HEAVY QUARK PHYSICS

AIP CONFERENCE PROCEEDINGS 196

RITA G. LERNER
SERIES EDITOR

HEAVY QUARK PHYSICS

ITHACA, NY 1989

EDITORS:

PERSIS S. DRELL & DAVID L. RUBIN
LABORATORY OF NUCLEAR STUDIES
CORNELL UNIVERSITY

American Institute of Physics   New York

Authorization to photocopy items for internal or personal use, beyond the free copying permitted under the 1978 US Copyright Law (see statement below), is granted by the American Insitute of Physics for users registered with the Copyright Clearance Center (CCC) Transactional Reporting Service, provided that the base fee of $3.00 per copy is paid directly to CCC, 27 Congress St., Salem, MA 01970. For those organizations that have been granted a photocopy license by CCC, a separate system of payment has been arranged. The fee code for users of the Transactional Reporting Service is: 0094-243X/87 $3.00.

Copyright 1989 American Institute of Physics.

Individual readers of this volume and non-profit libraries, acting for them, are permitted to make fair use of the material in it, such as copying an article for use in teaching or research. Permission is granted to quote from this volume in scientific work with the customary acknowledgment of the source. To reprint a figure, table or other excerpt requires the consent of one of the original authors and notification to AIP. Republication or systematic or multiple reproduction of any material in this volume is permitted only under license from AIP. Address inquiries to Series Editor, AIP Conference Proceedings, AIP, 335 E. 45th St., New York, NY 10017.

L.C. Catalog Card No. 89-081583
ISBN 0-88318-644-6
DOE CONF 8906204

Printed in the United States of America.

Contents

Preface ix

List of Participants x

THEORY OF HEAVY QUARK DECAYS

Models of Semileptonic Decays.. 3
    N. Isgur
Soccer in Indiana and Models for Nonleptonic Decays
of Heavy Flavours... 18
    I. I. Bigi
Phenomenology of the CKM Matrix 35
    Y. Nir
Predictions of Rare B Decays 46
    P. O'Donnell
Long Range QCD Effects in Nonleptonic Weak Decays 52
    M. Neubert
Non-Perturbative QCD in Weak Decays 59
    S. Sharpe
Symmetry, Topology and Helicity in $D_s \to \omega\pi$ and $D_s \to \rho\pi$ Decays........................ 72
    H. J. Lipkin

EXPERIMENTAL RESULTS ON B MESONS

Review of B-Meson Decay Results.. 79
    K. Schubert
Update on $B^0 \overline{B}^0$ Mixing... 97
    M. Schäfer
Semileptonic Decays of B Meson into Charmed Meson................................ 103
    N. Katayama
ARGUS Results on Rare B Decays 111
    A. Golutvin
Exclusive Semileptonic B Decays to Charm... 116
    R. Gläser
Charmless B Meson Decay at CLEO 122
    M. Procario
B Decay Studies with the TPC at PEP... 130
    G. Lynch
Argus Results on Baryonic Decays of B Mesons 136
    B. Spaan
Hadronic B Meson Decays 142
    Y. Kubota

EXPERIMENTAL RESULTS ON CHARM MESONS

Review of Charm Decay Results 157
    G. Gladding
Exclusive Semileptonic Decays of Charmed Mesons 172
    J. C. Anjos et al.
Charm Fragmentation 179
    G. Moneti
Recent Results on P-Wave Charmed Mesons at ARGUS 191
    R. Kutschke
Resonant Substructure in $K^-\pi^+\pi^+\pi^-$ and $\bar{K}^0\pi^+\pi^+\pi^-$ Decays of Charmed D Mesons 197
    F. De Jongh
Results on Hadronic Decays of D and D_s Mesons from Mark III 203
    P. Kim
Recent Results on D_s^+ Decays from E 691 210
    G. Punkar
NA14' Results on D_s Decays 219
    G. Wormser
Charmed Hadron Spectroscopy and Decay Results from CESR 226
    M. S. Alam

HADROPRODUCTION OF HEAVY QUARKS

Photo- and Hadroproduction of Charm and Beauty 239
    R. Morrison
Hadronic Production of Heavy Quarks in a Hybrid Emulsion Experiment 255
    R. Lipton
New Experimental Results in Hadronic Charm Production 262
    S. Kwan
Production of Prompt Electrons in the Charm P_t Region at $\sqrt{s}=630$ GeV 268
    O. Botner et al.
Shadowing and Nuclear Absorption in J/ψ Hadroproduction 274
    M. B. Gay Ducati
The New Hyperon Beam Experiment at CERN 280
    S. Paul
Charm Hadroproduction with an Impact Parameter Trigger 285
    D. Barberis

QUARKONIUM

Υ Spectroscopy: A Review of Recent Results 299
    M. Tuts
Radiative $\Upsilon(1S)$ Decays 311
    D. Besson

Study of ψ' Decays 317
    W. Toki
Two-Photon Production of the η_c 325
    T. Jensen
Inclusive Hadron Production at 10 GeV 331
    R. Waldi

SEARCH FOR TOP

Latest Results of the Search for Top at UA1 339
    M. Pimiä
Search for New Heavy Quarks at Tristan 345
    H. Sakamoto
Status of the Top Quark Search at CDF 354
    G. W. Foster

PREDICTIONS AND PROSPECTS FOR HQ PHYSICS AND *CP* VIOLATIONS

Predictions for *CP* Violation 367
    F. Gilman
Experimental Prospects for Observing *CP* Violation in *B*-Meson Decays 385
    T. Nakada
Prospects for *CP* Violation Experiments in Hadron Beams 401
    M. S. Witherell
B-Factory Plans at Cornell 409
    K. Berkelman
Detector for the *B* Factory at PSI 414
    J. Chauveau
The Status of the TPC/2γ Detector at PEP 421
    H. Marsiske
Tau Charm Factory Physics 428
    W. Toki
Multidimensional Analysis: *b* tagging at LEP 440
    Ch. de la Vaissiere and S. Palma-Lopes

Author Index 445

Preface

The 1989 International Symposium on Heavy Quark Physics took place in early June at Cornell University in Ithaca, New York. The invited speakers presented recent developments in the theory of heavy quark interactions and decays as well as the latest experimental results. There were reviews of recent measurements of beauty and charmed mesons, mixing in the B system, quarkonium spectroscopy, and of the relevant theoretical models.

In spite of our ever-growing knowledge of the details of quark interactions we were reminded that outstanding questions remain. The top quark, whose existence is unquestioned by most particle theorists, continues to elude the experimentalists. New mass limits for the top quark were reported. Speakers discussed the status of the theoretical predictions of rare decays of b quarks and CP violation in the decay of B mesons. The first evidence for b to u transitions was presented, but the CP symmetry of the b-quark system has yet to be investigated experimentally.

A special evening session was devoted to consideration of the physics accessible at a new generation of high luminosity e^+e^- colliders that operate in the $\Upsilon(4S)$ energy range. Evidently such B factories must deliver luminosities at least two orders of magnitude greater than has been achieved in existing storage rings. The prospects for measuring CP violation in hadron machines and the new electron storage rings were reviewed.

Outside of the formal sessions the symposium participants enjoyed the beautiful surroundings of the Finger Lakes region and the fine weather typical of late spring in central New York. The congenial setting helped us all to keep the physics of elementary particles and blue sky in proper perspective.

Many people worked very hard to make the conference a success. All of the organizers would like to thank Dale Held, Mary Wright, and Monica Norris for their outstanding efforts in organizing and administering the symposium, and the National Science Foundation and the U.S. Department of Energy for their support. The Cornell Physics Department generously provided the use of their facilities. The editors would like to thank all of those who contributed to the proceedings for their timely submission of manuscripts. We also gratefully acknowledge the assistance of Monica Norris and especially Pam Davis in the preparation of this volume.

P. Drell and D. Rubin

List of Participants

M. S. Alam
SUNY, Albany

Guido Altarelli
CERN

Ray Ammar
University of Kansas

Jeff Appel
Fermilab

Paul Avery
University of Florida

Steven Ball
University of Kansas

Dario Barberis
University of Heidelberg

Philip Baringer
University of Kansas

Karl Berkelman
Cornell University

Ikaras Bigi
University of Notre Dame

Elliott Bloom
SLAC

Bojan Bostjancic
Institut Jozef Stefan

Jacques Chauveau
LPNHE, Paris

Michael Church
Fermilab

Louis Culumovic
University of Western Ontario

David Cinabro
CERN

Michael Danilov
ITEP, Moscow

Fritz DeJongh
SLAC

Christian de la Vaissiere
LPNHE, Paris

D.-S. Du
Utah State University

Maria Ducati
Institute of Physics, Brazil

Richard Edelstein
Carnegie-Mellon University

Kenneth Edwards
Carleton University-Ottawa

Lars Owe Eek
University of Uppsala

Penny Estabrooks
Carleton University-Ottawa

John Fisher
(Independent Consultant)

Bill Foster
Fermilab

Soren Frederikson
Ohio State University

Bill Frisken
York University

Wolfgang Funk
University of Heidelberg

Lynn Garren
University of Florida

Chaogiang Geng
TRIUMF

Fred Gilman
SLAC

Gary Gladding
University of Illinois

Reiner Glaser
DESY

A. Golutvin
ITEP, Moscow

B. Govorkov
Lebdev Institute

Daniel Green
Fermilab

Tao Han
University of Wisconsin-Madison

David Hitlin
California Institute of Technology

Nathan Isgur
University of Toronto

Robert Jedicke
University of Toronto

A. Kamal
University of Alberta

Peter Kasper
Fermilab

Nobu Katayama
Cornell University

Gabrijil Kennel
University of Ljubljani

Shaukat Khan
DESY

Gregory Kilcup
Brown University

Peter Kim
SLAC

Robert Kowalewski
Carleton University-Ottawa

Robert Kutschke
University of Toronto

Nowhan Kwak
University of Kansas

Simon Kwan
CERN

Paolo Lariccia
CERN

Harry Lipkin
Weizmann Institute

Ronald Lipton
Carnegie-Mellon University

Jiang Liu
University of Michigan

Maurizio Loreti
University of Padova

Lee Lucking
Fermilab

Gerry Lynch
Lawrence Berkeley Lab

Yousef Makdisi
BNL/DOE

Helmut Marsiske
SLAC

David McFarlane
McGill University

Mac Mestayer
CEBAF

W. Metzger
Katholieke University

Giancarlo Moneti
Syracuse University

Rollin Morrison
U.C., Santa Barbara

Tatsuya Nakada
Paul Scherrer Institute

Charles Nelson
SUNY-Binghamton

Matthias Neubert
University of Heidelberg

Bogden Niczyporuk
CEBAF

Yosef Nir
SLAC

Patrick O'Donnell
University of Toronto

Roberto Onofrio
University of Rome

Stephan Paul
CERN

Martti Pimia
CERN

Francis Pipkin
Harvard University

James Prentice
University of Toronto

Greg Punkar
U.C., Santa Barbara

Natalie Roe
Lawrence Berkeley Lab

Michael Ronan
Lawrence Berkeley Lab

William Ross
Yale University

Thomas Ruf
University of Heidelberg

Hiroshi Sakamoto
KEK

Marion Schaefer
DESY

Henning Schroeder
DESY

Klaus Schubert
University of Karlsruhe

Daryl Scora
University of Toronto

Sally Seidel
University of Toronto

Stephen Sharpe
University of Washington

Edward Shibata
Purdue University

James Smith
University of Colorado

Xiaotong Song
Brookhaven National Lab

Amarjit Soni
UCLA

Bernhard Spaan
DESY/University of Dortmund

Joachim Spengler
DESY

C. R. Sun
SUNY, Albany

Walter Toki
SLAC

Michael Tuts
Columbia University

Fumiyo Uchiyama
Harvard University

Zenaida Uy
Millersville University

Anatoly Vorobiov
Institute of Nuclear Physics
Novosibirsk

Roland Waldi
University of Karlsruhe

Angel Wang
University of South Carolina

James Wiss
University of Illinois

Michael Witherell
U.C., Santa Barbara

Guy Wormser
LAL Orsay

Zhongxin Wu
Yale University

Rainer Wurth
DESY

Meiko Yamawaki
North Adams State College

A. R. Zhitnitsky
Institute of Nuclear Physics
Novosibirsk

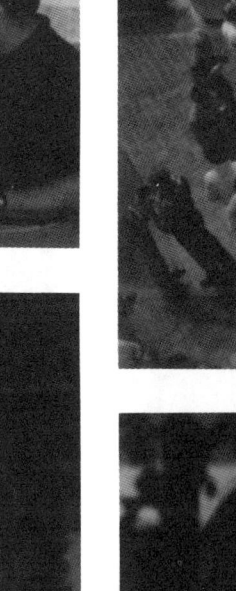

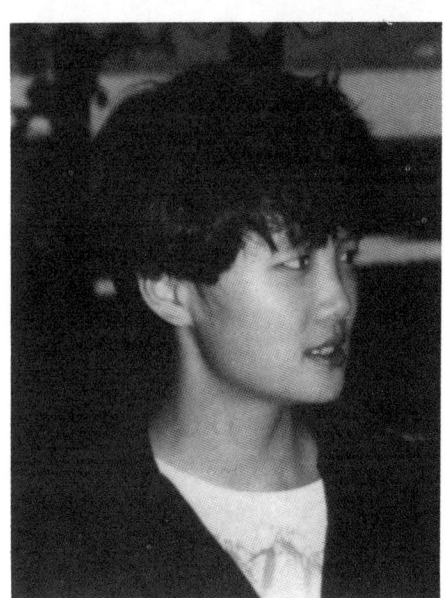

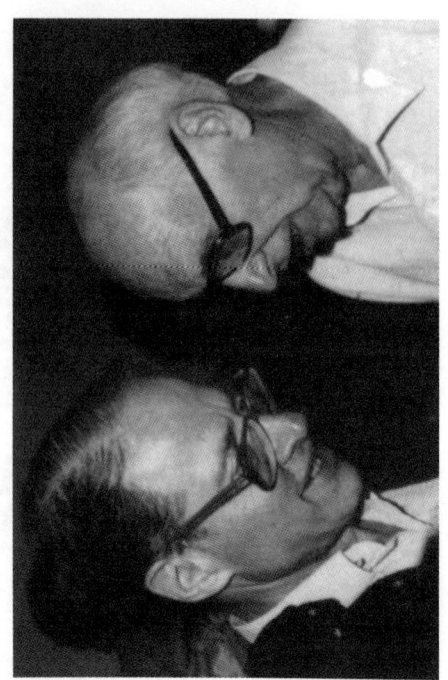

THEORY OF HEAVY QUARK DECAYS

MODELS OF SEMILEPTONIC DECAYS

Nathan Isgur
Department of Physics
University of Toronto
Toronto, Canada M5S 1A7

ABSTRACT

Semileptonic decays provide in principle one of the most direct methods for measuring Kobayashi-Maskawa angles, and one of the cleanest probes of hadronic structure. I describe and compare some models which have been developed for the hadronic matrix elements controlling such decays.

INTRODUCTION

Extracting the Kobayashi-Maskawa (K-M) matrix elements[1] from measurements of semileptonic decay rates necessarily involves theory since the rates depend on hadronic dynamics. Given our present rudimentary ability to calculate hadronic current matrix elements[2-8], the determination of the K-M angles will also involve an active interplay between experiment and theory which will test existing theoretical methods and help to define which measurements can be most useful for this purpose. The situation is sketched in Figure 1.

INCLUSIVE MODELS

As $m_Q \to \infty$ with $m_Q - m_q$ large, one would expect parton model ideas to be applicable to the decay $P_Q \to X_q e \bar{\nu}_e$ (we denote by P_α and V_α the ground state pseudoscalar and vector mesons with valence quark content $\alpha \bar{d}$, and by X_α an exclusive or inclusive final state containing the quark α). In this limit $m_Q \gg \Lambda_{QCD}$ so that Q is practically free and q is recoiling at high momentum as in deep inelastic scattering, so the inclusive decay rate should approach the rate for the free quark decay $Q \to q e \bar{\nu}_e$. This is the basis of the calculations of Altarelli, Cabibbo, Corbo, and Maiani[9] and other earlier work of this type[10]. These calculations are the most rigorous of all those we will discuss here, but they pay a certain price for this in having a limited range of validity: not only are they limited to inclusive processes, but also in the kinematic ranges over which they should be applied.

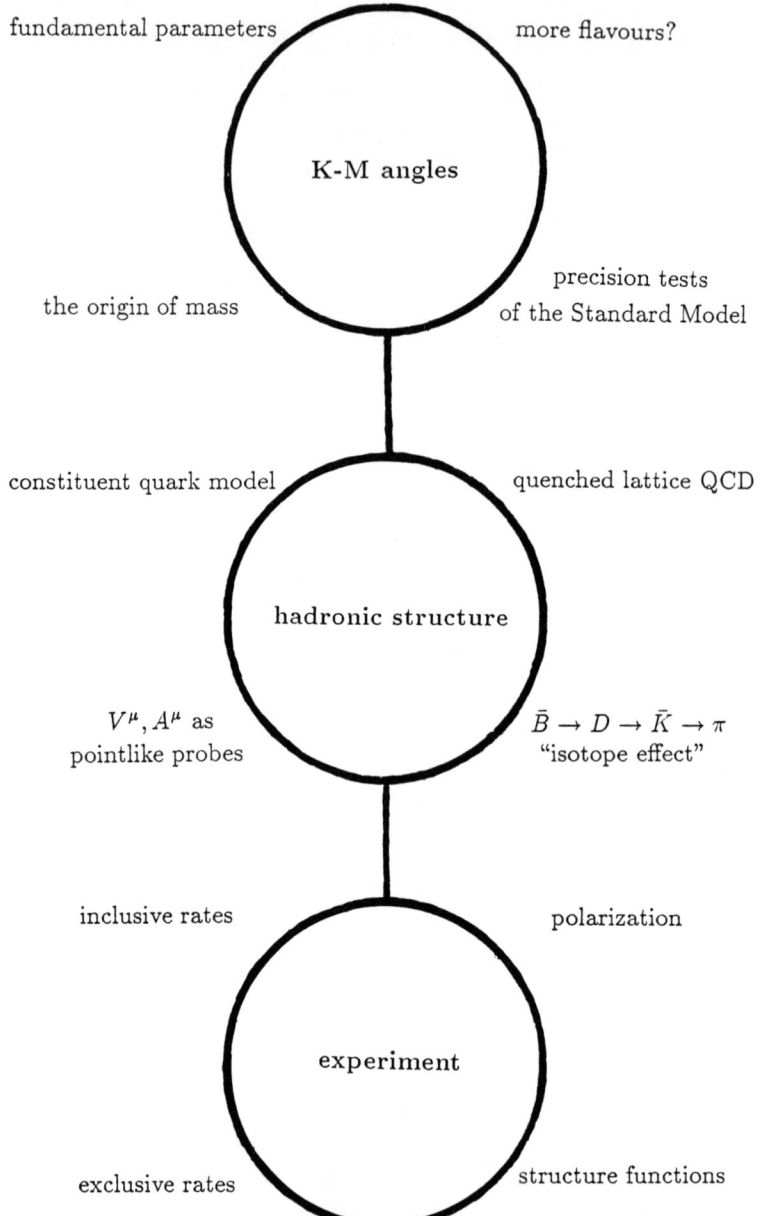

Figure 1: Determining the K-M angles requires an interplay between hadronic models and experiment.

In deep inelastic lepton-nucleon scattering, the hadronic recoil mass $m_X^2 \to 2M_N\nu(1-x) \to \infty$ in the Bjorken limit. However, in $P_Q \to X_q e \bar{\nu}_e$ the inclusive total rate Γ and electron spectrum $\frac{d\Gamma}{dE_e}$ receive contributions from hadronic recoil masses right down to threshold. This implies that the free quark decay calculations are not justified unless m_Q and $m_Q - m_q$ are sufficiently large that the threshold region makes a negligible contribution relative to the deep inelastic region.

Let me make this criticism more explicit[3,4]. Imagine a world in which pair creation by the strong interaction is suppressed. (This imaginary world corresponds to QCD in the large-N_c limit. Alternatively, we can view it as an approximation in which QCD is first solved in fixed sectors of Fock space with mixing between sectors treated as a perturbation.) Then the decay $P_Q \to X_q e \bar{\nu}_e$ will be saturated by hadronic systems X_q which are $q\bar{d}$ resonances. If, to simplify our discussion, we ignore the motion of the quarks in the decaying P_Q meson, then in the free-quark decay model the recoil hadronic mass $m_X^2 = (p_{P_Q} - p_e - p_{\bar{\nu}})^2$ will be continuously distributed from a minimum value $(m_q m_d)^2$ up to a maximum value $(m_q + m_d)^2 + (m_d/m_Q)(m_Q - m_q)^2$. On the other hand, in our illustrative world the quark q would be captured into one of the discrete eigenstates in which $q\bar{d}$ form a confined meson spectroscopy. In these circumstances, the free quark differential decay rates will approximately hold (in an energy average sense) only in those regions of phase space where the density of confined states has approached that of the continuum. This corresponds to the usual parton model condition that T_q, the kinetic energy of q, be large compared to 1 GeV. However, at low m_X the "true" hadronic recoil spectrum will be controlled by a sparse set of discrete states which will at best only remotely resemble the free spectrum: $d\Gamma/dm_X^2$ in this region will consist of a set of well-separated spikes. See Figure 2. We therefore see that in this limit (large N_c or treating pair creation as a perturbation), on purely kinematical grounds the free-quark decay model cannot possibly reproduce the differential spectrum $d\Gamma/dm_X^2$ at low m_X^2. While $d\Gamma/dm_X^2$ is not a very useful quantity experimentally, this failure reflects directly on the high energy end-point region of the electron spectrum $d\Gamma/dE_e$. The electron energy spectrum can be viewed as a superposition of m_X-dependent projections of the three-body Dalitz plots for $P_Q \to X_q e \bar{\nu}_e$ for fixed m_X, each with their corresponding end points $E_e^{\max} = (m_{P_Q}^2 - m_X^2)/2m_{P_Q}$, so that, as already stated, low m_X determines the high-E_e spectrum. Of course, one might hope that the parton model "averaging" of the confined spectrum will extrapolate right down to the minimum allowed hadronic mass m_X. This "averaging" would necessarily be rather crude, since as each new discrete m_X begins to contribute to the electron spectrum (as the energy drops below its end point), it has associated with it a threshold — with its prescribed threshold behavior — of which

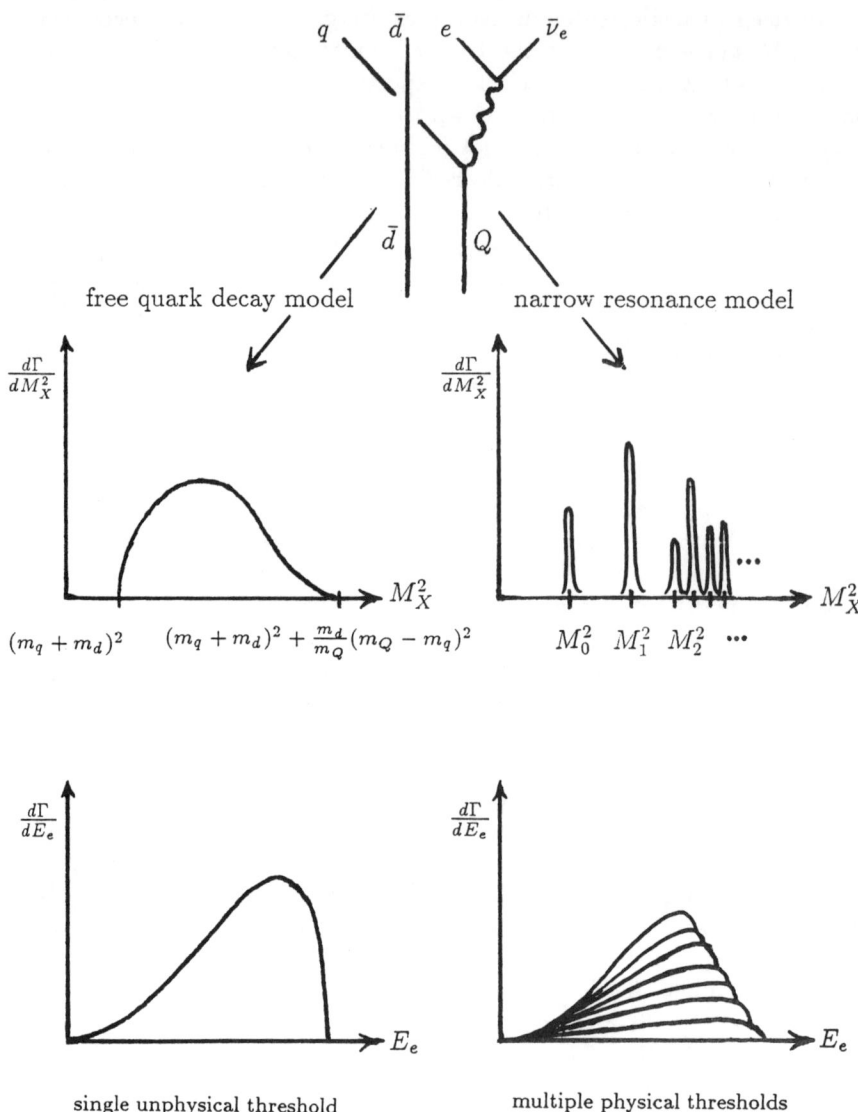

Figure 2: A comparison of the free quark decay model and the narrow resonance model for semileptonic decay spectra.

the free-quark decay model is completely ignorant. Even allowing for this failure, we can see no justification for such an optimistic view of the range of validity of the free-quark decay model. Its kinematic failures are, after all, a superficial reflection of the deeper dynamical problem of failing to take into account the strong, nonperturbative interactions which create the confined spectrum. In other words, in the low-m_X part of the spectrum there is no justification for the parton model mnemonic that "the short-distance physics determines the inclusive cross section; everything else happens with unit probability."

Despite this failure to describe the spectrum at high E_e, we should emphasize that in this picture the free-quark decay model (and the ancillary QCD jet picture) would hold as an approximation which would become increasingly accurate as the mass of the decaying quark Q increased. In particular, we note that the range of hadronic masses populated by the decay with a light-quark spectator leads to *all* the thresholds falling within a region δE just below the maximum energy, which becomes an increasingly small fraction of the total E_e range: as $m_Q \to \infty$,

$$\frac{\delta E}{E_e^{\max}} \simeq \frac{m_d}{m_Q} \tag{1}$$

(we once again neglect quark motion in the hadrons). This happens even though the highest mass m_X being produced is becoming large: as $m_Q \to \infty$,

$$m_X^{\max} \simeq (m_d m_Q)^{\frac{1}{2}} . \tag{2}$$

The crucial point is that the model does not produce states with m_X ranging up to $m_{P_Q} \simeq m_Q + m_d$ as allowed by pure kinematics, but rather (as $m_Q \to \infty$) only over an infinitesimal fraction of this range. Thus we see that there is, for example, no paradox in producing high recoil momentum through "soft" $q\bar{d}$ bound states: the highest momenta ($\sim \frac{1}{2} m_{P_Q}$) correspond to large recoils sustained by states which have a density approaching that of the continuum and masses around $(m_d m_Q)^{\frac{1}{2}} \ll m_{P_Q}$, in which the quarks each have mean center-of-mass momenta of $\frac{1}{2}(m_d m_Q)^{\frac{1}{2}}$ corresponding, in the original P_Q rest frame, to a recoil quark momentum of $\frac{1}{2} m_{P_Q}$.

Given these limitations on the range of validity of the free quark calculation, for which known processes can it be used? One might hope to apply it to $b \to c$ decays, but the rough estimates given above indicate that m_X is in this case dominated by the range within a few hundred MeV of threshold. Exclusive calculations[3,4] confirm this, so that the *usual* dual realization of the inclusive limit is inapplicable. Similar remarks apply to $c \to s$ decays. On the other hand, both types of models suggest that $b \to u$ transitions lead to m_X in the range from $m_\pi \leq m_X \lesssim 2$ GeV so that the inclusive calculation should work reasonably well in those parts of the Dalitz plot dominated by large m_X. Nevertheless, there is no good reason to expect the calculation to work in the high energy endpoint region crucial to the determination of V_{ub} since that region is dominated by a few low-massed hadronic states. It is, incidentally, simple to understand that the free quark decay calculation will overestimate the rate for $b \to u$ near the maximum electron energy E_e^{max} and therefore give limits on or values for V_{ub} which are too small. In the free quark calculation, if we neglect m_u, all collinear momenta of u and $\bar{\nu}_e$ recoiling against the electron will produce maximum recoil. In the hadronic calculation, high momentum u quarks typically produce large hadronic recoil masses and so, since $\frac{M_B^2 - M_X^2}{2M_B} < E_e^{max}$, cannot contribute to the endpoint.

There is a simple symptom of these problems. In the free quark decay model

$$\Gamma(Q \to q e \bar{\nu}_e) = \frac{G_F^2 |V_{qQ}|^2}{192\pi^3} m_Q^5 f(\frac{m_q}{m_Q}) \,. \tag{3}$$

Table 1, by way of some admittedly rather extreme variations, shows how strongly the factor $m_Q^5 f(\frac{m_q}{m_Q})$ depends on m_Q and m_q. This alone indicates that binding effects, through their effects on phase space, are still important.

Table 1: an illustration of the sensitivity of $\Gamma(Q \to q e \bar{\nu}_e)$ to variations of m_Q and m_q *

	m_b	m_c	m_u	$m_Q^5 f(\frac{m_q}{m_Q})$ (in 10^3 GeV^5)
b→c	5.3	1.4	-	2.5
	4.8	1.9	-	0.8
b→u	5.3	-	0.0	4.2
	4.8	-	0.5	2.3

*The minimum and maximum values of m_Q and m_q used in the table are meant to represent a typical spread of values from potential models. The maximum variation that could be obtained is displayed for emphasis, even though it might be more realistic to pair maximum and minimum values for Q and q.

EXCLUSIVE MODELS

The preceeding discussion strongly suggests that while inclusive calculations will suffice for semileptonic t quark decays, studies of $b \to c, b \to u, c \to s, c \to d$, and of course $s \to u$, require the development of theories of exclusive semileptonic decay. The history of such calculations (even if we restrict ourselves to those dealing with heavy quark decays) is a long one. The problem has been studied using the quark model (Refs. 11), current algebra (Refs. 12), and many other methods (Refs. 13). Here I will concentrate my remarks on three popular models.

The WBS Model. The model of Wirbel, Bauer, and Stech[5] begins with a decay at $t = m_{e\nu}^2 = 0$ and views it in an infinite frame (IMF) boosted along the direction opposite to the hadronic recoil momentum $\vec{p}_X$. This red shifts the lepton momenta to zero, leaving the calculation of the matrix elements

$$\langle X_q(\mathbf{P}\hat{z}, \lambda_x)|V^\mu - A^\mu|P_Q(\mathbf{P}\hat{z})\rangle \tag{4}$$

where $\mathbf{P}$ is the magnitude of the "infinite" momentum of the boosted hadrons. This frame is simple because one can presume that the hadronic wavefunctions have taken on a limiting form as $\mathbf{P} \to \infty$. This in turn means that all decays involving a given meson at t=0 involve this single limiting wavefunction, while calculations at (almost) every other kinematic point would require that one solve the extraordinarily difficult problem of boosting the meson state vector to an arbitrary momentum. WBS go on to adopt a simple model for these IMF wavefunctions, which gives them a way to calculate the required hadronic current matrix elements at t=0. They extrapolate the form factors associated with these matrix elements away from t=0 using the t-dependence expected from lowest vector, axial vector, scalar, or pseudoscalar meson dominance of the corresponding form factors.

This model, like the others we will discuss, has advantages and disadvantages. Its biggest advantage is that in principle it provides a Lorentz invariant description of the decay processes. It also leads to a simple interpretation of the decrease of the form factors at t=0 for fixed m_q with increasing m_Q (in the IMF the longitudinal momentum fractions carried by the light quark peak at $x \simeq m_d/m_q$ and m_d/m_Q, respectively, and so become increasingly mismatched), and a well-defined model for the t-dependence of all form factors. Most of the disadvantages of the WBS model are related to its advantages:

1) In practice, Lorentz invariance in the IMF is very difficult to obtain since total angular momentum (as opposed to J_z) can not be guaranteed by kinematical constraints, as it could be in the rest frame. Thus the WBS wavefunctions apparently do not correspond to good J^P eigenstates.

2) The wavefunction overlaps at t=0 (except for $m_q \simeq m_Q$) receive their support from the tails in x of the IMF wavefunctions. For example, in $B \to \pi$, the B valence quark wavefunction peaks at very low light quark x, so that the overlap probes the π valence quark wavefunction in the region where the light quarks have very asymmetric shares of the pion's momentum. It may be asking too much to expect the WBS wavefunctions to provide an accurate description of this region, which is related to the problem of accurately describing the high Q^2 behaviour of the pion's electromagnetic form factor.

3) The use of t-channel meson dominance of weak form factors is not an essential ingredient of the WBS model, but should rather be viewed as an initial parameterization. In its present form, however, it is certainly dubious: the region near t=0 seems unlikely to be controlled by a single distant pole when the t-channel spectral function is so densely populated by mesons on the scale of momentum transfers available in the physical region of heavy quark decay. A peculiar manifestation of this *ansatz* of the BWS model is that the $B \to \pi$ form factor nearly blows up as $t \to t_m$ since the B^* is so near in mass to t_m. If this feature of the model is unrealistic, it would mean that the BWS model's predictions for $B \to \pi$ and $B \to \rho$ are too large.

Despite these criticisms, one should bear in mind that the WBS model has, in its essentials, many of the correct physical ingredients to describe semileptonic heavy quark decay. The challenge will be to try to extend it from a rough first generation model into one that more fully realizes the advantages of the WBS approach.

The KS Model. The model of Körner and Schuler[8] resembles in many ways that of WBS. It also works in the IMF but, instead of calculating matrix elements in an explicit model for the mesons, relies on a helicity matching prescription to determine the matrix elements. The basic idea is that free quark decay determines a set of quark helicity amplitudes which should in some sense be "dual" to a set of hadronic helicity amplitudes. KS examine matching onto the states P_q and V_q. The matching process involves using the spin wavefunctions of P_Q, P_q, and V_q from the valence quark model plus the assumption that the spectator spin is preserved between the initial and final state. Having thus calculated at $t = 0$, the KS model uses meson dominance along with perturbative QCD scaling laws for the asymptotic t-dependence of the form factors to extrapolate to $t \neq 0$.

The main advantage of the KS model is that it is based on the fundamental principle of duality rather than explicit hadronic modelling. Its disadvantages are:

1) Because it is so general, the KS model can, at least in its present form,

only pick out the L=0 projections (*i.e.*, the P_q and V_q parts) of the full semileptonic decay spectrum. Within this sector, moreover, it can only give relations, and must rely on a more explicit model (in practice taken to be WBS) for absolute rate predictions.

2) The assumption that the spectator quark spin is preserved is related to the WBS problem with having good total angular momentum states. When the spectator quark is dragged away by the recoiling quark it will, even if the forces are spin preserving, have an amplitude for spin flip. This is because it in general has some $\vec{p}_\perp$ relative to the boost axis and as a result will experience a Wigner rotation. The KS matching method therefore seems difficult to justify in the IMF.

Despite these weaknesses, the KS model is important for its demonstration of the fundamental role played by quark-hadron duality in the decay process. Indeed, we will see below that in certain circumstances the analogue of the KS matching procedure in the P_Q rest frame would be completely justified.

The ISGW Model. The first point to be made here is that ISGW is *not* a Welsh rendition of my name, but is rather the acronym for the model of myself, Scora, Grinstein, and Wise[4]. The model was first developed by the GWI authors in Refs. 3; the new version of the model corrects an error made in the earlier version which led to the difficulties addressed by Altomari and Wolfenstein[6] and later discussed by Golowich *et al.*[14]. The ISGW model focuses not on the minimum value of t=0 but on t at its maximum value $t_m \equiv (m_{P_Q} - m_{X_q})^2$. This corresponds to zero recoil in the sense that in the rest frame of P_Q, the three momentum $\vec{p}_X = 0$; $t = 0$ corresponds to maximum recoil $|\vec{p}_X| = (m_{P_Q}^2 - m_{X_q}^2)/2m_{P_Q}$. The region of zero recoil is the natural habitat of the nonrelativistic quark model: here the heavy quark Q decays into a lighter quark q with most of the available energy $m_Q - m_q$ going into the leptons. Since X_q is recoiling slowly, or left at rest, the required matrix elements

$$\langle X_q(\vec{p}_X \simeq 0, s_x)|V^\mu - A^\mu|P_Q(\vec{0})\rangle \qquad (5)$$

can be computed with the usual rest frame quark model wavefunctions.

The main advantage of this model is that it builds upon the foundation of experience embodied in the nonrelativistic quark model. Thus, for example, the calculation of many matrix elements that one requires can be related to standard quark model calculations. A second advantage is that the model provides a clear physical picture of why the form factors peak at t_m: low t corresponds to high $\vec{p}_X$ which requires that the spectator quark be in the improbable high momentum tail of the rest frame wavefunction of P_Q. (It is then moving with the velocity $\vec{v}_X \simeq \vec{p}_X/m_X$ required to fit into the hadron X_q). As with the other models, the principal disadvantages of the ISGW model arise from it advantages:

1) The nonrelativistic quark model calculation of weak matrix elements can only be made Lorentz invariant in the weak binding limit where p_i/m_i is small (p_i and m_i are the momenta and mass of constitutent quark i) and where $m_{ij} \simeq m_i + m_j$ (m_{ij} is the mass of meson $q_i \bar{q}_j$). ISGW make a Lorentz invariant calculation in this limit, isolate the Lorentz scalar form factors, and then assume that their formulas for these form factors are valid even out of the weak binding regime. This makes their results Lorentz invariant as far as all kinematical effects are concerned, but, naturally, their form factor formulas do not take into account relativistic dynamics. At $t = t_m$ the resulting errors are thus those associated with the usual predictions of the nonrelativistic quark model for other current matrix elements like G_A/G_V and magnetic moments. In the region near $t = t_m$ their predictions for the $(t_m - t)$-dependence of form factors can be characterized by a slope which, being an effective "charge radius", is also a static quantity which should once again be predicted with typical quark model accuracy.

2) If we were interested only in a region near $\vec{p}_X = 0$, the ISGW model would be a good bet for a rough description of the data. However, as already emphasized in discussing the WSB model, $t_m - t$ can be large compared to typical hadronic scales, so that a modest error in an effective radius r at t_m can easily become a very large error in a form factor when $t_m - t \gg r^{-2}$. While not a big problem in $b \to c$, this flaw seriously compromises the ISGW model in $b \to u$. They try to compensate for it by adding a parameter which allows them to adjust their predicted effective radii to existing data, but it is difficult to know how well this prescription should work. (ISGW assign a factor of about two uncertainty to their $B \to \pi$ rate). Thus although the model gives a sensible picture of the $t_m - t$ dependence of the form factors, it cannot provide a reliable dynamical model for them at high $t_m - t$.

There is some hope for improving the ISGW model. The study of current matrix elements at zero recoil could very well point the way toward a systematic calculation of relativistic corrections to be model. (For example, the bag model suggests that the discrepancy between the quark model prediction $G_A/G_V = 5/3$ and experiment is due to the presence of lower components in the quark spinors. Such an effect can easily be incorporated into the ISGW model). Progress in accurately predicting the high $t_m - t$ behaviour of form factors seems a more remote possibility.

THE SHIFMAN-VOLOSHIN LIMIT

Although the interaction with experimental results will probably be the most important mechanism for improving models of heavy quark semileptonic decays, input from more fundamental theory can also have an impact. In particular, I would like to mention a theoretical constraint against which the models can be tested. This constraint originates with the work of Shifman and Voloshin[15] who pointed out that in a certain limit ("the S-V limit") the free quark calculation for $Q \to qe\bar{\nu}_e$ will predict exactly the same semileptonic rate as the sum over exclusive rates $P_Q \to X_q e\bar{\nu}_e$, with the latter exhausted by just two channels: $X_q = P_q$ and V_q. The idea is simple. If the free quark decay model is to be valid, then: i)the quarks must be approximately free, i.e. $\eta_i \equiv \Lambda_{QCD}/m_i \to 0$ for $i = Q, q$ (or, alternatively, $\eta_Q \to 0$ and $m_q/m_Q \to 0$); and ii)the recoil kinetic energy of q must be large, i.e., $x_m \equiv \frac{m_Q^2 - m_q^2}{2m_Q^2} >> (\Lambda_{QCD}/m_Q)^{\frac{1}{2}}$. On the other hand, it suffices to ensure that the system X_q will materialize entirely as the ground states P_q and V_q for the recoiling q to make a small invariant mass with the spectator quark; this can be translated into the condition $x_m << 1$. The crucial observation is that these conditions are all compatible: if x_m is sufficiently small, complete P_q, V_q dominance is assured, but for any fixed such x_m, as $m_Q \to \infty$ the conditions required for the accuracy of the free quark calculation will be met.

The arguments of Shifman and Voloshin can in fact easily be extended[16] to show that this duality is a strong one: it is local in the Dalitz plot

$$\frac{d^2\Gamma(Q \to qe\bar{\nu}_e)}{dxdt} = \frac{d^2\Gamma(P_Q \to P_q e\bar{\nu}_e)}{dxdt} + \frac{d^2\Gamma(P_Q \to V_q e\bar{\nu}_e)}{dxdt} \qquad (6)$$

and valid not only in the limit as η_i and $x_m \to 0$, but even once one includes corrections of order x_m. The existence of this limit i)assures us of the validity of the "sum over resonances" approach in at least some circumstances, ii)provides a constraint which must be satisfied by all models, and iii)gives us a method of assessing the model dependence of exclusive calculations in processes like $\bar{B} \to De\bar{\nu}_e$ which are sufficiently close to the S-V limit. Table 2 shows an explicit comparison between the predictions of the ISGW model and the S-V limit for the form factors and structure functions appearing in $\bar{B} \to D, D^*e\bar{\nu}_e$ and $D \to \bar{K}, \bar{K}^*e^+\nu_e$. Since we only have to rely on the model to predict the small *deviations* from the S-V limit, we can have more confidence than might at first seem warranted in, e.g., the extraction of V_{cb} from exclusive decay rates.

Table 2: The zero recoil values of some form factors and structure functions of the ISGW model compared to the S-V limit.

		f_+	$f/2m_{P_Q}$	g (GeV^{-1})	a_+ (GeV^{-1})	β_{++}^P	$\alpha^V/4m_{P_Q}^2$	β_{++}^V	γ^V
$\bar{B} \to D, D^*$	S-V limit	1	0.76	0.12	−0.12	1	0.53	1	2
	quark model	1.13	0.65	0.16	−0.15	1.28	0.42	0.99	2.23
$D \to \bar{K}, \bar{K}^*$	S-V limit	1	0.75	0.34	−0.34	1	0.52	1	2
	quark model	1.16	0.72	0.49	−0.36	1.34	0.52	1.00	2.64

LATTICE QCD

A discussion of constraints imposed on semileptonic decay models from other parts of theory would be incomplete without mentioning lattice QCD. This is even a case where, amusingly, we are most interested in the results of "unphysical" quenched calculations, since the models (at least the ones mentioned here) are also quenched. Such studies will in the relatively short term be able to directly "measure" some properties of quenched semileptonic form factors[17]. These can then be compared to model predictions and used to eliminate models.

COMPARISON TO EXPERIMENT

I will not attempt a detailed comparison to experiment[18] here, but will rather sketch the situation to set the stage for the discussion of a discrepancy between the models and experiment which has emerged in $D \to \bar{K}^* e^+ \nu_e$ decays.

In the $b \to c$ transitions, the ISGW model predicts that only 13% of the rate will go to resonances other than the D and D*. This dominance is consistent with the data. All three models (BWS, KS, and ISGW) are in fairly good agreement with each other and with the data on the $\bar{B} \to D e \bar{\nu}_e$ and $\bar{B} \to D^* e \bar{\nu}_e$ rates and on the longitudinal to transverse polarization ratio Γ_L/Γ_T in the latter decay. While the accuracy of the data has a long way to go, with the backing of the S-V limit I believe we can already deduce values for $|V_{cb}|$ which are accurate to about 20%.

The relationship between theory and experiment is not so harmonious for $c \to s$ transitions. There is some agreement: ISGW predicted that a very small fraction of the D decay would go to resonances other than the K and K^*, and although there was initially some indication to the contrary, it now seems likely that this is correct. The three models are in reasonable agreement with each other and experiment on the $D \to \bar{K} e^+ \nu_e$ rate. They

Table 3: The sensitivity of the transverse and longitudinal decay rates in $\bar{B} \to D^* e^- \bar{\nu}_e$ and $D \to \bar{K}^* e^+ \nu_e$ to variations in the normalizations of the f, g, and a_+ form factors from their calculated values.

f/f^{calc}	g/g^{calc}	a_+/a_+^{calc}	$\Gamma_\perp/\|V_{qQ}\|^2$		$\Gamma/\|V_{qQ}\|^2$		Γ_L/Γ_T	
			$\bar{B} \to D^*$ (10^{13} sec^{-1})	$D \to \bar{K}^*$ (10^{11} sec^{-1})	$\bar{B} \to D^*$ (10^{13} sec^{-1})	$D \to \bar{K}^*$ (10^{11} sec^{-1})	$\bar{B} \to D^*$	$D \to \bar{K}^*$
1	1	1	1.28	0.46	2.52	0.96	0.97	1.09
0.8	1	1	0.87	0.31	1.48	0.58	0.70	0.89
1	0.8	1	1.23	0.44	2.47	0.94	1.01	1.13
1	1	0.8	1.28	0.46	2.78	1.02	1.17	1.23

are also in agreement with each other on the $D \to \bar{K}^* e^+ \nu_e$ rate and polarizations, but no longer with a recent experiment[19] which finds the rate about half and Γ_L/Γ_T about twice that predicted.

There have been several recent attempts to tune the models to these new results. Bauer and Wirbel[20] have argued for an additional parameter z in the BWS model with which they can simultaneously fix the $D \to \bar{K}^* e^+ \nu_e$ rate and polarizations. The disadvantage of such a fix is that recent data on the $\bar{B} \to D^* e \bar{\nu}_e$ polarization leaves us with $z_D \neq z_B$, which is possible, but not very enlightening. Scora and I [21] have searched for a similar freedom in the ISGW model. Our results are summarized in Table 3 which shows that, with 20% uncertainties (typical of the quark model) in the three form factors controlling the $D \to \bar{K}^*$ matrix element, we could obtain substantial variations in the $D \to \bar{K}^* e^+ \nu_e$ rate, but could not reproduce the large value of Γ_L/Γ_T. We also studied the polarized decay rates in the S-V limit, and concluded that they too are dual locally in the Dalitz plot and immune to corrections of order x_m. (In the S-V limit, $\Gamma_L/\Gamma_T = 5/4$, rather close to the results of the models). Scora and I concluded that the ISGW model had no way to fix the Γ_L/Γ_T discrepancy so that, if this measurement is confirmed, it will kill the model[22].

CONCLUSIONS

The interplay between models and experiment required for accurate extraction of the K-M angles from heavy quark semileptonic decay has just begun. So far we can only conclude that the models are not hopelessly inadequate.

The era of comparing theory to measured rates is, nevertheless, nearly over. The reason is, paradoxically, that rates are too sensitive to elements of the physics over which the models have little control. Perhaps the best example of such an element is the t-dependences of the weak form factors. The models are all incapable of really predicting these dependences and, in one way or another, make guesses for them. We should bear in mind that, with good enough experiments, this is not an obstacle to realizing our principal ambitions in heavy quark semileptonic decays: experiment can, by detailed examination of the Dalitz plots and angular correlations for the exclusive decays, extract each contributing form factor (or structure function) at all values of t. To extract V_{cb}, for example, we then need only have a model which can reliably predict one exclusive form factor at one particular value of t. I have argued that the Shifman-Voloshin limit can be particularly useful in helping us to identify those quantities which can be most reliably predicted by theory.

The most difficult decays to study both theoretically and experimentally are, naturally, the most interesting: the $b \to u$ transitions. However, considering all the information on the analogous decays $b \to c, c \to s, c \to d$, and $s \to u$ that we will eventually have at our disposal, I believe the prospects are very good for reaching a satisfactory understanding of the whole family of weak semileptonic decays and their associated K-M angles.

REFERENCES

1. M. Kobayashi and T. Maskawa, *Prog. Theor. Phys* **49**, 652 (1973).
2. For a recent review see M. Wirbel, in *Prog. in Particle and Nuclear Physics* **21**, 33 (1988) and references therein.
3. B. Grinstein, M.B. Wise, and N. Isgur, *Phys. Rev. Lett.* **56**, 298 (1986); B. Grinstein, M.B. Wise and N. Isgur, Caltech preprint CALT-68-1311 (1985), and University of Toronto preprint UTPT-85-37 (1985).
4. N. Isgur, D. Scora, B. Grinstein, and M.B. Wise, *Phys. Rev.* **D39**, 799 (1989).
5. M. Wirbel, B. Stech, and M. Bauer, *Z. Phys.* **C29**, 637 (1985). See also Ref. 2.
6. T. Altomari and L. Wolfenstein, *Phys. Rev. Lett.* **58**, 1583 (1987); *Phys. Rev.* **D37**, 681 (1988).
7. B. Grinstein and M.B. Wise, *Phys. Lett* **B197**, 249 (1987).
8. J.G. Körner and G.A. Schuler, *Z. Phys.* **C38**, 511 (1988).
9. G. Altarelli, M. Cabibbo, G. Corbo, and L. Maiani, Nucl. Phys. **B207**, 365 (1982); N. Cabibbo, G. Corbo, and L. Maiani, *ibid.* **B155**, 93 (1979).

10. M.K. Gaillard, B.W. Lee, and J.L. Rosner, Rev. Mod. Phys. **47**, 277 (1975); J. Ellis, M.K. Gaillard, and D.V. Nanopoulos, Nucl. Phys. **B100**, 313 (1975).
11. I. Hinchliffe and C.H. Llewellyn Smith, Nucl. Phys. **B114**, 45 (1976); V. Barger, T. Gottschalk, and R.J.N. Phillips, Phys. Rev. **D16**, 746 (1977); G.K. Lane, Phys. Lett. **70B**, 227 (1977); M.B. Gavela, *ibid.* **83B**, 367 (1979); F.E. Close, G.J. Gounaris, and J. Paschalis, *ibid.* **149B**, 209 (1984); M. Suzuki, *ibid.* **155B**, 112 (1985); S. Nussinov and W. Wetzel, Phys. Rev. **D36**, 130 (1987); T. Altomari and L. Wolfenstein, Phys. Rev. Lett. **58**, 1583 (1987).
12. A. Ali and T.C. Yang, Phys. Lett. **65B**, 275 (1976); D. Fakirov and B. Stech, Nucl. Phys. **B133**, 315 (1978); S.C. Chao, G. Kramer, W.F. Palmer, and S.S. Pinsky, Phys. Rev. **D30**, 1916 (1984).
13. F. Bletzacker, M.T. Nieh, and A. Soni, Phys. Rev. **D16**, 732 (1977); Y. Kizikuri, Prog. Theor. Phys. **67**, 1598 (1982).
14. E. Golowich, F. Iddir, A. Le Yaouanc, L. Oliver, O. Pène, and J.C. Raynal, Phys. Lett. **B213**, 521 (1988).
15. M.B. Voloshin and M.A. Shifman, Yad. Fiz. **47**, 801 (1988) [Sov. J. Nucl. Phys. **47**, 511 (1988)]; M.A. Shifman, in *Lepton and Photon Interactions*, proceedings of the International Symposium on Lepton and Photon Interactions at High Energies, Hamburg, West Germany, 1987, edited by W. Bartel and R. Rückl [Nucl. Phys. B. Proc. Suppl. **3**, 289 (1988)].
16. N. Isgur, Phys. Rev. **D40**, 101 (1989).
17. See, for example, the talk by A. Soni in these proceedings.
18. For experimental results, see the appropriate reports in these proceedings.
19. See M. Witherell's talk on semileptonic charm decays in these proceedings.
20. M. Bauer and M. Wirbel, Univ. of Heidelberg preprint HD-THEP-88-22, 1988. See also Ref.2.
21. D. Scora and N. Isgur, Phys. Rev. **D40**, 1491 (1989).
22. It has been said that the ISGW model was a modified form of the original model of Ref. 3 designed to fit the $\bar{B} \to D^*$ polarization measurements. This is not the case. The original version of the ISGW model was never published because we realized (stimulated by the work of Refs. 6 and 7) that it had an error in it regarding the a_+ form factor. The model should not be blamed for errors made by theorists working on it! Assuming that there have been no other errors made in extracting the predictions of the model, a large Γ_L/Γ_T in $D \to \bar{K}^* e^+ \nu_e$ is inconsistent with the model.

SOCCER IN INDIANA
AND
MODELS FOR NON-LEPTONIC DECAYS OF HEAVY FLAVOURS *

I.I. Bigi

Department of Physics
University of Notre Dame du Lac
Notre Dame
IN 46556

ABSTRACT

Various descriptions of non-leptonic charm decays are reviewed and their relative strengths and weaknesses are listed. I conclude that it is mainly (though not necessarily solely) a destructive interference in nonleptonic D^+ decays that shapes the decays of charm mesons. Some more subtle features in these decays are discussed in a preview of future research before I address the presently confused situation in D_S decays. Finally I give a brief theoretical discussion of inclusive and exclusive non-leptonic decays of beauty mesons.

PROLOG

When I came to Notre Dame last summer, I continued a habit which I had formed many years ago as a graduate student, namely to play soccer on a regular basis with my colleagues. After the second game, one of the older players there took me aside and told me: "You run too much! We don't do that here. The game of soccer isn't played that way in Indiana!" This statement clearly contains a message about the state of Indiana, yet what this message really is is not so obvious.

An analogous situation is encountered in Heavy Flavour decays: we know the data contain important lessons, yet quite often we do not know what they really are.

I. INTRODUCTION

(1) Territory

Heavy flavour means I will not address K decays. Top will certainly qualify as a heavy flavour; yet it appears more and more likely that its mass will exceed even M_W. In that case top quarks will decay into another quark (b or s) and an on-shell W boson which will decay in turn. Thus there would be no *intrinsic non-leptonic* decays of top. Therefore, we are presumably dealing with a well-established territory when discussing non-leptonic decays of heavy flavours, namely the decays of charm and beauty mesons and baryons.

(2) Attitude

In Act I of "Dr. Faustus, Part I" a young man appears on stage declaring "Although I know a lot, I would like to know everything...". Anybody who teaches at a university will recognize immediately whom we are dealing with: it is a graduate student since only those are innocent enough to declare in public how much they know; unfortunately we are dealing with a graduate student who is not overly bright since otherwise he would not express the unrealistic desire to know everything. We have to keep a note of realism in mind when analyzing heavy flavour decays and always ask ourselves critically *"why do we want to know something and to which degree of accuracy do we have to know it ?"* I am referring to this attitude as the "Cui Bono?" principle.

(3) Goals

(a) It would be quite unrealistic to claim that we are *testing* QCD in the nonleptonic decays of heavy flavours. For the moment our goal can only be to *probe* QCD in a novel environment, namely the interface between the perturbative and non-perturbative regime: for the "mostly active" quark -- c and b -- is heavy

$$m_b \gg m_c \gg \Lambda_{QCD} \tag{1}$$

whereas the "mostly passive" light quarks -- u, d, s -- are not.

There is clearly an intrinsic intellectual merit in this goal, yet there is also another side to it as discussed next.

(b) A major driving force in studying rare decays, mixing and CP violation is the hope to uncover the presence of New Physics. Unfortunately, there is a certain complication that slows us down in this noble endeavor: the transitions that we can observe occur between hadrons and not quarks; thus at least some understanding of the hadronization process is required before progress can be made.

(4) Strategy

Since the Standard Model alleges that the intrinsic strength of all electro-weak forces in the charm sector are known (and controlled by the Cabibbo angle), it is quite natural to adopt the following strategy: learn the necessary lessons on the hadronization process in D decays and apply them subsequently in B decays. There are two caveats, though, that we have to keep in mind as will be explained later on:

- Information on hadronization in the D system cannot be carried over to the B system in a *mechanical* way.

- D decays carry a good potential for surprises, namely manifestations of New Physics.

(5) Outline

In Section II, I will give a rather detailed review of our present understanding of non-leptonic D decays. Section III deals, in a more tentative manner, with beauty decays. Section IV contains a summary and my outlook.

II. CHARM DECAYS -- A WELL ESTABLISHED BATTLE GROUND

(1) First Skirmish

Most battles are preceded by a skirmish, and that was the case here as well. It was suggested that the decays of charm hadrons can, to a rather good approximation, be described by the decay of the charm quark in close analogy to the decay of a heavy lepton while the other light quarks and antiquarks in the hadron act as interested, yet passive spectators. Accordingly, one can relate the charm lifetime to the muon lifetime:

$$\tau(\text{charm}) \cong \frac{1}{5} \tau_\mu \left(\frac{m_c}{m_\mu}\right)^5 \sim 7 \times 10^{-13} \text{ sec} \qquad (2)$$

for $m_c \sim 1.5$ GeV as inferred from K^o-$\overline{K}^o$ mixing. This simple prediction agrees surprisingly well with the observed *average* D lifetime:

$$\langle \tau(D) \rangle = \frac{1}{2}\left(\tau(D^o) + \tau(D^+)\right) \cong 7 \times 10^{-13} \qquad (3)$$

Such skirmishes tend to make the victor overconfident as if final victory were already in his grasp.

(2) The Ambush

Three experimental findings represent the original "charm decay puzzle"[1]:

(a) $$\frac{\tau(D^+)}{\tau(D^0)} \cong \frac{b_{SL}(D^+)}{b_{SL}(D^0)} \cong 2.5 \qquad (4)$$

in contrast to the simple theoretical expectation of a universal charm lifetime, as expressed in (2).

(b) $$\frac{BR(D^0 \to \overline{K}^0 \pi^0)}{BR(D^0 \to K^- \pi^+)} \cong 0.4 \qquad (5)$$

i.e. much larger than a value of $\frac{1}{40} - \frac{1}{18}$ as inferred from naive colour counting.

(c) $$\frac{BR(D^0 \to K^+ K^-)}{BR(D^0 \to \pi^+ \pi^-)} \sim 3\text{-}4 \qquad (6)$$

again different from the theoretical expectation of 1-1.4.

In addition, other unanticipated, and not immediately appreciated, findings emerged subsequently, albeit quite slowly, namely that non-leptonic decays of D^0 and D^+ mesons lead largely to two-body final states.

(3) Theorist's Responses

Theorists -- or at least those that cared about weak decays -- were surprised, yet their response was not slow in coming. It followed the conventional pattern that is adopted when faced by a surprise attack: a holding action followed by counter strikes that lead to a systematic counteroffensive. In the following, I will not describe these responses in any detailed way since there exists an extensive literature on that [2]; instead, I will give my personal evaluation on their merits -- successes and failures -- and their potential for improvements in the future.

(a) "Flexible defense": The Quark Diagrammatic Approach (= QDA).

The prescription one follows here is rather simple [3]:

(i) First one draws all quark diagrams with different topologies that involve the decay of a charm quark. There are six different types --denoted by M_i in the following-- namely spectator diagrams with or without colour matching, diagrams with W exchange in the s or t channel and two types of Penguin diagrams.

(ii) Next one expresses transition amplitudes as a linear combination of these basic matrix elements M_i:

$$T(D \to f) = \sum_i c_i^{(f)} M_i \qquad (7)$$

where the dependence on the specific channel f is contained in the coefficients $c^{(f)}$; phase shifts and absorption are also included in $c_i^{(f)}$ as best as possible.

(iii) The measured branching ratios of selected decay modes are then used to fit the matrix elements M_i.

(iv) Finally, one examines the agreement between "predictions" and data on other branching ratios.

My evaluation of this ansatz is as follows (where I have adopted the notation that *(+) denotes a strength, (-) a weakness and (+/-) a point of ambivalent value*):

(+) it provides quite a useful classification and thus represents a good *starting point* for analysis.

(+/-) it represents pure, rather unrefined phenomenology.

(-) final state interactions (hereafter referred to as FSI), namely phase shifts and absorption, are treated in a rather ad-hoc fashion allowing for a considerable amount of poetic license.

(-) There exists actually no good argument why FSI should be universal for one class of quark diagrams. Even if the same quark diagrams contribute to $D \to K\pi$ and $D \to K^*\rho$, say, the FSI will presumably be quite different in these two channels. Furthermore, rescattering and channel mixing has largely been ignored.

(-) I see little room for improvements within this ansatz, i.e. for overcoming these drawbacks.

(b) "First Counter Strike": The factorization ansatz.

This ansatz first put forward by Bauer and Stech and further developed by Bauer, Stech and Wirbel [4] consists of the following ingredients:

(i) QCD radiative corrections are included in the usual way, namely by expressing the effective Lagrangian in terms of two current-current operators O_+ and O_- :

$$\mathcal{L}_{eff}(\Delta C=1) \propto c_+ O_+ + c_- O_- \qquad (8)$$

(Semi-) Perturbative QCD is invoked to compute the c number coefficients c_+ and c_-.

(ii) One restricts oneself to describing two-body decay modes

$$D \to PP, PV, VV \qquad (9)$$

where P[V] stands for a pseudoscalar [vector] meson.

(iii) To determine the matrix elements for these decay modes one relies on a factorization ansatz. E.g.

$$T(D \to PP) \propto \langle P | J_\mu | D \rangle \langle P | J'_\mu | 0 \rangle \qquad (10)$$

(iv) W-exchange diagrams are ignored since their *dominant* contribution is *not* of the factorizable kind.

(v) A new non-perturbative parameter ξ is introduced which denotes the relative weight between a colour-mixed and a colour-matched diagram; ξ is treated as a free parameter, yet it is assumed -- and that is the crucial point -- to possess a *universal value* in all two-body decays.

(vi) FSI, namely phase shifts, absorption and some rescattering, are included only as much as phenomenologically needed.

This procedure allows to express all decay amplitudes in terms of two a priori unknown parameters a_1, a_2 which are defined as follows:

$$a_1 = \frac{1}{2}(c_+ + c_-) + \frac{\xi}{2}(c_+ - c_-) \qquad (11)$$

$$a_2 = \frac{1}{2}(c_+ - c_-) + \frac{\xi}{2}(c_+ + c_-) \qquad (12)$$

Accepting the numerical values for c_+ and c_- as they are inferred from perturbative QCD one is left with a *single* free parameter ξ (coupled with some poetic license in the inclusion of FSI as mentioned before).

More than twenty measured decay modes can be described in quite a satisfactory way with

$$a_1 |_{exp} \approx 1.2 \pm 0.1 \qquad (13)$$
$$a_2 |_{exp} \approx -0.5 \pm 0.1 \qquad (14)$$

Using the canonical values for c_+ and c_- one finds

$$a_1 |_{theor.} \approx 1.3 - 0.6\,\xi \qquad (15)$$
$$a_2 |_{theor.} \approx -0.6 + 1.3\,\xi \qquad (16)$$

Comparing in particular (14) and (16) shows that ξ has to be quite small: $\xi \ll 1$, $\xi < 1/3$. For the later discussion it is quite important to keep two things in mind:

(α) Many effects get lumped together into ξ; in addition to colour counting there are effects due to the soft FSI that enter here. Thus it poses no surprise at all that $\xi \neq 1/3$ is found. Yet these FSI effects should differ from channel to channel thus creating a strong channel dependance of ξ. It has to be viewed as quite amazing that a *universal*

value for ξ yields a quite satisfactory description (after some basic FSI have been factored out, in particular phase shifts in $D \to \overline{K}^{\circ} \pi^{\circ}$ vs. $D \to K^{-} \pi^{+}$).
(β) This universal value for ξ happens to be zero.

We will soon return to grapple with this apparent miraculum; for now I want to present my personal evaluation of this scheme:

(+) It represents phenomenology that is rather simple and quite successful.
(+) It actually made predictions that preceded the data – a feature that is not that common in this field.
(-) It makes several ad-hoc assumptions like factorization, a universal value for ξ, the add-on of FSI. These turn out to be successful assumptions – but they are ad-hoc nevertheless.
(-) I see little room for internal improvements of this scheme.

(c) "Systematic Counteroffensive": The 1/N Ansatz

The authors of ref.5 use the same effective $\Delta C = 1$ operator as before, namely (8). Very simple rules are followed to determine the colour weight of each decay diagram as a function of N, the number of colours. The transition amplitude is then expressed as an expression in 1/N

$$T(D \to f) = \sqrt{N} \left(b_0 + \frac{b_1}{N} + O\left(\frac{1}{N^2}\right) \right) \tag{17}$$

and the *leading term* b_0 is determined.
As said before, all the details can be found in the literature and I present only my evaluation here:

(+) One starts from a single basic rule, namely the retention of the leading term b_0 only. One is therefore dealing with a compact and internally consistent phenomenology.
(+) This basic rule has important implications:
• the factorization and valence quark ansatz are consequences, not independant assumptions; W exchange contributions have to be ignored by the same rational.
• Since $\xi = 1/N$ one realizes that $\xi_{eff} \sim 0$ is the *only possible* value on the leading level in 1/N; if, say, $\xi_{eff} = 1/3$ were found this scheme would have been a failure.

Another comment is in order he re to clear up a certain misconception: the data on D decays do not establish $\xi_{eff} = 0$; $\xi_{eff} \sim 0.15$ is allowed as well. Yet there is an important distinction to be kept in mind between a *fitted* and a *theoretically predicted or motivated* value: the latter does not have to represent the best fitted value as well, it only has to be consistent with it. In the same way one is not concerned at all that for example the predicted value of the W mass is not identical to the central value of the measured M_W as long as it is compatible with it.

(+) FSI do not contribute to the leading order in 1/N; considering the uncertainties that beset them one has to view this as a *theoretical* advantage.
(+) There appears room for internal improvement: at least in some cases it should be possible to estimate or even determine the next-to-leading correction expressed by b_1 in (17).
(+/-) We are dealing with a very smart phenomenology – yet phenomenology it still is in its determination of the numerical size of the various formfactors that enter.
(-) The basic rule, namely to retain the leading term b_0 only, is still ad-hoc.
(-) However unsavory it might be from a theoretical point of view (s. above) there is a clear phenomenological need for FSI, for example in $D \to K\pi$ decays or in $D^o \to \overline{K}^o \phi$ etc.

(d) The first theoretical description: QCD Sum Rules.

QCD sum rules represent the best theoretical technology that is presently available. Blok and Shifman [6] have applied them to $D \to PP$ and PV decays (the modes $D \to VV$ require much more work). Typical results are given in the following table:

	theoret. ratio	experim. ratio	pure W-exchange	naive Spectator Ansatz
$\dfrac{BR(D^o \to \overline{K}^o \pi^o)}{BR(D^o \to K^- \pi^+)}$	~ 0.25	~ 0.45	$\dfrac{1}{2}$	$\leq \dfrac{1}{18} = \dfrac{1}{2N_c^2}$
$\dfrac{BR(D^o \to \overline{K}^o * \pi^o)}{BR(D^o \to K^- * \pi^+)}$	~ 0.33	~ 0.49	$\dfrac{1}{2}$	$\leq \dfrac{1}{18}$
$\dfrac{BR(D^o \to \overline{K}^o \rho^o)}{BR(D^o \to K^- \rho^+)}$	~ 0.05	~ 0.07	$\dfrac{1}{2}$	$\leq \dfrac{1}{18}$
$BR(D^o \to \overline{K}^o \phi)$	~ 1.3%	~ 1 %	0.3 - 1%	$\leq 10^{-4}$

It was already said that the naive Spectator Ansatz is inconsistent with the data, as exemplified by this table; the same holds also for an ansatz assuming that W-exchange diagrams dominate, although this shows only in line 3 of this table. I have listed their predictions here mainly to illustrate that the QCD sum rules yield *non-trivial results in rough agreement with the data*. In the next, more sophisticated step one compares these results with those of the 1/N ansatz; doing that we realize two things:

(i) Whenever there is a leading term in 1/N the two treatments yield similar numbers. This implies that in these cases the various non-leading terms in 1/N cancel to a good degree.

(ii) When there is no leading term in 1/N -- for the moment the only example is $D^0 \to \overline{K}^0 \phi$ -- then the non-leading terms do not cancel to the same degree; instead they produce a number which is consistent with experiment.

My evaluation is as follows:

(+) The sum rules allow for a *systematic* inclusion of perturbative and non-perturbative features of QCD.
(+) A priori all possible contributions -- quark decay diagrams with and without colour matching, W exchange diagrams in the s and t channel -- have been included.
(+) Factorization has not been assumed.
(+) The sum rules provide good internal quality control via self-consistency requirements.
(+) There are *no free parameters* (in principle, though in practice there is some freedom in adopting numerical values for quark and gluon condensates etc.). This should be kept in mind before one overemphasizes the obvious fact that QCD sum rules do not yield an optimal fit to the data. We should also remember that from self-consistency requirements one estimates *the numerical uncertainties to amount to ~30-40%*.
(+) There is considerable room for improvement by including $SU(3)_{FL}$ breaking and by extending the anlysis to $D \to VV$ decays.
(-) At present it is unclear to which degree FSI have really been included or not.
(-) $SU(3)_{FL}$ breaking has been ignored so far.

(e) "The ultimate theoretical description": Lattice Monte Carlo calculations.

Like SDI they promise to take care of all our problems; like SDI they have not quite delivered yet on such promises; however unlike SDI they have a good chance to reach their stated goals and to do that within the foreseeable future. I actually believe that a world where SDI and Lattice Monte Carlo calculations shared one more feature, namely the funding level, would be a less evil place, irrespective of how such an equality were reached.

A note of caution: we already know that FSI -- rescattering etc. -- represent an important dynamical feature in charm decays. *The quenched approximation to lattice calculation will therefore not yield satisfactory results on the whole.*

(4) The Lessons:

There are four conclusions we can draw so far:
• It is mainly a *destructive interference in D^+ decays* that shapes the pattern of D^0 and D^+ decays.
• QCD sum rules are emerging as effective theoretical tools.
• Even so, our theoretical description contains at present a "~30% fuzziness". That is to say we cannot tell whether the ~30% uncertainties are due to improper or incomplete handling of FSI or an insufficient inclusion of W-exchange processes etc.
• *Comprehensive* data on branching ratios were crucial for the progress we have made in sorting out the actors in D decays. Twenty branching ratios measured with

20% accuracy yield much more insight than 2 branching ratios determined with 2% precison.

At this point it is tempting for theorists and also experimentalists to declare victory -- "we have conquered charm" -- and move on to greener pastures, namely B physics. Yet there are two things fundamentally wrong with such an attitude:

(i) It clearly reveals a lack of imagination. As Ronald Reagan declared after his reelection in 1984, "You haven't seen anything yet"! D^0-$\bar{D}^0$ mixing and CP violation in charm decays are excellent places to look for something exciting and new.

(ii) It represents hybris as well: apart from the fact that our description has sometimes been helped by a merciful imprecision in the data -- a kindness future data might not extend -- some new problems have emerged as discussed next.

(5) New Traps

It is expected that W-exchange contributions are more significant in Λ_c than in D^0 decays. Nevertheless, it is somewhat amazing that they shorten the Λ_c lifetime by a factor of two or so. I infer from (18) that two-body decay modes are significantly less important for Λ_c than D^0 decays. For lack of space I will not discuss this subject of growing topical interest any further here.

$$\frac{BR(D^0 \to K^+K^-)}{BR(D^0 \to \pi^+\pi^-)} = \begin{cases} 3.7 \pm 1.1 & \text{MK III} \\ 2.8 & \text{ARGUS} \\ 2.22 \pm 0.47 & \text{CLEO} \\ 1.4 & \text{theoret.} \end{cases} \quad (19)$$

The oldest piece of the original charm decay puzzle is thus still with us although more recent data seem to alleviate it.

$$\frac{BR(D^0 \to K^+\pi^-(\pi^0))}{BR(D^0 \to K^-\pi^+(\pi^0))} \sim \begin{cases} (9 \pm 5) \, tg^4\theta_c & \text{E691} \\ (0.2-2) \, tg^4\theta_c & \text{theor.} \end{cases} \quad (20)$$

$$BR(D_s^+ \to \phi\pi^+) \sim 2\% \quad (21)$$

(6) Future Campaigns

(i) Comparing the theoretical expectation with the data as stated in (19) we have to conclude that we do not understand FSI and/or $SU(3)_{FL}$ breaking.
So what?
We have to understand these decays before we can estimate reliably how much D^0-$\bar{D}^0$ mixing we expect in the Standard Model; the same holds for CP violation in $D^0 \to \pi^+\pi^-, K^+K^-$. There are further lines of attack we can choose:
(α) $SU(3)_{FL}$ breaking can be probed by comparing $BR(D^+ \to \pi^0\pi^+)$ with $BR(D^+$

$\to \bar{K}^0 \pi^+$). Yet even here allowance has to be made for the possibility of significant rescattering.

(β) It has not been established yet that $SU(3)_{FL}$ breaking is really as large as suggested by (19). For $SU(3)_{FL}$ symmetry relates the scattering amplitudes for $D^0 \to K\bar{K}$ with those for $D^0 \to \pi\bar{\pi}$ and not just T ($D^0 \to K^+ K^-$) with T ($D^0 \to \pi^+\pi^-$). Recent E400 and CLEO data suggest that BR ($D^0 \to K^0 \bar{K}^0$) is suppressed by a factor of five or so relative to BR ($D^0 \to K^+K^-$). It is *conceivable* that BR ($D^0 \to \pi^0\pi^0$) is actually equal to BR ($D^c \to \pi^+\pi^-$); in that case

$$\frac{|T(D^0 \to K^+K^-, K^0\bar{K}^0)|^2}{|T(D^0 \to \pi^+\pi^-, \pi^0\pi^0)|^2} \sim 1.4 \tag{22}$$

would hold -- in agreement with theoretical expectation. Data on $D^0 \to \pi^0\pi^0$ - while certainly not easy to obtain - are thus highly desirable.

(ii)   The doubly Cabibbo suppressed decays
$$D \to K^+ \pi's$$
deserve detailed studies [7].

(α) They provide sensitive tests for our understanding of non-leptonic D decays.

(β) They provide an annoying background to searches for $D^0 - \bar{D}^0$ mixing while at the same time representing an important ingredient for calculating the strength of mixing in the Standard Model. Looking at (20) shows that their observation is within reach.

(iii)   It would be quite instructive to analyze triple correlations in $D \to K_1 K_2 \pi_3 \pi_4$ decays, i.e., the expecation values $<\vec{p}_i (\vec{p}_2 \times \vec{p}_3)>$ where the $\vec{p}_i$ denote particle momenta.

(iv) D_S decays

D_S decays provided us with an early success, namely the approximate equality of $\tau(D_s^+)$ and $\tau(D^0)$ as obtained from V spin considerations. However our understanding has not progressed much further despite a considerable amount of data:

(α) The semi-leptonic branching ratio is still basically unknown.

(β) Actually no absolute D_S branching ratio has been determined directly; they are always related to the reference reaction $D_S \to \phi\pi^+$.

(γ) A recent CLEO analysis indirectly leads to
$$BR (D_s \to \phi\pi^+) \sim 2 \pm 1\% \tag{23}$$
whereas theoretically one predicts 3-4%. With one exemption no final state f has been found yet with

$$BR(D_s \to f) > BR(D_s \to \phi\pi) \qquad (24)$$

(δ) Adding up the observed as well as expected D_S decay modes and calibrating their branching ratios by (23) then suggests that up to half of all D_S decays are unaccounted for!

(ε) Data from MARK II and NA 14' suggest [8] one decay mode that could narrow down considerably this gap although there is contradictory evidence from MARK III:

$$\frac{BR(D_s \to \eta'\pi)}{BR(D_s \to \phi\pi)} = \begin{cases} 5 \pm 1.8 \pm 1.2 & \text{NA 14'} \\ \sim 1 & \text{th.} \\ \leq 1.9 & \text{Mk III} \end{cases} \qquad (25)$$

If the NA 14' number were to hold up, it would pose some intriguing theoretical questions:

• Maybe W-exchange is very important in D_S decays after all; it would enhance decays with an η', in particular if the latter contained a sizeable gluonic component in its wavefunction. Then we would also expect

$$BR(D_s \to \eta'\pi) \leq BR(D_s \to \eta'\rho) \qquad (26)$$

• Alternatively this large rate could be due to the (accidental) presence of a nearby *scalar* resonance with the appropriate internal quantum numbers: for this could contribute only to PP, but not PV final state. Accordingly one would expect

$$BR(D_s \to \eta'\rho) \ll BR(D_s \to \eta'\pi) \qquad (27)$$

There is some circumstantial evidence for such a scenario in other D_S decays:

$$\frac{BR(D_s \to \overline{K}^{0*}K^+)}{BR(D_s \to \phi\pi^+)} = \begin{cases} 1.05 \pm 0.17 & \text{CLEO} \\ 0.87 \pm 0.13 & \text{E691} \\ 1.44 \pm 0.37 & \text{ARGUS} \\ 1.03 \pm 0.3 & \text{MK III} \\ 0.77 \pm 30\% & \text{QCD SR} \end{cases} \qquad (28)$$

$$\frac{BR(D_s \to \overline{K}^0 K^+)}{BR(D_s \to \phi\pi^+)} = \begin{cases} 0.99 \pm 0.17 & \text{CLEO} \\ 0.92 \pm 0.28 & \text{MK III} \\ 0.93 \pm 0.27 & \text{ARGUS} \\ 0.43 \pm 30\% & \text{QCD SR} \end{cases} \qquad (29)$$

The apparent deficit in the theoretical expectation for the $D_S \to PP$ mode could be blamed on the presence of a scalar resonance.

In summary:

The situation in D_S decays is quite confused, if not even paradoxical at the moment:

– If one accepts the absolute value for BR $(D_S \to \phi\pi)$ given in (23) one is faced with the prospect of having roughly one half of the D_S decays unaccounted for.

– There is conflicting experimental evidence on whether $D_S \to \eta'\pi$ can provide a major part of the missing decays.

– The whole sitution is made even more puzzling by the fact that $\tau(D_S)$ and $\tau(D^o)$ agree within 10%, as expected.

Further experimental studies are urgently needed to clarify the situation. Once basic quantities like BR $(D_S \to \phi\pi, \eta'\pi, l\nu X)$ etc. have been determined one has to tackle more subtle issues:

• Does

$$\mathrm{BR}\,(D_s^+ \to \rho\pi^+)\,/\,\mathrm{BR}\,(D_s \to \omega\pi)\,/\,\mathrm{BR}\,(D_s \to \phi\pi) =$$

$$< 0.03\,/\sim 0.15\,/\,1 \tag{30}$$

hold as predicted by QCD sum rules?

• Are genuine multibody final states more important in D_S than in D^o decays?

• There is evidence for the $D \to VV$ transitions to be overestimated by the phenomenological models. Does this occur in D_S decays as well?

III BEAUTY DECAYS

(1)   Inclusive Rates

Of the several inclusive processes that have been measured in addition to $\tau(B)$ – BR $(B \to lX)$, BR $(B \to \psi X)$, BR $(B \to D_S X)$, BR $(B \to p/\Lambda X)$ – I will briefly discuss the first two only which are measured to be by ARGUS and CLEO [10]:

BR $(B \to l\nu X) = 10.3 \pm 0.3\,\%$ \hfill (31)

BR $(B \to \psi X) = 1.12 \pm 0.10 \pm 0.23\,\%$ \hfill (32)

To which degree can these numbers be reproduced by theory?

Even in the absence of a rigorous proof I accept the duality concept, namely that rates on the quark level are to be equated with rates on the hadron level asymptotically; e.g.,

$$BR([Q\bar{q}] \to l\, X) = BR(Q \to l\, v\, q') \quad \text{as } M_Q \to \infty \quad (33)$$

After all if top quarks are sufficiently heavy they decay as "free" quarks before they can hadronize and there should be a smooth transition to a phase where hadrons can still form [9]. This implies furthermore that

$$\xi_{asym} = \frac{1}{3} \quad (34)$$

will hold in this limit.

The real question is "how quickly is asymptotia reached"; more specifically, "is the beauty system already asymptotic?"

There is evidence that this is not quite the case. In particular to reproduce (32)

$$\xi_b \simeq 0 \quad (35)$$

has to be adopted! That means that at least the class of B decays that is driven by $b \to c\bar{c}s$ transitions is *not asymptotic yet*.

For the semi-leptonic branching ratio one finds accordingly

$$BR(B \to l\, X) \approx 12\text{-}14\,\% \quad \text{for } \xi \approx 0\text{-}1/3 \quad (36)$$

the lower value corresponding to $\xi \simeq 0$. The observed semileptonic branching ratio shows a ~ 20-30% deficit *relative to this expectation*.

Unless some unknown experimental uncertainties will close the gap between (31) and (36) there are three possible conclusions we can draw:

A) B lifetimes are not universal. E.g., if $b_{SL}(B^+) = 12\%$, $b_{SL}(B^0) = 8\%$ were to hold they would average out to (31); W exchange would be invoked to "explain" the reduced semileptonic branching ratio for B^0 mesons.

While I would not be surprised if

$$\frac{\tau(B^+)}{\tau(B^0)} = \frac{b_{SL}(B^+)}{b_{SL}(B^0)} \approx 1.1\,[1.2] \quad (37)$$

were to hold, I would find a ratio of 1.5 -- while conceivable in principle -- very hard to swallow theoretically. Yet there exists an even better argument now than my personal concern: by combining various analysis from CLEO and ARGUS it is concluded [10]

$$\frac{b_{SL}(B^+)}{b_{SL}(B^0)} \approx 1.0 \pm 0.2 \quad (38)$$

thus apparently closing this loophole.

B) The conservative conclusion would be to argue that the b quark is not sufficiently heavy to trust these simple calculation to better than, say, 20-30%.

C) One can consider the unlikely, yet conceivable and exciting possibility that there exists a neutral light Higgs H into which both B^+ and B^0 can decay [11]:
$$B \to HX \qquad (39)$$
The fact that the top mass has to exceed 76 GeV makes such a scenario less unlikely. It should be possible to identify such decays despite uncertainties about how the Higgs boson decays *if (39) indeed made up 20% of all B decays*.

(2) Exclusive Non-leptonic Decays

Two-body decay modes of B mesons can be treated like those of D mesons:
- Class I transitions -- in the terminology of ref. 4 -- are sensitive to the size of the coefficient a_1 (see (11)). The measured values [12] for BR(B→ $D^+\pi^-$, $D^{*+}\pi^-$, $D^{*+}\rho^-$, $D^{*+}a_1^-$) are quite consistent with the expectation
$$a_1 \approx 1.0 \qquad (40)$$
with little sensitivity to the value of ξ.
- Class II transitions depend on the coefficient a_2 and are thus very sensitive to the numerical size of ξ. From the observed branching ratios for B→ $\psi K^{(*)}$ one infers
$$|a_2| \approx 0.3 \qquad (41)$$
strongly suggesting (though not firmly establishing)
$$\xi \approx 0 \qquad (42)$$
at least for $b \to c\bar{c}s$ transitions. I find it quite remarkable (and somewhat miraculous) that the only value for ξ emerges that can make the 1/N ansatz relevant in B decays as well.
- This does not neccessarily mean (though it is tempting to conclude so) that the same value of ξ applies also for $b \to cud$ transitions or for B decays on the whole. The only information we have on that comes from a class III transition, namely $B^- \to D^0\pi^-$, where a_1 and a_2 terms interfere. The measured branching ratio appears to exhibit a preference for the value in (42), although it does not prove it within the present accuracy level.
- The factorization ansatz with the choice expressed in (40) appears to severely underestimate (by a factor of 4 or so) the observed size of BR(B→ D D_s etc.) [12]. While it might be premature to claim a definite discrepancy between data and theory, one should also keep in mind that these reactions take place fairly close to the charm-charm threshold where *pre-asymptotic effects might play a particularly significant role*.

IV SUMMARY AND OUTLOOK

A) Studies of charm decays represent a mature field: the available experimental technologies are excellent, the theoretical ones good. Yet such a situation is prone to breed complacency; the confused or at least confusing situation in D_S decays has reminded us again of this danger. We can make the reasonable claim that the main (though not all) features of non-leptonic D^0 and D^+ decays can be inferred from a destructive interference in D^+ decays. Yet at the same time we should feel obliged to exploit the excellent experimental technologies to the fullest and to transform the good theoretical technologies into excellent ones:
- Only then could we establish to have achieved a complete understanding of charm decays.
- *Dedicated searches for mixing and CP violation in D decays* are called for to exploit the substantial potential for discovering New Physics that exists in charm decays [13].

B) The physics of beauty decays is "exciting", i.e. "promising", i.e. not quite mature yet:
- Applying the dynamical lessons that we are learning in D decays to B decays represents more an art at present than a science.
- We are suffering from a stagnation in theoretical ideas or concepts for dealing with genuine multibody final states which appear to be dominating in B decays while being already significant in D decays.

EPILOG

There is a well-known saying:"When all is said and done, there is usually a lot more said than done!" This can serve as a characterization of our understanding of non-leptonic heavy flavour decays; it is certainly not very flattering -- yet at the same time not totally unfair either. What it points out is that the days of the "Gentleman Theorist" are over in this field: rather than the youthful elan of the dillettante -- however pleasing to the heart -- it is patient and plain hard work that is asked for! And that is also the way the game of soccer is supposed to be played in the state of Indiana.

ACKNOWLEDGEMENTS

I am grateful for interesting discussions with B.Gittelman, S.Stone, P.O'Donnell, N.Isgur and V.Khoze. I also benefitted from the stimulating ambience of Cafe Decadence.
This work was supported in part by the National Science Foundation under grant number PHY 89-09929.

REFERENCES

[1] For a very recent review see: R.Morrison and M.Witherell, preprint UCSB-HEP-89-01, to appear in Annual Review of Nuclear and Particle Science, Vol. 39.
[2] For recent reviews see: M.Wirbel, preprint DO-TH-88/2, to be published in Progress in Particle and Nuclear Physics;
I.I.Bigi, Nucl.Phys.B(Proc.Suppl.)7A(1989)318.
[3] L.-L.Chau and H.-Y.Cheng, Phys.Rev.D36(1987)137, with references to earlier work.
[4] M.Bauer, B.Stech and M.Wirbel, Z.Phys.C34(1987)103.
[5] A.J.Buras, J.-M.Gerard and R.Rückl, Nucl.Phys. B268(1986)16.
[6] B.Blok and M.A.Shifman, Yad.Fiz.45(1987)211,478,841; preprint ITEP 87-80.
[7] I.I.Bigi, Invited Lectures in: Proceedings of the XVIth SLAC Summer Institute, August 1988, SLAC Report No.336, E.Brennan (ed.).
[8] G.Wormser, these Proceedings.
[9] I.Bigi,Y.Dokshitzer, V.Khoze, J.Kühn and P.Zerwas, Phys.Lett.B181(1986)157.
[10]K.Schubert, these Proceedings.
[11]A.E.Blinov, V.A.Khoze and N.G.Uraltsev, preprint DESY 88-102.
[12]W.-Y.Chen et al., CLEO collab., preprint August 1989.
[13] I.I.Bigi, preprint UND-HEP-89-BIG01

PHENOMENOLOGY OF THE CKM MATRIX[*]

Yosef Nir
Stanford Linear Accelerator Center,
Stanford University, Stanford, CA 94309

ABSTRACT

The way in which an exact determination of the CKM matrix elements tests the Standard Model is demonstrated by a two-generation example. The determination of matrix elements from meson semileptonic decays is explained, with an emphasis on the respective reliability of quark level and meson level calculations. The assumptions involved in the use of loop processes are described. Finally, the state of the art of our knowledge of the CKM matrix is presented.

INTRODUCTION

The free parameters of the quark sector in the Standard Model (SM) are the quark masses and the mixing parameters. In the interaction basis, the charged gauge interactions are, by definition, diagonal:

$$\bar{d}_L^i M_d^I d_R^i + \bar{u}_L^i M_u^I u_R^i + \frac{g}{\sqrt{2}} W_\mu^+ \bar{u}_L^i \mathbf{1} \gamma^\mu d_L^i \quad . \tag{1}$$

For n generations, the mass matrices M_d^I and M_u^I are general $n \times n$ matrices, while $\mathbf{1}$ stands for the unit matrix. In the mass basis, the mass matrices are, by definition, diagonal:

$$\bar{d}_L^m M_d d_R^m + \bar{u}_L^m M_u u_R^m + \frac{g}{\sqrt{2}} W_\mu^+ \bar{u}_L^m V \gamma^\mu d_L^m \quad . \tag{2}$$

The charged gauge interactions, however, are no longer diagonal; the mixings are given by the unitary matrix V. The independent parameters are n eigenvalues of each mass matrix and $(n-1)^2$ parameters of the matrix V. At present we know of three quark generations, in which case V is the Cabibbo–Kobayashi–Maskawa (CKM) mixing matrix of *four* free parameters: three mixing angles and one phase.

If we have several independent measurements for a given CKM matrix element, or if we find the values of the nine entries, we will have the four mixing parameters *overdetermined*. Therefore, an *exact* determination of the CKM matrix elements provides us with a stringent test of the SM and with possible clues to physics beyond it. We explain this by showing what we can tell about the third generation from our present knowledge of the 2×2 Cabibbo matrix.

We survey the determination of different matrix elements from semileptonic meson decays. We explain the shortcomings of calculations at either the quark level or the meson level. We concentrate on the three above diagonal elements: $|V_{us}|$, $|V_{cb}|$ and $|V_{ub}|$.

---

[*] Work supported by Department of Energy contract DE–AC03–76SF00515.

© 1989 American Institute of Physics

Additional information can be derived from loop processes. The assumptions made are stronger. We explain these assumptions and show the constraints from $B - \overline{B}$ mixing and from the ϵ parameter.

THE FIRST TWO GENERATIONS

The Cabibbo mixing matrix for the first two generations is

$$V_C = \begin{pmatrix} V_{ud} & V_{us} \\ V_{cd} & V_{cs} \end{pmatrix} . \tag{3}$$

The value of $|V_{ud}|$ is calculated from the comparison of $0^+ \to 0^+$ superallowed Fermi β transitions:[1,2]

$$|V_{ud}| = 0.9747 \pm 0.0011 . \tag{4}$$

This is the most accurately determined of all CKM matrix elements. The calculation of the above value follows a continuous refinement of radiative corrections. The most recent one[2] takes into account $O(Z\alpha^2)$ corrections, and brings the eight accurately studied Ft values to agree within less than 1σ.

The value of $|V_{us}|$ is best determined from the measured rates of $K^+ \to \pi^0 e^+ \nu_e$ and $K_L^0 \to \pi^- e^+ \nu_e$, which give[3]

$$|V_{us}| = 0.220 \pm 0.002 . \tag{5}$$

The calculation cannot be carried out within the spectator quark model, because:
  a. The final spectrum is completely dominated by the single pion state, so that duality is not expected to hold.
  b. There are large QCD corrections as the relevant scale for $\alpha_s(\mu)$ is $\mu = O(m_s)$, but $m_s \sim \Lambda_{QCD}$ (the scale at which, by definition, $\alpha_s \sim 1$).
  c. There are large uncertainties in m_s: first, it is a running mass and we do not know the relevant scale and second, even if we knew the scale, the uncertainty in m_s is still about 30%.[4] This is significant, as the phase space for the decay depends on $(m_s)^5$.

Thus, the above value is derived from a phenomenological model:

$$\frac{BR(K \to \pi e \nu)}{\tau(K)} = C_K \left| f_+^K(0) \right|^2 |V_{us}|^2 , \tag{6}$$

where C_K includes factors with small uncertainties only. In general, the major difficulty is in the calculation of the form factor $|f_+(0)|$. In this case, however, only the three light quarks take part. In the $SU(3)$ symmetry limit ($m_u = m_d = m_s$) we have $|f_+(0)| = 1$. Deviations from the symmetry limit are second-order in the symmetry breaking parameter and calculable. Altogether we have a 1–1.5% error from experiment and about a 2% error in the theoretical calculations.

Our confidence in the above calculation of $|V_{us}|$ is supported by another independent measurement which gives a consistent value: a simultaneous fit to the rates of $\Lambda \to p e \nu$, $\Sigma^- \to n e \nu$ and $\Xi^- \to \Lambda e \nu$ gives[5] $|V_{us}| = 0.220 \pm 0.001 \pm 0.003$. Consistency with the meson decay data was achieved only after recoil corrections were taken into account.

There are two methods to determine $|V_{cd}|$ and $|V_{cs}|$. The first one is by using data from deep inelastic neutrino–nucleon scattering. One gets[6]:

$$|V_{cd}| = 0.21 \pm 0.03 \qquad (7)$$
$$|V_{cs}/V_{cd}| \geq 3.3 \ .$$

The bound on the ratio is derived with a mild assumption on the ratio of strange sea to antiquark sea in the nucleon, $2S \leq \overline{U} + \overline{D}$.

The second method is from D semileptonic decays. A reliable quark level calculation is still impossible due to the lightness of the c-quark; duality is questionable and QCD corrections may be large. However, the uncertainty in m_c at a given scale is small, so the question here is that of the relevant scale.

At the meson level, we have:

$$\frac{BR(D^0 \to X_q^- e^+ \nu)}{\tau(D^0)} = C_D \left| f_+^{D \to X_q}(0) \right|^2 |V_{cq}|^2 \ , \qquad (8)$$

where C_D includes factors with small uncertainties only. The uncertainty from the D^0 lifetime is common to both determinations,[7] $\tau(D^0) = 0.422 \pm 0.008 \pm 0.010$ psec. A large uncertainty comes from the calculation of the form factor. The charm quark is too heavy to make an $SU(4)$ symmetry useful for the calculation. Various calculations of the form factors, using quark models and QCD sum rules, give:

$$|f_+(0)| = \begin{cases} 0.6 \pm 0.1 & [8] \\ 0.75 - 0.82 & [9] \\ 0.75 \pm 0.05 & [10] \end{cases} \qquad (9)$$

The main difference between the determination of $|V_{cd}|$ and that of $|V_{cs}|$ comes from the experimental measurements. For $c \to s$, there are two measurements:

$$BR(D^0 \to K^- e^+ \nu) = \begin{cases} (3.8 \pm 0.5 \pm 0.6) \times 10^{-2} & [11] \\ (3.4 \pm 0.5 \pm 0.4) \times 10^{-2} & [12] \end{cases} \qquad (10)$$

With enough confidence in the models for the form factor, one may give a value for $|V_{cs}|$; e.g., $|V_{cs}| = 1.1 \pm 0.2$ for $|f_+(0)| = 0.7 \pm 0.1$. However, for $c \to d$, there is only one measurement and with large uncertainties[12]:

$$BR(D^0 \to \pi^- e^+ \nu) = (3.9^{+2.3}_{-1.1} \pm 0.4) \times 10^{-3} \ . \qquad (11)$$

The ratio $|V_{cd}/V_{cs}|$ is free of the uncertainties in $\tau(D^0)$. Moreover, it depends on the *ratio* $|f_+^{D \to \pi}/f_+^{D \to K}|$: this ratio is 1 in the $SU(3)$ limit, which is expected to hold within 10%. Thus, we get:

$$|V_{cd}/V_{cs}| = 0.25 \pm 0.06 \ . \qquad (12)$$

With present experimental errors and theoretical uncertainties, the more restrictive bounds come from deep inelastic scattering, but the measurements of D semileptonic decays give further confidence in these results.

To conclude, different direct measurements give the following range for the

Cabibbo matrix elements:

$$V_C = \begin{pmatrix} 0.9747 \pm 0.0011 & 0.220 \pm 0.002 \\ 0.21 \pm 0.03 & \geq 0.60 \end{pmatrix} . \qquad (13)$$

Now, suppose we knew about two generations only. Then unitarity would imply that the above matrix depends on one parameter only:

$$V_C = \begin{pmatrix} c_{12} & s_{12} \\ -s_{12} & c_{12} \end{pmatrix} . \qquad (14)$$

With the above measurements we have certainly *overdetermined* the Cabibbo angle. The test to the two-generation SM is the following: Can we find a range for the Cabibbo angle which is consistent with all measurements? The answer is positive; for $0.219 \leq s_{12} \leq 0.222$, we get the following ranges for the matrix elements:

$$V_C = \begin{pmatrix} 0.9750 - 0.9758 & 0.219 - 0.222 \\ 0.219 - 0.222 & 0.9750 - 0.9758 \end{pmatrix} , \qquad (15)$$

which is consistent with the measurements (13). Thus, the two-generation picture is still consistent and we could not tell that there is a third generation if not for its direct observation (or from CP violation). From our knowledge about $|V_{cb}|$ and $|V_{ub}|$, we know that the third-generation mixings would be probed only if we reached an accuracy level of 10^{-4} in the determination of $|V_{ui}|$ or 10^{-3} in the determination of $|V_{ci}|$ ($i = d, s$); this is well beyond the present level of accuracy. At present, the values in (13) imply only the following mild bounds on the possible mixings of a third generation:

$$V = \begin{pmatrix} \cdot & \cdot & \leq 0.07 \\ \cdot & \cdot & \leq 0.78 \\ \leq 0.14 & \leq 0.77 & \leq 1 \end{pmatrix} . \qquad (16)$$

Additional information on the parameters of the first two generations can be derived from indirect measurements, namely, SM loop processes. To extract useful information, we need to know all the significant contributions to such a process. Thus, we make two major assumptions:

  a. There are no additional generations. This assumption is unnecessary in the case of direct measurements.

  b. There are no significant "beyond-standard" contributions. For direct measurements, we assume that there are no beyond-standard processes which compete with the tree level SM processes, which is indeed the case for most "reasonable" models (with the possible exception of models with a light charged Higgs). For indirect measurements, we assume that there are no processes which compete with SM loop processes (which are suppressed by the high order in the weak interaction coupling and by the GIM mechanism). This is *not* the case in many extensions of the SM.

Finally, we note that as the GIM mechanism is in operation, the results have strong dependence on the masses of intermediate quarks.

The only loop process which does not a priori necessitate the existence of a third generation is ΔM_K, the mass difference between the two neutral K-mesons:

$$\frac{\Delta M_K}{N_K} \frac{(1-D)}{B_K} = \eta_1 \frac{m_c^2}{M_W^2} (s_{12})^2 \quad . \tag{17}$$

The N_K parameter is a known quantity, $N_K \equiv G_F^2 f_K^2 M_K M_W^2/(6\pi^2) = 2.1 \times 10^{-10}$ GeV. The long-distance contributions are given by $D \cdot \Delta M_K$. The B_K parameter gives the ratio between the short-distance contribution and its value in the vacuum insertion approximation. The η_1 parameter gives the QCD corrections, $\eta_1 = 0.7$. In the above, we used unitarity for two generations by putting

$$Re[(V_{cd}^* V_{cs})^2] \approx (s_{12})^2 \quad . \tag{18}$$

We note the strong dependence on m_c. When the original study of the K-$\overline{K}$ mixing[13] was performed, the c-quark was not yet experimentally discovered. Thus, one could use Eq. (17) to predict the mass of the c-quark. In the original calculation, the vacuum saturation approximation was used ($B_K = 1$), and neither long-distance contributions nor QCD corrections were taken into account ($D = 0$, $\eta_1 = 1$). This led, somewhat coincidentally, to the correct prediction: $m_c = 1.5$ GeV. With the full range of uncertainties in B_K and D, one gets:

$$8 \times 10^{-6} \le \frac{\Delta M_K}{N_K} \frac{(1-D)}{B_K} \le 5 \times 10^{-5} \quad , \tag{19}$$

which gives 1.3 GeV $\le m_c \le 3.2$ GeV. As we now know that $m_c \approx 1.4$ GeV, the two-generation picture is still self-consistent, even when information from the loop process ΔM_K is taken into account. Due to the very small mixings of the third generation, at present we could not find it from inconsistencies in the Cabbibo matrix.

THE ABOVE DIAGONAL ELEMENTS

In this section we concentrate on the determination of the three above diagonal elements:

$$V = \begin{pmatrix} \cdot & V_{us} & V_{ub} \\ \cdot & \cdot & V_{cb} \\ \cdot & \cdot & \cdot \end{pmatrix} \tag{20}$$

from semileptonic meson decays. The determination of $|V_{us}|$ was explained in the previous section: the s-quark is too light to allow a quark level calculation, but light enough to allow a reliable calculation of the form factor at the meson level.

The value of $|V_{cb}|$ is best determined from semileptonic B decays: $B \to X_c e \nu_e$. At the quark level, the process is $b \to c e \nu_e$. In this case:
 a. The dominant semileptonic modes are those with $X_c = D$, D^*. Duality should hold for the decay rate within about 10%.
 b. The relevant scale for QCD corrections is of order m_b. As $\alpha_s(m_b) \sim 0.2$, a first-order calculation should be fine to within 4% or so.
 c. The mass of the b-quark at a certain energy scale is known at the 2% accuracy level. Consequently, the crucial question is that of the relevant

energy scales. We will argue that there is no ambiguity of energy scales for m_c or, more accurately, in the ratio m_c/m_b. However, the question of energy scale for m_b in the $(m_b)^5$ factor is still open and remains the main source of uncertainty in the calculation. One possible way to overcome this difficulty is by fitting m_b to the leptonic spectrum. The fit is model-dependent, but if we use several models and let their parameters vary in a reasonable range, we may learn what is the uncertainty involved.[**]

Within the spectator quark model:

$$\frac{BR(b \to ce\nu)}{\tau_b} = \left[\frac{G_F^2}{192\pi^3}\right] m_b^5 F_{ps}(\rho_c) F_{QCD}(\rho_c) |V_{cb}|^2 \quad . \tag{21}$$

The experimantal quantities on the left-hand side are known with about 15% error, mainly from the b lifetime determination. The phase-space factor F_{ps} and the QCD correction factor F_{QCD} both depend on the mass ratio $\rho_c = m_c^2/m_b^2$. As mentioned, a priori there is an ambiguity, because quark masses are running, so that ρ depends on two scales:

$$\rho_c = \frac{[m_c(\mu_c)]^2}{[m_b(\mu_b)]^2} \quad . \tag{22}$$

The question is, what are the relevant scales μ_c and μ_b? The answer is[14] that to every choice of two scales, there corresponds a specific QCD correction factor. The modification of F_{QCD} is such that the product $F_{ps}(\rho) \cdot F_{QCD}(\rho)$ is independent of the choice of scales:

$$F_{ps}(\rho_c) F_{QCD}(\rho_c) = 0.46 \pm 0.04 \quad . \tag{23}$$

Various arguments suggest that the value of m_b should be taken as

$$m_b = 4.9 \pm 0.3 \text{ GeV} \quad . \tag{24}$$

As the decay width depends on $(m_b)^5$, this gives a 30% uncertainty. With the above values, we get:

$$|V_{cb}| = 0.046 \pm 0.008 \quad . \tag{25}$$

Various phenomenological models are, at present, in the stage of being tested against the experimental data. However, they all give $|V_{cb}|$ values which are somewhat higher than the spectator quark model value. To account for the model dependence of the calculation, we take:

$$|V_{cb}| = 0.048 \pm 0.009 \quad . \tag{26}$$

The value of $|V_{ub}|$ can be determined from semileptonic charmless B decays: $B \to X_u e \nu_e$. At the quark level, the process is $b \to u e \nu_e$. The calculation is subject to uncertainties similar to those of $|V_{cb}|$. It is advantageous to consider the ratio $q \equiv |V_{ub}/V_{cb}|$, rather than $|V_{ub}|$ itself:

$$\frac{BR(b \to ue\nu)}{BR(b \to ce\nu)} = \frac{F_{ps}(\rho_u)}{F_{ps}(\rho_c)} \frac{F_{QCD}(\rho_u)}{F_{QCD}(\rho_c)} q^2 \quad . \tag{27}$$

[**] We thank K. Schubert and G. Altarelli for discussions on this point.

The ratio is free of the uncertainties in $(m_b)^5$ and τ_b. Moreover, the ratio between the QCD correction factors does not depend [to $O(\alpha_s)$] on the choice of scale for α_s and, due to the lightness of the u-quark, $F_{ps}(\rho_u) = 1$ with no uncertainty. We get:

$$F_{ps}(\rho_u) F_{QCD}(\rho_u) = 0.85 \ . \tag{28}$$

The only theoretical uncertainty is then in $F_{ps}(\rho_c)$. We get:

$$q = (0.74 \pm 0.03) \left[\frac{BR(b \to u e \nu)}{BR(b \to c e \nu)}\right]^{1/2} . \tag{29}$$

Experiment does not provide us, at present, with $BR(b \to u e \nu)$, as there is no direct observation of charmless B decays. If one tried to subtract from the *measured* semileptonic rate the theoretically *calculated* charmed semileptonic decay rate, one would be left with zero and the $b \to u$ contribution "buried" within the large error bars.

Instead, $|V_{ub}|$ is determined from the electron energy spectrum. The spectator quark model is not appropriate for this analysis,[15] while various phenomenological models give very different results. The strongest experimental results with the weakest theoretical constraints give[16]:

$$q \leq 0.16 \ . \tag{30}$$

The CLEO collaboration recently reported[17] a measurement of $BR(b \to u e \nu) \neq 0$, but as the errors are still large and the result is not yet confirmed by other experiments we do not use it here.

To summarize: The above diagonal elements in the CKM matrix are best determined from semileptonic meson decays. For light mesons, or correspondingly light quarks, quark-meson duality does not hold because the spectrum is dominated by one final state. Moreover, even if the spectator quark model held, we would have practical difficulties in the calculation due to large QCD corrections and large uncertainties in the light quark masses. On the other hand, we are able to calculate rather accurately within phenomenological models, due to the approximate flavor symmetry. For heavier mesons, or correspondingly heavier quarks, the spectator quark model should give a reasonable description of the inclusive decay rate. QCD corrections are small and heavy quark masses are known rather well, though they remain the major source of uncertainty. In the case of heavy quarks, phenomenological models have no approximate symmetry to help control the hadronic matrix elements, and at this stage they should be tested against the experimental results rather than used to estimate the CKM matrix elements.

Direct measurements give:

$$|V_{us}| = 0.220 \pm 0.002 \ , \quad |V_{cb}| = 0.048 \pm 0.009 \ , \quad q \equiv \frac{|V_{ub}|}{|V_{cb}|} \leq 0.16 \ . \tag{31}$$

INDIRECT MEASUREMENTS

We now proceed in the same manner as in the two-generation case. We assume that there are only three generations. Unitarity implies that the following values for the CKM matrix elements:

$$V_{CKM} = \begin{pmatrix} 0.9747 \pm 0.0011 & 0.220 \pm 0.002 & \leq 0.009 \\ 0.21 \pm 0.03 & \geq 0.60 & 0.048 \pm 0.009 \\ \leq 0.14 & \leq 0.77 & \leq 0.9992 \end{pmatrix}, \quad (32)$$

should be consistent with a parametrization of four free parameters only:

$$V_{CKM} = \begin{pmatrix} c_{12} & s_{12} & s_{13}e^{-i\delta} \\ -s_{12}c_{23} - c_{12}s_{23}s_{13}e^{i\delta} & c_{12}c_{23} - s_{12}s_{23}s_{13}e^{i\delta} & s_{23} \\ s_{12}s_{23} - c_{12}s_{23}s_{13}e^{i\delta} & -c_{12}s_{23} - s_{12}c_{23}s_{13}e^{i\delta} & c_{23} \end{pmatrix}.$$
(33)

The above parametrization, recently adopted by the Particle Data Group,[18] is given here with the only approximation $c_{13} = 1$, which is good to $O(10^{-4})$—better than any of the experimental determinations.

Indeed, there is a range for the mixing parameters consistent with all data. It is simple to find it, as the values of the three mixing angles are equal to the absolute values of the above diagonal elements, which were derived in the previous section. Thus, the allowed ranges for the parameters is:

$$s_{12} = 0.220 \pm 0.002, \quad s_{23} = 0.048 \pm 0.009, \quad q \equiv \frac{s_{13}}{s_{23}} \leq 0.16. \quad (34)$$

Direct measurements do not constrain δ: $0° \leq \delta \leq 360°$.

Additional information on the matrix elements is derived from indirect measurements, namely loop processes. At present, we have no direct information on the mixings of the top quark:

$$V = \begin{pmatrix} \cdot & \cdot & \cdot \\ \cdot & \cdot & \cdot \\ V_{td} & V_{ts} & V_{tb} \end{pmatrix}. \quad (35)$$

The values of V_{ts} and V_{tb} are determined from unitarity, but V_{td} is still poorly determined:

$$|V_{tb}| \approx 1, \quad |V_{ts}| \approx |V_{cb}|, \quad |V_{td}| \leq 0.022. \quad (36)$$

The GIM mechanism implies a strong m_t dependence in loop processes. Thus, we will use the known values of five quark masses and of s_{12} and s_{23} to get constraints in the three-parameter space (m_t, q, δ).[19] To put constraints on the parameters from the indirect measurements, one assumes that

a. There are only three quark generations.

b. There are no significant contributions from any new physics.

The most useful measurements are those of the $B - \overline{B}$ mixing parameter x_d and the CP violating parameter ϵ. The x_d relation can be presented as follows:

$$N_B = \frac{x_d}{(\tau_b s_{23}^2)(B_B f_B^2)} \cdot F(m_t, q, \delta), \quad (37)$$

where N_B contains factors with small uncertainties only. There are large uncertainties in the quantities on the r.h.s of Eq. (37). We use:

$$x_d = 0.71 \pm 0.14$$
$$\tau_b s_{23}^2 = (4.1 \pm 1.0) \times 10^9 \text{ GeV}^{-1} \quad (38)$$
$$B_B f_B^2 = (0.15 \pm 0.05 \text{ GeV})^2 \quad ;$$

F is a function of the three unknown parameters (m_t, q, δ). We show the x_d bounds for either fixed m_t values (Fig. 1) or fixed q values (Fig. 2). The bounds correspond to the full range of parameters in Eq. (38). The ϵ relation can be presented as follows:

$$N_\epsilon = B_K \cdot G(m_t, q, \delta) \quad , \quad (39)$$

where N_ϵ contains factors with small uncertainties only. The only large uncertainty is in the B_K parameter. We use:

$$B_K = 0.7 \pm 0.3 \quad ; \quad (40)$$

G is a function of the three unknown parameters (m_t, q, δ). We show the ϵ bounds for either fixed m_t values (Fig. 1) or fixed q values (Fig. 2). The bounds correspond to the full range of B_K in Eq. (40) and of s_{23} in Eq. (34).

The final allowed range is that which lies within the direct bounds and within both the x_d-band and the ϵ-band. As the top mass becomes smaller, the allowed range in the (q, δ) plane becomes smaller. For $m_t \leq 47$ GeV, there is no allowed range, thus excluding this range for the top mass. For $m_t \sim 200$ GeV, almost all of the original range is allowed. For the mixing parameters, we get the following bounds from indirect measurements:

$$q \geq 0.015 \quad , \quad 12° \leq \delta \leq 178° \quad . \quad (41)$$

Within the three-generation SM, and using the unitarity conditions and all measurements (direct and indirect), we have:

$$V = \begin{pmatrix} 0.9750 - 0.9758 & 0.219 - 0.222 & 0.0008 - 0.009 \\ 0.217 - 0.223 & 0.9734 - 0.9753 & 0.039 - 0.057 \\ 0.006 - 0.020 & 0.037 - 0.057 & 0.9985 - 0.9993 \end{pmatrix} \quad . \quad (42)$$

The SM with three quark generations is still consistent with all measurements of the CKM matrix elements.

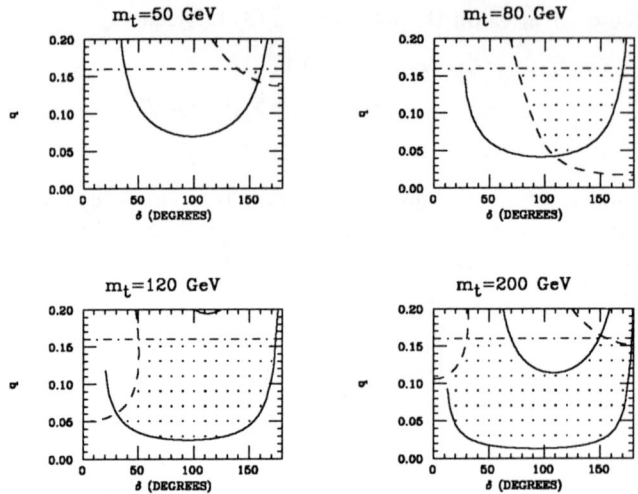

Figure 1: Allowed range of $q = s_{13}/s_{23}$ and δ for $m_t = 50, 80, 120$ and 200 GeV. The dot-dashed line gives the direct bound. The dashed lines give the x_d bounds. The solid lines give the ϵ bounds. The dotted area is the allowed range.

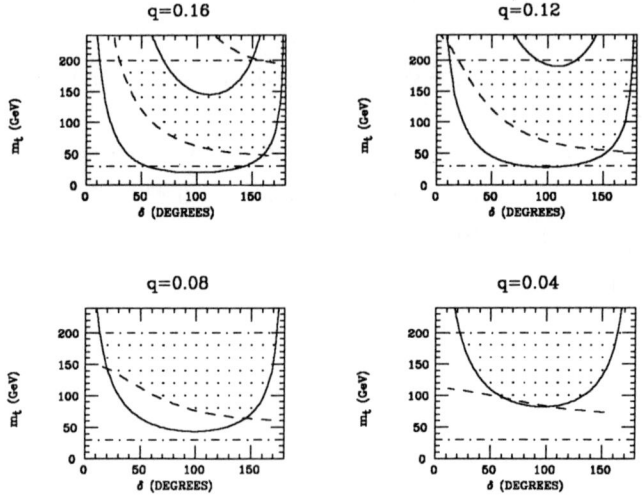

Figure 2: Allowed range of m_t and δ for $q = 0.16, 0.12, 0.08$ and 0.04. The dotdashed lines give the direct bounds. The dashed lines give the x_d bounds. The solid lines give the ϵ bounds. The dotted area is the allowed range.

REFERENCES

1. W. J. Marciano and A. Sirlin, Phys. Rev. Lett. **56**, 22 (1986).
2. A. Sirlin and R. Zucchini, Phys. Rev. Lett. **57**, 1994 (1986).
3. H. Leutwyler and M. Roos, Z. Phys. **C25**, 91 (1984).
4. J. Gasser and H. Leutwyler, Phys. Rep. **87**, 77 (1982).
5. J. F. Donoghue, B. R. Holstein and S. W. Klimt, Phys. Rev. **D35**, 934 (1987).
6. K. Kleinknecht and B. Renk, Z. Phys. **C34**, 209 (1987).
7. J. R. Raab *et al.*, Phys. Rev. **D37**, 2391 (1988).
8. T. M. Aliev *et al.*, Sov. J. Nucl. Phys. **40**, 527 (1984).
9. M. Wirbel, B. Stech and M. Bauer, Z. Phys. **C29**, 637 (1985).
10. C. A. Dominguez and N. Paver, Phys. Lett. **B197**, 423 (1987).
11. J. C. Anjos *et al.*, TPS Collaboration, Phys. Rev. Lett. **62**, 1587 (1989).
12. J. Adler *et al.*, MARK III Collaboration, Phys. Rev. Lett. **62**, 1821 (1989).
13. M. K. Gaillard and B. W. Lee, Phys. Rev. **D10**, 897 (1974).
14. Y. Nir, Phys. Lett. **B221**, 184 (1989).
15. N. Isgur, D. Scora, B. Grinstein, and M. B. Wise, Phys. Rev. **D39**, 799 (1989).
16. S. Behrends *et al.*, CLEO collaboration, Phys. Rev. Lett. **59**, 407 (1987).
17. M. Procario, these proceedings.
18. Particle Data Group, Phys. Lett. **B204**, 1 (1988).
19. Y. Nir, Nucl. Phys. **B306**, 523 (1988).

PREDICTIONS OF RARE B DECAYS

Patrick J. O'Donnell
University of Toronto, Toronto, Ont., Canada M5S 1A7

ABSTRACT

The rare decay modes of the B meson should soon be able to test the standard electroweak model; these rare decays may give a value for the top quark mass or signal the onset of new physics beyond the standard model. For this to be possible, an estimate of the expected background from the standard model is needed. It is shown how reliable these estimates can be in inclusive decays when the effects of QCD corrections taken into account.

INTRODUCTION

Much of what we know about the standard model comes from studies of the rare decays of particles, in particular, those of the K mesons. As is evident at this conference, there is now a possibility of using the B mesons as a new source of information. The rare decays of the B mesons (at present only in the form of upper limits) are exciting in that we shall have for the first time a system that replicates many of the details of the K system but which, due to differences in mass and couplings, has also very different characteristics. In particular, the couplings to the top quark, something which is highly suppressed in rare K decays, are expected to be a dominant feature in the B system. Consequently, the presence of, e.g., penguin diagrams, are much more likely to be evident in this new set of rare decays.

In this talk I shall consider two main questions, first, that of the sensitivity of the rare flavor-changing inclusive B decays to a putative value of the top quark mass and second, the degree of confidence that we may place in the decay rates of these processes in the presence of QCD corrections. The talk will summarize the results of three recent papers[1),2),3)] which have considered in some detail the consequences of QCD corrections to rare decays of the B.

RARE FLAVOR-CHANGING DECAYS

The rare decays involve the two processes $b \to s\gamma$ and $b \to sg$, where g denotes a gluon, which are absent at the tree level but arise when higher order electroweak interactions are taken into account. Such induced transition matrix elements are typical of the type occurring in the GIM mechanism[4)]. Indeed, if the top quark is sufficiently light, a GIM suppression will take place.

At this level of perturbation theory[5)] the matrix element for

the rare process b → sγ is rather sensitive to the value of the
top quark mass; this makes it an important way to determine the
value of this mass. If the top quark mass is greater than 50 GeV,
say, as reported as a possibility in this conference, then such a
mass becomes comparable to the scale set by the W boson, the other
intermediate particle in the induced process. We see here two very
different mass scales playing a role, the intermediate scale of
the order of the W mass and the external masses of the order of
the B meson. The QCD corrections to the basic lowest order matrix
element are expected to be of the order of $(\alpha_s(m_b)/\pi)\ln(m_t/m_b)$
which is of the same order of magnitude as the induced electroweak
process.

The one loop matrix element of the induced electroweak
process illustrates the sort of suppression mechanism described
above and how it is overcome whenever one of the quarks in the
intermediate state has a mass that differs considerably from the
others. The matrix element for the decay in the lowest order is

$$M = \frac{G_F}{2\sqrt{2}} \frac{e}{2\pi^2} \sum_i V_{ib} V_{is}^* F_2^i q^\mu \varepsilon^\nu \bar{s} \sigma_{\mu\nu} (m_b R + m_s L) b. \quad (1)$$

where L(R) is the left- (right-) projection operator and the sum
is over the charge +2/3 quark states u, c and t. The photon energy
is q^μ and F_2^i is a function of $x_i = (m_i/M_W)^2$, where m_i is the mass
of the u,c or t quark.

The function $F_2^i(x_i) \approx x_i$ for $x_i \ll 1$. Unitarity of the CKM[6]
matrix implies that the sum over the quark states in (1) may be
written as

$$\sum_i V_{ib} V_{is}^* F_2^i = V_{tb} V_{ts}^* (F_2(x_t) - F_2(x_u)) + V_{cb} V_{cs}^* (F_2(x_c) - F_2(x_u)). \quad (2)$$

In this form the GIM mechanism can readily be seen, for in
the case of all $x_i \ll 1$ the matrix elements will tend to zero. In
the b decay the c- and u- quarks satisfy the condition, $x_i \ll 1$,
leaving the terms involving the t quark. It is the dependence on
the t quark mass that makes this an interesting process.

QCD CORRECTIONS IN THE STANDARD MODEL

By the standard model we mean of course, the
$SU(3)_{color} \times (SU(2) \times U(1))$ model with a single complex Higgs
doublet, and six flavors of quarks. The heavy particles in the
b → s processes are the W bosons, the charged Higgs scalar $\phi^\pm$ of

the same mass as the W in the Feynman gauge with gauge parameter $\xi = 1$ and the top quark. Further details[7] are given in references 1, 2 and 3.

The important point is that in a perturbation expansion there will be terms of the form $\ln(M_W/\mu)$, where μ is the regularization mass scale. The obvious choice of the scale μ is that of the W mass in order to keep these logarithms small. A set of operators relevant to the $b \to s$ processes have their coefficients defined at this scale (and also all couplings and masses). Renormalization group equations are used to obtain the amplitudes at the physical scale relevant to the B decays. For the inclusive decays $B \to \gamma X_s$ ten operators are required[2] while the decays $B \to X_s e^+ e^-$ need two more operators[3].

In the calculations described here, the external particles are not put initially on the mass shell (which corresponds to using the equations of motion at an early stage of the calculation) since this may obscure the properties of the operators under renormalization. Also, the process $b \to s$ + gluon enters into the mixing of the set of operators. (These effects distinguish our calculations from the other complete calculation of these processes[8][9]).

Thus we seek an effective Lagrangian corresponding to the flavor changing $b \to s$ processes in which all heavy internal particles have been integrated out. Going from the scale $\mu \approx M_W$ down to the physical scale corresponding to the B system the renormalization group equations mix the operators. For example, if γ_{ij} are the elements of the

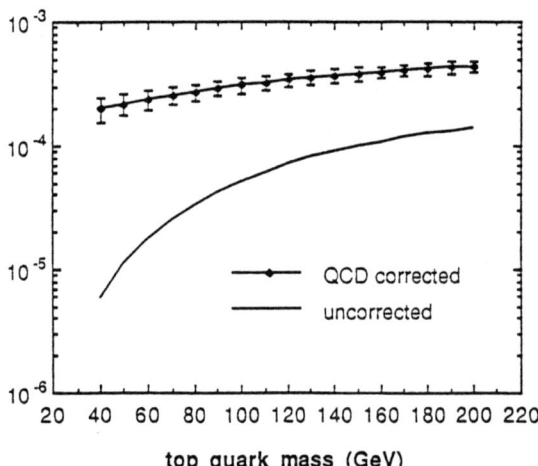

Fig. 1. The branching ratio for inclusive decays $B \to \gamma X_s$ as a function of the top quark mass showing the enhancement caused by QCD corrections. The error bars show the effect of varying Λ_{QCD}.

10 × 10 anomalous dimension matrix the renormalization group equations are:-

$$\mu \partial C_i / \partial \mu = (\alpha_s / 2\pi) \gamma_{ij} C_j (\mu) \tag{3}$$

The solution to these equations give rise to the inclusive branching ratio shown in Figure 1. Unlike the figures drawn in reference 2, this figure shows the ratio to the total rate rather than the semi-leptonic one.

(In the actual talk a different figure was shown corresponding to the controversy arising from the problem of definition of γ_5 in the dimensional regularization scheme. Since the conference it has been shown[10] that a method invented by Siegel[11] (dimensional reduction) to deal with this lack of definition in supersymmetry and explored by Altarelli et al[12] for use in V - A theories is in conflict with calculations done in four dimensions. Since the latter have a well defined γ_5 we conclude that the dimensional reduction scheme is not the correct method to use in two loop calculations such as those considered here. The four dimensional method does give agreement with the usual dimensional regularization method in which the γ_5 is defined to be a matrix which anti commutes with all γ's (as in the usual case in four dimensions) and is otherwise not defined in terms of the other γ's. This resolves the outstanding controversy between our results and those of Grinstein et al.[8]).

THE PROCESS $B \to X_s e^+ e^-$

In the standard model this process occurs in three distinct ways corresponding to penguin graphs with a single virtual photon, or a single (Z,ϕ) exchange or via box graphs involving the exchange of (W,ϕ). There are two further operators needed in addition to the ten for real photon emission[2].

In reference 2, the normalized decay spectrum is presented. The QCD corrections cause a suppression which becomes more pronounced if the mass of the top quark is at the lower end of the range we have been considering. This is due to the decrease in the contribution of the box diagrams and the increasing contribution coming from the charm threshold. However, at the smallest values of the invariant mass, $x = q^2/m_b^2$, of the dilepton pairs the contribution of the almost real photon dominates and, as we have seen, there is a significant enhancement. As a consequence, there is a mild enhancement to the integrated spectra as shown in Figure 2. That is, the contribution at low invariant mass overtakes the mild suppression at larger values of x. The branching ratio is down relative to the photon emission case by about two orders of magnitude.

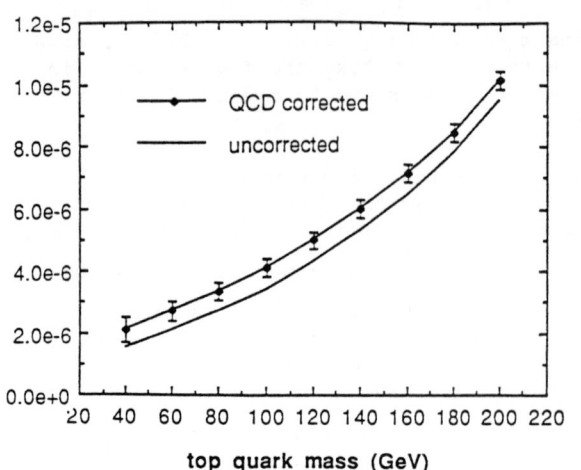

Fig. 2. The branching ratio for inclusive decays $B \to X_s e^+ e^-$ as a function of the top quark mass showing the enhancement caused by QCD corrections. The error bars show the effect of the variation of Λ_{QCD}.

NONLEPTONIC CHARMLESS B DECAYS

The rare decays which I have covered so far only involve the process $b \to sg$ in an indirect way, from the mixing of operators from the renormalization group scaling of the effective Lagrangian. Non-leptonic decays can proceed in a direct way through gluon emission. This makes them interesting since the effect of the QCD corrections in $b \to sg$ can be quite dramatic.

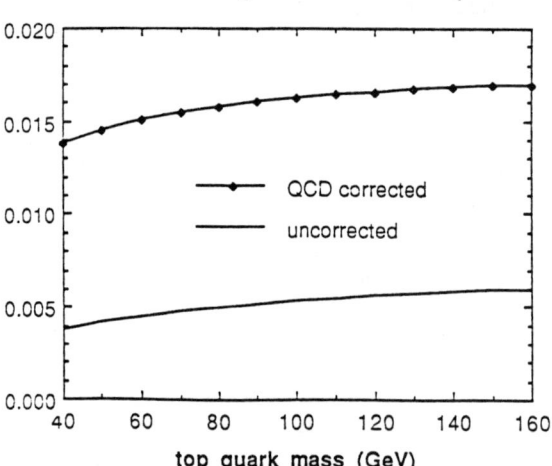

Fig. 3. The total of QCD corrections to the nonleptonic charmless B decay. The effects of the corrections to the sub-processes discussed in the text are to be found in ref. 3.

In the set of operators there are two types relevant here, involving the processes $b \to sg$ and $b \to sgg$. It was pointed out by Hou[13] that there are four types of charmless decays to consider. The most important ones are $b \to sgg$ and $b \to sq\bar{q}$ where the quarks can be u, d or s; the remaining ones are $b\bar{q} \to s\bar{q}$ and the direct single gluon emission b → sg. In the absence of QCD corrections the latter, as a function of the top quark mass, starts out for light masses as the smallest overall contribution but overtakes the b$\bar{q}$ → sq process at a top quark mass of about 120 GeV. The corrections give a dramatic change since the b → sg process is severely dampened by the QCD effects.

ACKNOWLEDGEMENTS

This work is the result of an enjoyable collaboration with Bob Grigjanis, Henri Navelet and Mark Sutherland. The research reported here was supported in part by NSERC (A3828), by NATO (248/88) and by CEN, Saclay.

REFERENCES

[1] R. Grigjanis, H. Navelet, P.J. O'Donnell and M. Sutherland, Phys. Lett. B213, 355 (1988).

[2] R. Grigjanis, H. Navelet, P.J. O'Donnell and M. Sutherland, Phys. Lett. B223, 239 (1989).

[3] R. Grigjanis, H. Navelet, P.J. O'Donnell and M. Sutherland, Phys. Lett. B224, 209 (1989).

[4] S.L. Glashow, J. Iliopoulos and L. Maiani, Phys. Rev. D2, 1285 (1970).

[5] B.A. Campbell and P.J. O'Donnell, Phys. Rev. D25, 1989 (1982).

[6] N. Cabibbo, Phys. Rev. Lett, 10, 531 (1963): M. Kobayashi and T. Maskawa, Prog. Theor. Phys. 49, 652 (1973).

[7] See also R. Grigjanis in Symmetry Violations in Subatomic Physics (World Scientific, Singapore,1988, ed. B. Castel and P.J. O'Donnell), p. 199, where a slightly more extended account is given of the techniques.

[8] B. Grinstein, R. Springer and M. Wise, Phys. Lett. B202, 138 (1988).

[9] B. Grinstein, M.J. Savage and M. Wise, Nucl. Phys. B319, 271 (1989).

[10] R. Grigjanis, H. Navelet, P.J. O'Donnell and M. Sutherland, UTPT preprint.

[11] W. Siegel, Phys. Lett. B84, 193 (1979).

[12] G. Altarelli, G. Curci, G. Martinelli and S. Petrarca, Nucl. Phys. B187, 461 (1981).

[13] W.-S. Hou, Nucl. Phys. B308, 561 (1988).

Long Range QCD Effects in Nonleptonic Weak Decays

Matthias Neubert[*)]

Institut für Theoretische Physik der Universität Heidelberg
Philosophenweg 16, D–6900 Heidelberg, West Germany

Abstract: Nonleptonic weak decays are very sensitive to nonperturbative quark-quark and quark-antiquark correlations. It is shown that effects of scalar diquarks dominate strange particle decays and are responsible for the drastic $|\Delta I| = \frac{1}{2}$ enhancement observed in these transitions. They also give rise to a large long range contribution to the $K_L - K_S$ mass difference. In more energetic processes, scalar diquarks are less important, but their contribution is still sizeable in inclusive D and D_s decays. The generation of virtual diquark pairs appears as an effective first step in the formation of baryons in B decays.

1. Introduction

Due to the strong influence of nonperturbative QCD, nonleptonic weak decays are notoriously difficult to calculate and have never been understood in a coherent way. From striking experimental observations like the famous $|\Delta I| = \frac{1}{2}$ enhancement in strange particle decays and the difference in the nonleptonic widths of D^0 and D^+ it is evident that QCD dynamics can cause drastic effects. As a consequence of such drastic manifestations of nonperturbative QCD, a precise extraction of Standard model parameters like V_{cb}, V_{ub}, and the CP violating phase angle from nonleptonic weak decays seems very difficult. On the other hand, these decays are ideally suited for the very study of nonperturbative QCD. The reason is that the initial state consists of a single isolated particle and that the transition operator is known (at the scale m_W) and has simple Lorentz and flavour structure. Furthermore, in decays of D and B mesons and heavy baryons an increasing number of open channels of great variety can be investigated.

We argue that the strong color forces between two quarks are as important as those between quark and antiquark. They lead to the formation of strongly correlated scalar two-quark states in color antitriplets, which dominate strange particle decays and, in particular, are responsible for the $|\Delta I| = \frac{1}{2}$ enhancement and for a large long range QCD contribution to $m_L - m_S$. The reason is that the effective weak Hamiltonian can directly generate and annihilate scalar diquark states with a sizeable amplitude. Effects of quark-quark correlations are also important in some D and D_s decays. Examples are decays with small energy release and inclusive decays. Also, diquarks seem to play an important role in baryonic B decays.

[*)] supported in part by the Bundesministerium für Forschung und Technologie, Bonn

2. Quark–Antiquark and Quark–Quark Correlations in Weak Decays

Short distance QCD corrections to the nonleptonic weak Hamiltonian give rise to an effective Hamiltonian consisting of local 4-quark operators multiplied by Wilson coefficients[1]. The conventional way is to write it as a product of color singlet charged and neutral currents. For the case of $|\Delta S| = 1$ strange particle decays one has

$$\mathcal{H}_{eff}^{|\Delta S|=1} = \frac{G_F}{\sqrt{2}} V_{us} V_{ud}^* \{ c_1(\mu) (\bar{d}\gamma_\nu(1-\gamma_5)u)(\bar{u}\gamma^\nu(1-\gamma_5)s) \\ + c_2(\mu)(\bar{u}\gamma_\nu(1-\gamma_5)u)(\bar{d}\gamma^\nu(1-\gamma_5)s)\} + \text{penguin operators} + \text{h.c.} \tag{1}$$

c_1 and c_2 are scale dependent coefficients compensating the scale dependence of the operators. At the scale where $\alpha_s(\mu) \simeq 1$ one has $c_1 \simeq 1.5, c_2 \simeq -0.9$, and the coefficients of penguin operators are still very small.

Long range QCD forces manifest themselves in hadronic matrix elements of $\mathcal{H}_{eff}$. To some extend, the attractive interaction between quarks and antiquarks shows up in the appearance of decay constants, which describe the direct generation or annihilation of mesons by quark currents. Since, for kinematical reasons, matrix elements of the (axial) vector current increase linearly with momentum, the direct generation of a final meson is the more important the more energy is available. At least part of these quark-antiquark forces can be accounted for by simply replacing the quark currents in $\mathcal{H}_{eff}$ by the associated hadronic field operators (in the following refered to as "new factorization"). With this approach, good semi-quantitative results for numerous energetic two-body decays have been obtained[2].

From the very existence of baryons it is evident that there are also strong attractive forces between two quarks in a color antitriplet state. The short distance contribution to this force is still one half of that between quark and antiquark in a color singlet. Clearly, bound two-quark states (diquarks) cannot exist as asymptotic particles because of their open color. They can, however, exist inside baryons and as intermediate virtual states. The essential point is that the major part of the weak Hamiltonian, i.e. that part involving the large coefficients c_1 and c_2, can be rewritten as a product of local (pseudo) scalar diquark currents[3]

$$\mathcal{H}_{eff}^{|\Delta S|=1} = \frac{G_F}{\sqrt{2}} V_{us} V_{ud}^* c_- \{ \epsilon_{kij}(\bar{u}_i^c(1-\gamma_5)d_j)^\dagger \epsilon_{klm}(\bar{s}_l^c(1-\gamma_5)u_m) + \text{h.c.}\} \\ + \text{product of color sextet currents} + \text{penguin terms}, \tag{2}$$

where we have explicitly written down color indices. $c_- = c_1 - c_2 \simeq 2.4$, and u^c etc. denote charge conjugate fields. The product of color antitriplet currents shown explicitly transforms like an $I = \frac{1}{2}$ operator, since the scalar (ud) diquark has necessarily $I = 0$. The product of sextet currents will be neglected in the following discussion of quark-quark correlations, since the corresponding short distance force is repulsive, and consequently a bound color sextet diquark state seems very unlikely.

The strong attractive quark-quark forces will manifest themselves in the fact that the diquark currents in $\mathcal{H}_{eff}$ can directly generate and annihilate correlated two-quark states. Although the associated couplings are not directly measurable, diquark decay constants g^D can be defined as the coupling of spatially extended two-quark states (as they exist in baryons) to the local currents present in the effective Hamiltonian. From a QCD sum rule analysis one obtains the following large numerical value at the scale where $\alpha_s(\mu) = 1$

$$g_2^D g_3^D \simeq 0.20 \times 0.17 \, \text{GeV}^4, \tag{3}$$

where the flavour indices 2 and 3 stand for (su) and (ud) diquarks, respectively[4]. A nice feature of the diquark model is that the product $(c_- g_2^D g_3^D)$, which arises in matrix elements of the effective Hamiltonian (2), is almost scale independent.

The sum rule analysis also indicates that diquarks of negative intrinsic parity are much heavier than 0^+ diquarks, as is expected from the mass splitting of mesons of different parity like π and $a_0(980)$. As a consequence, 0^- diquarks will generally give rise to smaller effects than 0^+ diquarks.

From the fact that matrix elements of scalar currents do not depend on the particle momentum, while those of vector currents increase linearly with momentum and hence become the more important the more energy is available, one gets the following general rule: If scalar diquarks can participate in a weak decay, this mechanism dominates over the direct emission of pseudoscalar mesons as long as $f_\pi m_q \ll g^D$, where m_q is the mass of the decaying quark[3]. This relation is well fulfilled in strange particle decays, where indeed diquarks play a dominant role and – as we will see – are responsible for the spectacular $|\Delta I| = \frac{1}{2}$ enhancement.

3. Diquarks and Strange Particle Decays

a) The $|\Delta I| = 1/2$ rule in K decays

In pure $|\Delta I| = \frac{3}{2}$ transitions like $K^- \to \pi^- \pi^0$ the relevant operators cannot directly generate scalar diquarks, and new factorization gives a relatively small amplitude in agreement with experiment. On the other hand, for decays like $\overline{K}^0 \to \pi^+ \pi^-$, where both $|\Delta I| = \frac{1}{2}$ and $\frac{3}{2}$ amplitudes contribute, the factorization result, which already takes into account quark-antiquark correlations, is far too small

$$\frac{1}{i} \mathcal{A}_{fac}(\overline{K}^0 \to \pi^+ \pi^-) \simeq 0.80 \times 10^{-7}\, \text{GeV}, \quad \text{but} \quad |\mathcal{A}_{exp}| \simeq 2.76 \times 10^{-7}\, \text{GeV}. \qquad (4)$$

From the previous discussion we expect long range quark-quark correlations to play a dominant role in these transitions. This is indicated in Fig. 1. Due to the generation of virtual diquark pairs the decay mechanism is considerably complex. However, quantitative calculations can be performed if one makes use of the stringent requirements of chiral symmetry.

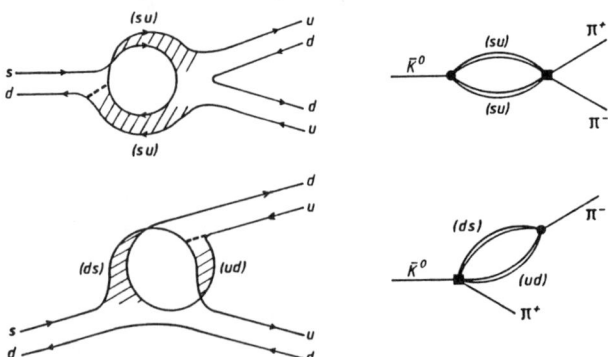

Fig. 1: Examples of quark-quark correlations in the decay $\overline{K}^0 \to \pi^+ \pi^-$. The W exchange is indicated by dashed lines. In the corresponding Feynman diagrams of the chiral diquark model the weak and strong vertices are indicated by full circles and squares, respectively

One represents the pseudoscalar meson octet P^a ($a = 1, ..., 8$) by the special unitary matrix

$$\Sigma(x) = \exp\left(\frac{i\sqrt{2}}{f_\pi} \lambda_a P^a(x)\right) \;, \; P_1 = \frac{1}{\sqrt{2}}(\pi^+ + \pi^-) \text{ etc.,} \tag{5}$$

which transforms linearly under the chiral group. The strongly correlated two-quark pairs are approximated by left and righthanded local diquark fields, $\Phi_L(x)$ and $\Phi_R(x)$, respectively. They transform like antitriplets with respect to both color and flavor SU(3). In the effective Hamiltonian (2) we replace the diquark currents by the local fields Φ_L multiplied by g^D, and obtain for the weak interaction part of the Lagrangian

$$\mathcal{L}_W = -\frac{G_F}{\sqrt{2}} V_{us} V_{ud}^* c_- \frac{2}{3} g_2^D g_3^D \Phi_L^\dagger \lambda_6 \Phi_L, \tag{6}$$

where $\frac{2}{3}$ is a color factor. The structure of the strong interaction Lagrangian is constrained by the requirements of chiral symmetry and the Gell-Mann theorem[5]. It is $SU(3)_L \times SU(3)_R$ invariant apart from terms involving the quark mass matrix m_q, which break the symmetry softly. The most general non-derivative diquark-to-meson coupling reads[6]

$$\mathcal{L}_{int} = \frac{m_1^2}{2}\left(\Phi_L^\dagger \Sigma^* \Phi_R + \Phi_R^\dagger \Sigma^T \Phi_L\right) + \frac{v'}{4}\left\{\Phi_L^\dagger\left(m_q^* \Sigma^T + \Sigma^* m_q^T\right)\Phi_L + \Phi_R^\dagger\left(m_q^T \Sigma^* + \Sigma^T m_q^*\right)\Phi_R\right\}. \tag{7}$$

Note that all the parameters of $\mathcal{L}_{int}$ are related to diquark masses. $m_1^2 = \frac{1}{2}(m_-^2 - m_+^2)$ determines the mass splitting between diquarks of different intrinsic parity, and v' controls the SU(3) mass splitting.

As a further approximation which greatly simplifies the model, we may remove the heavy 0^- diquarks from the theory by rescaling the Φ-fields

$$\psi_L = \zeta^\dagger \Phi_L \;, \; \psi_R = \zeta \Phi_R \text{ with } \zeta = (\Sigma^*)^{1/2}, \tag{8}$$

which makes the m_1-term a pure mass term. Then the limit $m_- \to \infty$ can be performed without complications.

By expanding ζ and Σ in powers of the meson fields P_a, the computation of the amplitude for the process $\overline{K}^0 \to \pi^+ \pi^-$ involves the evaluation of the diagrams shown in Fig. 1. We use a cutoff $\Lambda_\perp$ in the transverse momentum of the diquarks to regulate loop integrals. Such a cutoff appears very natural since clearly diquarks cease to be effective at large q^2. The loop integrals diverge only logarithmically with this cutoff. They depend on the masses of the (ud) and (su) diquarks, but it turns out that this mass dependence is almost compensated by the mass dependence of the diquark coupling constants[4], in practice leaving $\Lambda_\perp$ as the only free parameter. We shall consider values $\Lambda_\perp = 0.6 - 1.0$ GeV, corresponding to an average transverse momentum of the diquarks of $\langle p_\perp \rangle = 0.40 - 0.65$ GeV. Then

$$\frac{1}{i} \mathcal{A}_D(\overline{K}^0 \to \pi^+ \pi^-) = \frac{G_F}{\sqrt{2}} V_{us} V_{ud}^* c_- \frac{g_2^D g_3^D}{f_\pi^3} (m_K^2 - m_\pi^2) \frac{S(\Lambda_\perp, m_{ud})}{16\pi^2} = (1.75 \pm 0.43) \times 10^{-7} \text{ GeV}. \tag{9}$$

After adding the factorization contribution (4) (note that both have the same sign!), the total amplitude

$$\frac{1}{i}(\mathcal{A}_D + \mathcal{A}_{fact}) = (2.55 \pm 0.50) \times 10^{-7} \text{ GeV} \tag{10}$$

compares very favourably with the experimental number.

Our conclusion is therefore that the physics behind the $|\Delta I| = \frac{1}{2}$ rule, which has been a mystery since more than three decades, lies in a specific property of the weak Hamiltonian: It generates and annihilates with a sizeable amplitude correlated scalar two-quark states due to the strong attractive forces of QCD between two quarks in a color antitriplet state.

b) The $K_L - K_S$ mass difference

As a second application we study effects of quark-quark and quark-antiquark correlations in $|\Delta S| = 2$ transitions. Due to the nonvanishing $K^0 - \overline{K}^0$ mixing amplitude K_L and K_S have slightly different masses. The short distance contribution to the mass difference can be obtained from box diagrams as shown in Fig. 2a. Historically, this led to a semi-quantitative estimate of the charm mass[7]. The calculation in the six-quark Standard model, including short range QCD corrections, gives[8] (with $B_K = 0.6 \pm 0.2$)

$$\Delta m_K^{box} \simeq (0.95 \pm 0.35) \times 10^{-6} \text{ eV} \times \left(\frac{m_c}{1.35 \text{ GeV}}\right)^2. \tag{11}$$

Comparison with the experimental value $\Delta m_K^{exp} = (3.521 \pm 0.014) \times 10^{-6}$ eV immediately indicates that long range contributions to the mass difference must be of great importance. These contributions can be thought of as correlations between different fermion lines in the box diagrams. The strong quark-antiquark attraction leads to intermediate particle states, as indicated in Fig. 2b. In the spirit of the diquark picture, there will also be correlated two-quark intermediate states, which can be approximated by scalar diquarks. Typical graphs are depicted in Fig. 2c.

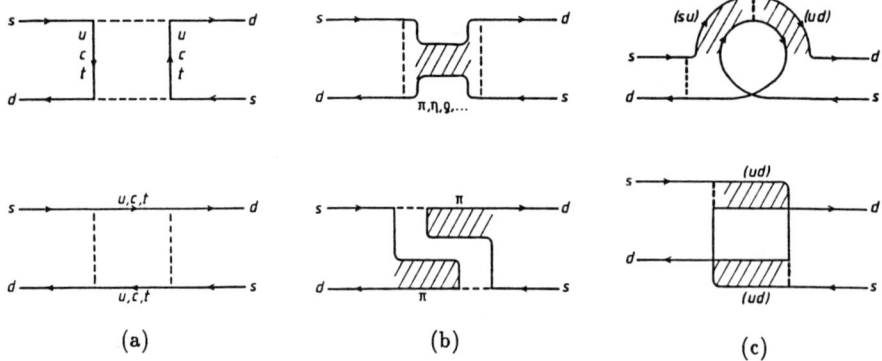

Fig. 2: Different contributions to $K^0 - \overline{K}^0$ mixing: (a) short distance contributions from box diagrams, (b) intermediate particle states due to quark-antiquark correlations, (c) intermediate diquark states due to quark-quark correlations

A straightforward calculation in the chiral diquark model gives a large positive contribution to the $K_L - K_S$ mass difference

$$\Delta m_K^D = \frac{2}{3} \frac{G_F^2}{m_K} (V_{us} V_{ud}^*)^2 c_-^2 \left(\frac{g_2^D g_3^D}{f_\pi}\right)^2 \frac{Q(\Lambda_\perp, m_{ud})}{16\pi^2} = (4.33 \pm 0.66) \times 10^{-6} \text{ eV}. \tag{12}$$

For the sum of all graphs the loop factor Q is convergent in the limit $\Lambda_\perp \to \infty$. However, we have again introduced a cutoff to account for the fact that diquarks are only effective at low q^2.

To get an estimate of the long range contributions due to intermediate particle states, we consider the single meson poles of π, η, η', ρ and ω (see Fig. 2b). We calculate the corresponding amplitudes in new factorization approximation and find $\Delta m_K^{pole} = -(1.5 \pm 0.5) \times 10^{-6}$ eV. The 2π intermediate state gives both negative and positive contributions to the mass difference, which are very small in factorization and partially cancel each other. We will neglect it here.

Adding to the short distance result the contributions due to quark-quark correlations and single meson intermediate states our theoretical estimate for the mass difference is

$$\Delta m_K^{th} = (3.88 \pm 0.90) \times 10^{-6} \text{ eV}, \tag{13}$$

which compares well with the experimental value. The theoretical uncertainties are quite large, however.

4. Diquark Effects in D and B Decays

In D decays, f_π times the mass of the decaying quark (m_c) is of about equal magnitude as g^D, indicating that diquark effects can still be important, but are no longer dominant. In energetic two-body decays the diquark loop contribution cannot compete anymore with a more direct meson generation. Indeed, factorization with asymptotic particle states gives a semi-quantitative description of many energetic two-body D decays[2]. On the other hand, for some specific channels diquark effects can still be sizeable, especially in some decays to pseudoscalar mesons and in decays with small energy release. They also have some influence on inclusive D decay rates. In fact, taking both quark-antiquark and quark-quark correlations into account one can qualitatively understand the nonleptonic width of D^0 and D^+ [6].

Since almost all B decays are very energetic, diquarks can in general only play a minor role. However, decays into baryons are an important exception because the formation of baryons is greatly facilitated if diquarks are generated in an intermediate step. It is very likely that at least almost all of those diquarks generated in the lowest binding states will finally turn into baryons. The Kobayashi–Maskawa matrix element V_{cb} allows the processes $b \to \bar{u}(cd)$, $b \to \bar{c}(cs)$, and from a simple scaling argument we expect $g_{cd}^D \simeq g_{cs}^D \simeq (0.35 \pm 0.05)$ GeV^2 for the corresponding diquark coupling constants to local currents. This gives[3,6]

$$\Gamma(b \to \text{ charmed } 0^\pm \text{ diquarks}) \simeq 4.4 \times 10^{10} \text{ s}^{-1}. \tag{14}$$

Contributions of excited diquark states could further enhance this estimate. Comparison with the experimental value

$$\Gamma_{exp}(B \to \text{ charmed baryons}) \simeq 6.5 \times 10^{10} \text{ s}^{-1} \tag{15}$$

indicates that the diquark mechanism is very effective and can in fact roughly account for the observed baryon yield.

Because of their connection to the matrix element V_{ub}, B decays into strangeness and charmless baryons are of particular interest. If baryon production proceeds via an intermediate diquark pair, a new $|\Delta I| = \frac{1}{2}$ rule must hold[6]. The current-generated (ud) diquark has necessarily isospin zero. Hence, $B \to N^* \bar{\Delta}$ can occur but not $B \to \Delta \bar{\Delta}$ (except through final state interactions). An estimate of the inclusive decay into an (ud) diquark in its lowest state gives a branching ratio

$$\text{Br}(b \to \bar{u}(ud)_{0+}) \simeq 4 \times 10^{-3} \left| \frac{V_{ub}}{V_{cb}} \right|^2. \tag{16}$$

This number is relevant for decays in which a proton or an N^* (decaying into $p\pi$) originates directly from the current-generated (ud) diquark in its lowest state. A three-particle decay mode such as $B^- \to p\bar{p}\pi^-$ is likely to be of this type. A branching ratio of $\sim 10^{-4}$ for this channel would therefore indicate that $|V_{ub}/V_{cb}|$ should be relatively large, i.e. $\gtrsim 0.15$.

5. Conclusions

Nonleptonic weak decays are ideally suited for the study of nonperturbative QCD. We have shown that strong long range color forces between two quarks are as important as those between quark and antiquark. Correlated two-quark states strongly affect weak transitions at low energy, because these states can directly be generated and annihilated by scalar diquark currents present in the effective Hamiltonian. In particular, diquarks are responsible for the $|\Delta I| = \frac{1}{2}$ enhancement in K decays and give a large long range contribution to the $K_L - K_S$ mass difference. The general rule in the low energy domain is that weak amplitudes are large if the appropriate transition operator can generate and annihilate scalar color antitriplet quark-quark states, and are small if not. In this latter case, amplitudes as obtained by new factorization are expected to give good results. Important examples are $|\Delta I| = \frac{3}{2}$ decays, B factors, and the quantity ϵ'.

Although diquark effects in D and B decays are in general much smaller than in strange particle decays, there exist some particular cases where they are still important. Examples are decays with a small energy release in the final state, and inclusive decay widths. Also, correlated two-quark states seem to play an important role in B decays into baryons, processes which can very effectively proceed via an intermediate pair of diquarks. This mechanism can roughly account for the observed yield of charmed baryons and gives rise to interesting selection rules. In particular, in decays to strangeness and charmless baryons a new $|\Delta I| = \frac{1}{2}$ rule is predicted.

Acknowledgements: This work was done in collaboration with B. Stech. It is a pleasure to thank M. Mistry and K. Berkelmann for organizing this stimulating Symposium. Lively discussions with H.G. Dosch, M. Jamin, and X.P. Xu are gratefully acknowledged.

References

1) M.K. Gaillard, B.W. Lee, Phys. Rev. Lett. **33**, 108 (1974)
   G. Altarelli et al, Phys. Lett. B**52**, 351 (1974)
   F.J. Gilman, M.B. Wise, Phys. Lett. B**93**, 129 (1980)
2) M. Bauer, B. Stech, M. Wirbel, Z. Phys. C **34**, 103 (1987)
3) B. Stech, Phys. Rev. D**36**, 975 (1987)
   B. Stech in: Diquarks, ed. by Anselmino and Predazzi, World Scientific 1989, 277
4) H.G. Dosch, M. Jamin, B. Stech, Z. Phys. C**42**, 167 (1989)
   M. Jamin, M. Neubert, in preparation
5) J.A. Cronin, Phys. Rev. **161**, 1483 (1967)
   M. Gell-Mann, Phys. Rev. Lett. **12**, 155 (1964)
6) B. Stech, Nucl. Phys. B (Proc. Suppl.) 7a, 106 (1989)
   M. Neubert, B. Stech, HD-THEP-89-11, to be publ. in New York Ann. of Phys.
7) M.K. Gaillard, B.W. Lee, Phys. Rev. D**10**, 897 (1974)
8) T. Inami, C.S. Lim, Prog. Theor. Phys. **65**, 297 (1981)
   F.J. Gilman, M.B. Wise, Phys. Rev. D**27**, 1128 (1983)

NON-PERTURBATIVE QCD IN WEAK DECAYS*

Stephen R. Sharpe
University of Washington, Seattle, WA 98195

ABSTRACT

The status of non-perturbative calculations of hadronic matrix elements is reviewed, with particular emphasis on lattice results. New results are presented for B_K and for the $K^+ \to \pi^+\pi^0$ decay amplitude.

INTRODUCTION

This talk gives a report on the status of non-perturbative QCD calculations as of June 1989. I will concentrate on developments which have occurred in the last year. I have tried to complement the presentation of Amarjit Soni.[1] He describes the status of calculations of amplitudes involving heavy quarks: f_D and f_B, the form factors in semileptonic decays such as $D \to K e\nu$, and the non-leptonic amplitudes $D \to K\pi$. I will only make some general comments on why it is difficult to do calculations for B mesons.

I should immediately confess that my talk is heavily weighted towards lattice QCD. This partly reflects my own expertise, partly lack of space and time, and partly my prejudice that the lattice method is the way of the future.

I will only discuss amplitudes involving exclusively light quarks. This is of interest to a Symposium on heavy quark physics for the following reasons. Much heavy quark physics involves the mesons containing a heavy quark and light anti-quark (or vica versa). While the physics of the heavy quark simplifies when $m_{\text{heavy}} \to \infty$, that of the light quark does not. The presence of the light quark makes calculations of amplitudes involving heavy-light mesons essentially as difficult as those involving only light quark mesons. Thus we can use the light quark mesons to judge the reliability of the calculations for heavy-light mesons.

The second reason for studying light quark amplitudes is that they are required for the indirect extraction of certain third generation parameters. An example, to be discussed below, is ϵ, which measures the amount of CP violation in $\overline{K}K$ mixing. ϵ is well measured, and is, in principle, predictable in the standard model in terms of m_t, V_{td} and V_{ts}. In practice, however, it is hard to calculate the QCD parts of the amplitude. Our ignorance is collected into the fudge factor B_K. If we could calculate B_K, then we would constrain the KM matrix and m_t.

Our inability to calculate is concentrated entirely in the QCD part of the standard model. In principle we aught to be able to calculate all strong interaction amplitudes in terms of m_u, m_d and m_s. A partial wish-list of simple quantities is:

- m_{proton}/m_ρ and other mass ratios
- f_π/m_ρ, f_K/f_π and other decay amplitudes
- electromagnetic form factors and structure functions
- Amplitudes for $K \to \pi\pi$ decay and $\overline{K}K$ mixing

---

* DOE Outstanding Junior Investigator. Work supported in part by DOE contract DE-AT06-88ER40423.

If we had a reliable method for doing such calculations we would proceed as follows. First, fix the overall scale using, say, m_ρ. Then fix the quark masses using m_π/m_ρ and m_K/m_ρ. Once this is done all other quantities, such as m_{proton}/m_ρ, are predictions. Furthermore, we would be able to predict the strength of α_s as measured in high-energy experiments.

I would not be giving this talk if we knew how to do such calculations. The difficulty occurs when the momenta of the gluons which are binding the quarks, or of the quarks themselves, become small. For then the effective coupling constant is large, and perturbation theory cannot be used. For low energy properties of mesons, such as masses or wavefunctions, the important physics is entirely outside the perturbative domain. One must use non-perturbative methods.

NON-PERTURBATIVE METHODS

Three approximate methods have been developed to do non-perturbative calculations in QCD: QCD sum rules, the large N_c (number of colors) approximation, and numerical lattice calculations. There are also models such as the non-relativistic quark model (NRQM) and bag model. I will not consider these models in this talk. This is because, unlike the three methods I will discuss, they do not have a direct connection to QCD. I am not leaving them out because they are less useful phenomenologically. On the contrary, the partially relativized quark model does a very good job of fitting the spectrum and decays of baryons and mesons.

An excellent account of the three methods is contained in the proceedings of the Ringberg Workshop.[2] Other useful reviews are Refs. 3 for sum rules, 4 for large N_c, and 5 for the lattice. Each method has its strengths and weaknesses. In my estimation, they are, at present, roughly equivalent in reliability. In the future the lattice method will be the most reliable, at least for the quantities listed in the Introduction.

In the sum rule method non-perturbative QCD is parameterized by a few vacuum expectation values. Quark confinement is assumed, and various sophisticated versions of duality are applied. The major advantages are that the approach is analytic, and that it can be applied to a wide range of amplitudes. The major drawbacks are that there are a number of different variants on the market, which do not always give the same results, and that the effect of higher order terms has not been systematically studied.

The large N_c approximation is narrower in its applicability, but more powerful when it can be applied. The most detailed applications are to the kaon decay amplitudes, to B_K and to the $\pi^+ - \pi^0$ mass splitting. The idea is to expand about $N_c = \infty$, in which limit all resonances become stable. Terms of order $1/N_c$, which are included, cause the resonances to decay. We cannot solve the theory in the large N_c limit, but we can use experimental data to fix certain parameters, and then make predictions. A simplified version of this approach has also been applied to D decays, with some success.[6] The advantages of the large N_c calculations are that they are analytic, and that, to a certain extent, they can be systematically improved. The disadvantages are, first, that it is hard to apply them to amplitudes not involving pions or kaons, second, that the higher order corrections are known to be large in certain cases (e.g. the η' mass), and, finally, that they have to use experimental data as input.

Lattice QCD is the only method based entirely on first principles. No assumption about confinement need be made, and, in fact, confinement has been demonstrated convincingly by the calculations. The approximations that must be made are, first, to formulate the theory on a grid of points with spacing a, and, second, to restrict the world to a finite box. Having made these approximations, the system is reduced to one with a finite number of degrees of freedom. At this point the quantum mechanics can be done on a computer using Feynman's path integral. The approximations can then be systematically removed, and, as this is done, one does recover continuum, infinite volume QCD. The important question is whether this recovery occurs for lattice sizes which can be put on present day computers.

The real answer to this question is: "not yet". We do have a good idea of what it will take, however, and, if the present rate of growth of computer power continues, computer simulations of QCD with few percent accuracy on mass ratios, decay constants, etc. are only 5-10 years away. Once we have the required computer power, the lattice method will be the preferred method for those quantities that it can calculate.

In the meantime, if we are to proceed, we must make another approximation: the QUENCHED approximation. This means we leave out all internal quark loops in diagrams, though all gluon loops are included. This has drastic consequences, such as making all resonances stable, and removing the pion cloud from around nucleons. This is similar to the large N_c approximation, which has no internal quark loops for $N_c \to \infty$. This is one of the reasons why I think of quenched QCD results as similar in reliability to large N_c calculations.

I should stress that the methodology for calculating quantities, whether mass ratios or kaon decay amplitudes, is the same in full QCD and in the quenched approximation. The theoretical work which has gone into figuring out how to calculate quantities will be directly applicable once we have the computer power to simulate the full theory. This means that, over the next few years, a large part of talks like this one will simply consist of a standard introduction followed by a list of results. The caveats will be gradually whittled away, mainly, I expect, due to increased computer power, rather than further theoretical work. In other words, what the future probably holds is a black box (labelled "non-perturbative QCD") which can calculate reliably a large number of amplitudes. I hope I am wrong about this, and that there will be a theoretical breakthrough in our ability to do calculations. But even if I am right, such a black box will be extremely useful, if not essential, in searching for physics beyond the standard model.

Since lattice calculations represent the probable way of the future, let me spend a little time explaining their strengths and limitations. My discussion will apply equally to the full theory and its quenched approximation. Using the quenched approximation simply introduces another systematic error, which is as yet poorly known.

The finite lattice spacing means that momenta greater than $\sim \pi/a$ are not allowed. For a small enough, however, such momenta can be dealt with perturbatively. Present day lattices have $a \geq 0.08$ fm, corresponding to $\pi/a \leq 8\,\text{GeV}$. This is either large enough, or nearly large enough, as has been deduced by comparing calculations at a variety of values of a.

A second problem due to finite lattice spacing is that one cannot use very heavy quarks. The Compton wavelength of a particle containing a heavy quark, such as a heavy-light meson, must be longer than the lattice spacing, or else the particle will see the effect of the grid of points. The condition is thus that $Ma < 1$, where M is the particle mass. Exactly how stringent the inequality must be can be determined by calculations. There is some evidence that $Ma \sim 1$ is sufficiently small.[7] If so, present lattices with $1/a = 2.0 - 2.5\,\text{GeV}$ do allow a direct study of charmed mesons, but not of bottom mesons. To study B mesons, one has to extrapolate from smaller quark masses. This is a tricky business which is in its early stages at present, but which will certainly improve as a is reduced. An alternative approach to study B mesons, due to Eichten and Le Page,[8] is to expand about infinite b quark mass. This is attractive in principle, but so far has not proved useful in practice.[9]

The finite box size has three effects. First, there is the direct effect of putting one or more particles into a box, rather than into infinite space. The energy and wavefunction of a particle are affected by the boundaries. The boxes I will discuss have linear dimensions of 1.6 – 2.4fm, just large enough for single particle finite size effects to be small. It is clearly impossible to perform a scattering experiment in such a box, for that would require asymptotic states consisting of two well separated particles. Instead, one has to deduce the scattering amplitude indirectly from the shifts in the energy of two particle states as the volume is changed. Thus one must be careful when looking at a two body decay such as $K \to \pi\pi$, for the attraction or repulsion of the final state pions can be substantial.[10] In principle, it is simple to test for such effects: simply vary the lattice size and see what happens. In practice, since computer time grows at least as fast as L^4, this is not so simple. One of the important steps forward in the last year has been that such finite volume tests have been performed.

The second effect of the finite box is to restrict the allowed momenta of particles: $|\vec{p}| = 0, 2\pi/La, \ldots$. For $1/a = 2\,\text{GeV}$ and $L = 24$, roughly the largest values used at present, the first non-zero momentum has a magnitude of $\sim 500\,\text{MeV}$. This makes it hard to do detailed studies of momentum dependence.

Finally, the finite box size puts a lower limit on the quark mass that can be used. The problem is due to the lightest hadron, the pion, the mass of which varies as $m_\pi^2 \propto m_q$ for small m_q. If the pion mass gets too small, its Compton wavelength becomes an appreciable fraction of the lattice size, and finite size effects alter its mass and properties. For these effects to be tolerable, one requires $m_q > m_s/3 - m_s/4$ with the present range of lattices sizes. This lower limit is also enforced by the fact that computer time grows as the inverse of m_q. I will argue, however, that this lower limit is low enough to see the small quark mass behavior of amplitudes.

I would sum all this up by saying that using lattice QCD one can calculate amplitudes involving a single, or possibly two, particles, at zero or small momentum, at moderately small quark mass. At present, this can be done well only when using the quenched approximation. In a number of years this restriction will be removed. The major drawbacks of the lattice method are the fact it is a numerical method, with associated statistical errors, and the impossibility of considering multi-particle states. For certain quantities one will always have to rely on other methods.

NEW RESULTS: (I) B_K

During the past year, new results for matrix elements have mainly come from lattice calculations, and I will concentrate on these. A summary of the status of all methods in 1988 is contained in Ref. 2. I will not talk at all about lattice results for the mass spectrum. A non-technical review which is almost current is given in Ref. 11.

The progress in lattice calculations has been due to an infusion of computer time. The Department of Energy, in addition to its regular allocations of time, held a competition for "Grand Challenge" proposals. One award, for 8000 hours on a Cray-2, was given to C. Bernard, R. Gupta, G. Kilcup, A. Soni and myself. Our aim has been to do reliable calculations of matrix elements in the quenched approximation. "Reliable" means that the systematic errors due to finite lattice spacing, finite volume and finite quark mass, should be reduced to a few percent level. Meanwhile, in Rome similar computations are underway using the home-built APE computers, which have a power similar to the Cray.

The kaon B parameter is defined by

$$\frac{8}{3}f_K^2 m_K^2 B_K(\mu) = \mathcal{M}(\mu) \equiv \langle \overline{K}_0|\bar{s}\gamma_\mu(1-\gamma_5)d\,\bar{s}\gamma_\mu(1-\gamma_5)d\,|K_0\rangle. \quad (1)$$

Here the normalization is $f_K = 1.22 f_\pi = 1.22 \times 135$ MeV. The matrix element on the right-hand side depends on the scale, μ. Roughly speaking, this means that, in the evaluation of the matrix element, quarks and gluons are restricted to have momenta with magnitude smaller than μ. Thus a lattice calculation of the matrix element, in which internal momenta are restricted to be less than π/a, corresponds to $\mu \sim \pi/a$.

The matrix element $\mathcal{M}$ in Eqn. (1) arises in the theoretical calculation of ϵ, the CP violating part of the $\overline{K}K$ mixing amplitude. In the standard model CP violation comes from box diagrams with intermediate charmed and top quarks. The third generation must be involved if the CP violating phase δ is to appear. Since the quarks in the box are heavy, one can use perturbation theory to calculate the box diagram. The end result it that the box gets replaced by the matrix element $\mathcal{M}$ times a perturbatively calculable coefficient:

$$\epsilon = (\text{known factors})g(m_t)s_3 s_\delta \eta_2(\mu)\mathcal{M}(\mu). \quad (2)$$

Here $g(m_t)$ is a known function, which monotonically increases with m_t. The sines s_3 and s_δ are from the Particle Data Booklet parametrization of the KM matrix. The coefficient $\eta_2(\mu)$ is calculable in perturbation theory. Thus it is known only for large enough μ, where perturbation theory is reliable. At 1-loop one finds $\eta_2(\mu) \propto \alpha_s(\mu)^{-2/9}$. The μ dependence of $\mathcal{M}$ exactly cancels that of η_2, so that the physical parameter ϵ is independent of the scheme being used to do the calculation.

Different methods of calculation correspond to different values of μ. Thus it is convenient to use a common standard to make comparisons. The conventional way is to use

$$\hat{B}_K = \alpha_s(\mu)^{-2/9} B_K(\mu) \propto \eta_2(\mu) B_K(\mu), \quad (3)$$

which does not depend on the scale. The only difficulty with this definition is that one has to know which value of μ to use in α_s. The lattice calculation corresponds to $\mu \sim \pi/a$, but should one use $\mu = \pi/2a$, $\mu = \pi/a$ or $\mu = 2\pi/a$? This makes a difference in $\hat{B}_K$. This questions can be resolved by higher order perturbative calculations, which have yet to be done. Because of this uncertainty, I prefer to use the "bare" $B_K(\mu)$ when comparing lattice results to each other, and turn to $\hat{B}_K$ only when comparing to results of other methods.

Chiral symmetry provides an important constraint on B_K. In particular, if one could vary the s and d quark masses, and thus vary m_K, then for small m_K one must find:

$$B_K = c_1 + c_1'\left(\frac{m_K}{4\pi f_K}\right)^2 \log(\frac{m_K^2}{\Lambda^2}) + c_2\left(\frac{m_K}{f_K}\right)^2. \quad (4)$$

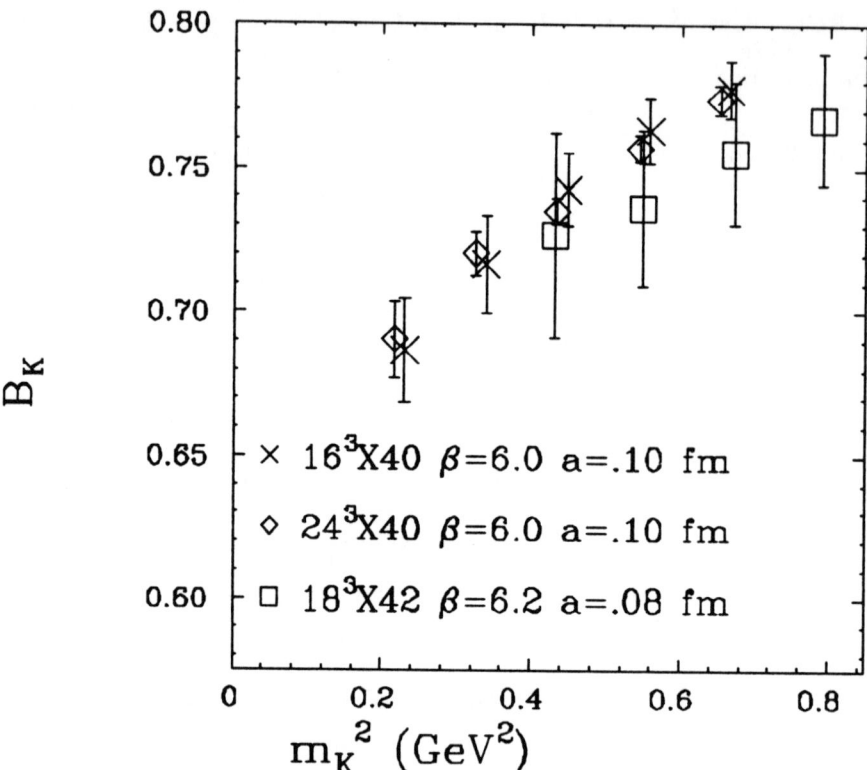

Figure 1. B_K for staggered fermions on various lattices.

Here c_1, c_1', c_2 and Λ are constants to be determined. The important point is that B_K should tend to a constant in the limit $m_K^2 \to 0$, rather than diverge.

It has taken some time to do a calculation which shows this behavior. The problem is caused by the difficulty in putting quarks on the lattice in such a way that chiral symmetry is not broken. The two computationally useful types of lattice quarks are staggered and Wilson fermions. For the former one can show[12] that appropriately defined lattice matrix elements must behave as in Eqn. (4). The trade-off is that the calculations are more complicated. For Wilson fermions the chiral behavior is not guaranteed, and one must make subtractions to enforce the correct limit.[13] This introduces greater errors into the calculation. Nevertheless, it is very important that both types or fermion are used, and that they agree on B_K, since this is a stringent test that a is small enough.

I show in Figure 1 the results from staggered fermions for B_K as a function of m_K^2.[14] Notice that the x-axis shows m_K^2 in physical units, so that the physical kaon appears at 0.25. This is within the range of the data points, so we can directly calculate B_K. Notice also that the y-axis is the bare B_K, and not $\hat{B}_K$. This means that there has been no massaging of our results between the

Table 1. Lattice B_K

$\beta = \frac{6}{g^2}$	a (fm)	Lattice	Number	Staggered	Wilson
5.7	0.2	$16^3 \times 32$	30	0.98 ± 0.02	$0.87 \pm .11 \pm .13$
6.0	0.1	$16^3 \times 40$	31	$\binom{0.71 \pm 0.02}{0.69 \pm 0.03}$	$0.70 \pm .10 \pm .14$
6.0	0.1	$24^3 \times 40$	15	0.70 ± 0.01	$0.61 \pm .09 \pm .03$
6.2	0.08	$18^3 \times 40$	27	0.69 ± 0.03	
6.0	0.1	$10^2 \times 20 \times 40$	30		$0.75 \pm .20$
6.2	0.08	$16^3 \times 48$	15		$0.60 \pm .15$

computer and the plot. The plot shows three sets of results: two sets at the same lattice spacing ($a \sim 0.1$fm) but with different sizes ($16^3 \times 40$ and $24^3 \times 40$); and a further set from a smaller lattice spacing $a \sim .08$ with size $18^3 \times 42$. The important points to note are:

- Correct chiral behavior. In particular, it appears that our lightest quark (with mass $m_q \sim m_s/3$) is light enough so see the $m_K^2 \to 0$ behavior.
- Small statistical errors. This is due in part to a quite large sample of configurations, but also due to the use of new methods.[15]
- Small finite volume effects. Even within the small errors the results on the two lattices with $a = 0.1$fm agree. The physical dimensions of the two lattices are ~ 1.6fm and 2.4fm. The latter is almost as large as is possible to run on present computers, and it is fortunate that it is large enough.
- Small finite lattice spacing corrections. This we deduce from a comparison of the data at the two lattice spacings. We expect B_K to be smaller for the smaller lattice spacing, because of the scale dependence given in Eqn. (3). However, this is a 0.5% effect, and thus smaller than the error bars. Because the $a = 0.08$fm, $18^3 \times 42$ lattice is relatively small in physical units (~ 1.4fm), we cannot go to such small kaon masses. This means that we have to extrapolate to calculate B_K at the physical kaon mass. Nevertheless, it is fairly convincing that the two lattice spacings are in agreement. We hope in the next year to check this by repeating the calculation on a large ($32^3 \times 48$) lattice with $a = 0.08$fm.

The results for B_K at the physical kaon mass are collected in Table 1, together with information about the lattices. β is related to the coupling constant: $\beta = 6/g^2$. As the table shows, we have made another check by doing two independent calculations on the smaller lattice with $a = 0.1$fm. The results agree within errors, giving us confidence in our error estimates. We also have a result at a larger lattice spacing, $a = 0.2$fm, on a 16^3 lattice. This is the largest lattice in physical units, permitting us to reach the smallest physical kaon masses. This allows us to check that the smooth behavior as a function of m_K^2 seen in Fig. 1 continues to smaller kaon masses.

Larger a corresponds to smaller μ. If μ is too small perturbation theory is inapplicable, and we expect the calculation of $\hat{B}_K$ to break down. According to Eq. (3), and using the coupling constant from $g^2 = 6/\beta$, we expect that $B_K(a = 0.2fm)$ should be 1% larger than $B_K(a = 0.1fm)$. In fact, we find a much larger change. For this reason, one should not trust quantitative calculations of $\hat{B}_K$ for $a > 0.1$fm. This is consistent with expectations from other calculations.

The table also includes, in the first three rows, the new results with Wilson fermions from the BS collaboration,[16] For Wilson fermions there is an extra systematic error due to the subtraction, and this is the second error in the BS data. The last two rows show old results from the European Lattice Collaboration (ELC), with only the statistical error.[17] There are also new ELC results,[18] but I do not know the details, so I do not wish to quote them. The agreement between Wilson and staggered fermion results is satisfactory, and provides additional evidence that the systematic errors due to finite lattice spacing are small.

The only remaining worry concerns the logarithmic terms in Eq. (4). These "chiral logs" come from loops of kaons. For example, the operator in $\mathcal{M}$ can produce an extra $\overline{K}K$ pair, in addition to converting the external K into a $\overline{K}$. The extra pair can then join to form a loop. It turns out the coefficient c_1' can be calculated in terms of c_1, and that, in full QCD, the chiral log term can be a large ($\sim 50\%$) contribution to B_K.[19] The important issue is whether lattice calculations properly include these chiral logs. If they do not, then this could be another, hidden, systematic error.

The first worry is that the quenched approximation either removes or reduces the chiral logs. This does not happen. The ratio c_1'/c_1 is very similar in the quenched and full theories.[14] The second worry is that the loops, which require intermediate states with three particles, will be greatly altered by the finite box. If this were so, then the result for B_K should vary with the volume. As Fig. 1 shows, however, there is little of no volume dependence. Thus it appears that the chiral logs are being calculated correctly. Further study, however, throws doubt on this conclusion, and suggests that the chiral logs are being underestimated.[14] If so, the result for B_K will increase, but by no more than 10%. Clearly, more work is needed to clarify this issue.

Although it is unfortunate that such worries remain, the fact that they involve detailed physical considerations indicate that QUENCHED lattice calculations have come of age.

I close this section with a comparison of different methods, using $\hat{B}_K$ defined in Eq. (3). I will use the staggered fermion result, as this has the smallest errors. The uncertainty in μ, or equivalently in α_s, means that a result $B_K(a = 0.1fm) = 0.70$ converts to $\hat{B}_K = 0.9 - 1.0$. Thus, with the possible exception of chiral logs, the statistical and systematic errors in B_K are swamped by the uncertainty in μ.

The lattice result is larger than those from the other methods. The large N_c approximation gives[4] $\hat{B}_K = .75 \pm .15$. Different implementations of the sum rules give various values, so I quote R. Decker[20] who gives an overall value $\hat{B}_K = 0.55 \pm 0.25$. The quenched lattice is thus consistent with large N_c, but may be inconsistent with the sum rules. As I said above, I do not think that one of these calculations stands out as clearly superior. Thus the message to someone attempting an analysis of the consequences of ϵ is to use the range $B_K = 2/3 \pm 1/3$.[21] The lattice result underscores the fact that values close to unity should not be ruled out. This is amusing, since $\hat{B}_K \sim 1$ is the value coming from the vacuum insertion approximation, first used in this context by Gaillard and Lee back in 1974.[22]

NEW RESULTS: (II) $K^+ \to \pi^+ \pi^0$

The amplitude $\mathcal{A}_{phys} = \mathcal{A}(K^+ \to \pi^+\pi^0)$ provides an opportunity to test the non-perturbative methods. We know the KM parameters in the quark-level decay, there are no loops of heavy quarks, and we know the experimental result. All we have to do is calculate the non-perturbative physics which converts from a quark decay to a meson decay. Furthermore, this amplitude is more simple to calculate than that for K^0 decay, because it is pure $\Delta I = 3/2$. Certain types of diagrams, the "eye diagrams", are absent.

The large N_c approximation has done a good job of understanding this decay. The result for the amplitude is too small by 5-30%, depending on the parameters used.[4] QCD sum rules are similarly successful,[23] finding a result 5% too large. The sum rule calculation, however, does not include the final state interactions which appear to be important in this amplitude, as discussed below. Thus the success may be partly illusory. Still, the challenge is for the quenched lattice is to reproduce the amplitude at the 30% level.

For technical reasons, a direct calculation of the on-shell amplitude is not possible. One must calculate the off-shell amplitude $\mathcal{A}_{off}$ in which the initial kaon and final pions are all at rest, and the u, d and s quarks are degenerate.[24] Lowest order current algebra then implies that $\mathcal{A}_{phys} \propto \mathcal{A}_{off}/m_K^2$. Here and in the following m_K^2 refers to the common mass of the lattice kaon and pions, and not to the physical kaon mass. Chiral symmetry implies a constraint for this ratio similar to that for B_K:

$$\mathcal{A}_{phys} \propto \frac{\mathcal{A}_{off}}{m_K^2} = d_1 + d_1' \left(\frac{m_K}{4\pi f_K}\right)^2 \log(\frac{m_K^2}{\Lambda^2}) + d_2 \left(\frac{m_K}{f_K}\right)^2 . \quad (5)$$

The actual calculation is more complicated than that for B_K, because there are two particles in the final state. On the other hand, the theoretical interpretation is cleaner, because there is no need for subtractions. For a review of the status at the end of 1988, see Ref. 25. The calculation has so far only been done with Wilson fermions,

Figure 2 shows data for $\mathcal{A}_{off}/m_K^2$, normalized so that the experimental amplitude is 1. I have only included data from lattices which have $a \leq 0.1$fm (i.e. $\beta \geq 6.0$). The new BS results[16] are from the same $a = 0.1$fm lattices as discussed in the previous section. The old ELC data are from the lattices listed in the last two rows of table 1.[17] Thus the ELC $a = 0.1$fm lattice is smaller than either of the BS lattices. New ELC data will presumably be appearing this summer.

The most striking feature is that all points lie considerably above the experimental result. An average of a subset of the accumulated data gives the ratio of lattice to experiment as 2.4 ± 0.5,[16] so there is a statistically significant discrepancy. If this were to represent a real disagreement between quenched QCD and experiment, one might expect 100% overestimates of other quenched

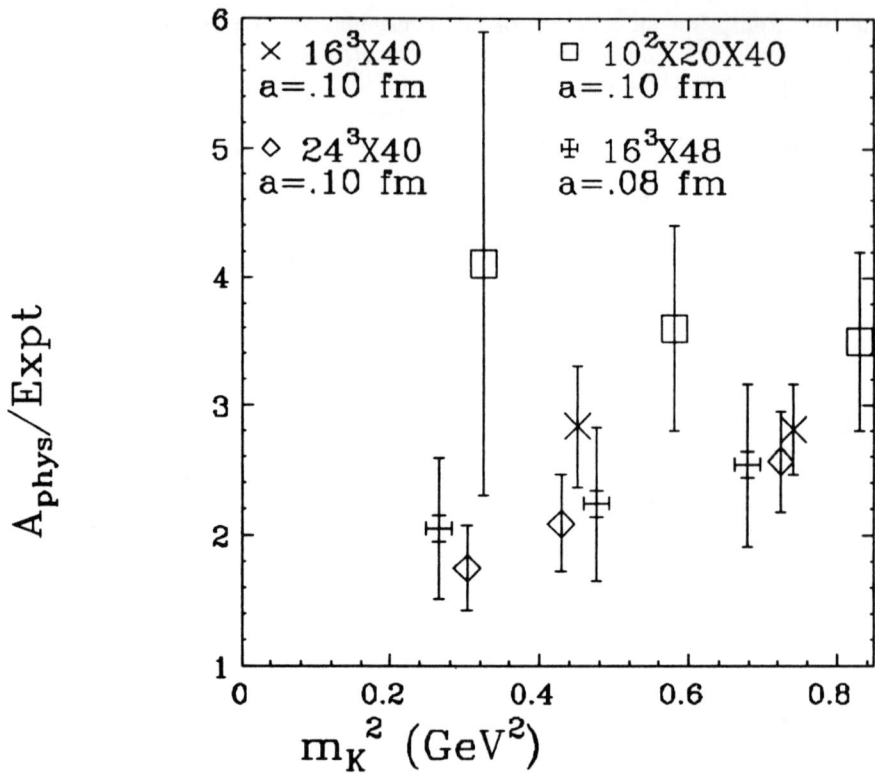

Figure 2. Wilson fermion data for the weak amplitude $\mathcal{A}(K^+ \to \pi^+\pi^0)$ in units of the experimental amplitude.

lattice results. In particular, the full QCD lattice result for $\hat{B}_K$ might be ~ 0.5 rather than ~ 1.0.

This unpleasant conclusion is consistent with the old result which relates $\hat{B}_K$ to the K^+ decay amplitude using current algebra and SU(3) symmetry.[26] This is a relationship between c_1 in Eq. (4) and d_1 in Eq. (5). If the leading terms are dominant, it predicts $\hat{B}_K \sim 0.3$. Conversely, if chiral logs and other higher order terms are small, the large lattice result for $\hat{B}_K$ converts into a result for $\mathcal{A}_{phys}$ larger than experiment by $2-3$. This is exactly what we find if we take the data in Fig. 2 at face value.

In my opinion, however, the systematic uncertainties are too large to draw this unpleasant conclusion. For a start, the physical amplitude contains isospin violating and electromagnetic contributions which are not included in the lattice calculation. These are notoriously hard to calculate, but may solve some of the problem. Secondly, comparing the three sets of data with $a = 0.1$fm suggests large finite volume effects. We may not yet have the infinite volume answers. Finally, the $a = 0.1$fm, $24^3 \times 40$ results decrease towards the chiral limit. Thus there may be significant corrections to the lowest order current algebra relation between $\mathcal{A}_{phys}$ and $\mathcal{A}_{off}$. Neither of the latter two trends can be definitively seen in the results, however, because the errors are too large. Better quality data is needed from larger lattices to determine exactly what the quenched lattice predicts. This should be forthcoming in the next year.

Clearly one should simply wait and see, but I cannot resist the temptation to speculate. My speculations are fueled by Isgur et al.,[10] who observe that the $\pi^+\pi^0$ (i.e. $I = 2$) final state interactions (FSI), which are repulsive, lead to a reduction in the K^+ decay amplitude by $1.5 - 2.0$. This conclusion is only mildly model dependent. This is a higher order effect which invalidates the lowest order relationship between $\hat{B}_K$ and $\mathcal{A}_{phys}$. Values of $\hat{B}_K$ greater than 0.33 are then expected. Such a strong FSI would have two effects on the lattice results. First, there would be a volume dependence in the amplitude. Second, the results would depend on m_K^2, i.e. the terms proportional to d'_1 and d_2 in Eq. (5) would not be small. In fact, the chiral logs (the d'_1 term) contain a large contribution from the FSI. It turns out that the $I = 2$ FSI, which are due to quark exchange, are fully present in the quenched approximation. Thus the results of Ref. 10 suggest the features of the data that appear to be observed. Much needs to be done, theoretically and numerically, to flesh out and then test this hypothesis.

CONCLUSIONS

From a pragmatic point of view, not much has changed in the last year. The work I have described has not reduced the uncertainty in $\hat{B}_K$.

Nevertheless, I am optimistic, because we understand the problems much more clearly than a year ago. We do understand how to go about calculating the interesting amplitudes on a infinite lattice with very small spacing. What we are learning is how to deal with the problems forced upon us by finite resources. One year's worth of a Cray-2 processor has revealed the important systematic errors, those on which we should concentrate next year. This all sounds very much like experimental physics.

Another result of this years running should be a better estimate of the matrix elements needed to calculate ϵ' in the standard model. The analysis is underway.

What should we expect from next year? I hope that we can pin down the K^+ decay amplitude, and get a first result for the K^0 amplitude. There should also be good results, at the 10% level, for the semileptonic D decay form factors, and an improved extrapolation from f_D to f_B. It is my prejudice that the quenched approximation is relatively unimportant for these quantities involving heavy-light mesons.

What about full QCD? At present it is possible to simulate, with a time scale of a year, up to a 16^4 lattice with $m_q \geq m_s$. This can be done given a Connection Machine-2 or an ETA-10, both running at roughly $500MFlop$. This is faster than the quenched QCD simulations on the Cray-2, the results of which I described above, which run at $100 - 200MFlop$. Quenched results show that the minimal requirements for beginning studies of complicated quantities such as matrix elements are a $16^3 \times 32$ lattice with quark masses down to $m_s/3$. This will require an increase in computer speed of roughly 100.

Such an increase should come in ~ 5 years. This claim is based upon the growth of speed in home-built computers, both completed and under construction, shown in Figure 3. The graph shows the sustainable speed with optimized

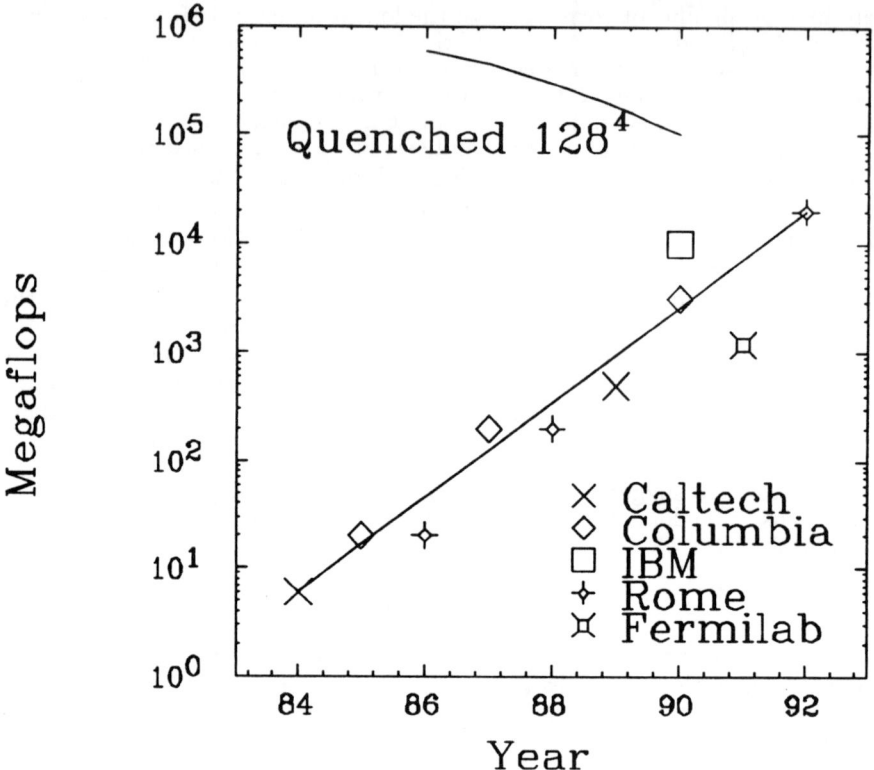

Figure 3. The growth of computer speed in home-built computers.

coding. Commercial machines lie on roughly the same curve. For example, the Connection Machine-2, on which lattice calculations only really began this year, would come in at about 2000 *MFlop*, using optimized codes.

Actually, we may not have to wait the full 5 years because of improvements in algorithms. As an example of such improvements, Fig. 3 also shows the speed needed to repeat our $24^3 \times 40$ calculations on a 128^4 lattice. This would be the ultimate quenched calculation. The curve decreases because we have improved (and know how to further improve) our inversion algorithm.

Of course, calculations in full QCD will be harder, because there will be resonances, FSI due to quark annihilation, etc.. On the other hand, these are the effects which will make the calculations interesting. We can understand them while we wait for the very large full QCD lattices. For such lattices, which will constitute the black box I described earlier, the wait will be comparable to that for physics from the SSC or a B factory. Then we will have all the tools in hand to find the physics beyond the standard model.

ACKNOWLEDGEMENTS

I thank Greg Kilcup for help in preparing the talk, and Claude Bernard, Rajan Gupta, Greg Kilcup, Nathan Isgur, Kim Maltman, Apoorva Patel and Amarjit Soni for many discussions.

REFERENCES

1. A. Soni, these proceedings.
2. "Hadronic Matrix Elements and Weak Decays", A.J. Buras, J.-M. Gerard and W. Huber eds, *Nucl. Phys.* **B(Proc. Suppl.) 7A** (1989).
3. S. Narison, CERN-TH5354/89, to be published by the World Scientific Lecture Notes Series.
4. A.J. Buras, Lectures at XIX Int. Seminar on Theor. Phys., Jaca, Spain, 6/88, preprint MPI-PAE/PTh 68/88;
   J.M. Gerard, Lectures at the 10th Autumn School of Physics, Lisbon, Portugal, 10/88, preprint MPI-PAE/PTh 2/89.
5. R. Gupta, Proc. 1st CCAST workshop/symposium on *"Lattice Gauge Theory using Parallel Processors"*, Beijing, China, 1987, Gordon and Breach;
   S. Sharpe, Proc. 16th Slac Summer Institute (1988) 271, E. Brennan ed.
6. For a discussion, see I. Bigi, these proceedings.
7. C. Bernard, T. Draper, G. Hockney and A. Soni, *Phys. Rev.* **D38** (1988) 3540.
8. E. Eichten, *Nucl. Phys.* **B(Proc. Suppl.)4** (1988) 170 ;
   P. LePage, *ibid* 199.
9. Ph. Boucaud, O. Pene, V. Hill, C. Sachrajda and G. Martinelli, *Phys. Lett.* **220B** (1989) 219.
10. N. Isgur, K. Maltman, J. Weinstein and T. Barnes, Toronto preprint UTPT-89-10.
11. S. Sharpe, AIP Conf. Proc. **185** (1989) 55, Suh-Urk Chung, ed.
12. G. Kilcup and S. Sharpe, *Nucl. Phys.* **B283** (1987) 493.
13. See B. Gavela *et al.* (p. 228) and C. Bernard *et al.* (p. 277) in Ref. 2.
14. G. Kilcup, S. Sharpe, R. Gupta and A. Patel, University of Washington preprint 40423-10, 8/89 and work in preparation.
15. S. Sharpe (p. 255) in Ref. 2;
    G. Kilcup, talk at the Symposium on Lattice Field Theory, Fermilab, September 1988, to be published in *Nucl. Phys.* **B(Proc. Suppl.)** (1989).
16. C. Bernard and A. Soni, private communication.
17. M. Gavela, L. Maiani, S. Petrarca, F. Rapuano, G. Martinelli, O. Pene and C. Sachrajda, *Nucl. Phys.* **B306** (1988) 677.
18. G. Martinelli, private communication.
19. J. Bijnens, H. Sonoda and M.B. Wise, *Phys. Rev. Lett.* **53** (1984) 2367.
20. R. Decker (p. 180) in Ref. 2.
21. The consequences of this range are discussed by G. Altarelli and Y. Nir in these proceedings.
22. M.K. Gaillard and B.W. Lee, *Phys. Rev.* **D10** (1974) 897.
23. B. Guberina, A. Pich and E. de Rafael, *Phys. Lett.* **163B** (1985) 198.
24. C. Bernard, T. Draper, G. Hockney and A. Soni, *Nucl. Phys.* **B(Proc. Suppl.) 4** (1988) 483.
25. C. Bernard, talk at the Symposium on Lattice Field Theory, Fermilab, September 1988, to be published in *Nucl. Phys.* **B(Proc. Suppl.)** (1989).
26. J. Donoghue, E. Golowich and B. Holstein, *Phys. Lett.* **119B** (1982) 412.

SYMMETRY, TOPOLOGY AND HELICITY IN $D_s \to \omega\pi$ AND $D_s \to \rho\pi$ DECAYS*

Harry J. Lipkin
Department of Nuclear Physics
Weizmann Institute of Science, Rehovot 76100, Israel
and
School of Physics and Astronomy
Raymond and Beverly Sackler Faculty of Exact Sciences
Tel Aviv University, Tel Aviv, Israel
and
High Energy Physics/Physics Divisions
Argonne National Laboratory, Argonne, IL 60439

ABSTRACT

Diagrams contributing to $D_s \to \omega\pi$ and $D_s \to \rho\pi$ decays are classified by topology as (a) OZI-forbidden or disconnected, (b) OZI-allowed or connected; by helicity as (a) helicity-conserving, (b) with helicity flips on both outgoing quark lines. The $D_s \to \omega\pi$ decay has an exotic final state and is forbidden for all annihilation diagrams which go via an intermediate state containing a single quark-antiquark pair and less than two additional gluons. It is also forbidden for helicity-conserving OZI-allowed diagrams, which are expected to be dominant and are allowed for $D_s \to \rho\pi$ decays.

INTRODUCTION

We clarify the confusion regarding the $D_s \to \omega\pi$ and $D_s \to \rho\pi$ decays, suggested as candidates for checking the presence of annihilation diagram contributions, by pointing out and proving the following general results:

1. The contribution to the $D_s \to \omega\pi$ decay vanishes for all diagrams in which there is an intermediate state consisting of a single quark-antiquark pair and less than two additional gluons; i.e. it is possible to cut the diagram with a continuous line intersecting only with a single quark-antiquark pair and less than two additional gluons. This is a generalization of a result previously noted[1].

2. The relevant diagrams have two different quark-line topologies.

   (a) "Disconnected diagrams" of the type normally called "OZI-forbidden" in which the quark-antiquark pair forming the vector meson is created from the vacuum by at least three gluons.

---

*Supported in part by the U.S. Department of Energy, Division of High Energy Physics and Division of Nuclear Physics, Contract W-31-109-ENG-38.

(b) "Connected diagrams" of the type normally called "OZI-allowed" in which the lines of the quark-antiquark pairs forming the vector and pseudoscalar mesons are connected together. This also includes the doubly-Cabibbo-suppressed W-exchange diagram, in which the connection is via the W.

3. The $D_s \to \rho\pi$ decay is described by a connected diagram. It cannot be described by a disconnected diagram, because the gluons have isospin zero and cannot create an isovector meson.

4. The $D_s \to \omega\pi$ decay is described by a disconnected diagram. It cannot be described by a connected diagram if quark helicity is conserved at the gluon vertices.

5. Suggested[2] constructive interference between related diagrams in $D_s \to \omega\pi$ and cancellation in $D_s \to \rho\pi$ cannot occur unless diagrams with at least two quark helicity flips play a dominant role.

6. Since connected diagrams with helicity conservation are expected to be dominant and OZI-violating diagrams and double helicity flip diagrams suppressed, $D_s \to \omega\pi$ is suppressed relative to $D_s \to \rho\pi$.

We now prove and explain these results.

THE G-PARITY-EXOTIC SELECTION RULE

$D_s \to \omega\pi$ cannot go via an intermediate state containing only a single light quark-antiquark pair, because the final state must have total angular momentum $J = 0$, the $\omega - \pi$ state has even G parity and a $J = 0$ state with even G is exotic and cannot be made from a quark-antiquark pair. All intermediate states in any diagram; i.e. any combination of light quarks and gluons which cross a single continuous line drawn through the diagram, must have $J = 0$ and even G parity, since angular momentum is conserved and G parity is conserved by the strong interactions which act after the weak interaction turns the charmed-strange pair into a light pair. $D_s \to \omega\pi$ via simple annihilation has been shown to require a second class current[1]. The present selection rule which uses the exotic classification[3] is stronger, since it is valid even if second class currents exist.

GENERAL TOPOLOGICAL FEATURES OF $D_s \to VP$ DECAYS

The most general diagram consists of a black box with two external lines carrying $c\bar{s}$ flavor numbers going in from the initial D_s and four lines carrying nonstrange flavor numbers going out. Inside the box there can be any number of gluons, quark-antiquark pairs, vertices and loops. However, since flavor is conserved at all vertices except for two vertices coupled to a single W boson somewhere in the box, we obtain the following topological features:

1. The two incoming lines must be connected somewhere in the diagram via a W to eliminate the c and s flavors. The box can contain charmed and strange quark loops, but no charm nor strangeness can escape except in the two incoming lines.

2. The four outgoing nonstrange quark lines must be connected somewhere in the diagram. There are two possible topologies:

    (a) The quark and antiquark lines in the *same* meson are connected. This gives a disconnected or OZI-forbidden diagram.

    (b) The quark and antiquark lines in different mesons are connected, either directly or through a W. This gives a connected or OZI-allowed diagram.

3. Since electric charge is transferred from the initial state to the final state one of the two quark lines producing the final state must have a charge change somewhere along its length at or through a W. It can also have an arbitrary number of additional gluon vertices. The other quark line which retains its electric charge is produced only by gluons and has only gluon vertices.

4. Since $D_s \to \omega\pi$ cannot go via an intermediate state containing only a single light quark-antiquark pair, some gluons must connect the initial $c - \bar{s}$ pair with light quarks produced. For the annihilation diagram a minimum of two gluons are needed to leave the $c - \bar{s}$ pair in a color singlet state which can annihilate at a W vertex. The doubly-Cabibbo-suppressed exchange diagram can go with one gluon.

IMPLICATIONS OF ISOSPIN AND HELICITY CONSERVATION

1. Isospin invariance requires the diagrams to come in pairs with the charge conserving quark line carrying the flavors u and d. The amplitudes from two members of such a pair must be equal and add with a positive relative phase.

2. Since vector interactions conserve helicity in the limit of zero quark mass, the light quark helicity is expected to be conserved along the quark line. The outgoing quark and antiquark connected by the same quark line inside the black box then have opposite helicity.

3. A vector meson recoiling against a spinless pion in a $J = 0$ state must have helicity zero to conserve angular momentum; i.e. the quark and antiquark have opposite helicity. This selection rule is trivially satisfied in a disconnected diagram with helicity conservation. In the connected diagram with helicity conservation the quarks in both final state mesons must have the same helicity and similarly for the antiquarks. Thus, even if helicity is not conserved, diagrams with an odd number of flips are forbidden and connected diagrams with flips must have flips on *both* quark lines.

4. A $q\bar{q}$ state where the quark and antiquark each have definite and opposite helicities is a linear combination of vector and pseudoscalar states with a relative phase depending upon the helicity and the phase convention. For the connected diagram with helicity conservation both final state mesons are in the same state, which is either V+P or V-P. If there are helicity flips on both quark lines, a final state is produced with one meson in the V+P state and the other in the V-P state. We assume that such double flip diagrams are suppressed.

EXPLICIT CALCULATION FOR CONNECTED DIAGRAMS

We are now in a position to determine the implications of the above symmetry and topological conditions for the connected diagrams.

Let $|V_u\rangle$ and $|V_d\rangle$ denote the vector meson states which are composed respectively of $u\bar{u}$ and $d\bar{d}$ pairs; i.e.

$$|V_u\rangle = (1/\sqrt{2})(|\omega\rangle + |\rho^\circ\rangle) \tag{1a}$$

$$|V_d\rangle = (1/\sqrt{2})(|\omega\rangle - |\rho^\circ\rangle) \tag{1b}$$

and similarly for the pseudoscalar states denoted respectively by $|P_u\rangle$ and $|P_d\rangle$.

Since isospin invariance requires the connected diagrams to come in pairs, one having a neutral meson composed of a $u\bar{u}$ pair and the other composed of a $d\bar{d}$ pair, and the amplitudes must add with a positive phase, one might jump to the conclusion that the diagrams add for the ω and cancel for the ρ. However, phases are tricky, and one must be sure to make an even number of mistakes in sign; an odd number gives the wrong answer. The interchange of the $u\bar{u}$ and $d\bar{d}$ pairs also interchanges the charged and neutral mesons. Since the final state is a p-wave, this introduces an additional negative phase. Now the diagrams add for the ρ and cancel for the ω. This seems more reasonable, since the G-parity exotic selection rule requires pairs of diagrams with an intermediate state containing a single quark-antiquark pair to cancel for the ω if they do not both vanish. This argument is essentially correct, but it is not fool proof.

The $u\bar{u}$ to $d\bar{d}$ interchange also interchanges the vector and pseudoscalar mesons; i.e. it relates the amplitude for a neutral vector and a π^+ to the amplitude for a ρ^+ and a neutral pseudoscalar; i.e. a trivial isospin relation. But we must also consider helicity amplitudes, which relate vector and pseudoscalar amplitudes, since gluons and W's create quark-antiquark pairs in helicity eigenstates.

We now present an explicit calculation with all the meson interchanges and helicity constraints, being careful to stop after an *even* number of mistakes to get the right answer. Let $\langle M_1(\vec{k}); M_2(-\vec{k})|T_C|D_s\rangle$ denote the total contribution of all connected diagrams to the transition matrix element for the decay of a D_s at rest into two meson states with momenta $\vec{k}$ and $-\vec{k}$ respectively. Then the isospin symmetry condition gives the following relation for helicity conserving transitions,

$$\langle [V_u(\vec{k}) \pm P_u(\vec{k})]; [\rho^+(-\vec{k}) \pm \pi^+(-\vec{k})]|T_C|D_s\rangle =$$
$$\langle [\rho^+(\vec{k}) \pm \pi^+(\vec{k})]; [V_d(-\vec{k}) \pm P_d(-\vec{k})]|T_C|D_s\rangle \tag{2a}$$

For connected diagrams with helicity flips on both quark lines,

$$\langle [V_u(\vec{k}) \pm P_u(\vec{k})]; [\rho^+(-\vec{k}) \mp \pi^+(-\vec{k})]|T_C|D_s\rangle =$$
$$\langle [\rho^+(\vec{k}) \pm \pi^+(\vec{k})]; [V_d(-\vec{k}) \mp P_d(-\vec{k})]|T_C|D_s\rangle = \epsilon_1(\vec{k}) \pm \epsilon_2(\vec{k}) \tag{2b}$$

where we assume that the connected diagrams with helicity flips on both quark lines are small and denote the small quantities by $\epsilon_1(\vec{k}) \pm \epsilon_2(\vec{k})$. Then combining eqs. (2a) and (2b) to eliminate P_u, P_d and ρ^+ gives

$$\langle V_u(\vec{k}); \pi^+(-\vec{k})|T_C|D_s\rangle = \langle \pi^+(\vec{k}); V_d(-\vec{k})|T_C|D_s\rangle + \epsilon_2(\vec{k}) \tag{3a}$$

Since this holds for all values of $\vec{k}$, we can substitute $-\vec{k}$ for $\vec{k}$ to obtain

$$\langle \pi^+(-\vec{k}); V_d(\vec{k}) | T_C | D_s \rangle + \epsilon_2(-\vec{k}) = \langle V_u(-\vec{k}); \pi^+(\vec{k}) | T_C | D_s \rangle \tag{3b}$$

Since the final state is required by angular momentum conservation to be an odd-parity p-wave,

$$\langle \pi^+(-\vec{k}); V_f(\vec{k}) | T_C | D_s \rangle = - \langle V_f(-\vec{k}); \pi^+(\vec{k}) | T_C | D_s \rangle \tag{3c}$$

where f denotes either u or d. Combining eqs. (3) and substituting (1) gives

$$\langle \omega(\vec{k}); \pi^+(-\vec{k}) | T_C | D_s \rangle = (1/\sqrt{2})\epsilon_2(\vec{k}) \tag{4}$$

Thus the $D_s \to \omega\pi$ amplitude is of order ϵ and vanishes when double helicity flip amplitudes are neglected. There is no similar relation for $D_s \to \rho\pi$.

We therefore conclude that the $D_s \to \omega\pi$ decay can be produced only via diagrams with no intermediate state containing a single quark-antiquark pair and less than two gluons, and that the allowed diagrams must be either disconnected OZI-violating or connected with at least two helicity-flip vertices. If either $D_s \to \omega\pi$ or $D_s \to \rho\pi$ goes via the annihilation diagram, $D_s \to \rho\pi$ should be much stronger.

We also note that $D_s \to \omega\pi$ can go via an OZI-forbidden disconnected spectator diagram, via the relatively strong $D_s \to \phi\pi$ decay mode and $\omega - \phi$ mixing. Thus a small branching ratio observed for $D_s \to \omega\pi$ at the level expected from the mixing diagram cannot be taken as any test for an annihilation contribution. On the other hand, the failure to observe $D_s \to \rho\pi$ is not evidence against the annihilation diagram, since it is not obvious that annihilation should produce mainly the vector-pseudoscalar final state.

REFERENCES

1. A. N. Kamal, N. Sinha and R. Sinha, Zeits. f. Physik **C41**, 207 (1988) and Alberta preprint Thy-29-88, to be published in Phys. Rev. D.

2. G. Punkar, These Proceedings.

3. Edmond L. Berger and Harry J. Lipkin, Classification and J^{PG} Selection Rules for Weak Currents, Phys. Rev. Lett. **59**, 1394 (1987).

EXPERIMENTAL RESULTS ON B MESONS

REVIEW OF B-MESON DECAY RESULTS

K. R. Schubert

Institut für Experimentelle Kernphysik, Universität Karlsruhe
Postfach 6980, D - 7500 Karlsruhe 1

INTRODUCTION

This review covers the present status of experimental results on B-meson decays. It concentrates on recent results and omits all searches for decay modes which are not expected in the frame of the standard electroweak theory. Most information in this review originates from the two experiments ARGUS at DORIS-II and CLEO at CESR. Table 1 summarizes their main properties. End of 1988, the produced number of B-mesons in the two experiments had reached a total of one million.

Table 1: ARGUS and CLEO figures of merit

	$\Upsilon(4S)$ since	$\int \mathcal{L}dt$ until 1988	$\sigma(\Upsilon 4S)$	$N(B)$	$\sigma(\sqrt{s})$
ARGUS	1983	$181/pb$	$(0.80 \pm 0.06)nb$	290 K	8 MeV
CLEO	1982	$118 + 212/pb$	1.11 nb	730 K	3.5 MeV

1. MASSES OF B^0 AND B^+

The 1987 status of B-meson masses, based on the mass scale $m(\Upsilon 4S) = (10580 \pm 3.5)$ MeV [1], is given in the upper part of table 2; the average mass difference was $\Delta m = m(B^0) - m(B^+) = (+2.0 \pm 1.0)$ MeV. New results are given in the lower part of table 2, those of ARGUS from $B^0 \to D^-\pi^+$, $D^-\rho^+$ and $B^+ \to \overline{D}^0\pi^+$, $\overline{D}^0\rho^+$ [5], and those of CLEO from seven selected channels in their new data [6]. The final result, the average of rows 2, 3, and 4, is given in row 5 of the table.

Table 2: B-meson mass results

	$m(B^0)$ in MeV	$m(B^+)$ in MeV	Δm in MeV
CLEO [2]	$5280.6 \pm 0.8 \pm 2$	$5278.6 \pm 0.8 \pm 2$	$2.0 \pm 1.1 \pm 0.3$
ARGUS [3, 4]	$5280.1 \pm 0.8 \pm 3$	$5278.1 \pm 1.1 \pm 3$	$2.0 \pm 1.4 \pm 1.0$
ARGUS [5]	$5280.8 \pm 1.6 \pm 3$	$5279.4 \pm 1.7 \pm 3$	$1.4 \pm 2.3 \pm 1.0$
CLEO [6]	$5279.1 \pm 0.5 \pm 2$	$5279.3 \pm 0.5 \pm 2$	-0.2 ± 0.7
average	$5279.4 \pm 0.4 \pm 2$	$5279.1 \pm 0.4 \pm 2$	$+0.3 \pm 0.6$

The mass difference of B^0 and B^+ is relevant for the decay fractions $f_0 = \mathcal{B}(\Upsilon 4S \to B^0\overline{B}^0)$ and $f_+ = \mathcal{B}(\Upsilon 4S \to B^+B^-) = 1 - f_0$ for the following reason: The two-body decay rates are given by

$$\Gamma = \frac{\mathcal{M}^2}{12\pi} \cdot \frac{p}{M^2} \, , \qquad (1)$$

where M is the mass of the decaying meson and p the momentum of each of the decay products. For a $J^P = 1^- \to 0^-0^-$ decay, the matrix element $\mathcal{M}$ must be of the form

$$\mathcal{M} = K \cdot \epsilon^\mu q_\mu \cdot f(q^2) \, , \qquad (2)$$

where K is a constant, ϵ^μ the polarization vector, and $q = 2p$. With a constant form factor $f(q^2)$, the $\Upsilon(4S) \to B\overline{B}$ rate would be proportional to p^3. A mass difference of 2 MeV would lead to $f_0/f_+ = 0.75$, and $\Delta m = (0.3 \pm 0.6)$ MeV to $f_0/f_+ = 0.96 \pm 0.08$.

In a coupled-channel calculation of Eichten [7], $f(q^2)$ is a falling function of q^2 and has a zero at $p = q/2 = 0.7$ GeV/c. Multiplied by p^2, this leads to a rate with $d\Gamma/dp = 0$ around $p = \sqrt{2m_B(M_{\Upsilon 4S}/2 - m_B)}$, i.e. to $f_0/f_+ = 1.00$. The conclusion

$$f_0/f_+ = 1.00 \pm 0.08 \, , \quad f_0 = 0.50 \pm 0.02 \, , \qquad (3)$$

seems to be very conservative and should be used for future B-meson decay results obtained on the $\Upsilon(4S)$ resonance.

2. MASS DIFFERENCE BETWEEN B_1^0 AND B_2^0

The phenomenology of $B^0\overline{B}^0$ oscillations is presented by G. Altarelli [8] at this Conference. The relevant parameters are M_{12} and Γ_{12} in the $B^0\overline{B}^0$ mass matrix. A situation with $\Gamma_{12} = 0$ and $M_{12} \neq 0$ leads to nonexponential decay laws as illustrated in fig.1. This time evolution has not been observed so far; but the observed time-integrated oscillation quantity χ, defined as the fraction of all produced B^0 which decay as a $\overline{B}^0$, is related to M_{12} and Γ_{12} in the following way:

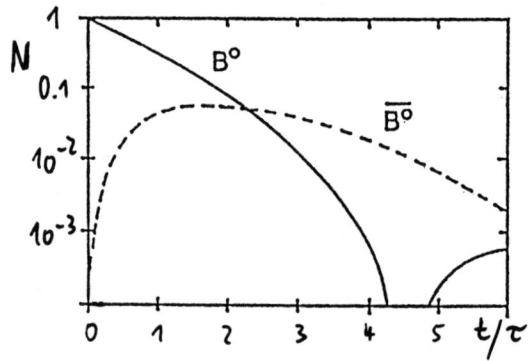

Fig. 1: Intensity of B^0 and $\overline{B}^0$ after production of B^0 at $t = 0$.

$$\chi = \frac{N(B^0 \to \overline{B}^0 \to decay)}{N(B^0 \, produced)} = \frac{x^2 + y^2}{2 + 2x^2} \, , \qquad (4)$$

$$x = \frac{-2M_{12}}{\Gamma} = \frac{m(B_1) - m(B_2)}{\Gamma} \, , \quad y = \frac{-\Gamma_{12}}{\Gamma} = \frac{\Gamma(B_1) - \Gamma(B_2)}{2\Gamma} \, , \qquad (5)$$

where $\Gamma = \Gamma_{11} = 1/\tau_b$. Since B^0 and $\overline{B}^0$ have very few common decay channels, one expects $y \ll 1$ and

$$\chi = \frac{x^2}{2+2x^2} . \qquad (6)$$

Another time-integrated oscillation quantity is r, related to χ by

$$r = \frac{\chi}{1-\chi}, \quad \chi = \frac{r}{1+r} \pm \frac{\sigma_r}{(1+r)^2} . \qquad (7)$$

Only ARGUS and CLEO determine $\chi(B^0)$; experiments at higher energies determine a linear superposition $\chi = f(B^0) \cdot \chi(B^0) + f(B_s) \cdot \chi(B_s)$ which will be discussed in chapter 12. On the $\Upsilon(4S)$, the final $B^0\overline{B}^0$ state is a coherent two-particle state with $C = P = -1$, and $\chi(B^0)$ becomes easily measurable as

$$\chi = \chi(B^0) = \frac{N(l^\pm l^\pm)}{N(ll \text{ from } B^0\overline{B}^0)} = \frac{N(l^\pm l^\pm)}{N(ll)} \cdot \left(1 + \frac{f_+ \cdot \mathcal{B}_+^2}{f_0 \cdot \mathcal{B}_0^2}\right) , \qquad (8)$$

where $\mathcal{B}_+$ and $\mathcal{B}_0$ are the inclusive decay fractions of B^+ and B^0 into leptons.

Table 3: Pairs of like- and opposite-sign leptons in the two $\chi(B^0)$ experiments

	$N(l^{\pm\pm})$	$N(l^+l^-)$
ARGUS [10], obs.	70	403
background	35.4 ± 5.9	21.7 ± 7.6
signal	34.6 ± 10.8	381 ± 22
CLEO [11], obs.	71	470
background	38.5 ± 5.2	44.0 ± 8.2
signal	32.5 ± 10.0	426 ± 23

A preliminary 1989 update of the ARGUS result [9] was presented by M. Schäfer [10] at this Conference. Table 3 summarizes the update, from a combination of old and new ARGUS data, together with the CLEO result [11]. Using $f_+\mathcal{B}_+^2/f_0\mathcal{B}_0^2 = 1.2 \pm 0.2$ consistently, leads to

$$\chi(B^0) = \begin{cases} 0.18 \pm 0.06 & \text{ARGUS,} \\ 0.16 \pm 0.06 & \text{CLEO,} \\ 0.17 \pm 0.04 & \text{average.} \end{cases}$$

Additional information from ARGUS on D^*l and D^*ll coincidences [10] confirms this result but does not improve it. With $y = 0$, the above average gives

$$|x| = 0.72 \pm 0.13 , \qquad (9)$$

$$|m(B_1^0) - m(B_2^0)| = (0.4 \pm 0.1) \, meV = (8 \pm 2) \cdot 10^{-14} \cdot m(B) . \qquad (10)$$

This mass difference is 100 times larger than $m(K_2^0) - m(K_1^0)$, and even the relative mass diffence is 10 times larger than in the neutral Kaon system. For the consequences see ref. [8].

3. B-MESON MEAN LIFE TIME

No news on τ_b have been presented at this Conference. Table 4 summarizes the August 1988 status [12], the average of which was

$$\tau_b = (11.8 \pm 1.4) \cdot 10^{-13} \ s \ , \tag{11}$$

taking care of correlations in the systematic errors. This result is a mean life time of B^0, B^+, B_s, and Λ_b, and the determination of separate life times remains a subject for the future. Mark-II reported last year four reconstructed B^0 decays with separately measured production and decay vertices [13], resulting in $\tau(B^0) = (13^{+12}_{-6}) \cdot 10^{-13} \ s$.

Table 4: B Lifetime results; see [12] for references

Experiment	τ_b in $10^{-13} \ s$
JADE 1986	$18.0 \ ^{+5.0}_{-4.0} \pm 4.0$
MAC 1987	$12.9 \pm 2.0 \pm 2.1$
HRS 1987	$10.2 \ ^{+4.2}_{-3.9}$
DELCO 1988	$11.7 \ ^{+2.7}_{-2.2} \ ^{+1.7}_{-1.6}$
MARK-II 1988	$9.8 \pm 1.2 \pm 1.3$
TASSO 1988	$13.5 \pm 1.0 \pm 2.4$
JADE 1988	$14.6 \ ^{+2.2}_{-2.1} \pm 3.4$

4. SEPARATE MEAN LIVES FOR B^+ AND B^0

Direct measurements of $\tau(B^+)$ and $\tau(B^0)$ might be obtained at LEP in a few years from now. However, ARGUS and CLEO have some indirect means to determine the ratio $\tau(B^+)/\tau(B^0)$. Semileptonic decays, either measured inclusively or exclusively, should have the same absolute rates for B^+ and B^0, leading to

$$\frac{\tau(B^+)}{\tau(B^0)} = \frac{\mathcal{B}(B^+ \to l^+\nu F_i^0)}{\mathcal{B}(B^0 \to l^+\nu F_j^-)} \ , \tag{12}$$

where F_i, F_j is either a pair of exclusive isospin partners or just anything $F_i = F_j = X$. Four methods have been used so far to determine the ratio of charged and neutral semileptonic decay fractions:

1. Tagging of inclusive lepton decays with reconstructed B^0 candidates, i.e.

$$\mathcal{B}(B^0 \to l^+\nu X) = \frac{N(B^0_{rec} + l^\pm)}{N(B^0_{rec})} \ , \tag{13}$$

where B^0_{rec} is a fully reconstructed B^0 or $\overline{B}^0$, determines the inclusive semileptonic branching fraction of the B^0 independent of f_0. Comparing this result to $\mathcal{B} = f_0 \cdot \mathcal{B}_0 + f_+ \cdot \mathcal{B}_+ = 0.109 \pm 0.006$, as obtained in chapter 9, determines $f_0 + f_+ \cdot \tau(B^+)/\tau(B^0)$.

2. The ratio of $D^*l\nu$ decays on the $\Upsilon 4S$, abbreviating $\tau_+ = \tau(B^+)$ and τ_0 accordingly,

$$\frac{N(B^+ \to D^{*0}l^+\nu)}{N(B^0 \to D^{*-}l^+\nu)} = \frac{f_+ \cdot \tau_+}{f_0 \cdot \tau_0}.$$

3. Counting lepton pairs from B decays on the $\Upsilon 4S$ resonance,

$$\frac{N(ll)}{N(l)^2} = F\left(\frac{\tau_+}{\tau_0}; \frac{f_+}{f_0}\right) \approx F\left(\frac{\tau_+}{\tau_0}\right).$$

4. Counting the number of D^0l, D^+l, and $D^{*+}l$ coincidences on the $\Upsilon 4S$, and assuming that semileptonic B decays are dominated by $Dl\nu$ and $D^*l\nu$, results in

$$\frac{N(D^0l^-) - N(D^{*+}l^-, D^{*+} \to D^0\pi^+)}{N(D^+l^-) + N(D^{*+}l^-, D^{*+} \to D^0\pi^+)} = \frac{f_+ \cdot \tau_+}{f_0 \cdot \tau_0}, \quad (14)$$

using the fact that D^{*+} mesons decay into $D^0\pi^+$ and $D^+\pi^0$, $D^+\gamma$, but D^{*0} mesons decay only into $D^0\pi^0$, $D^0\gamma$.

The following results have been obtained with the help of these methods:

$$\begin{aligned}
\tau_+/\tau_0 &= 1.0^{+1.0}_{-0.5}, & \text{method 3, CLEO 1987 [14]}, \\
f_0 + f_+ \cdot \tau_+/\tau_0 &= 1.16 \pm 0.25, & \text{method 1, CLEO 1989 [15]}, \\
f_+\tau_+/f_0\tau_0 &= 0.85 \pm 0.27, & \text{method 2, CLEO 1989 [15,16]}, \\
f_+\tau_+/f_0\tau_0 &= 1.14 \pm 0.44, & \text{method 4, ARGUS 1989 [17]}.
\end{aligned}$$

Exclusive hadronic decays may also be used for a determination of τ_+/τ_0. With the expectations $\Gamma(B^+ \to \overline{D}^0 D_s) = \Gamma(B^0 \to D^- D_s)$, $\Gamma(B^+ \to J/\Psi K^+) = \Gamma(B^0 \to J/\Psi K^0)$, $\Gamma(B^+ \to J/\Psi K^{*+}) = \Gamma(B^0 \to J/\Psi K^{*0})$ and the results [6] $\overline{D}^0 D_s/D^+ D_s = 1.77 \pm 1.33$, $\Psi K^+/\Psi K^0 = 1.33 \pm 0.95$, $\Psi K^{*+}/\Psi K^{*0} = 1.30 \pm 1.04$ we obtain

$$f_+\tau_+/f_0\tau_0 = 1.4 \pm 0.6. \quad (15)$$

We do not include $\overline{D}^0\pi^+/D^-\pi^+$ since these two rates are not expected to be equal and their theoretical ratio of 0.67 [18] has a large uncertainty. The average of the four semileptonic results and the hadronic result in eq. 15, together with f_0 from eq. 3, is

$$\tau(B^+)/\tau(B^0) = 1.03 \pm 0.19. \quad (16)$$

5. NONLEPTONIC $b \to c$ DECAYS

Ten B^+ and ten B^0 channels of exclusive nonleptonic B decays have been observed so far; their decay fractions are summarized in table 5. The CLEO 89 numbers are obtained with $f_0 = 0.50$, those of ARGUS with 0.45, and those of CLEO 87 with 0.43. The agreement between the three columns is good with the only exception of the new CLEO value for $B^0 \to J/\Psi K^{*0}$.

CLEO has also new preliminary results on $B \to J/\Psi + X$ and $B \to \Psi 2S + X$ [19]. The J/Ψ mesons are well reconstructed in their e^+e^- and $\mu^+\mu^-$ decay modes. The new

Table 5: Exclusive B decay fraction results, in percent

	ARGUS [3,4]	CLEO 87 [2]	CLEO 89 [6]
$B^0 \to D^{*-}\pi^+$	0.35 ± 0.22	0.31 ± 0.17	0.30 ± 0.10
$D^{*-}\pi^+\pi^0(\rho)$	2.0 ± 1.4	–	1.9 ± 1.6
$D^{*-}\pi^+\pi^+\pi^-$	4.3 ± 2.3	–	1.5 ± 1.1
$D^-\pi^+$	0.31 ± 0.16	0.59 ± 0.33	0.24 ± 0.11
$D^-\pi^+\pi^0(\rho)$	2.2 ± 1.5	–	–
$J/\Psi K^0$	(1 ev.)	–	0.06 ± 0.04 (3 ev.)
$J/\Psi K^{*0}$	0.33 ± 0.18	0.41 ± 0.19	0.10 ± 0.04
$J/\Psi K^-\pi^+_{nonres.}$	–	–	0.11 ± 0.05
$\Psi 2S \overline{K}^{*0}$	–	–	0.13 ± 0.08
$D^- D_s^+$	–	–	3.6 ± 1.8
$B^+ \to D^{*-}\pi^+\pi^+$	0.66 ± 0.50	0.20 ± 0.15	–
$D^{*-}\pi^+\pi^+\pi^0$	5.7 ± 3.8	–	–
$\overline{D}^0\pi^+$	0.19 ± 0.12	0.47 ± 0.18	0.31 ± 0.08
$\overline{D}^0\pi^+\pi^0(\rho)$	2.1 ± 1.2	–	–
$D^-\pi^+\pi^+$	–	$0.25^{+0.47}_{-0.25}$	–
$J/\Psi K^+$	0.07 ± 0.04	0.09 ± 0.06	0.08 ± 0.02
$J/\Psi K^{*+}$	–	–	0.13 ± 0.09
$J/\Psi K^+\pi^+\pi^-$	0.11 ± 0.07	–	0.14 ± 0.05
$\Psi 2S K^+$	0.22 ± 0.17	–	–
$\overline{D}^0 D_s^+$	–	–	6.4 ± 3.2

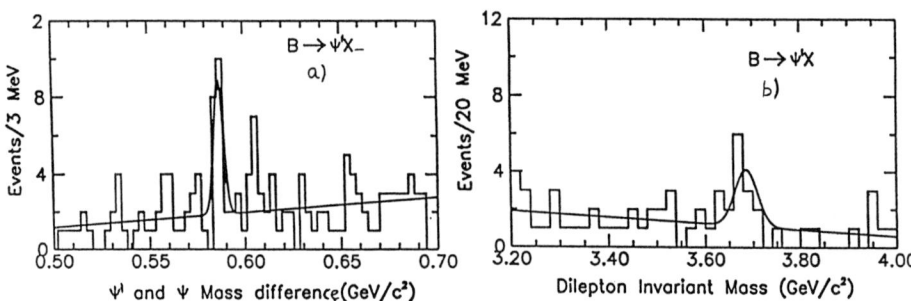

Fig. 2: New CLEO results on $B \to \Psi 2S\, X$,
a: $\Psi 2S \to J/\Psi \pi^+\pi^-$, $J/\Psi \to e^+e^-$, $\mu^+\mu^-$, b: $\Psi 2S \to e^+e^-$, $\mu^+\mu^-$.

result is $\mathcal{B}(B \to J/\Psi + X) = (1.12 \pm 0.10 \pm 0.15)\%$ which confirms the old results of ARGUS, $(1.07\pm0.16\pm0.22)\%$ [4], and of CLEO, $(1.09\pm0.16\pm0.21)\%$ [20]. The inclusive $\Psi 2S$ decay is seen in the $\Psi 2S \to J/\Psi \pi^+\pi^-$ and in the $\Psi 2S \to e^+e^-$, $\mu^+\mu^-$ decay modes as shown in fig. 2. The combined result, $\mathcal{B}(B \to \Psi 2S + X) = (0.40 \pm 0.10 \pm 0.15)\%$ confirms the 1987 result of ARGUS, $(0.46\pm0.17\pm0.11)\%$ [4]. The momentum spectrum of J/Ψ mesons in inclusive $B \to J/\Psi + X$ decays had shown a hint for a structure with two components in the ARGUS data, see fig. 3. This is clearly absent in the new CLEO data as shown in fig. 4.

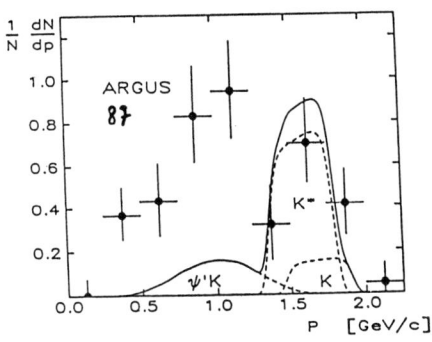
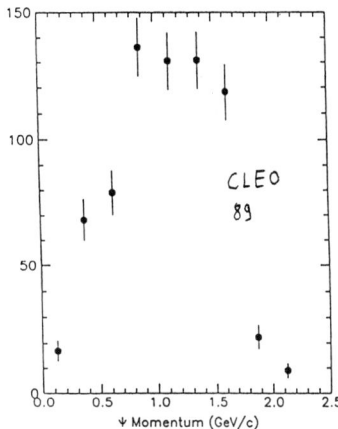

Fig. 3: Momentum spectrum of J/Ψ mesons in $B \to J/\Psi\, X$ decays [4].

Fig. 4: The same spectrum in the new CLEO data.

6. BARYONIC B-MESON DECAYS

Both ARGUS [21] and CLEO [6] see a clear signal for Λ_c production in B-meson decays. Fig. 5 shows the $\Lambda_c \to pK^-\pi^+$ signal on the $\Upsilon 4S$ for $x_p < 0.5$; there is no continuum contribution as seen in the histogram in fig. 5. The Λ_c momentum spectrum is shown in fig. 6; it is much softer than expected from $\overline{B} \to \Lambda_c \bar{p}$ and seems to be dominated by high multiplicity decays. The two observations agree with each other, ARGUS finds $(0.30 \pm 0.12 \pm 0.06)\%$ and CLEO $(0.28 \pm 0.06 \pm 0.05)\%$. The average is

$$\mathcal{B}(\overline{B} \to \Lambda_c X) \cdot \mathcal{B}(\Lambda_c \to pK^-\pi^+) = (0.29 \pm 0.07)\% \ . \qquad (17)$$

The decay fraction $\mathcal{B}(\Lambda_c \to pK^-\pi^+)$ is not well known. There are two measurements, Mark-II [22] finds $(2.2 \pm 1.0)\%$ and EHS $> 4.4\%$ [23]. The B-decay result in eq. 17 would add important information if one would find a way to determine $\mathcal{B}(\overline{B} \to \Lambda_c X)$ separately. This determination requires some assumptions, and we will try to present here a coherent analysis of the available data, based on a minimal set of assumptions.

We assume that the production of all baryons in B decays is dominated by the decays $\overline{B} \to \Lambda_c \bar{p} X$, $\Lambda_c \bar{n} X$, and $\Lambda_c \overline{\Lambda} X$. In detail, this requires

1. $b \to u$ and $b \to s \ll b \to c$, also in baryonic decays.

2. $\mathcal{B}(\overline{B} \to \Lambda_c X) \gg \mathcal{B}(\overline{B} \to \Xi_c^0, \Xi_c^+, \Omega_c\, X)$.

3. $\mathcal{B}(\overline{B} \to \Lambda_c X) \gg \mathcal{B}(\overline{B} \to DN\overline{N}X)$, i.e. production of baryon pairs in the W-current hadronisation in negligeable compared to baryon production in the c-quark fragmentation.

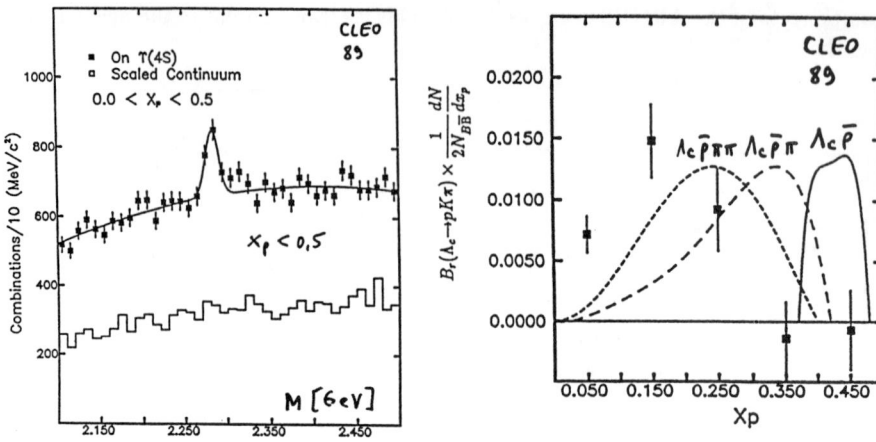

Fig. 5: CLEO results on $B \to \Lambda_c X$, invariant mass spectrum of $pK^-\pi^+$.

Fig. 6: Momentum spectrum of Λ_c and expectations for three decay modes

The third assumption does not have strong experimental support until now. The second assumption, however, is fulfilled to better than 10%. CLEO observes Ξ^- production on the $\Upsilon(4S)$, the origin is either $\overline{B} \to \Xi_c X$ or $\Lambda_c \to \Xi^- X$. The two extreme cases lead to

$$\mathcal{B}(\overline{B} \to \Xi_c X) < 0.10 \cdot \mathcal{B}(\overline{B} \to \Lambda_c X) \, , \quad \mathcal{B}(\Lambda_c \to \Xi X) < 6\% \, . \tag{18}$$

Table 6: Inclusive baryon decay fractions of B mesons, in percent

	CLEO [6]	ARGUS [24]	ARGUS [25]
$\mathcal{B}(B \to p \text{ or } \overline{p} + X)$	8.4 ± 0.9	8.2 ± 1.3	
$\mathcal{B}(\overline{B} \to p\overline{p} + X)$	2.6 ± 0.4	2.5 ± 0.3	
$\mathcal{B}(\overline{B} \to \Lambda \text{ or } \overline{\Lambda} + X)$	3.9 ± 0.6	4.2 ± 0.8	
$\mathcal{B}(\overline{B} \to \Lambda\overline{p} \text{ or } p\overline{\Lambda} + X)$	2.3 ± 0.5	2.3 ± 0.5	
$\mathcal{B}(\overline{B} \to \Lambda\overline{\Lambda} + X)$	< 0.45	< 0.88	
$\mathcal{B}_\Lambda$ from $l^+\Lambda$			29^{+13}_{-10}
f_Λ from $l^+\overline{\Lambda}$			19^{+10}_{-6}
$\mathcal{B}_p$ from $l^+\Lambda, l^+p$			27^{+11}_{-6}
f_p from $l^+\overline{\Lambda}, l^+\overline{p}$			40 ± 11
$\mathcal{B}_B$ from $l^+p\overline{p}$			$8.2^{+2.5}_{-1.6}$

There are three sets of observations: inclusive baryon rates and baryon-antibaryon coincidences from ARGUS [24] and CLEO [6], and lepton-baryon coincidences from ARGUS [25], where a lepton momentum cut ensures that leptons and baryons come from two different B-mesons. Table 6 gives the experimental results. The observations can be described by 5 parameters, $\mathcal{B}_B = \mathcal{B}(\overline{B} \to \Lambda_c X)$, $\mathcal{B}_\Lambda = \mathcal{B}(\Lambda_c \to \Lambda X)$, $\mathcal{B}_p =$

$\mathcal{B}(\Lambda_c \to pX, p \text{ not from } \Lambda), f_\Lambda =$ fraction of $\overline{\Lambda}$ in the associated production, i.e. $f_\Lambda = \mathcal{B}(\overline{B} \to \Lambda_c \overline{\Lambda} X)/\mathcal{B}(\overline{B} \to \Lambda_c X)$, and $f_p =$ fraction of $\overline{p}$ not originating from $\overline{\Lambda}$ defined in the same way. The observations are related to these parameters by

$$\mathcal{B}(\overline{B} \to p \text{ or } \overline{p} + X) = \mathcal{B}_B \cdot (\mathcal{B}_p + 0.64\mathcal{B}_\Lambda + f_p + 0.64 f_\Lambda) , \quad (19a)$$
$$\mathcal{B}(\overline{B} \to p\overline{p} + X) = \mathcal{B}_B \cdot (\mathcal{B}_p + 0.64\mathcal{B}_\Lambda) \cdot (f_p + 0.64 f_\Lambda) , \quad (19b)$$
$$\mathcal{B}(\overline{B} \to \Lambda \text{ or } \overline{\Lambda} + X) = \mathcal{B}_B \cdot (\mathcal{B}_\Lambda + f_\Lambda) , \quad (19c)$$
$$\mathcal{B}(\overline{B} \to \Lambda \overline{p} \text{ or } \overline{\Lambda} p + X) = \mathcal{B}_B \cdot [\mathcal{B}_\Lambda (f_p + 0.64 f_\Lambda) + f_\Lambda (\mathcal{B}_p + 0.64 \mathcal{B}_\Lambda)] , \quad (19d)$$
$$\mathcal{B}(\overline{B} \to \Lambda \overline{\Lambda} X) = \mathcal{B}_B \cdot \mathcal{B}_\Lambda \cdot f_\Lambda , \quad (19e)$$
$$N(l^+ p)/N(l^+) = \mathcal{B}_B \cdot (\mathcal{B}_p + 0.64 \mathcal{B}_\Lambda) , \quad (19f)$$
$$N(l^+ \overline{p})/N(l^+) = \mathcal{B}_B \cdot (f_p + 0.64 f_\Lambda) , \quad (19g)$$
$$N(l^+ p\overline{p})/N(l^+) = \mathcal{B}_B \cdot (\mathcal{B}_p + 0.64 \mathcal{B}_\Lambda) \cdot (f_p + 0.64 f_\Lambda) , \quad (19h)$$
$$N(l^+ \Lambda)/N(l^+) = \mathcal{B}_B \cdot \mathcal{B}_\Lambda , \quad (19i)$$
$$N(l^+ \overline{\Lambda})/N(l^+) = \mathcal{B}_B \cdot f_\Lambda , \quad (19j)$$

An overall least-square fit to the 15 observations leads to the following results:

$$\mathcal{B}_B = \mathcal{B}(\overline{B} \to \Lambda_c X) = (7.9 \pm 0.9)\% , \quad (20)$$
$$\mathcal{B}_\Lambda = \mathcal{B}(\Lambda_c \to \Lambda X) = (37 \pm 7)\% , \quad (21a)$$
$$\mathcal{B}_p = \mathcal{B}(\Lambda_c \to pX, p \text{ not from } \Lambda) = (31 \pm 8)\% , \quad (21b)$$
$$\mathcal{B}_n = \mathcal{B}(\Lambda_c \to nX, n \text{ not from } \Lambda) = (32 \pm 11)\% , \quad (21c)$$
$$f_\Lambda = \mathcal{B}(\overline{B} \to \Lambda_c \overline{\Lambda} X)/\mathcal{B}(\overline{B} \to \Lambda_c X) = (14 \pm 4)\% , \quad (22a)$$
$$f_p = \mathcal{B}(\overline{B} \to \Lambda_c \overline{p} X)/\mathcal{B}(\overline{B} \to \Lambda_c X) = (47 \pm 8)\% , \quad (22b)$$
$$f_n = \mathcal{B}(\overline{B} \to \Lambda_c \overline{n} X)/(\overline{B} \to \Lambda_c X) = (39 \pm 9)\% , \quad (22c)$$

where the result in eq. 22c follows from $f_n = 1 - f_\Lambda - f_p$ and that in eq. 21c from $\mathcal{B}_n = 1 - \mathcal{B}_\Lambda - \mathcal{B}_p$. The combination of eq. 20 with eq. 17 finally leads to

$$\mathcal{B}(\Lambda_c \to pK^-\pi^+) = (3.7 \pm 1.0)\%. \quad (23)$$

It should be added that the uncertainty from assumption 3 can only decrease the result in eq. 20 and therefore only increase that in eq. 23.

7. NONLEPTONIC $b \to u$ DECAYS

The ARGUS result [26] on charmless nonleptonic B decays, obtained in 1987 with $103/pb$,

$$\mathcal{B}(B^+ \to p\overline{p}\pi^+) + \mathcal{B}(B^0 \to p\overline{p}\pi^+\pi^-) = (11.2 \pm 2.4 \pm 4.1) \cdot 10^{-4} \quad (24)$$

was not confirmed by CLEO. In 1988, they found [27]

$$\mathcal{B}(B^+ \to p\overline{p}\pi^+) + \mathcal{B}(B^0 \to p\overline{p}\pi^+\pi^-) < 5.1 \cdot 10^{-4} , \quad (25)$$

where the limit in eq. 25 is for the extrapolation into the full phase space as for ARGUS. The 1988 ARGUS data with $70/pb$ do not show any signal in the $p\overline{p}\pi^+$ and $p\overline{p}\pi^+\pi^-$ channel, and more data are taken now in order to understand the origin of the number in eq. 24.

Both ARGUS and CLEO have searched for a large number of other nonleptonic $b \to u$ channels. The higher integrated luminosity of CLEO leads to the smaller upper limits, the most important of these are [28] with 90 % CL:

$$\begin{aligned}
\mathcal{B}(B^0 \to p\bar{p}) &< 4 \cdot 10^{-5}, \\
\mathcal{B}(B^0 \to \pi^+\pi^-) &< 9 \cdot 10^{-5}, \\
\mathcal{B}(B^+ \to \rho^0\pi^+) &< 1.5 \cdot 10^{-4}, \\
\mathcal{B}(B^0 \to \rho^0\rho^0) &< 3.4 \cdot 10^{-4}, \\
\mathcal{B}(B^+ \to \pi^+\pi^+\pi^-) &< 1.7 \cdot 10^{-4}.
\end{aligned} \qquad (26)$$

Only the $\pi^+\pi^-$ channel gives an interesting limit on the CKM matrix element V_{ub}. Bauer et al. [18, 29] give

$$\Gamma(B^0 \to \pi^+\pi^-) = 0.17 \cdot a_1^2 \cdot |V_{ub}/V_{cb}|^2 \cdot 10^{10}/s,$$

which, together with $a_1 = 1.1$, leads to

$$|V_{ub}/V_{cb}| < 0.19 \ (90\% CL). \qquad (27)$$

8. NONLEPTONIC $b \to s$ DECAYS

No new data on penguin-graph B-meson decays have been presented at this Conference. All searches for radiative penguins, like $\mathcal{B}(B \to K^*892\gamma) < 2.4 \cdot 10^{-4}$ (90%CL) [30], and for hadronic penguins have been negative so far. The upper limits obtained, however, are not in conflict with theoretical expections.

9. INCLUSIVE SEMILEPTONIC DECAYS

Fig. 7 shows the 1988 results of ARGUS [31] for the momentum spectrum of inclusive B decays into electrons and muons. The data are well described by a variety of models. Using the quark decay model of Altarelli et al. [32], where b, c, and u quarks are treated as quasi-free partons, the main results are

$$\mathcal{B}(B \to l\nu X) = (10.3 \pm 0.7)\%, \qquad (28)$$

$$<m_b> = (4.94 \pm 0.06) \ GeV, \qquad (29)$$

$$<m_b - m_c> = (3.38 \pm 0.02) \ GeV, \qquad (30)$$

$$|V_{ub}/V_{cb}| < 0.14 \ (90\% CL).$$

The meson decay model of Isgur et al. [33] gives

$$\mathcal{B}(B \to l\nu X) = (9.7 \pm 0.6)\%,$$

$$|V_{ub}/V_{cb}| < 0.16 \ (90\% CL). \qquad (31)$$

The two branching ratio determinations — the first is essentially for $b \to l\nu c$, the second for $B \to l\nu(D + D^* + 13\% \ more)$ — are rather close to each other. The two limits on

V_{ub} differ since there is much more agreement between the two models for the $b \to c$ decays than for the $b \to u$ decays. For the analysis in chapter 13, we will use the more conservative limit in eq. 31. As decay fraction $\mathcal{B}(B \to l\nu X)$ we will use the quark decay result in eq. 30, averaged with the results of CLEO, $(10.1 \pm 0.5)\%$ [15], and of Crystal Ball, $(12.0 \pm 0.5)\%$ [34], both also obtained from of the Altarelli model [32]:

$$\mathcal{B}(B \to l\nu X) = (10.9 \pm 0.6)\% \ . \tag{32}$$

The agreement between the three experiments is not good at all, the average in eq. 32 contains a "scale factor" [1] of $S = 2$.

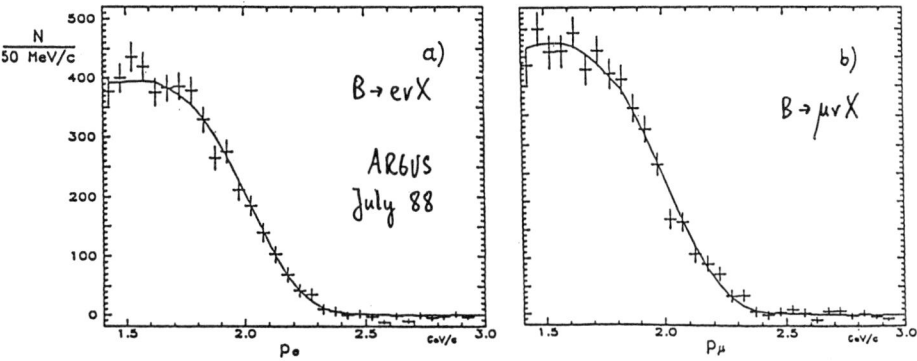

Fig. 7: ARGUS results on lepton momentum spectra in $B \to l\nu X$

As mentioned in chapter 4, CLEO [15] has obtained a result for $\mathcal{B}(B^0 \to l\nu X)$ by tagging inclusive leptons with reconstructed $\overline{B}^0 \to D^{*+}\pi^-$ and $\overline{B}^0 \to D^{*+}l^-\nu$ decays, the result is $(9.4 \pm 2.0)\%$ using Altarelli et al. [32].

The $|V_{ub}/V_{cb}|$ analysis of the new CLEO data with $212/pb$ has been presented for the first time at this Conference; for details see M. Procario [35]. In the momentum spectrum of inclusive electrons and muons from the $\Upsilon(4S)$, there is an excess of events between 2.3 and 2.6 GeV/c with a significance of 2.2 standard deviations. This momentum region should have no leptons from $b \to c$ decays, and the observed excess corresponds to $|V_{ub}/V_{cb}| = 0.07 \pm 0.02$ in the quark decay description [32] and to 0.12 ± 0.03 in the meson decay description [33]. This level of V_{ub} is just the expected one if CP violation in the K^0 system has its origin in the standard electroweak theory with three fermion families, see chapter 13. More data and new search methods for V_{ub} are now urgently needed.

10. EXCLUSIVE SEMILEPTONIC DECAYS

There are ARGUS and CLEO results on decay fraction and D^* polarization of $B^0 \to D^{*-}l^+\nu$, and decay fractions of $B^0 \to D^{**-}l^+\nu$ and $B^+ \to \overline{D}^0l^+\nu$, $\overline{D}^{*0}l^+\nu$, $\overline{D}^{**0}l^+\nu$.

Table 7: Exclusive semileptonic B-meson decay fractions, *) after rescaling

	ARGUS	CLEO
$B^0 \to D^{*-}l^+\nu$	$(7.0 \pm 1.2 \pm 1.9)\%$	$(4.6 \pm 0.5 \pm 0.7)\%$
	$(5.7 \pm 1.0 \pm 1.5)\%$ *)	$(4.9 \pm 0.5 \pm 0.7)\%$ *)
$B^0 \to D^{**-}l^+\nu$	$< 1\%$	$(2.2 \pm 1.3)\%$
$B^+ \to \overline{D}^0 l^+\nu$	$-$	$(2.4 \pm 0.8 ^{+0.7}_{-0.8})\%$
$B^+ \to \overline{D}^{*0} l^+\nu$	$-$	$(3.9 \pm 0.8 ^{+1.1}_{-0.8})\%$
$B^+ \to D^{**0} l^+\nu$	$-$	$(1.9 \pm 1.0)\%$

In this list, D^{**} is not meant to be one state, but includes all $c\bar{q}$ states different from 1^1S_0 and 1^3S_1.

The $B^0 \to D^{*-}l^+\nu$ decay is selected by the missing mass of $D^{*-}l^+$ assuming $|\vec{p}_B| = 0$. ARGUS, who pioneered this method [36], found $\mathcal{B}(B^0 \to D^{*-}l^+\nu) = (7.0 \pm 1.2 \pm 1.9)\%$, CLEO [15,16] finds $(4.6 \pm 0.5 \pm 0.7)\%$. If we rescale both results using common parameters $f_0 = 0.50$ and $\mathcal{B}(D^{*+} \to D^0\pi^+) = (54 \pm 5)\%$, where the last number is the average between the Particle Data Group value [1] and a recent MARK-III result [37], the agreement between them is not as bad, see the summary in table 7. The two groups do not agree on the amount of D^{**} in the missing mass spectrum, see also table 7. This disagreement is not significant, and much more data are needed before the 1.3% expectation for $\mathcal{B}(B \to D^{**}l\nu)$ in the model of Isgur et al. [33] can be tested with reasonable precision.

The D^* polarisation in $B^0 \to D^{*-}l^+\nu$ decays has been measured by ARGUS [38] and CLEO [16]. Both groups agree on a small longitutinal polarisation, $\mathcal{B}_L/\mathcal{B}_T = 0.85\pm0.45$, which also agrees with the expectation of the newest version of the Isgur et al. model [33] and with other form factor models [39,40].

The $B^+ \to \overline{D}^0 l^+\nu$ and $B^+ \to \overline{D}^{*0}l^+\nu$ results of CLEO in table 7 are obtained from the $D^0 l$ missing mass spectrum [16] again assuming $|\vec{p}_B| = 0$. The ratio $\overline{D}^{*0}l^+\nu/D^{*-}l^+\nu$ has been used for a determination of $\tau(B^+)/\tau(B^0)$ as presented in chapter 4. Do the exclusive decay fractions in table 7 exhaust the inclusive value of eq. 32 ? The average of the three D^* fractions 5.7%, 4.9%, and 3.9% is $(4.6\pm0.6)\%$. Adding the two CLEO results $(2.4\pm1.1)\%$ for $Dl\nu$ and $(2.0\pm0.8)\%$ for $D^{**}l\nu$ to this value leads to $(9.0\pm1.5)\%$ which disagrees with the inclusive value of $(10.9\pm0.6)\%$ by only 1.2 standard deviations. The present precision is really not sufficient for claiming a discrepancy here.

ARGUS has just finished a new search [41] for the exclusive V_{ub}-induced semileptonic decays $B^+ \to \rho^0 l^+\nu$ and $B^0 \to \pi^- l^+\nu$. The full $\Upsilon(4S)$ sample of 181 events/pb was used to investigate the missing mass spectrum of $\rho^0 l^\pm$ and $\pi^\mp l^\pm$ combinations assuming $|\vec{p}_B| = 0$. Fig. 8a shows the missing mass of $(\rho^0 l^\pm)$ for the data together with the expectation for a genuine $B^+ \to \rho^0 l^+\nu$ signal if its decay fraction would be 1%. Since no signal appears in the data, the following selection criteria have been applied:

1. $|\cos\theta(\vec{p}_{\rho l}, \vec{T}_{rest})| < 0.7$, where $\vec{p}_{\rho l}$ is the momentum vector of the $\rho^0 l^\pm$ pair, and $\vec{T}_{rest}$ is the direction of the thrust axis as found from all particles in the event with the

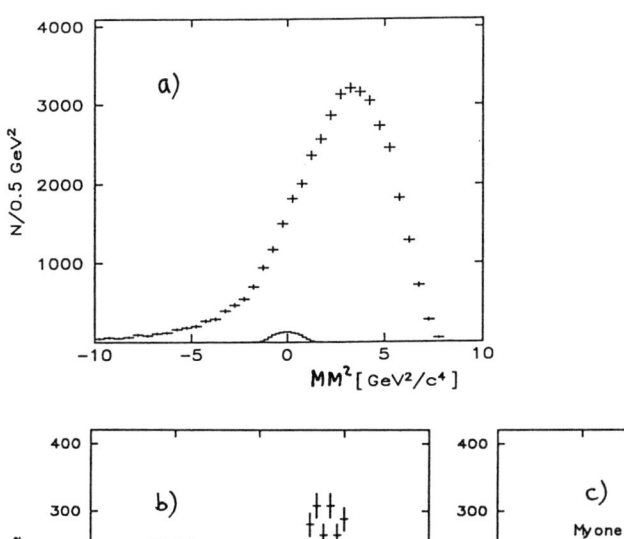

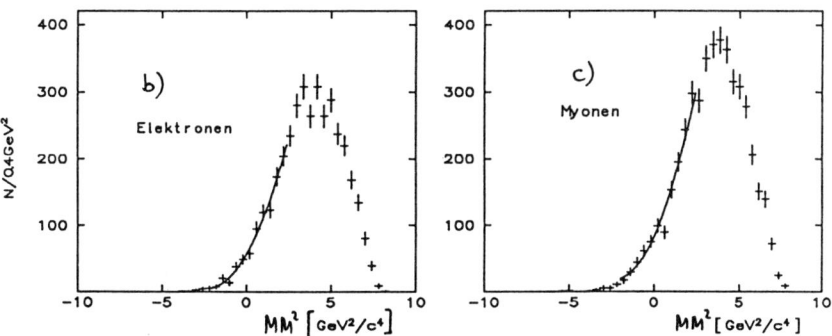

Fig. 8: ARGUS results [41] on $B^+ \to \rho^0 l^+ \nu$. a: spectrum of missing mass squared for all $\rho^0 l^\pm$ combinations (the histogram shows the Monte Carlo signal for $\mathcal{B}(B^+ \to \rho^0 l^+ \nu) = 1\%$), b: for $\rho^0 e^\pm$, and c: for $\rho^0 \mu^\pm$ after the cuts as described in the text.

exception of ρ^0 and $l^\pm$.

2. $\cos\theta(\vec{p}_\rho, \vec{p}_l) < -0.5$.

3. $|\sum_{i \neq \rho, l} \vec{p}_i| < 2.5~GeV$, where the sum runs over any particle combination in the event with the exception of ρ^0 and $l^\pm$.

The resulting missing mass distributions are shown in figs. 8b and 8c. The absence of a signal leads to

$$\mathcal{B}(B^+ \to \rho^0 l^+ \nu) < 1.0 \cdot 10^{-3}~(90\% CL)~. \tag{33}$$

A similar selection procedure for the pion mode yields

$$\mathcal{B}(B^0 \to \pi^- l^+ \nu) < 1.0 \cdot 10^{-3}~(90\% CL)~. \tag{34}$$

11. THE CKM MATRIX ELEMENTS V_{ub} AND V_{cb}

The CKM matrix does not need an introduction here, for recent reviews see refs. [42, 43, 44, 12]. The modulus of V_{ub} is bound by the experiments on $B^0 \to \pi^+ \pi^-$, $B \to$

$l\nu X$, and $B^+ \to \rho^0 l^+ \nu$, $B^0 \to \pi^- l^+ \nu$. The $\pi\pi$ result in eqs. 26, 27 was

$$|V_{ub}/V_{cb}| < 0.19 \,, \quad |V_{ub}| < 0.01 \,.$$

The $l\nu X$ result in eq. 31 from ARGUS was

$$|V_{ub}/V_{cb}| < 0.16 \,, \quad |V_{ub}| < 0.008 \,,$$

and the 2.2σ effect of CLEO [35] translates into just this same upper limit. The upper limits on $B^+ \to \rho^0 l^+ \rho$ and $B^0 \to \pi^- l^+ \nu$ are not easily transformed into an upper limit on $|V_{ub}|$, since the different meson decay models [33, 39, 40] disagree in the expected rates by nearly a factor of 4. See N. Isgur [45] for comments and for the hope that this disagreement will desappear when real events and their q^2 spectrum have been observed. The most conservative upper limit on $|V_{ub}|$ from $\rho l\nu$ and $\pi l\nu$ is 0.013 [41] which is not better than the inclusive lepton limit in eq. 31.

There are three methods for a determination of V_{cb}:

$$\Gamma(B \to l\nu X) = \frac{G_F^2 V_{cb}^2}{192\pi^3} \cdot m_b^5 \cdot \phi(m_c/m_b) \cdot \eta_{QCD} \,, \tag{35}$$

$$\Gamma(B^0 \to D^{*-} l^+ \nu) = V_{cb}^2 \cdot (2.2 \pm 0.5) \cdot 10^{13}/s \,, \tag{36}$$

$$\Gamma(B \to l\nu X) = V_{cb}^2 \cdot (2.2 \pm 0.5 + 1.0 \pm 0.2 + 0.4 \pm 0.1) \cdot 10^{13}/s \,. \tag{37}$$

The partial widths Γ are obtained from the B lifetime in eq. 11 and the relevant decay fractions — eq. 32 for the inclusive fraction, and the average of ARGUS and CLEO, $(4.6 \pm 0.6)\%$, for $D^* l\nu$. The first method requires the quark masses m_b and m_c. Since the Altarelli model [32] with the parameters as given in eqs. 29 and 30 describes the lepton spectrum very well, we assume that also the total rate is well governed by these parameter values. Using eq. 29 (with 3 times englarged error) and eq. 30 and taking properly into account the anticorrelation between m_b^5 and the phase space integral $\phi(m_c/m_b)$ for fixed difference $m_b - m_c$ leads to

$$V_{cb}(inclusive) = 0.047 \pm 0.005 \,. \tag{38}$$

The second method leads to

$$V_{cb}(exclusive) = 0.042 \pm 0.006 \,, \tag{39}$$

where the numerical value for the coefficient in eq. 36 is an average over four different form factor models [33, 39, 40, 46]. The third method, where the inclusive experimental rate is compared to the sum of three expected meson decay rates — D^* (as above), D [33, 39, 40, 46] and D^{**}[33] — leads to

$$V_{cb}(\text{``mixed''}) = 0.051 \pm 0.005 \,. \tag{40}$$

The agreement between the three methods is very good; we conclude

$$V_{cb} = 0.047 \pm 0.005 \,. \tag{41}$$

12. ANYTHING ON B_s DECAYS ?

No B_s meson has been reconstructed so far. There is some indirect evidence for B_s and B_s^* production on the $\Upsilon(5S)$ in the CUSB data, as discussed by M. Tuts at this Conference [47]. The standard electroweak theory expects a large $B_s\overline{B}_s$ oscillation rate with $\chi(B_s)$ close to its maximal value of 0.5. Fig. 9 summarizes the oscillation results obtained so far. The three high-energy e^+e^- experiments MARK-II [48], MAC [49], JADE [50] and the $p\overline{p}$ experiment UA1 [51] are sensitive to $\chi = f_0 \cdot \chi(B^0) + f_s \cdot \chi(B_s)$ with $f_0 \approx 0.38$ and $f_s \approx 0.15$. For some more details, see ref. [52]. The only oscillation evidence with more than 2σ in these four experiments comes from UA1. But their result is compatible with the ARGUS/CLEO value for $\chi(B^0)$ and any value between 0 and 0.5 for $\chi(B_s)$. The JADE result is obtained from the hemisphere asymmetry of $b\overline{b}$ jets as expected from γ/Z^0 interference. A recent analysis of R.B. Hurst [53] on this asymmetry in all high-energy e^+e^- experiments results in $\chi = 0.15 \pm 0.10$. Unfortunately this does not add much information, and our conclusion on $B\overline{B}$ oscillations remains

$$\chi(B^0) = 0.17 \pm 0.04 \ , \quad \chi(B_s) = any\ value\ within\ \pm 1\sigma \ . \tag{42}$$

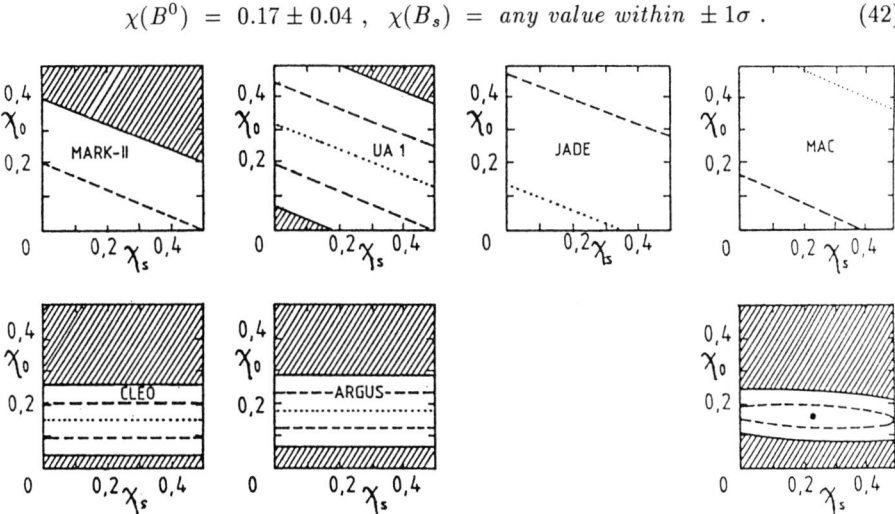

Fig. 9: Summary of $B\overline{B}$ oscillation results. The dotted curves are central values, the dashed curves correspond to one and the solid curves to two standard deviations. The four results from $b\overline{b}$ jets are all assumed to be $0.38 \cdot \chi_0 + 0.15 \cdot \chi_s$. The last part of the figure is the Gaussian average of the six experiments.

13. CKM MATRIX FIT AND CONSEQUENCES

The two results of chapter 11 are important inputs for our knowledge of the CKM matrix. Combining them with the four other matrix element determinations leads to the conclusion that unitarity is well fulfilled experimentally,

$$V_{ud}^2 + V_{us}^2 + |V_{ud}|^2 = 0.9977 \pm 0.0022 \ , \tag{43a}$$

$$|V_{cd}|^2 + |V_{cs}|^2 + V_{cb}^2 = 1.05 \pm 0.18 , \qquad (43b)$$

$$\sum V_{ui}^* \cdot V_{ci} = 0.005 \pm 0.035 , \qquad (43c)$$

where the numerical inputs are those of ref. [12]. A fit to the six measured moduli $|V_{ij}|$ with $i = u, c$ and $j = d, s, b$, imposing unitarity, leads to the following result:

$$V_{CKM} = \begin{pmatrix} 0.9753 \pm 0.0005 & 0.2207 \pm 0.0020 & 0.0000 \pm 0.0055 \\ -0.2204 \pm 0.0021 & 0.9742 \pm 0.0005 & 0.0470 \pm 0.0050 \\ 0.0108 \pm 0.0055 & -0.0458 \pm 0.0050 & 0.9989 \pm 0.0002 \end{pmatrix}$$

$$+ i \cdot \begin{pmatrix} 0 & 0 & 0.0000 \pm 0.0055 \\ 0 \pm 0.0003 & 0 \pm 0.0001 & 0 \\ 0 \pm 0.0054 & 0 \pm 0.0012 & 0 \end{pmatrix} . \qquad (44)$$

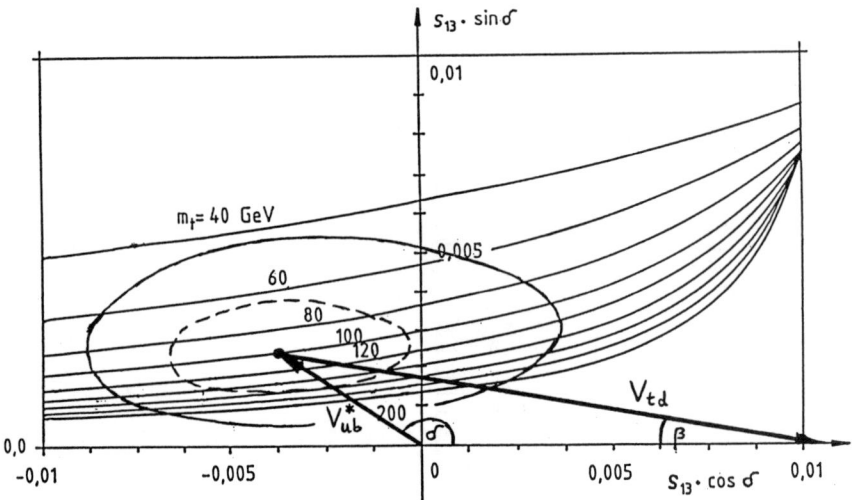

Fig. 10: The "CP triangle" of the CKM matrix with approximate one- (dashed ellipse) and two- (solid ellipse) standard deviation contours [12]. The solid lines relate V_{ub}^* to the observed value of $\epsilon(K^0)$ for different m_t values [57].

The most uncertain elements are V_{ud} and V_{td}. They are, however, connected by unitarity,

$$V_{ud}^* V_{td} + V_{us}^* V_{ts} + V_{ub}^* V_{tb} = 0 , \qquad (45a)$$

$$0.975 \cdot V_{td} + 0.999 \cdot V_{ub}^* = 0.010 \pm 0.001 , \qquad (45b)$$

a relation which is usefully presented as a triangle in the complex plane [54, 55]. The best fit result for this "CP Triangle" is shown in fig. 10, where the fit inputs are the six moduli already used for the fit leading to eq. 44, and in addition [12] $\chi(B^0)$ with $f_B = (140 \pm 25) \, MeV$, $\epsilon(K^0)$ with $B_K = 2/3 \pm 1/6$, $Re(\epsilon'/\epsilon)$ with the experimental result of NA31 [56] and $H = -0.27 \pm 0.09$ from Buras and Gerard [57], and $m_t < 200\ GeV$ from radiative corrections to m_W/m_Z [58]. The consequences of this fit, which uses all experimental information and assumes that $\epsilon(K^0)$, $Re(\epsilon'/\epsilon)$, and $\chi(B^0)$ have their origin completely in the standard electroweak theory with threee quark families, are the following:

$$|V_{ub}| = 0.005 \pm 0.002, \qquad (46)$$

$$\delta_{CKM} = (140 ^{+\ 20}_{-\ 50})^o, \qquad (47)$$

$$m_t = (110 \pm 35)\ GeV, \qquad (48)$$

$$sin(-2argV_{td}) = 0.41^{+\ 0.20}_{-\ 0.10}, \qquad (49)$$

$$x(B_s) = 10 \pm 5. \qquad (50)$$

The result in eq. 49 is relevant for the expected CP violation in B^0, $\overline{B}^0 \to J/\Psi K_s^0$ [59]. Its observation, together with a measurement of the other unknown observables in eqs. 46, 48, and 50 will show — hopefully in the not too far future — if the standard theory needs an extension or if CP violation in both the K and the B meson sector is completely described by the present version of the theory.

ACKNOWLEDGEMENT

I thank my colleagues in ARGUS and CLEO for providing me with newest results and for useful discussions. I thank S. Weseler for his contributions to preparing this talk and G. Dresel and W. Weißmann for their typesetting help.

REFERENCES

[1] Review of Particle Properties, Particle Data Group, Phys.Lett.B 204(1988)1
[2] C.Bebek et al. (CLEO), Phys.Rev.D 36(1987)1289
[3] H.Albrecht et al. (ARGUS), Phys.Lett.B 185(1987)218, updated
[4] H.Albrecht et al. (ARGUS), Phys.Lett.B 199(1987)451
[5] H.Albrecht et al. (ARGUS), Phys Lett.B 215(1988)424
[6] Y.Kubota (CLEO), these Proceedings
[7] E.Eichten et al., Phys. Rev.D 17(1978)3090
    E.Eichten, Phys.Rev.D 22(1980)1819
[8] G.Altarelli, these Proceedings and references therein
[9] H.Albrecht et al. (ARGUS), Phys.Lett.B 192(1987)245
[10] M.Schäfer (ARGUS), these Proceedings
[11] M.Artuso et al. (CLEO), Cornell preprint CLNS 89/889 (1989)
[12] K.R.Schubert, Karlsruhe preprint IEKP-KA/88-4, to be published in
    Proc.Conf.Phenomenology in High Energy Phys., ICTP Trieste 1988
[13] S.R.Wagner (MARK-II), SLAC-PUB-4635 (1988)
[14] A.Bean et al. (CLEO), Phys.Rev.Lett. 58(1987)183
[15] S.Stone (CLEO), presented at the Int. Workshop on Weak Interactions and
    Neutrinos, Ginosar/Israel, April 1989
[16] N.Katayama (CLEO), these Proceedings

[17] R.Gläser (ARGUS), these Proceedings
[18] M.Bauer, Dr.rer.nat. Thesis, Heidelberg University, 1987
[19] S.Stone (CLEO), private communication
[20] M.S.Alam et al. (CLEO), Phys.Rev.D 34(1986)3279
[21] H.Albrecht et al. (ARGUS), Phys.Lett.B 210(1988)263
[22] G.S.Abrams et al. (SLAC-LBL), Phys.Rev.Lett. 44(1980)10
[23] M.Aguilar-Benitez et al. (EHS), Phys.Lett.B 199(1987)462
[24] H.Albrecht et al. (ARGUS), Z.Physik C 42(1989)519
[25] A.I.Golutvin (ARGUS),
      Proc.XXIV.Int.Conf.High Energy Phys., Munich 1988, p.553
[26] H.Albrecht et al. (ARGUS), Phys.Lett.B 209(1988)119
[27] D.L.Kreinick (CLEO),
      Proc.XXIV.Int.Conf.High Energy Phys., Munich 1988, p.511
[28] D.Bortoletto et al. (CLEO), Cornell preprint CLNS 89/887(1989)
[29] M.Bauer, B.Stech, and M.Wirbel, Z.Physik C 34(1987)103
[30] H.Albrecht et al. (ARGUS), Phys.Lett.B 210(1988)258
[31] H.Albrecht et al. (ARGUS), "Search for $b \to u$ Transitions in Semileptonic
      B-Meson Decays, submitted to XXIV.Int.Conf.High Energy Phys., Munich 1988
[32] G.Altarelli et al., Nucl.Phys. B 208(1982)365
[33] N.Isgur, D.Scora, B.Grinstein, and M.B.Wise, Phys.Rev.D 39(1989)799
[34] K.Wachs et al. (Crystal Ball), Z.Physik C 42(1989)33
[35] M.Procario (CLEO), these Proceedings
[36] H.Albrecht et al. (ARGUS), Phys.Lett.B 197(1987)452
[37] J.Adler et al. (MARK-III),Phys.Lett.B 208(1988)152
[38] H.Albrecht et al. (ARGUS), Phys.Lett.B 219(1989)121
[39] J.G.Körner and G.A.Schuler, Z.Physik C 38(1988)511
[40] M.Wirbel, B.Stech, and M.Bauer, Z.Physik C 29(1985)637
[41] T.Ruf (ARGUS), Dr.rer.nat. Thesis, Karlsruhe University, 1989
[42] Y.Nir, these Proceedings
[43] H.Harari, Proc.Int.Conf. on CP Violation, Blois (France), 1989
[44] K.Kleinknecht, Proc.XXIV.Int.Conf.High Energy Phys., p.98
[45] N.Isgur, these Proceedings
[46] T.Altomari and L.Wolfenstein, Phys.Rev.D 37(1988)681
[47] M.Tuts (CUSB), these Proceedings
[48] T.Schaad et al. (MARK-II), Phys.Lett. 160B(1985)188
[49] H.R.Band et al. (MAC), Phys.Lett.B 200(1988)221
[50] W.Bartel et al. (JADE), Phys.Lett. 146B(1984)437
[51] C.Albajar et al. (UA1), Phys.Lett.B 186(1987)247
[52] K.R.Schubert, Progress in Nuclear and Particle Physics 21(1988)3
[53] R.B.Hurst (MAC), to be published in Proc.24.Rencontre de Moriond, 1989
[54] J.D.Bjorken, presented at the Workshop on Experiments, Detectors,
      and Experimental Areas for the Supercollider, Berkeley 1987
[55] C.Jarlskog and R.Stora, Phys.Lett.B 208(1988)268, and references therein
[56] H.Burkhardt et al. (NA31), Phys.Lett.B 206(1988)169
[57] A.J.Buras and J.-M.Gerard, Phys.Lett.B 203(1988)272, and references therein
[58] U.Amaldi et al., Phys.Rev.D 36(1987)1385
[59] I.Dunietz and T.Nakada, Z.Physik C 36(1987)503, and references therein

Update on B^0-$\overline{B}^0$ Mixing

Marion Schäfer
DESY, Hamburg, Germany

Abstract

Using the ARGUS detector at the e^+e^- storage ring DORIS II at DESY an update of the B^0-$\overline{B}^0$ mixing measurement has been performed including the new 1988 data. With a total of $\int Ldt = 172pb^{-1}$ on the $\Upsilon(4S)$ resonance and $55pb^{-1}$ in the nearby continuum the dilepton analysis yields a mixing parameter of $r = 0.22 \pm 0.07 \pm 0.06$. We have also investigated charge correlations between a D^{*+} and one or two leptons. Combining the results of all three methods we obtain $r = 0.22 \pm 0.07$.

1 Introduction

In the Standard Model transitions between B^0 and $\overline{B}^0$ are mediated by second order weak interactions through the Box diagrams [1], which are practically saturated by top quark exchange. Consequently B^0-$\overline{B}^0$ mixing leads to predictions for the CKM matrix element $|V_{td}|$ and the top quark mass [2]. Observable effects of B^0-$\overline{B}^0$ oscillations depend on the parameter $x = \Delta M / \Gamma$, where ΔM is the mass difference of the CP eigenstates and Γ is the mean B decay width. Due to the short lifetime of B mesons, one is experimentally restricted to measurements of time integrated quantities, such as the probability that a particle produced as B^0 decays as a $\overline{B}^0$ divided by the probability that it decays as a B^0:

$$r = \frac{Prob(B^0 \to \overline{B}^0)}{Prob(B^0 \to B^0)} = \frac{x^2}{2 + x^2}.$$

In the case of $\Upsilon(4S)$ decays, the parameter r, which is a measure of the strength of B^0-$\overline{B}^0$ oscillations is simply given by the ratio of events containing two identical B's to events with $B^0\overline{B}^0$ pairs:

$$r = \frac{N(B^0 B^0) + N(\overline{B}^0 \overline{B}^0)}{N(B^0 \overline{B}^0)}.$$

In 1987 ARGUS succeeded in fully reconstructing one B^0B^0 event ,thereby establishing the existence of B^0-$\overline{B}^0$ mixing [3]. Combining different methods we found an unexpectedly large value of $r = 0.21 \pm 0.08$. Since then another 67 $pb^{-1}\Upsilon(4S)$ data and 7 pb^{-1} in the nearby continuum have been collected ,and the analysis presented here is based on a total of $\int Ldt = 172 pb^{-1}$ on the resonance and 55 pb^{-1} off the resonance.The results derived from this data sample are preliminary.

2 Methods to observe $B^0\overline{B}^0$ Mixing

2.1 Dilepton Analysis

The traditional method to search for B^0-$\overline{B}^0$ oscillations is lepton tagging in semileptonic B decays,where the flavor of the B 's can be identified by the charge of the primary leptons.Thus in events with two B^0's or two $\overline{B}^0$'s one observes uncorrelated lepton pairs of the same charge ,while opposite sign leptons are produced in $B\overline{B}$ events.Since the $\Upsilon(4S)$ resonance decays to charged as well as neutral B's,the calculation of the mixing parameter r requires a correction factor λ [3] to account for l^+l^- pairs from B^+B^- events :

$$r = \frac{(N_{l+l+} + N_{l-l-})(1+\lambda)}{N_{l+l-} - (N_{l+l+} + N_{l-l-})\lambda}, \quad \text{with } \lambda = \frac{f^\pm (Br_{sl}^\pm)^2}{f^0 (Br_{sl}^0)^2} = 1.22.$$

The dominant background in this analysis arises from continuum $q\overline{q}$ events,secondary charm decays $(B \to \overline{D} \to l^-\nu X)$,and hadron tracks that are misidentified as leptons (fakes). These background sources can be effectively suppressed by applying the following cuts :

- total multiplicity $N_{tot} = N_{charge} + N\gamma/2 \geq 7$

- opening angle between the two leptons $-0.85 < cos\vartheta_{ll} < 0.95$

- lepton momentum 1.4 GeV/c $< p_l <$ 2.4 GeV/c .

After determination of the remaining background we obtain the dilepton rates quoted in Table 1.

From the observed 64 like-sign leptons ,29.4 are due to background, leaving a signal of 34.6 events for mixing. Comparing this to 381.3 unlike-sign candidates ,we obtain a mixing parameter of $r = 0.22 \pm 0.07 \pm 0.06$, where the first error is statistical and the second systematic.

Mixed Dilepton Candidates	$e^{\pm}e^{\pm}$	$\mu^{\pm}\mu^{\pm}$	$e^{\pm}\mu^{\pm}$
$\Upsilon(4S)$	15	21	34
Continuum	-	1	1
$\Upsilon(4S)$ (direct)	15 ± 4.9	18.1 ± 5.4	31.1 ± 6.5
Background			
Fake Leptons	1.8	4.0	6.7
Secondary Charm Decays	4.6	2.9	7.2
Converted Photons	0.3	-	0.3
J/ψ Decays	0.6	0.3	0.8
Sum :	7.7 ± 4.9 ± 2.6	10.9 ± 5.4 ± 2.1	16.0 ± 6.5 ± 3.9
Unmixed Dilepton Candidates	e^+e^-	$\mu^+\mu^-$	$e^{\pm}\mu^{\mp}$
$\Upsilon(4S)$	102	97	204
Continuum	2	3	2
$\Upsilon(4S)$ (direct)	96.1	88.2	198.1
J/ψ Correction	16.2 ± 4	15.0 ± 3	
Corrected $\Upsilon(4S)$ Rate	112.3 ± 10.9	103.2 ± 11.1	198.1 ± 14.9
Background			
Fake Leptons	3.6	7.0	13.2
Secondary Charm Decays	1.9	1.2	3.0
Converted Photons	0.3	-	0.3
J/ψ Decays	0.6	0.3	0.8
Sum :	105.9 ± 10.9 ± 1.8	94.7 ± 11.1 ± 2.5	180.7 ± 14.9 ± 4.9

Table 1. Preliminary dilepton rates

2.2 D^*ll Charge Correlations

An alternative way to measure B^0-$\overline{B}^0$ mixing exploits the fact that a partial reconstruction of $\overline{B}^0$ decaying into $D^{*+}l^-\overline{\nu}$ can be performed using the "recoil mass technique" .This technique has been used to determine the exclusive branching ratio for this decay [4] and relies on the fact that B mesons produced at the $\Upsilon(4S)$ resonance decay almost at rest .The effective neutrino mass can then be inferred from

$$M^2_{recoil} = [E_{beam} - (E_{D^*} + E_{lep})]^2 - (\vec{p}_{D^*} + \vec{p}_{lep})^2 .$$

Thus by reconstructing one $\overline{B}^0$ and tagging the other B with a fast lepton , mixed events produce $D^{*+}l^-l^-$ combinations ,while $D^{*+}l^-l^+$ combinations arise in $B^0\overline{B}^0$ events.

The D^{*+} has been reconstructed in the decay chain $D^{*+} \to D^0\pi^+$,followed by $D^0 \to K^-\pi^+, K^-\pi^+\pi^-\pi^+, K^-\pi^+\pi^0, K^0_s\pi^+\pi^-$. The resulting M^2_{rec} distribution for events with an additional fast l^+ or l^- with $p > 1.4$ GeV/c is shown in Fig.1 ,and the corresponding rates are listed in Table 2.

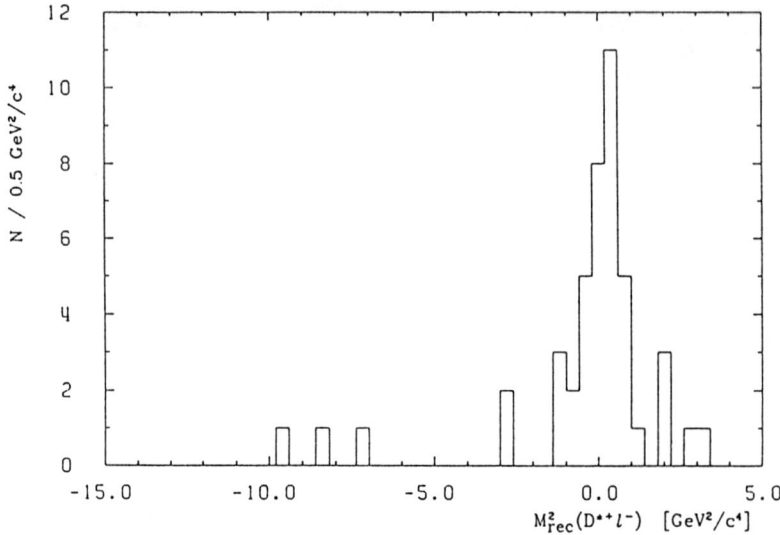

Fig.1 : M^2_{rec} for $D^{*+}l^-$ combinations in events with an additional lepton of $p_l > 1.4$ GeV/c

	$D^{*\pm}l^{\mp}l^{\mp}$	$D^{*\pm}l^{\mp}l^{\pm}$
$\Upsilon(4S)$	8 ± 2.8	27 ± 5.9
Background	1.9 ± 0.5	1.7 ± 0.4
Signal:	$6.1 \pm 2.8 \pm 0.5$	$25.3 \pm 5.9 \pm 0.4$

Table 2 : $D^{*+}ll$ rates with $|M^2_{rec}| < 1.5$ GeV^2/c^4

Since there is no contamination from charged B decays using this technique ,one obtains for the mixing parameter r

$$r = \frac{N_{D^{*+}l^-l^-}}{N_{D^{*+}l^-l^+}} = 0.24 \pm 0.12 \pm 0.02.$$

2.3 D^*l Charge Correlations

A third approach to observe B^0-$\overline{B}^0$ transitions is based on the assumption that B mesons decay exclusively via spectator processes [5]; consequently charged D^*'s are produced only in decays of neutral B's .Thus one can tag the flavor of one B with a $D^{*\pm}$,and the other with a lepton. Mixing manifests itself in the observation of $D^{*+}l^-$ pairs whereas $B^0\overline{B}^0$ events produce $D^{*+}l^+$ combinations.

To guarantee that the $D^{*+}l^-$ does not originate from the decay of a single B meson, we require the recoil mass to be less than $-2.5 GeV^2/c^4$. The $D^{*\pm}$ reconstruction has been performed in the channel $D^* \to D^0\pi^+$ followed by $D^0 \to$

$K^-\pi^+$, and we require a lepton momentum of greater than $1.0 GeV/c$. The M^2_{rec} distributions are shown in Fig.2a and 2b, and the rates are summarized in Table 3.

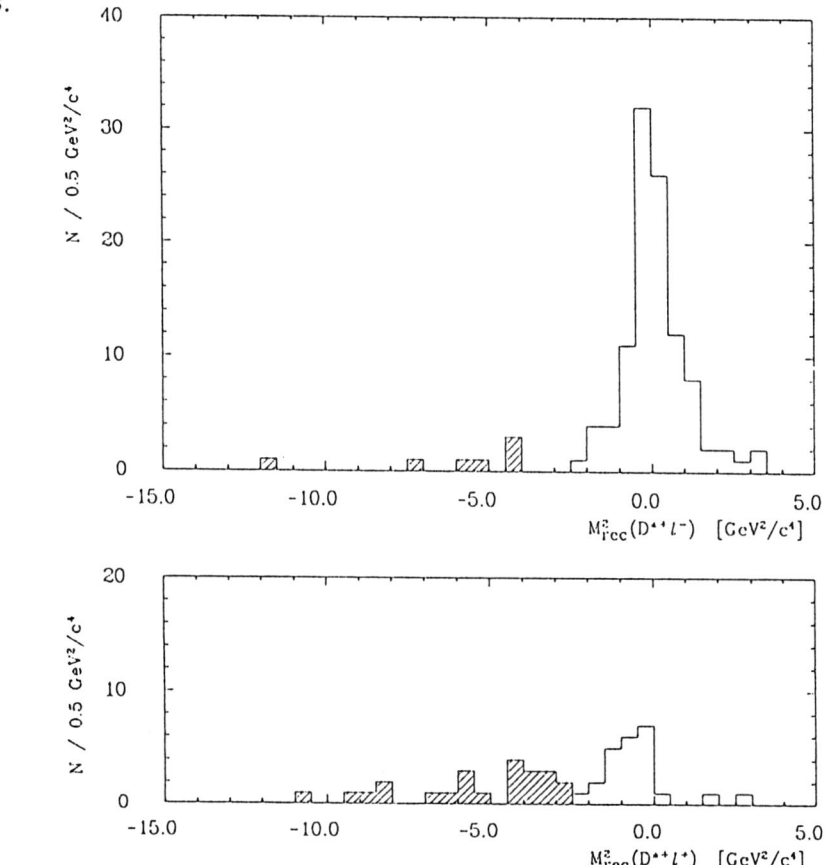

Fig.2 : M^2_{rec} distributions for (a) $D^{*\pm}l^{\mp}$ and (b) $D^{*\pm}l^{\pm}$.

	$D^{*\pm}l^{\mp}$	$D^{*\pm}l^{\pm}$
$\Upsilon(4S)$	7 ± 3.1	23 ± 5.5
Background	2.0 ± 0.5	2.3 ± 0.5
Signal:	$5.0 \pm 3.1 \pm 0.5$	$20.7 \pm 5.5 \pm 0.5$

Table 3 : $D^{*+}l$ rates with $M^2_{rec} < -2.5$ GeV^2/c^4

Subtracting the remaining backgrounds ,which arises mostly from cascade decays we observe 5 $D^{*+}l^-$ and 20.7 $D^{*+}l^+$ pairs ,resulting in $r = 0.24 \pm 0.16 \pm 0.02$. Despite the large error ,the good agreement between this value for r and the first two results is an indication that charged D^*'s are predominantly produced in neutral B decays.

3 Summary

Combining the three results on the mixing parameter r and taking correlations among the data samples into account we arrive at a mean value of $r = 0.22 \pm 0.07$. This is in excellent agreement with our first publication. The result has meanwhile been confirmed by the CLEO Collaboration [6], who find using the dilepton analysis $r = 0.19 \pm 0.06 \pm 0.06$.

As the value for r has not changed since its first announcement [3] ,the consequences remain essentially the same as discussed in [3]. A more precise determination of the Standard Model parameters $|V_{td}|$ and m_{top} depends strongly on solving the theoretical uncertainties in the quantity $B_B^{\frac{1}{2}} f_B$ [7] ,which is very difficult to measure.

References

[1] M.K.Gaillard and B.W.Lee, Phys.Rev. **D10** (1974) 897

[2] J.S.Hagelin, Phys.Rev. **D20** 2893 (1979)

[3] H.Albrecht et al (ARGUS collaboration), Phys.Lett. **192B** 245 (1987)

[4] H.Albrecht et al (ARGUS collaboration), Phys.Lett. **197B** 452 (1987)

[5] M.Wirbel ,DoTh 89/6 , May 89

[6] M.Artuso et al (CLEO collaboration), Cornell preprint CLNS 89/889 February 1989

[7] G.Altarelli , in the Proceedings of this Symposium

SEMILEPTONIC DECAYS OF B MESON INTO CHARMED MESON *

Nobuhiko Katayama

Laboratory of Nuclear Study, Cornell University, Ithaca, N.Y. 14853

ABSTRACT

Recent results on semileptonic decays of B meson into charmed meson using the CLEO detector at CESR are summarized. Ratios of the inclusive semileptonic branching fractions, $B(B^-\to D^0 X\ell^-\bar\nu)$, $B(\bar B^0\to D^+ X\ell^-\bar\nu)$, and $B(\bar B^0\to D^{*+} X\ell^-\bar\nu)$ to the average B meson semileptonic branching fraction are reported. The branching fractions of the exclusive final states, $B^-\to D^0\ell^-\bar\nu$, $B^-\to D^{*0}\ell^-\bar\nu$, and $\bar B^0\to D^{*+}\ell^-\bar\nu$ are measured. The ratio of $B(B^-\to D^{*0}\ell^-\bar\nu)$ to $B(B^-\to D^0\ell^-\bar\nu)$ and the polarization of the D^{*+} are obtained and compared with theoretical models. The value of $|V_{cb}|$ is calculated from the exclusive branching fractions for various models. Finally, the lifetime ratio of the charged and neutral B's (τ^+/τ^0) is determined from the ratio of $B(B^-\to D^{*0}\ell^-\bar\nu)$ to $B(\bar B^0\to D^{*+}\ell^-\bar\nu)$ to be $0.85 \pm 0.20^{+0.22}_{-0.16}$.

INTRODUCTION

Semileptonic decays of B's are the simplest B decays that can be described theoretically. Model predictions for D and D^* exclusive channels and D^* polarization are available.[1-4] Until recently, however, there has been little experimental data available on semileptonic B decays.[5-7] Here I report a detailed study of inclusive and exclusive semileptonic B decays into D^0, D^+, D^{*0} and D^{*+}. The data sample used in this study was collected with the improved CLEO detector at the Cornell Electron Storage Ring (CESR); It consists of $212 pb^{-1}$ at the $\Upsilon(4S)$ resonance, containing 242,000 $B\bar B$ pairs, and $102 pb^{-1}$ at energies just below the $B\bar B$ threshold (referred as continuum data).

INCLUSIVE SEMILEPTONIC DECAYS OF B MESON

In order to look for the semileptonic decay of a B meson into charmed mesons, $b\to cW^-$, $W^-\to\ell^-\bar\nu$, we study the "Yield of D mesons from B decays[8]" in events with an identified lepton.[9,10] The candidate lepton is required to have momentum p_l between 1.4 and 2.4 GeV/c: The lower momentum cut suppresses leptons that are not primary B decay products, while the upper momentum cut is close to the kinematic limit for B decay into charmed particles.

We observe the D^0 by searching for its decay into $K^-\pi^+$, the D^+ searching for $K^-\pi^+\pi^+$, and the D^{*+} by searching for $D^0\pi^+$ with $D^0\to K^-\pi^+$ or $D^0\to K^-\pi^+\pi^-\pi^+$. (Throughout this paper charge conjugate modes are implied.) The invariant-mass distributions for D^0 and D^+ candidates in events with an identified ℓ^- are shown in Fig. 1. The mass distributions of D^0, D^+ and D^{*+} with ℓ^- (right sign combination) are fitted to a signal Gaussian and polynomial background functions. The numbers of $D\ell^-$ candidates after continuum subtraction are given in Table I; the values shown are corrected for D detection efficiency ($\epsilon_{D\ell^-}$, see below) but not for lepton identification efficiency.

The number of D's per semileptonic decay is calculated as :

$$R_{D\ell^-} \equiv \frac{B(\bar B \to D\,X\ell^-\bar\nu)}{B(\bar B \to X\ell^-\bar\nu)} = \frac{N_{D\ell^-}}{N_\ell \epsilon_{D\ell^-}},$$

where $N_{D\ell^-}$ is the number of candidate events, N_ℓ is the number of leptonic events, and

* Research funded by the National Science Foundation.

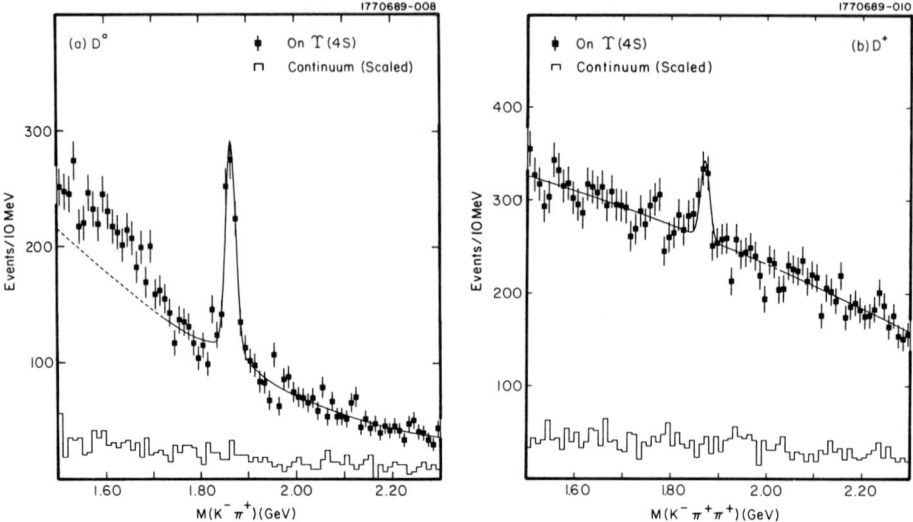

Figure 1 Invariant mass distributions for (a) D^0 and (b) D^+ candidates in events containing ℓ^-. The data points are for data taken on the $\Upsilon(4S)$ while the histograms are for data taken at energies just below resonance and have been scaled to account for the difference in luminosities and crosssections.

Table I Number of $D\ell^-$ candidates and # of D's per B decay

# of $D\ell^-$ candidates		# of D's per B decay	
D	$D\ell^-$	Semileptonic	Generic
D^0	810 ± 65	$0.73 \pm 0.09 \pm 0.09$	$0.52 \pm 0.02 \pm 0.09$
D^+	889 ± 156	$0.32 \pm 0.08 \pm 0.04$	$0.29 \pm 0.03 \pm 0.05$
D^{*+}	240 ± 40	$0.40 \pm 0.10 \pm 0.06$	$0.31 \pm 0.03 \pm 0.06$

$\epsilon_{D\ell^-}$ is the efficiency to reconstruct a D when a lepton is found. This ratio is independent of the lepton identification efficiency, assuming that the momentum spectra of leptons is the same in the two decays ($\overline{B} \to DX\ell^-\overline{\nu}$ and $B \to X\ell^-\overline{\nu}$). $B^0\overline{B}^0$ mixing[10,11] ($\overline{B}^0 \to DX$, $B^0 \to \overline{B}^0 \to Y\ell^-\overline{\nu}$), cascades in which the lepton comes from D decay will cause this technique to overestimate the number of D's per semileptonic B decay. $R_{D\ell^-}$ is given in Table I, where the first error is statistical and the second systematic. $B(D^{*+} \to D^0\pi^+) = (57 \pm 4 \pm 4)\%$, $B(D^0 \to K^-\pi^+) = (4.2 \pm 0.4 \pm 0.4)\%$ and $B(D^0 \to K^-\pi^+\pi^-\pi^+) = (9.1 \pm 1.3 \pm 0.4)\%$[12] are used and the systematic error includes uncertainties of 13% and 10% for the D and D^{*+} branching fractions respectively. The sum of $R_{D^0\ell^-}$ and $R_{D^+\ell^-}$, $1.05 \pm 0.12 \pm 0.10$, is consistent with B decay being 100% $b \to c$. The update of the number of D's per generic B decay[8] (hadronic & semileptonic) is also given in Table I.

DECAY $\overline{B}^0 \to D^{*+}\ell^-\overline{\nu}$

The $\overline{B}^0 \to D^{*+}\ell^-\overline{\nu}$ candidate events are identified by finding D^{*+} and ℓ^- and calculating the missing mass squared (MM^2) given by

$$MM^2 = (E_{beam} - (E_{D^*} + E_\ell))^2 - (\vec{p}_B - (\vec{p}_{D^*} + \vec{p}_\ell))^2,$$

where we set $\vec{p}_B$ equal to zero (this broadens the widths of the MM^2 distributions.) and use the fact that the beam energy E_{beam} is equal to the B energy. The MM^2 distributions for $D^{*+}\ell^-$ and $D^{*+}\ell^+$ are shown in Fig. 2. The prominent peak at $MM^2 \approx 0$ in Fig. 2a) is evidence for $\overline{B}^0 \to D^{*+}\ell^-\overline{\nu}$.[13]

Non-negligible sources of background in the MM^2 distribution of right sign events include non-resonant e^+e^- annihilation, fake leptons, fake D^{*+}, mixed events, and the decays $B \to D^*(2420)\ell^-\overline{\nu}$ where $D^*(2420) \to D^{*+}\pi$ as well as other resonant or non-resonant $B \to D^{*+}\pi\ell^-\overline{\nu}$'s (Hereafter D^{**} represents both excited D^* and non-resonant $D^*\pi$.[13]) and cascades. In Fig. 3a) we compare the MM^2 distributions expected for the background contributions with the one for $\overline{B}^0 \to D^{*+}\ell^-\overline{\nu}$. The signal (solid line) is well described by a Gaussian function. The D^{**} background (dashed line) gives rise to a structure peaking at small positive value of MM^2 that is well described by a skewed Gaussian. The wide background structure due to mixing is the dot-dashed line. The background due to cascade leptons is similar, and we use the mixing shape to parametrize both backgrounds.

The result of the fit is shown as a histogram in Fig. 2a). The number of $\overline{B}^0 \to D^{*+}\ell^-\overline{\nu}$ events is 108 ± 12. The D^{**} component is 18 ± 10 events. We determine the branching fraction $B(\overline{B}^0 \to D^{*+}\ell^-\overline{\nu})$ using the formula :

$$B(\overline{B}^0 \to D^{*+}\ell^-\overline{\nu}) = \frac{N(D^{*+}\ell^-\overline{\nu})}{\epsilon_{D^{*+}\ell^-}\epsilon_\ell\epsilon_b\epsilon_s f_{00} N_B}, \tag{1}$$

where $N(D^{*+}\ell^-\overline{\nu})$ is the number of candidate events, N_B is the number of B's in the data sample, f_{00} is the fraction of neutral B's produced at the $\Upsilon(4s)$, ϵ_ℓ is the efficiency to detect a lepton with $p_l > 1.4$ GeV/c (0.62 ± 0.02 for electrons and 0.41 ± 0.01 for muons), ϵ_s is the fraction of the leptons with $p_l > 1.4$ GeV/c (0.63 in the ISGW model), and ϵ_b is a correction for the charm decay branching fractions. We determine $N(D^{*+}\ell^-\overline{\nu})$ by subtracting from the number of events found by the fitting procedure the contribution due to continuum and lepton fakes. The efficiencies are estimated using the ISGW model[1] and the CLEO Monte Carlo simulation (MC). The efficiency $\epsilon_{D^{*+}\ell^-}$ depends upon the lepton species and the D^0 decay mode considered. Assuming that f_{00} is 50 % (here and throughout this paper) we obtain $B(\overline{B}^0 \to D^{*+}\ell^-\overline{\nu}) = (4.6 \pm 0.5 \pm 0.7)\%$. The observed branching ratio is in agreement with the ARGUS result[5,14] The D^{**} component can be a weak evidence for observing $\overline{B} \to D^{**}\ell^-\overline{\nu}$, or $D^*\pi\ell^-\overline{\nu}$ decays. If we assume $B(\overline{B}^0 \to D^{*+}\ell^-\overline{\nu})$ is equal to $B(B^- \to D^{**0}\ell^-\overline{\nu})$, the branching fraction, $B(\overline{B}^0 \to D^{**+}\ell^-\overline{\nu})$ is $(1.8 \pm 1.1 \pm 0.3)\%$.

The polarization of the D^{*+} in the decay $\overline{B}^0 \to D^{*+}\ell^-\overline{\nu}$ can be used to distinguish among models of semileptonic B decay. The D^{*+} decay angular distribution can be parametrized as :

$$\frac{1}{N}\frac{dN}{d\cos\Theta^*} = 1 + \alpha \cos^2 \Theta^*, \tag{2}$$

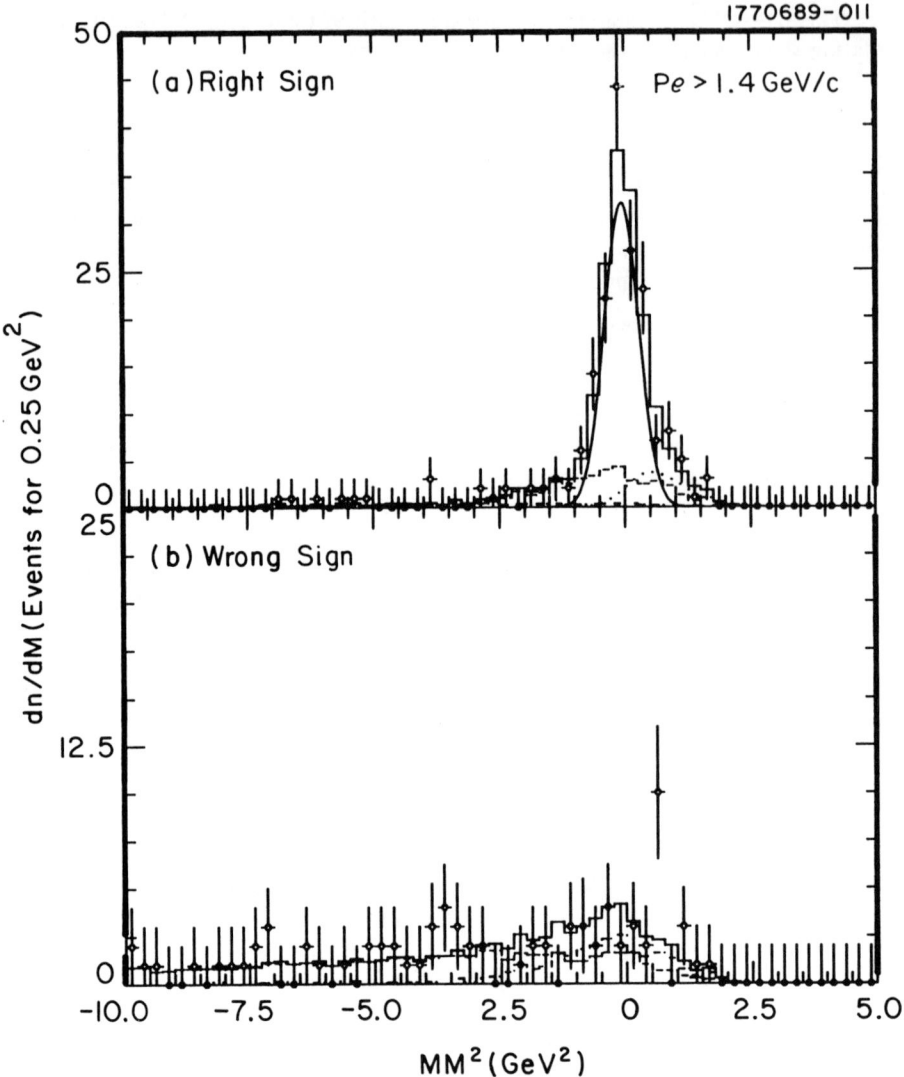

Figure 2 Missing mass squared distribution for :

a) $\overline{B}^0 \to D^{*+} \ell^- \overline{\nu}$ (right sign); the circles are the data points, the solid histogram is the result of the fit with a Gaussian for the signal and backgrounds. The background due to mixing and cascades is the dot-dashed histogram, the background due to fake D^*'s is the dashed histogram and the background due to D^{**}'s is the dotted line.

b) $D^{*+}\ell^+$ (wrong sign); the circles are the data points, the solid histogram is the result of the fit that contains the contribution from fake D^{*+}'s (dashed histogram) and mixed events and cascades (dot-dashed line).

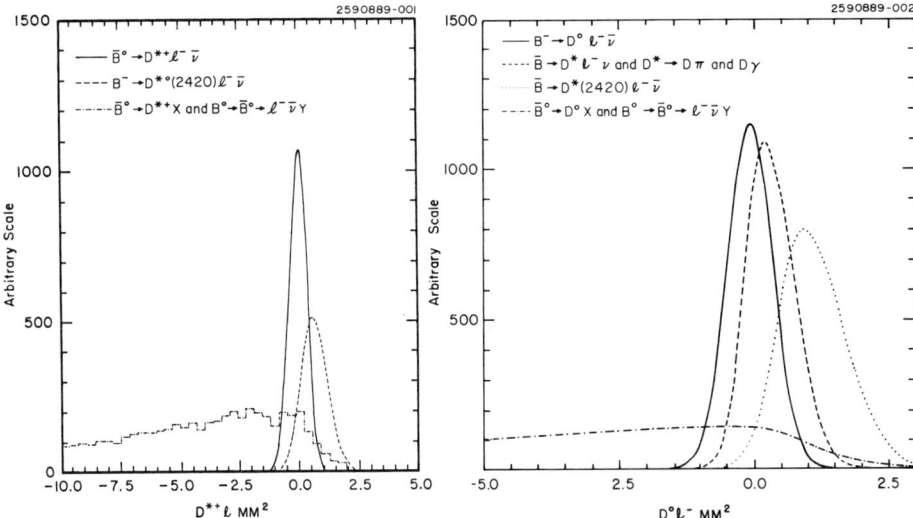

Figure 3 MM^2 generated by the CLEO Monte Carlo simulation for :

a) $D^{*+}\ell^-$: $\overline{B}^0 \to D^{*+}\ell^-\overline{\nu}$ (solid line); $B^- \to D^{**}\ell^-\overline{\nu}$ (dotted line); mixed events and cascade leptons (dot-dashed line).

b) $D^0\ell^-$: $B^- \to D^0\ell^-\overline{\nu}$(solid line); $\overline{B} \to D^*\ell^-\overline{\nu}$(dashed line); $\overline{B} \to D^{**}\ell^-\overline{\nu}$ (dotted line); mixed events and cascade leptons (dot-dashed line).

where Θ^* is the angle of the D^0 in the D^{*+} rest frame with respect to the D^{*+} direction in the laboratory frame. α relates the amount of longitudinally polarized states to the amount of transverse, $\alpha = \frac{2\Gamma_L}{\Gamma_T} - 1$. Events with $p_\ell > 1.4$ GeV/c with MM^2 consistent with zero ($MM^2 < 1$ GeV^2) are divided into $\cos\Theta^*$ bins. The number of $\overline{B}^0 \to D^{*+}\ell^-\overline{\nu}$ candidates is extracted in each bin through a fit of the D^0 invariant mass spectrum. The decay angular distribution thus obtained is fitted to the function given in Eq. (2). The result of the fit is $\alpha = 0.65 \pm 0.66 \pm 0.25$ for $P_\ell > 1.4$GeV/c. The small value of α confirms a previous measurement by the ARGUS collaboration.[5]

DECAYS $B^- \to D^0\ell^-\overline{\nu}$ AND $B^- \to D^{*0}\ell^-\overline{\nu}$

In order to extract B$(B^- \to D^0\ell^-\overline{\nu})$ and B$(B^- \to D^{*0}\ell^-\overline{\nu})$ we use the same missing mass technique but reconstruct only the D^0 in leptonic events. MC distributions for various processes that could result in the observation of a D^0 and an ℓ are shown in Fig. 3b); shown are (1) $B^- \to D^0\ell^-\overline{\nu}$, (2) D^0 coming from D^* decays; $B^- \to D^{*0}\ell^-\overline{\nu}$ and $\overline{B}^0 \to D^{*+}\ell^-\overline{\nu}$, (3) $\overline{B} \to D^{**}\ell^-\overline{\nu}$, and (4) mixing and cascade background. The MM^2 distribution for $D^0\ell^-$ candidate events is shown in Fig. 4a). There is a clear peak near $MM^2 = 0.5 GeV^2$ along with a large amount of background. The contribution from fake D^0s can be subtracted using the sideband of the D^0 mass peak (See Fig. 1.); The D^0 signal (sideband) region is between 1.85 and 1.88 GeV/c (1.7 and 1.8 GeV/c and 1.95 and 2.05 GeV/c). The histogram in Fig. 4a) shows the MM^2 distribution of the sideband events. Both data points and histogram are continuum subtracted. The MM^2 distribution after sideband subtraction is shown in Fig. 4b).

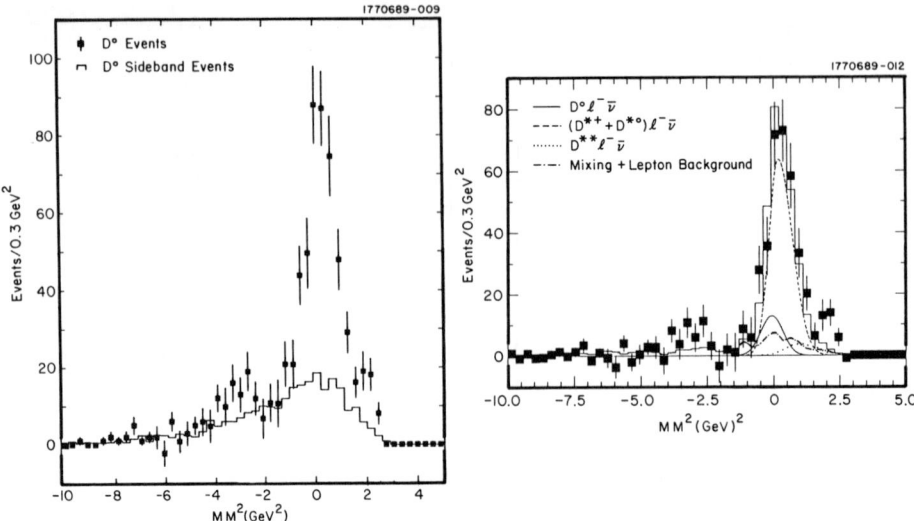

Figure 4 The MM^2 spectrum from :

a) $K^-\pi^+\ell^-$. The data have been continuum subtracted. The data points are those for which the $K^-\pi^+$ are close to the D^0 mass while the histogram is for D^0 sideband.

b) $D^0\ell^-$ (data points). The histogram is the fit to the $B^-\to D^0\ell^-\bar{\nu}$ and $\overline{B}\to D^*\ell^-\bar{\nu}$ components as indicated on the figure plus mixing, cascades, and $\overline{B}\to D^{**}\ell^-\bar{\nu}$ contributions which are estimated from the $\overline{B}^0\to D^{*+}\ell^-\bar{\nu}$ analysis and the MC.

To extract the number of $B^-\to D^0\ell^-\bar{\nu}$ and $B^-\to D^{*0}\ell^-\bar{\nu}$ events we use the results of the previous measurement. The ratio of the D^{**} and D^* contribution to the $D^0\ell^-$ MM^2 distribution is fixed to be 17%. The amount of $\overline{B}^0\to D^{*+}\ell^-\bar{\nu}$ contribution is also fixed by the previous measurement of $B(\overline{B}^0\to D^{*+}\ell^-\bar{\nu})$. We use our measurement of the "wrong sign" ($D^0\ell^+$) MM^2 distribution and our measurement of the $B^0\overline{B}^0$ mixing[10] to determine the expected amount of mixed events. Cascade lepton contribution is calculated using the result of the fit of the measured lepton momentum spectrum from B decay[6]. The MM^2 distribution for $D^0\ell^-$ is then fitted using the shapes shown in Fig. 3b) with only the area of the $B^-\to D^0\ell^-\bar{\nu}$ and $B^-\to D^{*0}\ell^-\bar{\nu}$ distributions allowed to float (shown as a histogram in Fig. 4b)). The number of $B^-\to D^0\ell^-\bar{\nu}$ ($B^-\to D^{*0}\ell^-\bar{\nu}$) events is 58 ± 20 (133 ± 26). Using equations similar to Eq. (1) we find $B(B^-\to D^{*0}\ell^-\bar{\nu}) = (3.9\pm0.8\pm0.9)\%$ and $B(B^-\to D^0\ell^-\bar{\nu}) = (2.4\pm0.8^{+0.7}_{-0.8})\%$. The statistical error here includes the correlated error between the D^0 and D^{*0} components of the fit. The systematic error is dominated by the uncertainty in the magnitude of the contribution from D^{**} and $(D\pi)_{nr}$ states.

DISCUSSIONS AND CONCLUSIONS

The ratio of vector to pseudoscalar decay widths is obtained from $B(B^-\to D^{*0}\ell^-\bar{\nu})$ and $B(B^-\to D^0\ell^-\bar{\nu})$:

$$\frac{\Gamma_{SL}(D^*)}{\Gamma_{SL}(D)} = 1.6^{+1.2}_{-0.6}{}^{+0.7}_{-0.5}.$$

This result, together with the measurement of the D^{*+} polarization, restricts the free parameter J_B in a recent model by Wirbel and Bauer to be $0.62 < J_B/J_B^o < 1.08$ at 90% C.L. These results are in agreement with KS and ISGW models. The value of $|V_{cb}|$ from the exclusive branching fractions is calculated[15] for various models and shown in Table II. The average B lifetime of 1.17 ± 0.14 ps is used.[16] Finally the lifetime ratio of the charged and neutral B's is determined to be

$$\frac{\tau^+}{\tau^0} = 0.85 \pm 0.20^{+0.22}_{-0.16}$$

from the ratio of $B(B^- \to D^{*0}\ell^-\bar{\nu})$ to $B(\overline{B}^0 \to D^{*+}\ell^-\bar{\nu})$. The major component of the systematic error is the uncertainty in the D^{**} production. Lowering the B^0 fraction (f_{00}) from 50% would decrease this ratio. Since theoretically τ^0 is expected to be less than or equal to τ^+, this would push the result in the wrong direction.

Table II $|V_{cb}|$ from exclusive decays

Mode	ISGW	WBS	KS
$B^- \to D^0\ell^-\bar{\nu}$	0.043 ± 0.007	0.051 ± 0.009	0.051 ± 0.009
$\overline{B}^0 \to D^{*+}\ell^-\bar{\nu}$	0.039 ± 0.004	0.043 ± 0.005	0.039 ± 0.004
$B^- \to D^{*0}\ell^-\bar{\nu}$	0.036 ± 0.005	0.039 ± 0.005	0.036 ± 0.005

In conclusion ratios of the inclusive semileptonic branching fractions, $B(B^- \to D^0 X\ell^-\bar{\nu})$, $B(\overline{B}^0 \to D^+ X\ell^-\bar{\nu})$, and $B(\overline{B}^0 \to D^{*+} X\ell^-\bar{\nu})$ to the average B meson semileptonic branching fraction, the absolute branching fractions of the exclusive final states, $B^- \to D^0\ell^-\bar{\nu}$, $B^- \to D^{*0}\ell^-\bar{\nu}$, and $\overline{B}^0 \to D^{*+}\ell^-\bar{\nu}$ and the D^{*+} polarization in the decay $\overline{B}^0 \to D^{*+}\ell^-\bar{\nu}$ have been measured and various quantities derived from the measurements compared with theoretical models.

REFERENCES AND FOOTNOTES

1) N. Isgur et al., *Phys. Rev.*, **D39**, 799 (1989).

2) J.G. Körner and G.A.Schuler, *Z. Phys.* **C 38**, 511 (1988).

3) M.Bauer and M.Wirbel, *Form factor effects in exclusive D and B decays*, Dortmund Preprint **HD-THEP 88-22**, (1988).

4) T. Altomari and L.Wolfenstein, *Phys. Rev. Lett* **58**, 1583 (1987).

5) A measurement of $D^{*+}\ell^-\bar{\nu}$ branching ratio and polarization is reported by ARGUS in H. Albrecht et al., *Phys. Lett.* **197B**, 452 (1987) and *Phys. Lett.* **219B**, 121 (1989).

6) R.V. Kowalewski (CLEO collaboration), *Ph.D. Thesis, Cornell University* (1988) measures $B(B \to X\ell^-\bar{\nu}) = (10.0 \pm 0.2 \pm 1.2)\%$.

7) H. Schröder, plenary talk at the *XXIV International Conference on High Energy Physics*, Munich (1988) gives the most recent ARGUS collaboration measurement $B(B \to X\ell^-\bar{\nu}) = (10.0 \pm 0.8)\%$.

8) D. Bortoletto et al., (CLEO collaboration), "Inclusive B-Meson Decays into Charm," Phys. Rev. D **35**, 19 (1987).

9) K. Chadwick et al., (CLEO collaboration), *Phys. Rev.* D **27**, 475 (1983).

10) M. Artuso et al., (CLEO collaboration), *Phys. Rev. Lett.* **62**, 2233 (1989).

11) H. Albrecht et al., *Phys. Lett.* **192B** (1987) 245.

12) J. Adler et al., *Phys. Rev. Lett.* **60**, 89 (1986); ibid., Phys. Lett. **208B** 153 (1988).

13) D. Bortoletto et al., (CLEO collaboration), CLNS 89–922.

14) The ARGUS measurement[5] is $B(\bar{B}^0 \to D^{*+}\ell^-\bar{\nu}) = (7.0 \pm 1.2 \pm 1.9)\%$ which becomes $(5.4 \pm 0.8 \pm 1.3)\%$ taking into account that we use $B(D^{*+} \to D^0\pi^+) = (57 \pm 4 \pm 4)\%$ [12] and $f_{00} = 0.50$ while ARGUS uses $B(D^{*+} \to D^0\pi^+) = (49 \pm 7 \pm 7)\%$ and $f_{00} = 0.45$.

15) S. Stone, Heavy quark decay session summary at XII International Workshop on Weak Interaction and Neutrinos, Ginosar, Sea of Galilee, Israel, April 1989.

16) S. L. Wu, "e^+e^- Interaction at High Energies," in Proceedings of 1987 Int. Symp. on Lepton and Photon Interactions at High Energies, ed W. Bartel and R. Ruckl, North Holland, Amsterdam (1987).

ARGUS Results on Rare B Decays

Andrei Golutvin
ITEP, Moscow, USSR

1 Introduction

Between 1983 and 1988, a sample of 152 000 $\Upsilon(4S)$ decays was obtained with the ARGUS detector at the e^+e^- storage ring DORIS II. Using this data we have searched for charmless B decays both in inclusive and exclusive decay modes.

In the Standard Model of electroweak interactions, the b quark decays via the charged current to either a c or a u quark. The decays of B mesons are dominated by $b \to c$ transitions. This can be demonstrated by the inclusive yields of D^0, D^+, D_s, Λ_c and J/ψ (Table 1).

	Branching Ratio [%]
$B \to D^0 X$	$46.6 \pm 7.1 \pm 6.3$
$B \to D^+ X$	$23.2 \pm 5.3 \pm 3.5$
$B \to D_s X$	$16 \pm 4 \pm 3$
$B \to \Lambda_c' X$	$7.6 \pm 1.4 \pm 1.8$
$2 \cdot B \to J/\psi X$	4.2 ± 1.0
Σ	$98 \pm 10 \pm 8$

Table 1. Inclusive B decays into charmed particles from ARGUS

The sum over all inclusive channels is lower than the expected value of 1.20 c quark per B decay [1], where the additional 20% come from charm production in the fragmentation of $W \to c\bar{s}$. This leaves enough room for non $b \to c$ processes, which can proceed through $b \to u$ or $b \to s$ transitions. It is however not possible to obtain any sensitive limits on the corresponding branching ratios from charm counting.

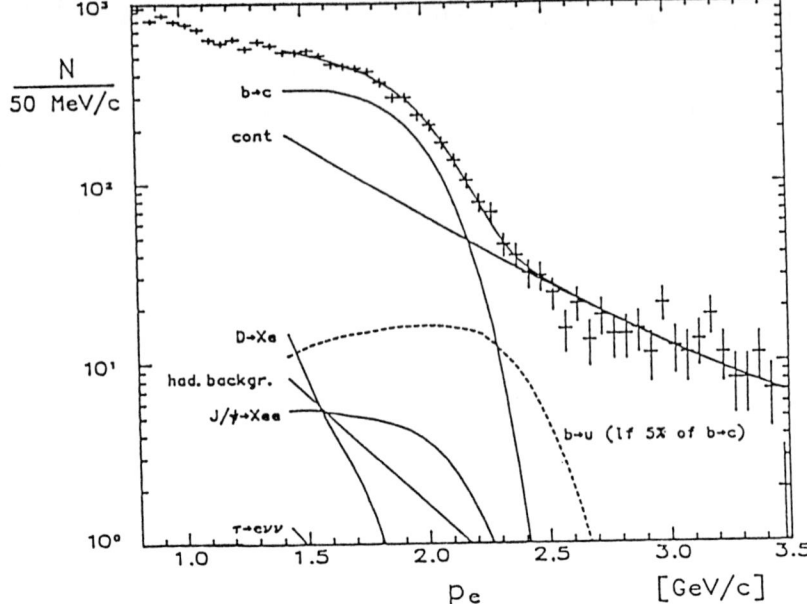

Fig. 1: Inclusive electron spectrum for $\Upsilon(4S)$ decays.

2   Searches for $b \to u$ transition

At the moment the most sensitive limit on the strength of $b \to u$ transition comes from the analysis of the inclusive lepton spectrum in B decays near the end-point. The signature for a decay via a $b \to u$ transition would be seen as an excess of leptons with momenta above the kinematic limit for $b \to c$ decays. The momentum spectrum for electrons from $\Upsilon(4S)$ decays is shown in Figure 1. This spectrum has been obtained for the old data sample which corresponds to an integrated luminosity of 84 pb^{-1}. Clearly there is a substantial background from continuum lepton sources, which must be subtracted in order to obtain the direct B decay contribution. This measurement is limited by the subtraction of a large amount of continuum. To increase the sensitivity, the continuum was fitted with a smooth function and then subtracted. However this method has associated systematic errors which are difficult to reliably estimate.

From the absence of a significant signal beyond the kinematical limit for $b \to c\ell\nu$ transitions ($p = 2.3 GeV/c$), a limit is obtained on the ratio of the Kobayashi-Maskawa matrix elements: $|V_{ub}/V_{cb}| < 0.16$ at 90% CL [2].

The systematic problems inherent in the analysis of inclusive lepton spectra can be partially overcome by looking for exclusive semileptonic $b \to u$ decays. In

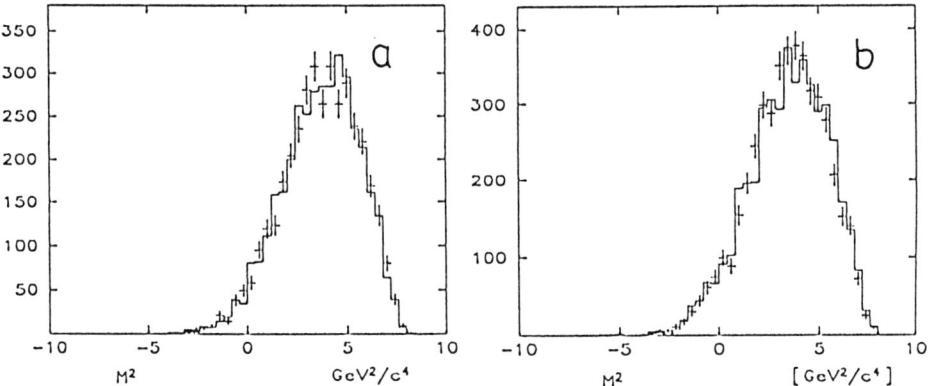

Fig. 2: Recoil mass squared against the $\rho^0 l^-$ system for electrons (a) and muons (b). The background shape is shown with solid histogram.

this context, the channel $B^- \to \rho^0 l^- \bar{\nu}$ seems particularly suitable, especially in the light of the success with $\bar{B}^0 \to D^{*+} l^- \bar{\nu}$. For this reason, ARGUS undertook a study of the decay $B^- \to \rho^0 l^- \bar{\nu}$, using the recoil mass technique. The resulting spectra for the mass recoiling against the $\rho^0 l^-$ system are shown in Figure 2 for electrons (2^a) and muons (2^b). No signal is observed at zero recoil mass. The backgrounds are substantially larger for this analysis than for $\bar{B}^0 \to D^{*+} l^- \bar{\nu}$ because of the larger combinatorial background under the broad ρ^0, as well as the substantial number of ρ^0 mesons from higher mass states produced in $b \to c$ or $b \to u$ transitions. From the fit one obtains an upper limit on the number of detected $B^- \to \rho^0 l^- \bar{\nu}$ decays. The same analysis has been performed for the decay $\bar{B}^0 \to \pi^+ l^- \bar{\nu}$. To calculate an upper limit on the ratio $|V_{ub}/V_{cb}|$ we have used different models. The conservative limit for the combined result of the $B^- \to \rho^0 l^- \bar{\nu}$ and $\bar{B}^0 \to \pi^+ l^- \bar{\nu}$ decays is $|V_{ub}/V_{cb}| < 0.29$ at 90% CL [3]. This result is not quite at the level obtained from the full inclusive lepton spectrum, but brings with it less systematic uncertainty.

It is also possible to search for charmless B decays in the exclusive hadronic decay modes. In this analysis ARGUS tried to reconstruct B mesons in the channels $B \to n\pi (n = 1, ..., 6)$ and $B \to p\bar{p}\pi(\pi)$. The mass of the B mesons is determined from its momentum by an energy constraint fit which uses the fact that the energy of a B meson has to coincide with the beam energy, e.g.:

$$M_{n\pi}^2 = E_{Beam}^2 - p_{n\pi}^2$$

No signal is found in the channels $B \to n\pi$. The corresponding 90 % C.L.

upper limits are shown in Table 2 together with the theoretical predictions [4,5] and upper limits obtained by CLEO. The theoretical predictions have been calculated assuming $|V_{ub}/V_{cb}| = 0.1$. The model [5] predicts some enhancement for multipion final states with π^0. Nevertheless for each particular decay mode the experimental upper limits are substantially higher than the theoretical predictions and less sensitive than the one derived from the endpoint of the lepton spectrum in B decays. In future it may be possible to increase the sensitivity of this method by summing over all charmless hadronic decay modes.

Decay	ARGUS	CLEO	Theory ($|v_{bu}/v_{bc}| = 0.1$)
$\pi^\pm \pi^0$	$5.0 \cdot 10^{-4}$	$2.6 \cdot 10^{-3}$	$0.6(1.3) \cdot 10^{-5}$
$\pi^+ \pi^-$	$1.9 \cdot 10^{-4}$	$0.8 \cdot 10^{-4}$	$2.0(2.5) \cdot 10^{-5}$
$\pi^\pm \pi^+ \pi^-$	$8.0 \cdot 10^{-4}$	$1.9 \cdot 10^{-4}$	$6 \cdot 10^{-5}$
$\rho^0 \pi^\pm$	$1.9 \cdot 10^{-4}$	$1.7 \cdot 10^{-4}$	$2 \cdot 10^{-6}$
$\pi^+ \pi^- \pi^0$	$1.8 \cdot 10^{-3}$	–	$2 \cdot 10^{-4}$
$\rho^0 \pi^0$	$4.3 \cdot 10^{-4}$	–	$2 \cdot 10^{-6}$
$\pi^+ \pi^+ \pi^- \pi^-$	$1.0 \cdot 10^{-3}$	–	$1 \cdot 10^{-4}$
$\pi^\pm \pi^+ \pi^- \pi^0$	$5.4 \cdot 10^{-3}$	–	$4 \cdot 10^{-4}$
$\pi^+ \pi^- \pi^0 \pi^0$	$5.7 \cdot 10^{-3}$	–	$5 \cdot 10^{-4}$
$\rho^+ \rho^-$	$4.2 \cdot 10^{-3}$	–	$5 \cdot 10^{-5}$
$\pi^\pm 2\pi^+ 2\pi^-$	$1.2 \cdot 10^{-3}$	–	$2 \cdot 10^{-4}$
$3\pi^+ 3\pi^-$	$3.3 \cdot 10^{-3}$	–	$2 \cdot 10^{-4}$

Table 2: Limits on charmless B decays from ARGUS

In the decays $B \to p\bar{p}\pi^+(\pi^-)$ ARGUS observed in 1987 a signal of 25 ± 8 events[6]. A preliminary analysis of the new data, corresponding to 65 % of the previous sample, does not show the expected signal. New data, which ARGUS is taking in 1989, will help to clarify the situation.

3 Searches for $b \to s$ transition

In recent years there has been considerable interest in penguin decays of B mesons, either with radiated photon or gluon ($b \to s\gamma$ or $b \to sg$ respectively). Such transitions probe the electroweak interaction at the one-loop level and provide a possible window on physics well beyond the directly accessible mass scale. The searches for $b \to s$ transitions were performed through the reconstruction of B meson in the decays $B \to $ "K"π^+, "K"ρ^0, "K"ϕ and $K^*\gamma$. "K" represents K_s^0, K^+, K^{*0} and K^{*+}. No signals are seen, leading to the limits shown in Table 3 [7]. Some of them are close to the theoretical predictions [8].

Decay mode	N_{events} (90% CL)	BR (90% CL)
$B^0 \to K^+\pi^-$	< 9.4	< $1.8 \cdot 10^{-4}$
$B^+ \to K_S^0\pi^+$	< 3.2	< $1.0 \cdot 10^{-4}$
$B^0 \to K^{*+}(892)\pi^-$	< 2.3	< $6.2 \cdot 10^{-4}$
$B^+ \to K^{*0}(892)\pi^+$	< 3.6	< $1.7 \cdot 10^{-4}$
$B^0 \to K_S^0\rho^0$	< 2.3	< $1.6 \cdot 10^{-4}$
$B^+ \to K^+\rho^0$	< 6.4	< $1.8 \cdot 10^{-4}$
$B^0 \to K^{*0}(892)\rho^0$	< 7.7	< $4.6 \cdot 10^{-4}$
$B^+ \to K^{*+}(892)\rho^0$	< 3.4	< $9.0 \cdot 10^{-4}$
$B^0 \to K_S^0\phi$	< 2.3	< $3.6 \cdot 10^{-4}$
$B^+ \to K^+\phi$	< 2.9	< $1.8 \cdot 10^{-4}$
$B^0 \to K^{*0}(892)\phi$	< 2.3	< $3.2 \cdot 10^{-4}$
$B^+ \to K^{*+}(892)\phi$	< 2.3	< $1.3 \cdot 10^{-3}$
$B^0 \to K^{*0}(892)\gamma$	< 7.9	< $4.2 \cdot 10^{-4}$
$B^+ \to K^{*+}(892)\gamma$	< 2.3	< $5.2 \cdot 10^{-4}$

Table 3: Limits on "penguin-type" B decays from ARGUS

In addition, these upper limits may serve to constrain some models involving more exotic phenomena. There can be no doubt that the experimental and theoretical study of these decays will continue to test the Standard Model and provide a window into physics beyond.

References

[1] R.Rückl, Habilitationsschrift,Universität München, Munich (1984).

[2] H. Schröder, Proc. of the 24th International Conference on High Energy Physics, Munich (1988).

[3] T.Ruf (ARGUS collaboration), Ph.D.thesis, Universität Karlsruhe, Germany, in Preparation.

[4] M.Bauer et al. Z. Phys **C34** (1987) 103.

[5] A.V.Dobrovolskaya et al. Preprint ITEP 108-89, Moscow (1989).

[6] H.Albrecht et al (ARGUS collaboration), Phys.Lett. **209B** 119 (1988).

[7] J.Swain (ARGUS collaboration), Ph.D. thesis, University of Toronto, Toronto, Canada, in Preparation.

[8] M.B.Gavela et al, Phys.Lett. **154B** 425 (1985), L.-L.Chau and H.Y.Cheng, Phys.Rev.Lett. **59** 958 (1987), N.G.Deshpande et al, Phys.Rev.Lett. **59** 183 (1987).

Exclusive Semileptonic B Decays to Charm

Reiner Gläser

DESY, Hamburg, Germany

Abstract

Using the ARGUS detector at the e^+e^- storage ring DORIS II at DESY we have investigated the exclusive decays $B^0 \to D^{*-}l^+\nu$ and $B^0 \to D^-l^+\nu$ and have determined the CKM matrix element $|V_{cb}|$.
By measuring the yields of $\overline{D}^0$, D^0, D^- and D^{*-} mesons originating from semileptonic $\overline{B}$ decays, the lifetime ratio of charged and neutral B mesons, $\tau(B^+)/\tau(B^0)$, is derived.

Introduction

The investigation of exclusive semileptonic B decays can provide important information on the CKM matrix elements $|V_{ub}|$ and $|V_{cb}|$, on the interplay between strong and weak interactions in B decays and on the currently unmeasured lifetime differences between weakly decaying b-flavoured hadrons.

For B_d^0 and B_u^+ mesons [1] one expects [1–6] the semileptonic width to be dominated by the decays into the lowest-lying charmed vector (D^*) and pseudoskalar (D) mesons. Since B mesons from $\Upsilon(4S)$ decays are produced almost at rest, exclusive semileptonic decays can be efficiently reconstructed, therefore providing a dedicated laboratory for our understanding of weak B decays. In particular, the CKM matrix element $|V_{cb}|$ can be derived with few model uncertainties and the lifetime ratio of charged and neutral B mesons can be determined.

1 The Decay $\overline{B}^0 \to D^{*+}l^-\overline{\nu}$

The decay $\overline{B}^0 \to D^{*+}l^-\overline{\nu}$ was first measured by ARGUS in 1987 [7]. The reconstruction of exclusive semileptonic B decays is possible, since the unobserved neutrino can be inferred from an approximation ($\vec{p}_B = 0$) of its effective mass:

$$M_{rec}^2 = [E_{beam} - (E_{D^*} + E_l)]^2 - [\vec{p}_{D^*} + \vec{p}_l]^2 \approx M_\nu^2 \qquad (1)$$

Fig. 1a) shows the M_{rec}^2 distribution of $D^{*+}l^-$ pairs ($l^- = e^-$, μ^-) with $x_p(D^{*+}) < 0.5$ and $p_{l^-} > 1\,GeV/c$. The prominant peak at $M_{rec}^2 = 0$ contains 73 ± 11 decays $\overline{B}^0 \to D^{*+}l^-\overline{\nu}$ on a background of 23 ± 5 events.

Scaling our branching ratio for $\overline{B}^0 \to D^{*+}l^-\overline{\nu}$ [7] to the MARK III value of $BR(D^{*+} \to \pi^+D^0) = (57 \pm 4 \pm 4)\%$ [9] and using $BR(\Upsilon(4S) \to B^0\overline{B}^0) = 45\%$, we obtain

---

[1] References to a specific charged state imply the charged conjugate state also.

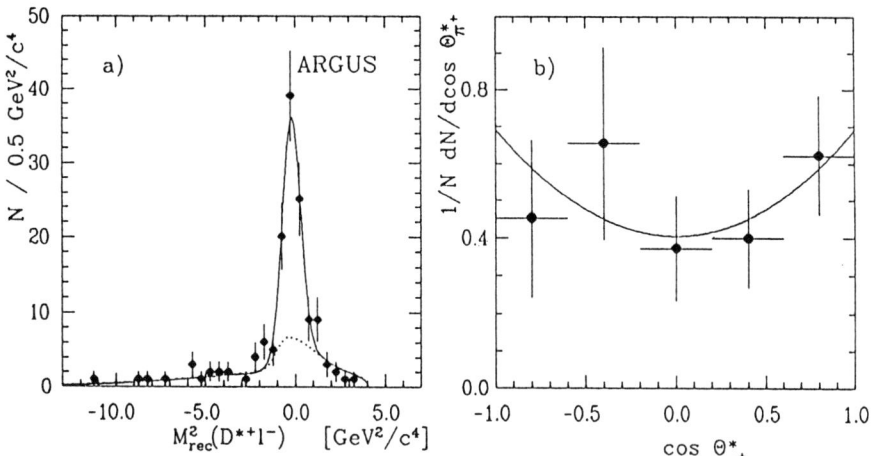

Fig. 1: a) M^2_{rec} distribution of $D^{*+}l^-$ pairs. b) $\cos\theta^*_{\pi+}$ distribution of the decay $\overline{B}^0 \to D^{*+}l^-\overline{\nu}$.

$$BR(\overline{B}^0 \to D^{*+}l^-\overline{\nu}) = (6.0 \pm 1.0 \pm 1.4)\% . \tag{2}$$

Thus, semileptonic $\overline{B}$ decays to D^* mesons represent about 60% of the total semileptonic rate, $BR(B \to l^+\nu X) = (10.9\pm0.6)\%$ [8]. In view of the theoretical uncertainties in the determination of $|V_{ub}|/|V_{cb}|$ from the shape of the lepton spectrum, it is highly desirable to obtain detailed information on the decay mechanism of the reaction $\overline{B}^0 \to D^{*+}l^-\overline{\nu}$.

The helicity structure of the decay is of particular interest. The width is given by three helicity components

$$\Gamma(\overline{B}^0 \to D^{*+}l^-\overline{\nu}) = \Gamma_{T_-} + \Gamma_{T_+} + \Gamma_L = \Gamma_T + \Gamma_L , \tag{3}$$

reflecting the degree of transverse and longitudinal polarization of the D^{*+} meson. The average polarization of the D^{*+} meson can be measured by analyzing the strong decay $D^{*+} \to \pi^+ D^0$. The distribution in $\cos\theta^*_{\pi+}$, where $\theta^*_{\pi+}$ is the polar angle of the slow pion in the D^{*+} rest frame with respect to the D^{*+} boost direction, is given by

$$\frac{1}{N}\frac{dN}{d\cos\theta^*_{\pi+}} = \frac{1 + \alpha\cos^2\theta^*_{\pi+}}{2(1+\alpha/3)} , \tag{4}$$

where α measures the ratio of the longitudinal and transverse helicity components of the width:

$$\alpha = \frac{2\Gamma_L}{\Gamma_T} - 1 \tag{5}$$

Fig. 1b) shows the measured $\cos\theta^*_{\pi+}$ distribution, from which we derive [10,11]

$$\alpha = 0.7 \pm 0.9 \implies \Gamma_L/\Gamma_T = 0.85 \pm 0.45 . \tag{6}$$

This result is in good agreement with the theoretical predictions of [1,5,6], whereas the models [2,3], giving values of $\Gamma_L/\Gamma_T > 2$, can be ruled out with more than three standard deviations.

The ratio of Γ_L/Γ_T is also reflected in the shapes of the lepton and q^2 spectra, which are dominated at high values by the transverse component Γ_{T_-} [5]. The measured distributions are displayed in Fig. 2 and are well described by those models which predict values of $\Gamma_L/\Gamma_T \approx 1$ [1,5,6].

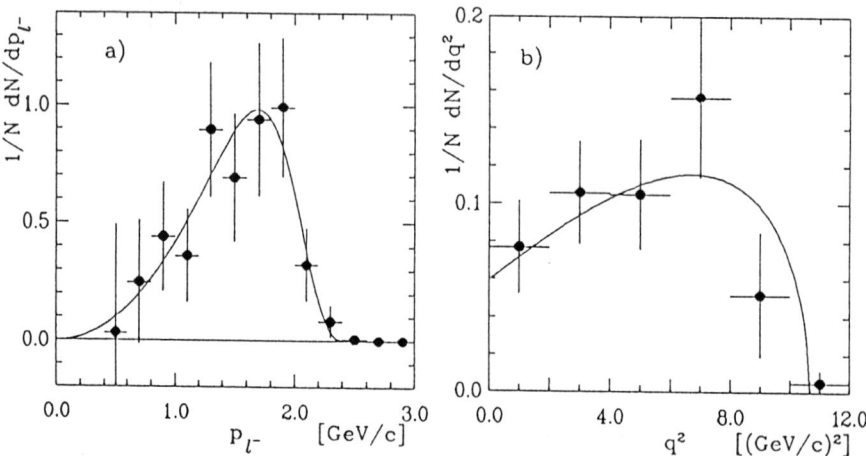

Fig. 2: a) Lepton spectrum of the decay $\overline{B}^0 \to D^{*+} l^- \overline{\nu}$. b) q^2 distribution of the reaction $\overline{B}^0 \to D^{*+} e^- \overline{\nu}$. The solid lines are the predictions from [5].

The results of these investigations together with the B hadron lifetime allow the determination of $|V_{cb}|$ by exploiting the relation

$$\Gamma_{sl}(\overline{B}^0 \to D^{*+}) = BR(\overline{B}^0 \to D^{*+} l^- \overline{\nu})/\tau_b = |V_{cb}|^2 \cdot \hat{\Gamma}_T \cdot (1 + \Gamma_L/\Gamma_T) \ . \quad (7)$$

$\hat{\Gamma}_T$ can be reliably computed in the formfactor approach [1,5,6]. Using $\hat{\Gamma}_T = 12 \times 10^{12} \, s^{-1}$, $\tau_b = (1.15 \pm 0.14) \, ps$ [12] and our values for $BR(\overline{B}^0 \to D^{*+} l^- \overline{\nu})$ and Γ_L/Γ_T, we obtain

$$|V_{cb}| = 0.048 \pm 0.010 \ . \quad (8)$$

The model dependence of this result is small compared to the statistical error [11].

2 The Decay $B^0 \to D^- l^+ \nu$

The decay $B^0 \to D^- l^+ \nu$ ($l^+ = e^+, \mu^+$) is, in principle, best suited for the determination of $|V_{cb}|$, since the matrix element contains only one substantial formfactor. Therefore, the width can be reliably predicted by theoretical models [1,5,6].

ARGUS performed the first measurement of $B^0 \to D^- l^+ \nu$ using a data sample of $172\, pb^{-1}$ on the $\Upsilon(4S)$ resonance, corresponding to about 150000 B meson pairs. Electrons and muons were selected in the momentum intervall $1.0 < p_{l+} < 2.5\, GeV/c$. D^- mesons were reconstructed in their decay $D^- \to K^+ \pi^- \pi^-$, where the $K^+ \pi^- \pi^-$ combinations were required to have $1.5 < p(K^+ \pi^- \pi^-) < 2.5\, GeV/c$. Here, the lower momentum cut suppresses heavily the combinatorial background in the D^- mass region and the contribution from D^- mesons, originating from the decay sequence $B^0 \to D^{*-} l^+ \nu$, $D^{*-} \to D^- (\pi^0, \gamma)$.

In order to derive the observed number of decays $B^0 \to D^- l^+ \nu$, the $K^+ \pi^- \pi^-$ mass distribution was fitted in bins of $M^2_{rec}(K^+ \pi^- \pi^- l^-)$, where M^2_{rec} is calculated according to eqn. (1). After subtraction of backgrounds from continuum events, uncorrelated $D^- l^+$ pairs from $\Upsilon(4S)$ decays and fake leptons the M^2_{rec} distribution of Fig. 3a) is obtained. The remaining background (dashed line) are $D^- l^+$ pairs from the cascade decay $B^0 \to D^{*-} l^+ \nu$, $D^{*-} \to D^- (\pi^0, \gamma)$. The contribution of this background source can be determined by studying $\overline{D}^0 l^+$ pairs from the decay $B^0 \to D^{*-} l^+ \nu$, $D^{*-} \to \pi^- \overline{D}^0$ (Fig. 3b).

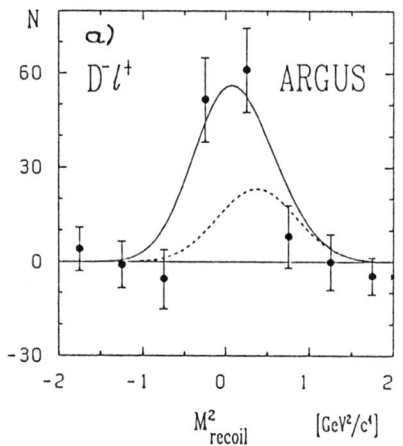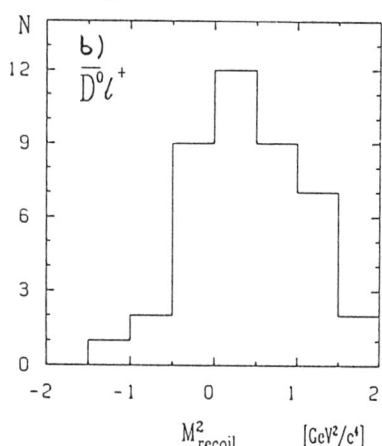

Fig. 3: a) Distribution of $M^2_{rec}(D^- l^+)$. The dashed curve shows the contribution from $B^0 \to D^{*-} l^+ \nu$, $D^{*-} \to D^- (\pi^0, \gamma)$. b) $M^2_{rec}(\overline{D}^0 l^+)$ spectrum for $\overline{D}^0$ mesons originating from decays $B^0 \to D^{*-} l^+ \nu$.

The fit to the M^2_{rec} distribution of Fig. 3a) gives a signal of 82 ± 27 $B^0 \to D^- l^+ \nu$ decays. Using the model [1] for efficiency correction due to the limited phase space, $BR(D^- \to K^+ \pi^- \pi^-) = (9.1 \pm 1.3 \pm 0.4)\%$ [13] and $BR(\Upsilon(4S) \to B^0 \overline{B}^0) = 45\%$, we obtain [14]:

$$BR(B^0 \to D^- l^+ \nu) = (1.8 \pm 0.6 \pm 0.5)\% \qquad (9)$$

From this result $|V_{cb}|$ is determined to be

$$|V_{cb}| = 0.044 \pm 0.009 , \qquad (10)$$

where $\Gamma(B \to \overline{D}l^+\nu) = 8.1 \cdot |V_{cb}|^2 \times 10^{12} s^{-1}$ [1] and $\tau_b = (1.15 \pm 0.14)\, ps$ [12] have been used. As in the case of $B^0 \to D^{*-}l^+\nu$, the model dependence of $|V_{cb}|$ is considerably smaller than the experimental error [14].

From the ARGUS measurements of $B^0 \to D^{*-}l^+\nu$ and $B^0 \to D^-l^+\nu$, one derives the ratio

$$\frac{BR(B^0 \to D^{*-}l^+\nu)}{BR(B^0 \to D^-l^+\nu)} = 3.3 \pm 1.7 , \qquad (11)$$

which is in good agreement with the expected value of about 3 [1,5]. Furthermore, one can conclude that semileptonic decays of B mesons into the lowest-lying pseudoscalar and vector $\overline{D}$ mesons constitute about 80% of the total semileptonic rate, $BR(B \to l^+\nu X) = (10.9 \pm 0.6)\%$ [8].

3 Measurement of the lifetime ratio $\tau(B^+)/\tau(B^0)$

The ratio of the B^+ and B^0 lifetimes can be extracated from a reconstruction of semileptonic B decays with a $\overline{D}^0$, D^- or D^{*-} meson in the final state. The expectation that the total semileptonic decay rate is nearly saturated by decays into the lowest-lying charmed pseudoscalar or vector mesons is confirmed by the ARGUS measurements of $B^0 \to D^{*-}l^+\nu$ and $B^0 \to D^-l^+\nu$. Hence,

$$\frac{\tau(B^+)}{\tau(B^0)} = \frac{BR(B^+ \to \overline{D}^0 l^+\nu) + BR(B^+ \to \overline{D}^{*0} l^+\nu)}{BR(B^0 \to D^- l^+\nu) + BR(B^0 \to D^{*-} l^+\nu)} . \qquad (12)$$

Since $\overline{D}^{*0}$ decays do not produce D^- mesons, all D^- mesons in semileptonic decays originate from B^0 decays, while semileptonic B^+ decays always yield a $\overline{D}^0$ meson. Thus the charge of the final state D meson serves as a tag for the charge of the B, except for $\overline{D}^0$ mesons produced in the chain $B^0 \to D^{*-}l^+\nu$, $D^{*-} \to \overline{D}^0 \pi^-$. The production ratio times the lifetime ratio of charged and neutral B mesons is therefore given by

$$\frac{f_+\, \tau(B^+)}{f_0\, \tau(B^0)} = \frac{N(\overline{D}^0 l^+) - r_* N(D^{*-}l^+, D^{*-} \to \overline{D}^0 \pi^-)}{r_- N(D^- l^+) + r_* N(D^{*-}l^+, D^{*-} \to \overline{D}^0 \pi^-)} , \qquad (13)$$

where N denotes the observed number of $\overline{D}l^+$ ($D^{*-}l^+$, $D^{*-} \to \overline{D}^0 \pi^-$) combinations and r_* and r_- account for the relative reconstruction efficiencies of D^{*-} and D^- decays with respect to $\overline{D}^0$ decays. Note that expression (13) does not contain D^{*-} branching ratios or the relative strength of semileptonic decays into vector and pseudoscalar mesons.

$\overline{D}$ mesons were reconstructed in the channels $\overline{D}^0 \to K^+\pi^-$, $D^- \to K^+\pi^-\pi^-$ and $D^{*-} \to \overline{D}^0\pi^-$ with $\overline{D}^0 \to K^+\pi^-$. The lepton momentum was required to be greater than $1.2\, GeV/c$.

After background subtraction, we obtain $325 \pm 28 \pm 9$, $183 \pm 37 \pm 12$ and $58 \pm 9 \pm 3$ events from the decays $B \to \overline{D}^0 l^+ X$, $B \to D^- l^+ X$ and $B \to D^{*-} l^+ X$, respectively. Correcting for the relative efficiencies leads to the following result for the production ratio f_+/f_0 times the lifetime ratio:

$$\frac{f_+}{f_0} \frac{\tau(B^+)}{\tau(B^0)} = 1.00 \pm 0.23 \pm 0.14 \ . \tag{14}$$

Thus, for a production ratio of about one (ARGUS used up to now $f_+/f_0 = 55:45$), the lifetime ratio $\tau(B^+)/\tau(B^0)$ is consistent with one within the experimental errors.

This result is insensitive to the effect of possible decays $B \to \overline{D}_J^* l^+ \nu$, where $\overline{D}_J^*$ denotes any excited charmed meson state, if one assumes isospin conservation in the decays of $\overline{D}_J^*$ mesons into the lower-lying $\overline{D}$ and $\overline{D}^*$ states.

References

[1] M.Wirbel, M.Bauer und B.Stech, Z. Phys. **C29** (1985) 637;
M.Bauer und M.Wirbel, Z. Phys. **C42** (1989) 671.

[2] H.Pietschmann und F.Schöberl, Europhys.Lett. **2** (1986) 583.

[3] B.Grinstein, M.B.Wise und N.Isgur, Phys. Rev. Lett. **56** (1986) 298.

[4] M.Shifman and M.Voloshin, ITEP-64 (1987).

[5] J.G.Körner und G.Schuler, Z. Phys. **C38** (1988) 511.

[6] N.Isgur, D.Scora, B.Grinstein und M.B.Wise, Phys. Rev. **D39** (1989) 799;
N.Isgur, D.Scora, UTPT-89-12 (1989).

[7] H.Albrecht et al. (ARGUS Collab.), Phys. Lett. **197B** (1987) 452.

[8] K.R.Schubert, these proceedings.

[9] J.Adler et al. (MARK III Collab.), Phys. Lett. **208B** (1988) 152.

[10] H.Albrecht et al. (ARGUS Collab.), Phys. Lett. **219B** (1989) 121.

[11] R.Gläser, Ph.D. thesis, University Hamburg (1989).

[12] H.Schröder, Proceedings of the *XXIV International Conference on High Energy Physics*, Munich (1988).

[13] J.Adler et al. (MARK III Collab.), Phys. Rev. Lett. **60** (1988) 89.

[14] H.Albrecht et al. (ARGUS Collab.), DESY 89-082 (1989), submitted to Phys. Lett.

CHARMLESS B MESON DECAYS AT CLEO

Michael Procario

Harvard University, Cambridge, Massachusetts 02138

Abstract

We have measured the lepton momentum spectra in semileptonic B meson decays using data taken with the CLEO detector at CESR. We find an excess of leptons in the region above that kinematically allowed for decays containing charm. Ascribing these leptons to charmless semileptonic decays yields $|V_{ub}/V_{cb}|$ of approximately 0.1.

1 Introduction

A central feature of the standard model of weak interactions is the 3×3 quark-mass mixing matrix first proposed by Kobayashi and Maskawa to explain CP violation in K meson decays.[1] If any of the nine matrix elements is exactly zero, however, the quark mixing will not give any CP violation[2]. It is important to show the matrix element between the b quark and the u quark (V_{ub}) is larger than zero. Prior to this work there was no direct, confirmed evidence for a nonzero value of (V_{ub})[3].

The lepton spectrum from B decays has been experimentally investigated before. The value of $B(B \to Xl^-\overline{\nu})$ is known to be approximately 10% where l is either an electron or a muon, and upper limits on $|V_{ub}/V_{cb}|$ have been determined[4]. (In this paper, statements about particular reactions are also true for the charge conjugate reactions.) We have found that the best way of looking for the $b \to u$ transition is to concentrate in the region where the lepton momentum p_l is near its maximum allowed value[5]. Figure 1 shows the expected momentum spectra due to $B \to X_c l\nu$ and $B \to X_u l\nu$ in the Altarelli model[6] for $\Gamma(b \to u) = 0.02\Gamma(b \to c)$. Since all realistic models of $b \to c$ semileptonic decays predict that our data sample would have less than one lepton above 2.4 GeV/c from $b \to c$ decayss, a search for leptons with momentum between 2.4 and 2.6 GeV/c should give a measurement of $|V_{ub}|$, which is highly independent of the choice of $b \to c$ model. However, to increase our acceptance, we will search from 2.3–2.6 GeV/c. This region is expected

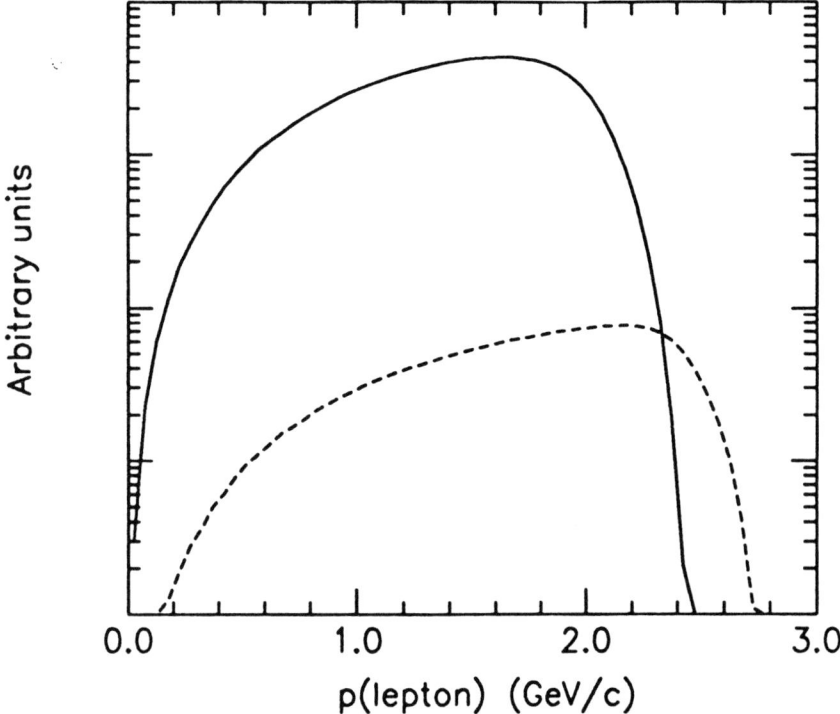

Figure 1: Comparison of Altarelli model predictions for the $b \to c$ (solid line) and $b \to u$ (dashed line) lepton momentum spectra for $\Gamma(b \to u) = 0.02\Gamma(b \to c)$.

to have almost double the $b \to u$ rate of 2.4–2.6 GeV/c and the $b \to c$ rate is comparable to the $b \to u$ rate to which we are sensitive.

2 Lepton sample

Data were accumulated with the CLEO detector located at the CESR e^+e^- storage ring. The sample consists of 212 pb^{-1} taken at the peak of the $\Upsilon(4S)$ resonance, which yielded 480,000 B mesons, and 101 pb^{-1} taken on the continuum at center-of-mass energies 30 MeV below the $\Upsilon(4S)$ peak. CESR was cycled so that data were taken for two-day periods on the $\Upsilon(4S)$ peak and one day off resonance.

The CLEO detector has been improved with a new 51-layer central drift chamber, which operates with a 50%–50% mixture of argon-ethane gas. The dE/dx resolution is 6.5%, which gives approximately three standard deviations separation between electrons and pions for much of momentum range 1.4 to 3.0 GeV/c. The momentum resolution is $(\delta p/p)^2 = (0.23\%p)^2 + (0.7\%)^2$, where p is in Gev/c.

Events that pass our standard event selection criteria[7] and have five or more charged tracks are searched for leptons. Muon identification in CLEO has been described in detail elsewhere[8]. The overall muon identification efficiency, including geometrical acceptance, is 60%, and the misidentification probability is 1.1% in the endpoint region. The misidentification probability is determined in two ways. A Monte Carlo simulation of the detector calculates the shape of the momentum spectrum for fakes and the spectrum is normalized with data on the lepton-poor $\Upsilon(1S)$ resonance. We also directly measure the misidentification probability by using pions from K_s^0 decays.

Electrons that pass through the electromagnetic calorimeter are identified by combining measurements from the electromagnetic calorimeters, time-of-flight scintillators, outer dE/dx counters and the central drift chamber[9]. The solid angle for electron identification in the electromagnetic calorimeter is $0.47(4\pi)$. Using this information, electrons are identified with 90% efficiency from 1.5 to 3.0 GeV/c, and the probability of misidentifying a hadron as a lepton is 0.3%. The misidentification probability is determined from data taken on the lepton-poor $\Upsilon(1S)$.

Electrons outside the the electromagnetic calorimeter's solid angle but with $|\cos\theta| \leq 0.8$ are identified using only the dE/dx measurement in the drift chamber. The solid angle for these electrons is $0.32(4\pi)$, and the electrons in the endpoint region are identified with 65% efficiency, while misidentification probability is 0.8%. The misidentification probability for these electrons is measured with tracks found to be hadrons in the electromagnetic calorimeter.

We measure the lepton momentum spectrum due to B decays by subtracting the spectrum of leptons from the continuum of non-$B\overline{B}$ production from spectrum taken on the $\Upsilon(4S)$. We estimate the continuum lepton spectrum with the data taken off resonance. The momentum is scaled by the beam energy difference between the on-resonance and off-resonance samples, and the number of leptons is scaled by the ratio of luminosities and the ratio of cross-sections between the the on-resonance and off-resonance samples. The luminosities were measured with wide-angle bhabha events. The scaling is 2.08 and is expected to be accurate to 1/2%, The scaling agrees with that found by subtracting the momentum spectrum of all tracks in hadronic events above 2.8 GeV, which is accurate to 1.7%.

The continuum produces a large fraction of the leptons in the endpoint region and, therefore makes a large contribution to the uncertainty on the number of leptons from B decays in the endpoint region. In order to reduce amount of continuum we need to subtract, we use the Fox-Wolfram event-shape parameters[10] to reject jet-like events. We require $R2$ be less than 0.4, where $R2 = H2/H0$. We reject about 70% of continuum lepton events with lepton momentum in the endpoint region, and accept 65% of the $b \to u$ events with high momentum leptons, which proceed through the mixture of exclusive final states as given in the ISGW model[12].

A second method to reduce the error from the continuum subtraction is to fit the continuum to a smooth function and use the increased statistical power of the additional events outside the endpoint region to make a more accurate estimate of the number of leptons that need to be subtracted. The lepton continuum samples are fitted to a series of smooth functions, which are exponential and polynomial functions and mixtures of the two. The statistical error is calculated from the fit and we use the spread of values from the different functions for the systematic error. The $\Upsilon(4S)$, continuum, and fitted continuum momentum spectra are shown in Figure 2.

There are sources of high momentum leptons in B decay in addition to those that might come from $b \to u$ transitions. The largest of these come from leptonic decays of J/ψ's from B decay. In order to reduce the systematic and statistical errors due to subtraction of the J/ψ contribution from the lepton spectrum, we search for J/ψ candidates and eliminate them. If a lepton can be combined with any other track in the event and have a mass within $\pm 60 \text{MeV}/c^2$ of the J/ψ mass, then that lepton is rejected. From Monte Carlo studies, we estimate that 90% of leptons from J/ψ's are rejected this way. There are also leptons from ψ' decays. From the measurement of $B \to \psi' X$[11], we estimate that the number of leptons from ψ' is approximately the same as the leptons from ψ that we fail to reject.

3 Momentum resolution

The momentum resolution of the tracking system is ~ 20 MeV/c in the endpoint region. If the momentum resolution were truly Gaussian, only a small and calcualbe number of leptons from $b \to c$ transitions would appear above 2.3 GeV/c. However, there is always the possibility that some fraction of the tracks will reconstruct significantly different from the true momentum. In order to insure that this does not happen, we require stricter track quality for leptons than for other tracks. Lepton candidates are rejected if the rms deviation between track position and hit position is greater than 250 microns. Tracks that fail to have at least 30 hits out of 51 layers in the central drift chamber or 5 hits out of 13 layers in the combination of the two inner drift chambers are also rejected. These requirements reject tracks where the pattern recognition has failed to find the correct set of hits to make a track.

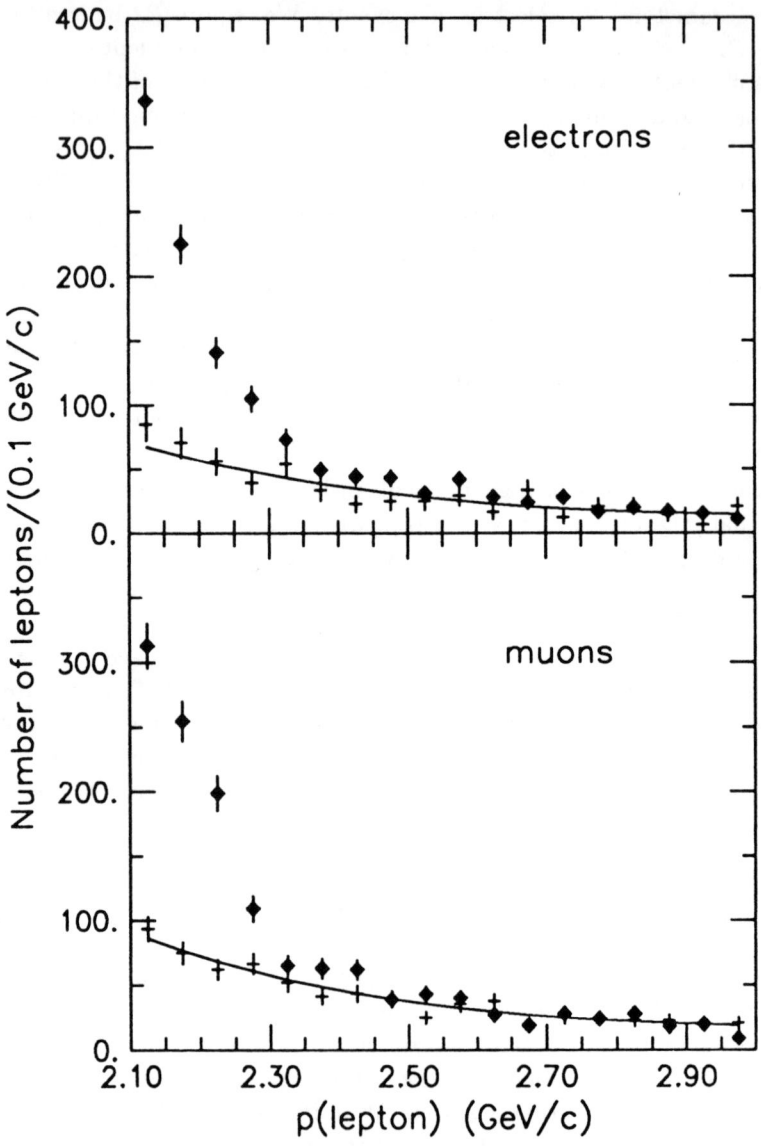

Figure 2: Lepton momentum spectra from $\Upsilon(4S)$ (diamonds), the continuum (crosses), and the fitted continuum (line) for electrons and muons.

The effect of the requirements are checked for isolated tracks and for tracks in complicated events with overlapping tracks. For the isolated tracks μ pairs were used. The entire sample of μ pairs from the $\Upsilon(4S)$ and continuum running were searched for tracks with momentum above the beam energy. We also embed well-measured isolated tracks into randomly selected hadronic events, and remeasure their momenta to check for differences in the momentum measurement due to complicated event topology. After imposing the strict track quality cuts, neither sample show non-Gaussian momentum shifts.

4 Backgrounds

We have considered several other possible sources of real leptons in the 2.3–2.6 GeV region. These include the decay chain $B \to DX \to D \to Yl^-\nu$, dileptonic decays of vector mesons, electrons from π^0 Dalitz decay and mismeasured muon tracks caused by lower momentum pion or kaon decay. The sum of all these sources is less than one event.

A known source of leptons that does contribute in the endpoint region is the tail of $B \to D(D*)l\nu$ decays. In order to estimate this contributions, we smeared the lepton spectra from ISGW and Altarelli models by our momentum resolution. We then fitted these spectra to our data in the momentum region 1.5–2.2 GeV/c, and used the extrapolation of these spectra into 2.3–2.6 GeV/c for our $b \to c$ background estimate. There is a 20% variation in the extrapolation due to different semileptonic decay models, but but the extrapolation is not sensitive to changes in fitting procedure.

Table I summarizes the lepton yields and the backgrounds. The different momentum regions are given to check the momentum dependence of the signal. In most models of semileptonic B decay the $b \to u$ lepton spectrum is expected to be approximately equal in the two regions. Results are given for both fitting the continuum before subtracting and a direct subtraction. Since the two results are compatible, we chose to use the fitted results because of their smaller errors. We have 4 ± 18 events in the 2.3–2.4 GeV/c region and 56 ± 21 events in the 2.4–2.6 GeV region, for a total of 60 ± 28 events. We have not been able to account for these leptons from any known source. The most natural explanation is that they arise from semileptonic decays proceeding through the $b \to u$ transition.

$2.3 < p_l \leq 2.4$	electrons		muons	
	direct	fitted	direct	fitted
Leptons from $B\overline{B}$	34 ± 17	$37 \pm 11 \pm 3$	35 ± 18	$26 \pm 12 \pm 2$
Fakes	4 ± 1		11 ± 4	
Leptons from J/ψ	1.5 ± 1		1.5 ± 1	
Leptons from ψ'	1.5 ± 1		1.5 ± 1	
$B \to X_c l \nu$	16 ± 4		20 ± 4	
$B \to X_u l \nu$	$10 \pm 17 \pm 4$	$13 \pm 11 \pm 5$	$0 \pm 18 \pm 4$	$-9 \pm 13 \pm 5$
$2.4 < p_l \leq 2.6$	electrons		muons	
	direct	fitted	direct	fitted
Leptons from $B\overline{B}$	53 ± 19	$38 \pm 14 \pm 5$	41 ± 22	$36 \pm 15 \pm 2$
Fakes	1.2 ± 0.7		8.0 ± 4	
Leptons from J/ψ	1.5 ± 1		1.5 ± 1	
Leptons from ψ'	2 ± 1		2 ± 1	
$B \to X_c l \nu$	1 ± 1		1 ± 1	
$B \to X_u l \nu$	$47 \pm 19 \pm 1$	$32 \pm 14 \pm 5$	$29 \pm 22 \pm 1$	$24 \pm 15 \pm 2$

Table I: Lepton yields and backgrounds in the endpoint region

If we attribute all the excess leptons to $b \to u$, we can use various theoretical models of semileptonic $b \to u$ decays to extract a value for the ratio of KM matrix elements $|V_{ub}/V_{cb}|$. First we determine the average of the e^- and μ^- cross-section in the 2.3–2.6 Gev region to be 0.37 ± 0.17pb. The efficiency we use is determined by generating the proper proportion of exclusive final states in the endpoint region using the ISGW model. The final states are $Xl^-\nu$, where $X = \pi^+, \pi^0, \rho^+, \rho^0$, and ω^0. The overall efficiency is a product of the trigger efficiency (100%) with the event selection efficiency including the R2 cut (65%) and the lepton detection efficiency (electron 64% and muon 60%). Then we use the relationship

$$\frac{|V_{ub}|^2}{|V_{cb}|^2} = \frac{\sigma(p)}{\sigma_0} \frac{1}{d}$$

where $\sigma(p)$ is the above determined cross-section, σ_0 is our previously measured total lepton cross-section of 236 ± 14 pb[5], and d is a model dependent factor that includes the relative probabilities for producing charmless and charmed final states and the fraction of the charmless momentum spectrum between 2.3 and 2.6 GeV. For example, in the ISGW model $d = 0.09$, and in the Altarelli model $d = 0.24$. We find $|V_{ub}|/|V_{cb}| = 0.13 \pm 0.03$ in the ISGW model and 0.08 ± 0.02 in the Altarelli model.

In conclusion, we have observed an excess of leptons in semileptonic B meson decay in the momentum interval $2.3 \,\text{GeV}/c \leq p_l \leq 2.6 \,\text{GeV}/c$, which can be interpreted as positive evidence for the $b \to u$ transition. The value of $|V_{ub}/V_{cb}|$ is approximately 0.1.

References

1. M. Kobayashi and T. Maskawa, Prog. of Theor. Phys. **49**, 652(1973).

2. C. Jarlskog, Z. Phys. **C29**, 491 (1985).

3. There is a published claim by ARGUS of seeing the final states $p\bar{p}\pi$ and $p\bar{p}\pi\pi$, H. Albrecht et. al., Phys. Lett. **B209**, 119 (1988). However this result is contradicted by CLEO in C. Bebek et. al., Phys. Rev. Lett. **62**, 8 (1989). Furthermore, recent ARGUS data do not show this effect, H. Schultz, "B Physics at ARGUS", in *Proceedings of the Symposium on the Fourth Family of Quarks and Leptons*, Santa Monica, Cal. (1989).

4. B. Gittelman and S. Stone, "B Meson Decay", in High Energy Electron-Positron Physics, ed. by A. Ali and P. Söding, World Scientific, Singapore, p. 275 (1988).

5. S. Behrends, et al., Phys. Rev. Lett. **59**, 407 (1987).

6. G. Altarelli, et al., Nucl. Phys. **B208**, 365 (1982).

7. Hadronic events selection criteria are discussed in S. Behrends, et al., Phys. Rev. **D31**, 2161 (1985).

8. K. Chadwick, et al., Phys. Rev. **D27**, 475 (1983).

9. D. Andrews, et al., Nucl. Inst. and Meth. **211**, 47 (1983).

10. G. C. Fox and S. Wolfram, Phys. Rev. Lett. **41**, 1581 (1978).

11. H. Albrecht, et al., Phys. Lett. **B199**, 451 (1987); W.-Y. Chen, "B decays to Charm: Exclusive Decays and Inclusive Rates", paper submitted to 1989 Lepton-Photon Conference.

12. N. Isgur, D. Scora, B. Grinstein and M. Wise, Phys. Rev **D39**, 799 (1989).

B DECAY STUDIES WITH THE TPC AT PEP

Gerald R. Lynch
Lawrence Berkeley Laboratory, Berkeley, CA. 94720

INTRODUCTION

In the past the TPC/2γ collaboration [1] at PEP has not been in position to contribute much to B decay studies because it did not have good position resolution at the origin. Now that we have built and tested a new Vertex Chamber (VC), we are ready to proceed with a B-physics program. Last November we had a short run at a beam energy of 13.6 GeV. The TPC and VC worked well. We are using a sample of 13 pb^{-1} of analyzed events from this run to test our system.

THE VERTEX CHAMBER

Our Vertex Chamber is a straw chamber that was built at SLAC. It has 14 layers of straws with a radius of 4.15 mm filled with 4 atmospheres of Argon-CO_2. It extends from 4.3 cm to 15.8 cm in radius. With 14 layers the VC has enough redundancy to allow excellent track finding. This is illustrated in Figure 1.

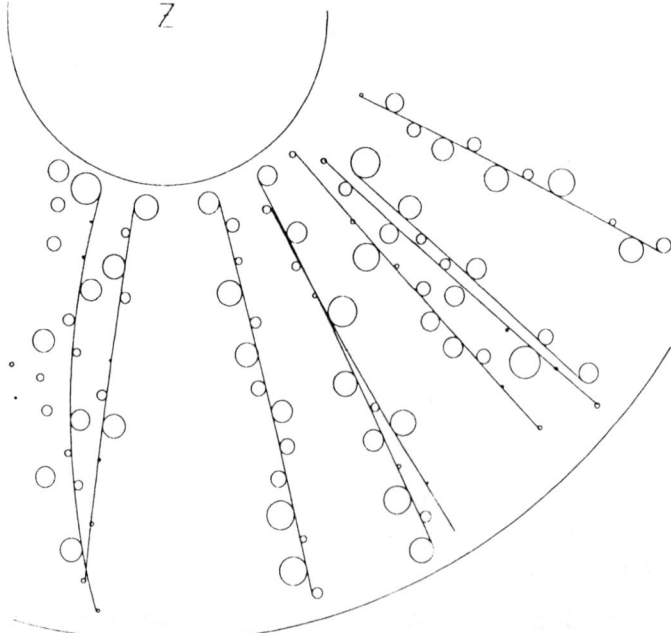

Figure 1. A part of a multi-hadron event in the Vertex Chamber. The measured points are represented by circles centered on the wire and with a radius equal to the calculated drift distance. A track goes tangent to the circles.

Figure 2 shows the resolution that has been achieved with this chamber. The average local resolution for drift distances greater than 2 mm is 30 microns. This resolution was reduced from 70 μ to near 30μ by using Bhabha electrons to determine six parameters for each channel, one to correct time, one to correct drift velocity, three to correct position, and one to correct for the wire not being in the center of the straw.

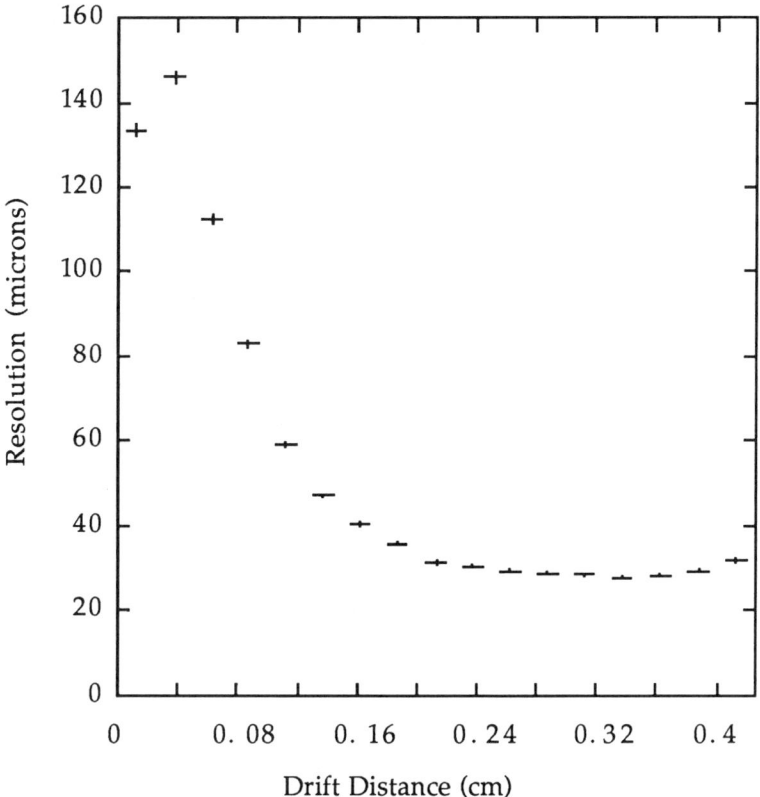

Figure 2. The resolution for the Vertex Chamber as a function of drift distance. The point at the smallest drift distance is biased low.

We now have done a fairly good job of aligning the VC and TPC. This alignment involved evaluating five effects, each of which contributed misalignments between the VC and TPC of about 1 mm.
1. A translation(of 1.3 mm).
2. A rotation about the z-axis (of about 3 mr).
3. A rotation about an axis normal to the z-axis (also about 3 mr).
4. An angle (of 1.3 mr) between the E-field and the B-field in the TPC.
5. A grid potential correction to the electric field in the TPC.

With these corrections we can fit to tracks using both VC and TPC points to get a momentum resolution of $(dp/p^2)^2 = (0.0065)^2 + (0.015/p)^2$ and an impact parameter resolution of $(dx)^2 = (54\mu)^2 + (100\mu/p)^2$.[2] With our position resolution we ought to be able to improve from 0.0065 to <0.004 in dp/p^2 and from 50μ to <40μ in impact parameter resolution. We expect that better TPC distortion corrections will improve the these resolutions.

B LIFETIME STUDIES

To test our capabilities for measuring the B lifetime using the impact parameter measurements of leptons from B decays, we have selected an enriched sample of B events with high p_t leptons as follows:

1. Choose events that pass the standard multi-hadron event selection.
2. Require Thrust>0.75.
3. Require the thrust axis to be more than 45° from the beam line.
4. Require a well identified lepton, either
    a. An electron identified in both the the hexagonal calorimeter and in the TPC, or
    b. A muon identified in the barrel muon chambers.
5. Require the lepton to be found in both the TPC and the VC.
6. Require the lepton to have a momentum P>2 GeV/c and to have P_t > 1 GeV/c relative to the thrust axis.
7. Require the lepton direction to be more than 0.1 radians from the thrust axis in the x-y plane.

The 55 leptons that are found by this selection are estimated to be a 70% pure BB sample. The distribution for the impact parameters relative to the measured position of the interaction region for these 55 tracks is shown in Figure 3. Some properties of this distribution are:

1. The full width at half maximum is at most 400 microns. This is a manifestation of our good impact parameter resolution and the small size of the beam at PEP. For most runs the interaction region at PEP had an RMS of about 225 microns in the horizontal direction and less than 60 microns in the vertical direction.
2. The average impact parameter is 195±63 microns (statistical errors only). The magnitude is close to what Monte Carlo studies say that it should be.
3. The statistical error is about 32% of the effect, suggesting that with about 10 times the data we could have the statistical errors comparable to the best previous experiments.

Figure 4 shows one of these events, the one with an impact parameter of about 1.9 mm.

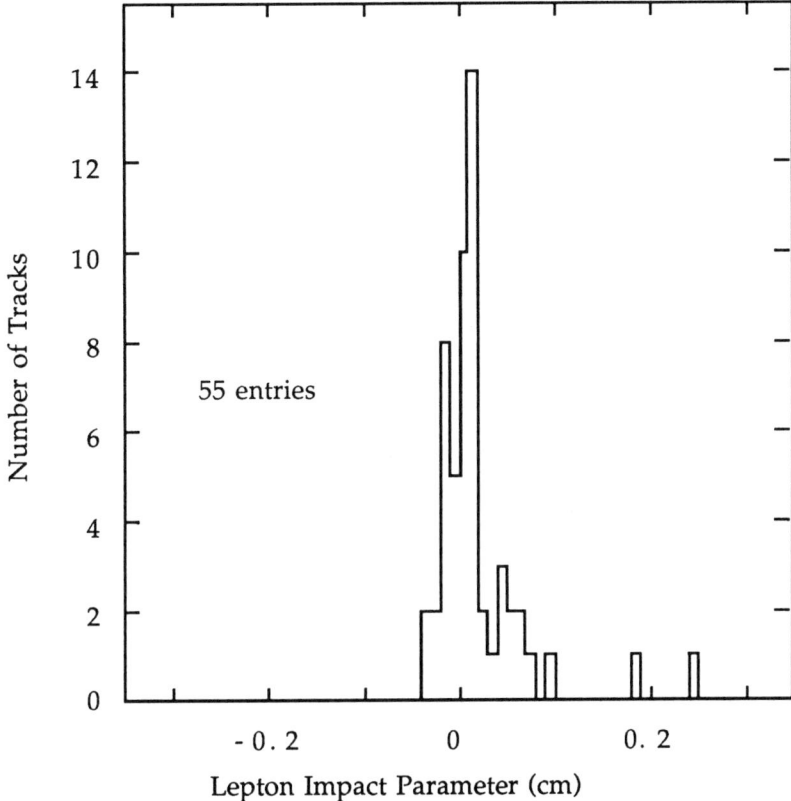

Figure 3. The distribution of measured impact parameters for the high p_t leptons that were selected. The sign of the impact parameter is determined from the intersection point of the lepton and the thrust axis.

CONCLUSIONS

With our new Vertex Chamber and the upgraded PEP, the TPC at PEP is ready to study B decays. With the 300 pb^{-1} that we plan to get in the next 18 months, we should be able to measure the B lifetime with an accuracy equal to the present world average. With the one fb^{-1} that we plan to get in the next three years, we should be able to measure $B\bar{B}$ mixing with an accuracy of 6 or 7 standard deviations, from which one would get a useful constraint on the magnitude of $B_s\bar{B}_s$ mixing.

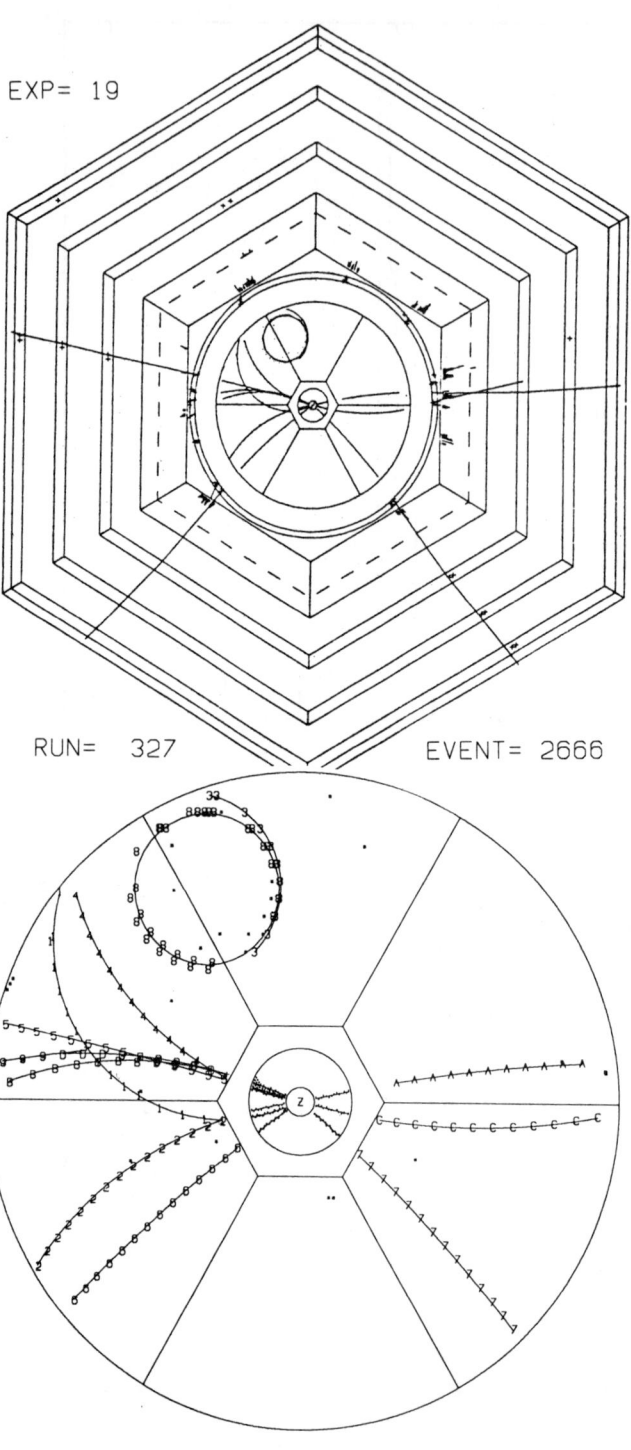

Figure 4. This figure has four views of one event that is very probably an event with a B⁻ that has a long decay length.

Figure 4a. An end view of the entire detector. The lines in the outer part of the detector are where the track seen in the TPC would go if it were a muon. The event has two muons.

Figure 4b. A view of the TPC and VC.

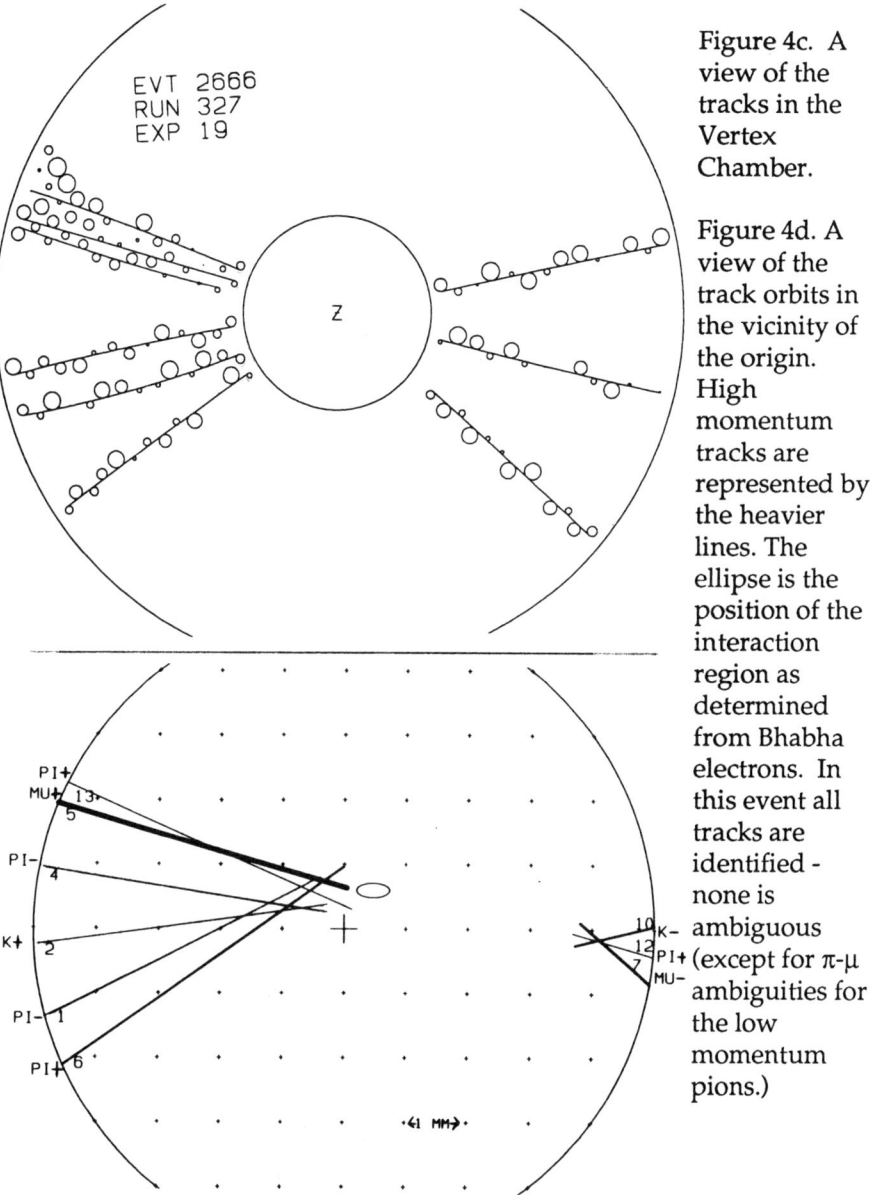

Figure 4c. A view of the tracks in the Vertex Chamber.

Figure 4d. A view of the track orbits in the vicinity of the origin. High momentum tracks are represented by the heavier lines. The ellipse is the position of the interaction region as determined from Bhabha electrons. In this event all tracks are identified - none is ambiguous (except for π-μ ambiguities for the low momentum pions.)

1. Descriptions of this experiment can be found in H. Aihara, et al., I.E.E.E. Trans. Nucl. Sci. NS30, 63, 67, 76, 117, 153, 162 (1983), and H. Aihara et. al. LBL23737 (1988)
2. This impact parameter resolution, as well as some other resolutions quoted in this report, are substantially smaller than the ones that were reported at the conference - the result of work done in the interim.

ARGUS RESULTS ON BARYONIC DECAYS OF B MESONS

Bernhard Spaan[*]

Institut für Physik, Universität Dortmund[**], Germany

Abstract

The experimental ARGUS results on baryonic B decays are summarized. The branching ratio of B mesons into baryons has been measured to be Br(B → baryons) = (7.6 ± 1.4)%. It appears that baryonic B decays are dominated by charmed baryons and multibody final states.

1 INTRODUCTION

From the analysis of B meson decays several "fundamental" parameters of the standard model are accessible. The extraction of the basic physical quantities often is complicated by the influence of non-perturbative strong interaction phenomena. This is especially true for B meson decays where baryons appear in the final state, since the three-quark structure of baryons gives additional freedom to the models. Therefore from the analysis of baryonic B decays on gains experimental information which can be used as input for models. Here two questions are of relevance, namely:

- How large is the baryonic branching ratio of B mesons?

- By which decay mechanism are baryons produced?

---

[*] Representing the ARGUS collaboration.
[**] Supported by the German Bundesministerium für Forschung und Technologie, under contract number 054DO51P.

Table 1: Ratios of baryonic branching ratios revealing information on baryon/antibaryon correlations.

$\frac{2 \cdot \text{Br}(B \to p\bar{p}X)}{\text{Br}(B \to pX)}$	0.60 ± 0.09
$\frac{\text{Br}(B \to \Lambda\bar{p}X)}{\text{Br}(B \to \Lambda X)}$	0.54 ± 0.15
$\frac{\text{Br}(B \to \Lambda\bar{\Lambda}X)}{\text{Br}(B \to \Lambda X)}$	< 0.42 (90% cl)

The latter question arises since for baronic B decays already in the pure spectator model there exist three different decay topologies, from which for pure mesonic decays one is colour suppressed and another (W hadronizing into baryons) not existing.

Furthermore the branching ratio of B mesons into charmed baryons is needed for the so called charm counting, which tells whether there is room for others than $b \to c$ transitions.

2 B DECAYS INTO P AND Λ

From B decays into SU(3) ground state baryons up to now only the inclusive spectra of both p and Λ have been measured. Recent results from the ARGUS collaboration, obtained after subtracting the continuum contribution, are shown in fig. 1. The baryon spectra appear to be rather soft, thus indicating that baryonic B decay channels are dominated by multibody final states. Integrating the inclusive spectra yields the branching ratios for $B \to pX$ and $B \to \Lambda X$:

$$\text{Br}(B \to pX) = (8.2 \pm 0.5^{+1.3}_{-1.0})\%$$
$$\text{Br}(B \to \Lambda X) = (4.2 \pm 0.5 \pm 0.6)\%.$$

In order to obtain the total baryonic branching ratio of B mesons, assumptions have to be made about the neutron abundance. From isospin symmetry one would expect that protons and neutrons have equal rates, whereas the small asymmetry from Λ decays can be taken into account by the known branching ratio of $B \to \Lambda X$. Further (small) asymmetries in the p and n yield can occur through the weak decays of Σ and Λ_c^+ baryons.

Indirect measurements of the neutron rate in B decays are obtained from the analysis of $p\bar{p}$ or $\Lambda\bar{p}$ correlations. The ratios of branching ratios $2 \cdot \frac{\text{Br}(B \to p\bar{p}X)}{\text{Br}(B \to pX)}$

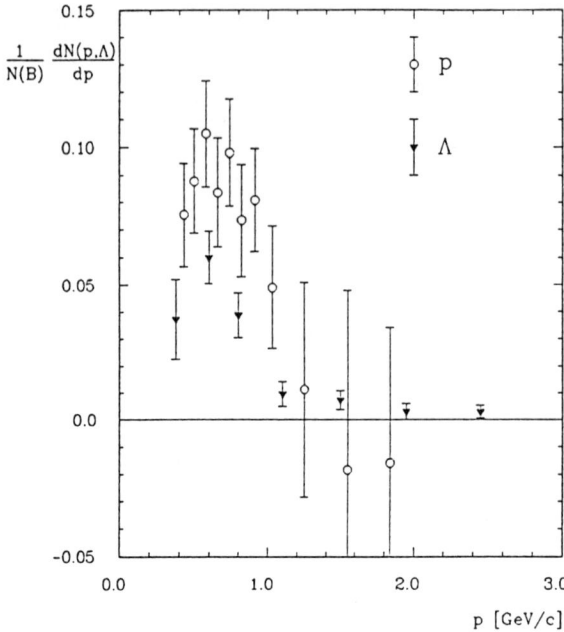

Figure 1: Momentum spectra of p and Λ particles from B decays as measured by the ARGUS collaboration [1].

and $\frac{Br(B \to \Lambda \bar{p} X)}{Br(B \to \Lambda X)}$ given in table 1 reflect that the probability for a p or Λ to be accompanied by an antiproton is not very different from 50%, hence supporting the hypothesis of about equal proton and neutron rates. No indication has been found for joined $\Lambda \bar{\Lambda}$ production so far, which could be explained if all Λ baryons are produced from Λ_c^+ decays.

From these measurements we are able to derive a total baryonic branching ratio for B mesons:

$$Br(B \to \text{baryons}) = (7.6 \pm 1.4)\%.$$

3 B DECAYS INTO CHARMED BARYONS

The question still remains, whether baryonic B decays are dominated by transitions to charmed baryons. This can be naturally expected, since the additional production of a baryon/antibaryon pair in $B \to D^*$ or $B \to D$ decays is likely suppressed by phase space. Indirect evidence for a charmed baryon production

Table 2: Dependence of the branching ratio B → Λ_c^+ on the Λ_c^+ branching ratio into $pK^-\pi^+$.

Br(B → Λ_c^+X)[%]	Br($\Lambda_c^+ \to pK^-\pi^+$)[%]
11.5	2.6
7.5	4
5	6

is given also by the large Λ/p ratio of 0.5 ± 0.1 in B decays.

In 1987 the ARGUS collaboration observed for the first time a Λ_c^+ signal from B decays using the Λ_c^+ decay channel $\Lambda_c^+ \to pK^-\pi^+$[2]. The measured product of the B and Λ_c^+ branching ratio is:

$$\text{Br}(B \to \Lambda_c^+ X) \cdot \text{Br}(\Lambda_c^+ \to pK^-\pi^+) = (0.30 \pm 0.12 \pm 0.06)\%\,.$$

The Λ_c^+ momentum spectrum in fig. 2 has again a soft behaviour as already observed for p and Λ. A comparison with the predictions from a phase space model shows that multibody decays seem to dominate in baryonic decays.

The determination of the branching ratio B → Λ_c^+ requires the knowledge of the Λ_c^+ branching ratios. Using e.g. the branching ratio Br($\Lambda_c^+ \to pK^-\pi^+$) from ref. [4] of $2.6 \pm 0.9\%$ yields Br(B → Λ_c^+) = 11.5% which is already larger than the total baryonic B branching ratio. Therefore one can conclude that the branching ratio $\Lambda_c^+ \to pK^-\pi^+$ is larger than given in ref. [4]. This is additionaly supported by the measurement from ref. [3], which gives a lower limit of Br($\Lambda_c^+ \to pK^-\pi^+$) > 4.4% at the 90% confidence level. The dependence of B → Λ_c^+X on the Λ_c^+ branching ratio is demonstrated in table 2. Assuming that baryonic B decays are saturated by the decays into Λ_c^+ baryons, a Λ_c^+ branching ratio into $pK^-\pi^+$ of order 4% is obtained. Even using a somewhat larger value for this branching ratio, baryonic B decays are dominated by decays to charmed baryons.

4 Λ LEPTON CORRELATIONS

Correlations between leptons and Λ hyperons reveal information on extra strangeness production in B decays, i.e. strangeness which is not generated by the b → c → s

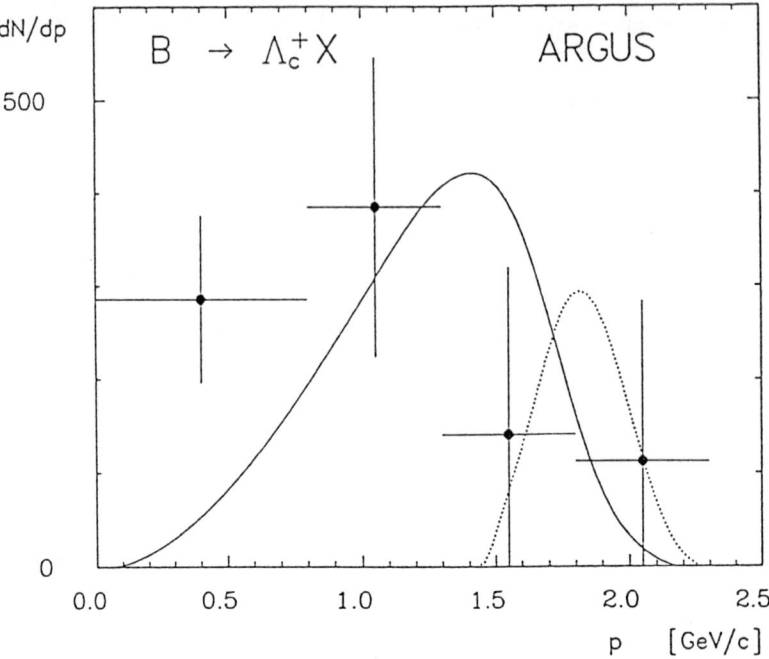

Figure 2: Momentum spectrum of Λ_c^+ baryons from B decays. Also shown are the expected form of the contribution from two-body (dotted line) and three-body (solid line) B-decays.

transition. From this one gains a partly better understanding of the hadronization on the spectator quark side, which might also be helpful for the reconstruction of exlusive channels. On the other hand Λ lepton correlations offer the chance to determine an absolute value for inclusive branching ratio $\Lambda_c^+ \to \Lambda X$.

The basic idea of this analysis is outlined in the following. The flavour of one B meson is tagged by the lepton charge, whereas a cut on the lepton momentum around 1.5 GeV/c ensures that the Λ originates from the other B meson. Neglecting for the moment $B^0 \overline{B^0}$ mixing, Λ hyperons in Λl^+ combinations will likely come from Λ_c^+ decays, while a signal in Λl^- combinations provides evidence for additional strangeness production.

We have searched for the corresponding correlations, where leptons are either electrons or muons. After continuum and background subtraction we find 37 ± 10 entries in the Λl^+ combinations respectively 23 ± 8 in the Λl^- combinations [5]. The considered background sources are lepon-hadron misidentification, secondary leptons from charm decays, J/ψ-decays, asymmetrical conversion and Λ baryons produced in charge-exchange reactions in the inner detector material. While a correlation between Λ and l^+ is expected from the observation of Λ_c^+ production, the observed Λl^- Signal cannot be explained due to $B^0\overline{B^0}$ mixing, which amounts to about 10% of the Λl^+ signal. Therefore this result implies a strong Λl^+ contribution from extra strangeness production in B decays.

5 B DECAYS INTO Ξ^-

Additional evidence for an extra strangeness production can be obtained from the observation of Ξ particles in B decays. Ξ particles can originate either from decays of Ξ_c or Ω_c states or from the spectator side.

Since the Ξ^- decays via $\Lambda\pi^-$ this should be an easy measurement. We have searched for inclusive Ξ^- production [1], where the scaled Ξ^- momentum was restricted to the interval $0.1 < x_p < 0.2$ in order to reduce the continuum background. After continuum subtraction there remains a signal of 54 ± 27 Ξ^- from B decays, which translates into an upper limit for Ξ^- production of

$$\text{Br}(B \to \Xi^- X) < 0.51\%$$

at 90% cl.

REFERENCES

References

[1] H. Albrecht et al., Z.Phys. **C42** (1989) 519

[2] H. Albrecht et al., Phys.Lett. **210B** (1988) 263.

[3] M. Aguilar-Benitez et al., Phys.Lett. **199B** (1987) 462.

[4] Particle Data Group, Phys.Lett. **204B** (1988) 1.

[5] H. Albrecht et al. (ARGUS Collaboration), Contributed paper to the XXIV International Conference on High Energy Physics in Munich (1988).

HADRONIC B MESON DECAYS

Yuichi Kubota
Cornell University, Ithaca, NY. 14853

Abstract

Using the CLEO detector at CESR, we have measured branching ratios for various inclusive charmed hadron production including the Λ_c. We have also measured the exclusive branching ratios of several decay modes of the B meson. $\bar{B}^0 \to D^{*+}D_s^-$ and $D^+D_s^-$, and $B^- \to D^0D_s^-$ as well as $\bar{B}^0 \to D^{*+}\rho^-$ have been observed for the first time. The mass difference between $\bar{B}^0$ and B^- has been measured to be 0.4±0.6 MeV.

INTRODUCTION

In this article we summarize our measurements of branching ratios of many fully reconstructed B meson decays. The B mesons are produced from decays of the $\Upsilon(4S)$ resonance at the Cornell Electron Storage Ring (CESR) in e^+e^- annihilations. The exclusive B-decay modes provide an extensive set of data to test theoretical models of hadronic B meson decays. Although several such models exist for two-body D meson decay,[1] only two models have been applied to the B meson system. Theoretical predictions are not yet available for D or B meson decays into more than two bodies. Furthermore, exclusive decays provide the only means of measuring the masses of the neutral and charged B mesons. A precision measurement of their difference gives a better understanding of the electromagnetic and strong interactions between a heavy quark and a light quark. In this paper we present new measurements of the branching ratios for inclusive charmed hadron production in B decays and the branching ratios for exclusive modes including at least one charmed hadron. We also report new measurements of the $\bar{B}^0 - B^-$ mass difference.[2]

The data sample used in this study was collected in 1987 by the CLEO detector at Cornell Electron Storage Ring (CESR). It consists of 212 pb^{-1} at the $\Upsilon(4S)$ resonance containing 240,000 $B\bar{B}$ pairs, and 102 pb^{-1} at energies below $B\bar{B}$ threshold. The CLEO detector, described in detail elsewhere,[3] has undergone major improvements to the central tracking system.[4] Charged particle tracking is done inside a superconducting solenoid of radius 1.0 m which produces 1.0 T magnetic field. Three nested cylindrical drift chambers measure momenta and specific ionization (dE/dx) of charged particles. The momentum resolution achieved by this system is $(\delta p/p)^2 = (0.23\%p)^2 + (0.7\%)^2$, where p is in GeV/c. The main drift chamber measures dE/dx to 6.5% accuracy. Photons are detected in barrel calorimeters which cover 47% of the solid angle. The energy resolution is $\sigma_E/E = 21\%/\sqrt{E}$, where E is in GeV, and the angular resolution is 10 mrad.

π^0 candidates are formed from 2 photons each of which has an energy greater than 150 MeV and which does not lie on the path of any charged track in the event. They are required to have an invariant mass within 65 MeV of the π^0 mass. Because of the relatively good angle measurement, a kinematic fit constraining the candidate mass to that of the π° improves the energy resolution by more than a factor of 2. The π° detection efficiency was 10% for π° momenta above 1 GeV/c.

INCLUSIVE CHARMED PARTICLE DISTRIBUTIONS

All exclusive B meson decays considered here result from the dominant b→c transition and thus have at least one charmed quark in the final state. When the reactions proceed through the chain: b→cW$^-$, W$^-$→cs there are two charmed quarks in the final state which can hadronize either as two charmed particles or as hidden charm (for example ψ or ψ'). Otherwise there is only one charmed particle in the final state.

To find the exclusive decays, we look first for the charm or hidden charm signals. The momentum range is limited since B decays at the Υ(4S) cannot produce particles with more than half the beam energy. We define $x = p/p_{max}$, where p is the measured particle momentum and p_{max} is the maximum possible momentum the particle could have if produced in continuum e^+e^- annihilations. Thus, x should be less than 0.5. Signals for D°, D^{*+}, D^+ and D_s^+ are sought using the following signatures. The D° is searched for in the $K^-\pi^+$ or $K^-\pi^+\pi^-\pi^+$ modes. In order to reduce combinatoric background, particles assigned to be charged kaons are required to have specific ionization measurements consistent with that expected for kaons. The signal in the $K^-\pi^+$ mode is shown for both on the Υ(4S) and continuum data in Fig. 1. The D^{*+} is required to decay to $D^\circ\pi^+$; the signal for the $D^{*+} \to D^\circ\pi^+$, $D^\circ \to K^-\pi^+$ mode is shown in Fig. 2. The D^+ is searched for in the $K^-\pi^+\pi^+$ and $K^-\pi^+\pi^+\pi^-$ modes. The signal in the $K\pi\pi$ mode is shown in Fig. 3. The inclusive spectra after continuum subtraction are shown as a function of x in Fig. 4. Charmed mesons produced in two-body decays in conjunction with a light meson have momenta close to the kinematic limit (x=0.5).

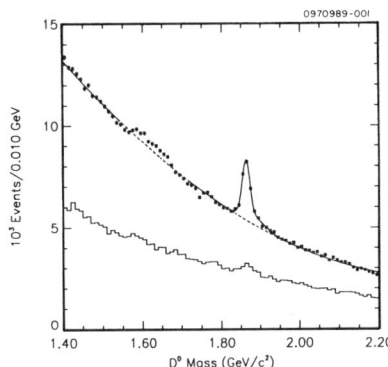
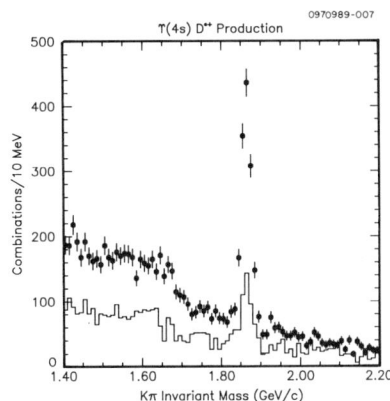

Fig. 1 $K^-\pi^+$ mass distribution showing D°

Fig. 2 $K^-\pi^+$ mass distribution showing D^{*+}

The D_s^- are found in the decay modes $\phi\pi^-$, $K^{*\circ}K^-$, $K^{*-}K^{*\circ}$, $K_s K^-$, $K^{*-}K_s$, or $\phi\pi^-\pi^-\pi^+$. The $\phi\pi^-$ mass spectrum is shown in Fig. 5, and the continuum subtracted x distribution from this mode is shown in Fig. 6. This spectrum peaks at large x values, possibly indicating a large contribution from two-body B decays into a D and a D_s, for which x should be between 0.25 and 0.36. The inclusive branching ratios for decay into charmed mesons, averaged over the mixture of $\bar{B}^0$ and B^- at the Υ(4S), are given in Table I. The D_s contribution is listed assuming that the $D_s \rightarrow \phi\pi$ branching ratio is 1.5% as given by a recent estimate.[5]

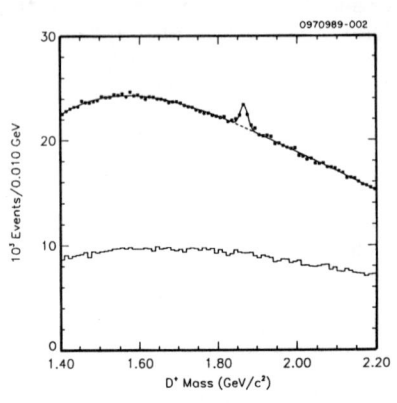

Fig. 3 $K^-\pi^+\pi^+$ mass distribution showing D^+

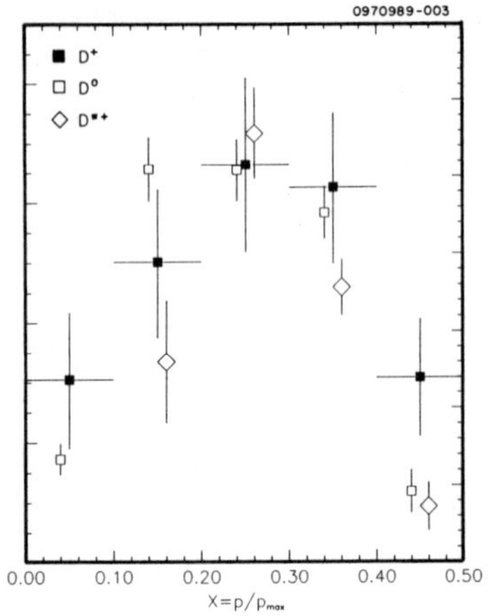

Fig. 4 D^{*+}, D^0 and D^+ momentum spectra; x = p/p_{max}

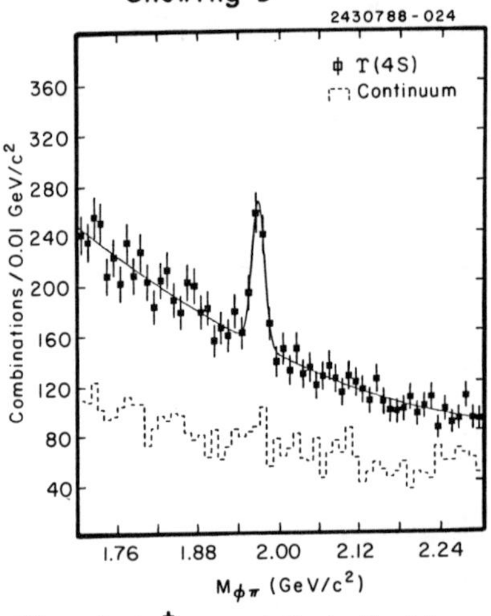

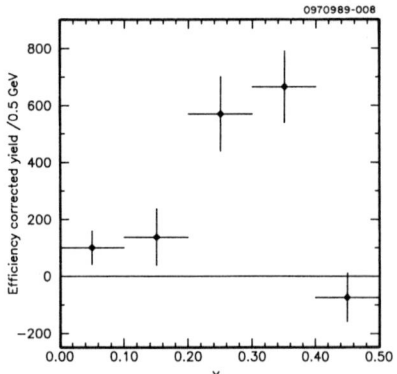

Fig. 5 $\phi\pi^+$ mass distribution showing D_s

Fig. 6 D_s momentum spectrum $x = p/p_{max}$

We have also searched for inclusive ψ and ψ' coming from B decay. A ψ is detected through its decay to electron or muon pairs. The dilepton invariant mass distributions, showing strong ψ signals are shown in Fig. 7. The x distribution for B→ψX decay is shown in Fig. 8. The relative flatness of this spectrum implies that most of the ψ's do not arise from two-body B decays into a ψ and another meson. The shape of the two body contributions ψK and ψK^* are also shown along with the contribution from ψ' to ψ decay that we now will discuss.

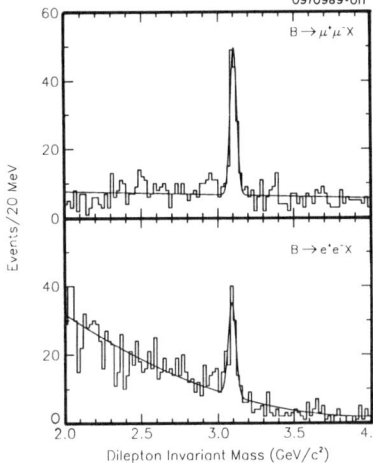

Fig. 7 $\ell^+\ell^-$ mass   Fig. 8 ψ momentum spectrum

We see ψ' decay both through the cascade decay, $\psi' \to \psi\pi^+\pi^-$, and via its direct decay to a lepton pair. To search for the cascade decay we first detect ψ via the dilepton decay and then consider the invariant mass difference between the ψ plus a $\pi^+\pi^-$ pair and the original ψ. We insist that the $\pi^+\pi^-$ invariant mass be greater than 450 MeV, which includes 88% of the $\psi' \to \psi$ decays. The resultant distribution, shown in Fig. 9a, has clear peak at 586.6 MeV, evidence for production of the ψ'. There are 15±4.4 events in the peak. Fig. 9b illustrates the search for the direct decay of ψ' to dileptons. It shows dilepton mass for dilepton momentum less than 1.6 GeV/c, the maximum allowed for ψ' coming from B's. There is a clear peak of 9±4 events at the ψ' mass. The ψ and ψ' branching ratios are given in Table I. Our measurement of $B \to \psi' X$ decays confirms a statistically weaker observation made by ARGUS.[6]

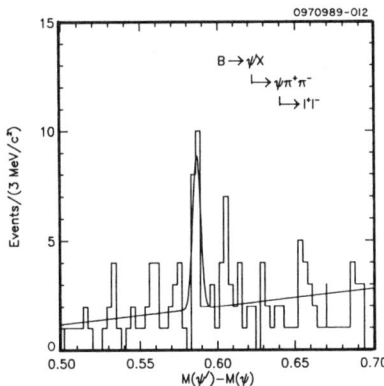 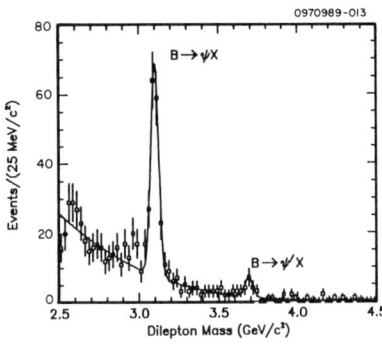

Fig. 9a $\psi\pi\pi$ ψ mass difference   Fig. 9b $\ell^+\ell^-$ mass

We observe the signal for B decays into Λ_c via its $pK^-\pi^+$ decay, as shown in Fig. 10. In Fig. 11. we show the x distribution. It is peaked toward low values of momentum suggesting that the multiplicity in these decays is often high and there is little contribution from two-body decays.

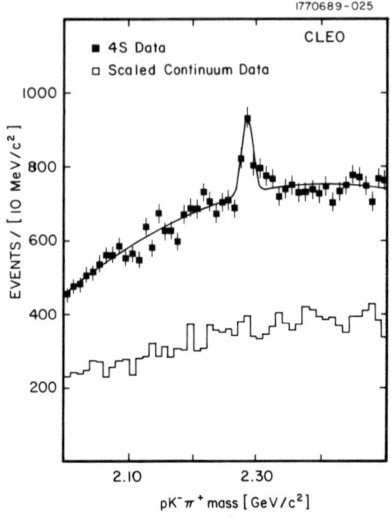

Fig. 10 $pK^-\pi^+$ mass   Fig. 11 Λ_c momentum spectrum

Table I Inclusive Branching Ratios

$B \to D^{*+}X$	$31 \pm 3 \pm 6$ %
$B \to D^0 X$	$52 \pm 2 \pm 8$ %
$B \to D^+ X$	$26 \pm 2 \pm 4$ %
$B \to D_s^+ X$	20 ± 4 % assuming $BR(D_s^+ \to \phi\pi^+) = 1.5$%[6]
$B \to \psi X$	$1.12 \pm 0.10 \pm 0.23$ %
$B \to \psi' X$	$0.33 \pm 0.08 \pm 0.12$ %
$B \to \Lambda_c X$	7.4 ± 2.3 %[7] assuming that all baryons from B decays are either from the Λ_c or its partner baryon.

The first error is the estimate of our combined statistical and systematic uncertainties. The second error, when quoted, is the uncertainty due to the charm or ψ branching ratio. The rates for all open charm modes represent the sum of B (or $\bar{B}$) to charmed and anticharm modes. The cascade decays from the D^{**} are not subtracted from D^0 or D^+ rates. Neither are ψ' cascades subtracted from the inclusive ψ rate. For $B \to D^{**}X$, the momentum spectrum is extrapolated to zero to obtain the rate since the detection efficiency below 500 MeV/c is poor.

To find the fraction of charm per B decay, we sum the D^0, D^+, Λ_c and ψ rates. The ψ rate is multiplied by 4 to take into account the fact that there are two charmed quarks per ψ and that there are unseen χ states. The summed rate is 110±15%, where the error includes the uncertainty on the D branching ratios.[9] This is in agreement with naive expectations of approximately 115%,[8] if the b→c transition accounts for all b-quark decays. The large systematic errors, however, leave considerable room for a b→u component.

B-MESON RECONSTRUCTION

After finding heavy meson candidates as described above, we require their mass to be within two standard deviations (σ) of their known masses.[9] The mass resolution for various particles is listed in Table II. In order to improve the momentum measurement, a kinematic fit is performed using the known particle mass as a constraint. Then the candidate is combined with other tracks in the event to form a B candidate, requiring the difference (ΔE) between the measured energy of the B candidate (E_m) and the beam energy (E_b) to be consistent with zero. The experimental resolution in ΔE varies between 15 and 25 MeV for most of the decay modes, as listed in Table II. It is 125 MeV for $\bar{B}^0 \to D^{*+}\rho^-$. We require $|\Delta E|$ to be less than twice the resolution. The tight ΔE cut excludes B candidates which are genuine B decays but have an additional particle which is missed in the analysis. It also is sensitive enough to reject candidates containing tracks which have been given wrong mass assignments.

For particle combinations which meet these requirements, we compute the B candidate mass from the relation

$$M^2 = E_b^2 - \vec{p}_b^{\,2}, \qquad (1)$$

where $\vec{p}_b^{\,2} = [\Sigma \vec{p}_i]^2$ and $\vec{p}_i$ is the 3-momentum vector of the i-th daughter particle. The technique of using E_b instead of E_m improves the resolution (σ_M) in M by an order of magnitude.

The dependence of the resolution, σ_M, on the measured B momentum $\vec{p}_b$ can be found by differentiating equation (1). The result is:

σ_M (due to error in $\vec{p}_b$) = $(|\vec{p}_b|/M)$ (error in measurement of $\vec{p}_b$).

Since ($|\vec{p}_b|/M$) is only 0.06, the resulting contribution to σ_M is actually smaller than the contribution to the error from the spread in beam energy. The beam energy uncertainty caused by synchrotron radiation amounts to 2.0 MeV. The net resolution σ_M is 2.6 MeV. These procedures are similar to those we have used previously.[3]

Other selection criteria are applied to reduce the remaining background from both continuum and $B\bar{B}$ sources. Because the B has spin 0, the distribution of cosine of the angle (θ_B) between the B direction and the beam axis is proportional to $\sin^2\theta_B$. Since the background distribution is flat in this variable, rejecting candidates with $|\cos\theta_B|>0.8$ eliminates 20% of the background but only 5% of the signal. (This cut is not used in the DD_s or $D^{*+}\rho^-$ analyses.) In cases where it is necessary to reduce background from continuum events, the event shape was examined. Because the B's are produced nearly at rest, the decay products of the two B's have very little angular correlation. Therefore, the distribution of the angle (θ_{sp}) between the sphericity axes calculated from charged tracks forming the B candidate and from those in the rest of the event should be isotropic for real B candidates. The continuum background events fall mostly near $\cos\theta_{sp} = \pm 1$. For example, eliminating events with $|\cos\theta_{sp}|>0.8$ reduces background substantially while losing 20% of the signal. The actual value of the cut is varied from mode to mode.

When a spin 0 particle decays into spin 1 and spin 0 particles, such as $B \to \psi K$ or $D^*\pi$, the helicity of the spin 1 particle is 0. If one defines θ_d to be the angle between the helicity axis and the direction of the decay product of the vector particle, the decay rate is proportional to $\cos^2\theta_d$ if the daughters have spin 0, and $\sin^2\theta_d$ if the vector particle decays electromagnetically into lepton-antilepton pairs. For the ψK case we require $|\cos\theta_d|<0.85$.

Table IIa - Experimental Parameters in Exclusive B⁻ Reconstruction

B⁻ Decay Mode	Charm Decay Mode	Energy Difference Resolution (MeV)	ψ/D Mass Resolution (MeV)	Efficiency*	Number of Events	Branching Ratio (%)
$D^0\pi^-$	$D^0 \to K^-\pi^+$	21	11.0	0.35	11	0.30 ± 0.10
	$D^0 \to K^-\pi^+\pi^+\pi^-$	25	8.2	0.25	16	0.30 ± 0.08
$D^{*+}\pi^-\pi^-$	$D^{*+} \to D^0\pi^+$		0.7			
	$D^0 \to K^-\pi^+$	23	11.0	0.19	1.5 ± 1.7	< 0.4
	$D^0 \to K^-\pi^+\pi^+\pi^-$	22	8.2	0.11	1.8 ± 2.1	
ψK^-	$\psi \to \mu^+\mu^-$ or e^+e^-	20	25	0.41	11	0.08 ± 0.02
ψK^{*-}	$\psi \to \mu^+\mu^-$ or e^+e^-	20	25	0.047	2	0.13 ± 0.09
$\psi K^-\pi^+\pi^-$	$\psi \to \mu^+\mu^-$ or e^+e^-	20	25	0.14	6 ± 3	0.12 ± 0.06
$\psi' K^-$	$\psi' \to \mu^+\mu^-$ or e^+e^-	20	26	0.53	0	< 0.1
	$\psi' \to \psi\pi^+\pi^-$		3			
	$\quad\hookrightarrow \mu^+\mu^-$ or e^+e^-	20	25	0.23	0	< 0.09
$\psi' K^{*-}$	$\psi' \to \mu^+\mu^-$ or e^+e^-	15	26	0.075	0	< 0.7
	$\psi' \to \psi\pi^+\pi^-$		3			
	$\quad\hookrightarrow \mu^+\mu^-$ or e^+e^-	15	25	0.03	0	< 0.7
$D^0 D_s^-$	$D^0 \to \begin{cases} K^-\pi^+ \\ \bar{K}^0\pi^+\pi^- \end{cases}$	22	11	0.2	5	3.7 ± 1.9
	$D_s^- \to \begin{cases} \phi\pi^- \\ \phi\pi^-\pi^+\pi^- \\ K^{*0}K^{*-}, K^{*0}K^- \\ K^0K^{*-}, K^0K^- \end{cases}$		10			

* This includes detector and identification efficiency. For the ψ and ψ' modes, branching ratios of K and K* to observable final states are included.

Table IIb - Experimental Parameters in Exclusive B^0 Reconstruction

B^0 Decay Mode	Charm Decay Mode	Energy Difference Resolution (MeV)	ψ/D Mass Resolution (MeV)	Efficiency*	Number of Events	Branching Ratio (%)
$D^+\pi^-$	$D^+\to \bar{K}^0\pi^+$	26	10	0.16	3	0.69 ± 0.41
	$D^+\to K^-\pi^+\pi^+$	25	10	0.28	12	0.24 ± 0.08
$D^{*+}\pi^-$	$D^{*+}\to D^0\pi^+$		0.7			
	$D^0\to K^-\pi^+$	27	11.0	0.44	7 ± 3	0.29 ± 0.14
	$D^0\to K^-\pi^+\pi^+\pi^-$	24	8.2	0.26	9	0.30 ± 0.11
$D^{*+}\rho^-$	$D^{*+}\to D^0\pi^+$		0.7			
	$D^0\to K^-\pi^+$	23	11.0	0.024	2	$1.9\pm 0.9\pm 1.3$
	$D^0\to K^-\pi^+\pi^+\pi^-$	22	8.2	0.016	4	
$D^{*+}\pi^-\pi^+\pi^-$	$D^{*+}\to D^0\pi^+$		0.7			
	$D^0\to K^-\pi^+$	20	11.0	0.11	9	1.7 ± 0.7
	$D^0\to K^-\pi^+\pi^+\pi^-$	19	8.2	0.076	11 ± 4	1.4 ± 0.6
$\psi \bar{K}^0$	$\psi\to\mu^+\mu^-$ or e^+e^-	20	25	0.15	3	0.06 ± 0.03
$\psi \bar{K}^{*0}$	$\psi\to\mu^+\mu^-$ or e^+e^-	15	25	0.2	7 ± 3	0.11 ± 0.05
$\psi K^-\pi^+$	$\psi\to\mu^+\mu^-$ or e^+e^-	20	25	0.2	7 ± 3	0.10 ± 0.04
$\psi'\bar{K}^0$	$\psi'\to\mu^+\mu^-$ or e^+e^-	15	26	0.18	0	<0.3
	$\psi'\to\psi\pi^+\pi^-$		3			
	$\hookrightarrow\mu^+\mu^-$ or e^+e^-	15	25	0.07	0	<0.3
$\psi'\bar{K}^{*0}$	$\psi'\to\mu^+\mu^-$ or e^+e^-	15	26	0.25	2	0.19 ± 0.13
	$\psi'\to\psi\pi^+\pi^-$		3			
	$\hookrightarrow\mu^+\mu^-$ or e^+e^-	15	25	0.10	1	0.09 ± 0.09
$D^+D_S^-$	$D^+\to\begin{cases}\bar{K}^0\pi^+\\ K^-\pi^+\pi^+\end{cases}$	20	10	0.2^*	4	2.1 ± 1.1
	$D_S^-\to\begin{cases}\phi\pi^-\\ \phi\pi^-\pi^+\pi^-\\ K^{*0}K^{*-},K^{*0}K^-\\ \bar{K}^0K^{*-},\bar{K}^0K^-\end{cases}$		10			
$D^{*+}D_S^-$	$D^{*+}\to D^0\pi^+$		0.7			
	$D^0\to\begin{cases}K^-\pi^+\\ K^-\pi^+\pi^+\pi^-\\ \bar{K}^0\pi^+\pi^-\end{cases}$	20	8-11	0.2^*	3	3.2 ± 1.7
	$D_S^-\to\begin{cases}\phi\pi^-\\ \phi\pi^-\pi^+\pi^-\\ K^{*0}K^{*-},K^{*0}K^-\\ \bar{K}^0K^{*-},\bar{K}^0K^-\end{cases}$		10			

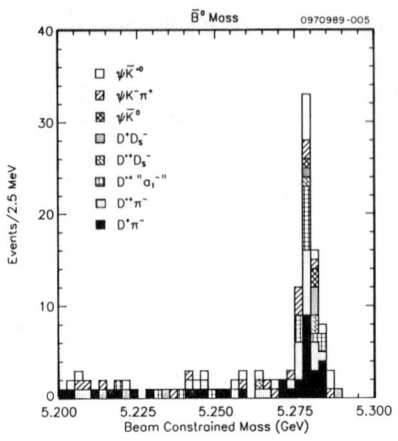 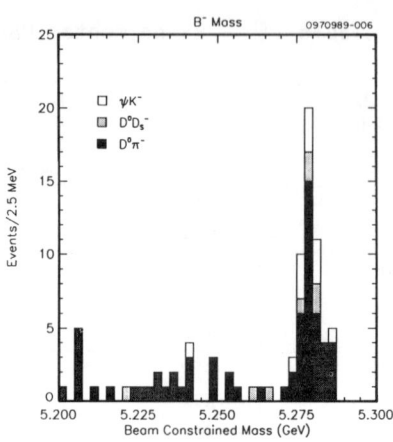

Fig. 12 mass distribution for B^0 and B^- candidates

To obtain the $\bar{B}^0$ and B^- meson masses, we fit the B mass distributions in Fig. 12, for all B candidates in modes containing only charged particles. The entry labeled $D^{*+}"a_1^-"$, which is included in the plot, is a subset of $D^{*+}\pi^+\pi^-\pi^-$ candidates where the 3 pion invariant mass is required to be consistent within ±300 MeV of the a_1 mass of 1260 MeV and one of the $\pi^+\pi^-$ mass combinations is within ±115 MeV of the ρ^0 mass. We fit these distributions with a Gaussian signal with the expected r.m.s. resolution of 2.6 MeV and a flat background. The resulting $\bar{B}^0$ mass is 5279.3±0.4 MeV and B^- mass, 5278.9±0.4 MeV, after correcting for the effect of initial state radiation. The dominant systematic error in these measurements is due to the 5 MeV uncertainty in the CESR beam energy. However, the difference between the two masses of 0.4±0.6 MeV has very small systematic error since the beam energy error cancels out. This mass difference is consistent with our previous measurement of 2.0±1.1 MeV and the ARGUS measurement of 1.9±1.6 MeV.[3] Averaging the two CLEO results, we obtain a mass difference of 0.8±0.5 MeV. This mass difference is consistent with several theoretical predictions.[10]

To calculate branching ratios it is important to know what fraction of the $\Upsilon(4S)$ decays are to charged or neutral B's. Formerly, we assumed a p^3 dependence of the phase space for $\Upsilon(4S) \to B\bar{B}$. Using the measured mass difference, we inferred that the production ratio B^+/B^0 is 57/43. However, the phase space argument has been contested.[11] In light of the new measurement and theoretical uncertainty, we assume this ratio to be 50/50 throughout this paper.

By fitting the mass distribution of B candidates in each decay mode with a flat background and a Gaussian at the above-measured mass and with a standard deviation of 2.6 MeV, we obtain the number of B decays in that mode. The efficiency for detecting a particular decay mode is calculated using a Monte Carlo simulation of the CLEO detector. These efficiencies are listed in Table II. We correct the number of observed events with the detection efficiency and relevant intermediate particle branching ratios to obtain the branching fractions given in Table III[12]

Table III - B Branching Ratios (%)

Mode	CLEO 1987	CLEO 1985[3†]	ARGUS[3†]	Bauer, et al. Model[2]
$B^- \to D^0 \pi^-$	$0.30 \pm 0.06 \pm 0.04$	$0.54 \pm 0.17 \pm 0.11$	$0.21 \pm 0.11 \pm 0.07$	$0.48(a_1 + 0.75a_2)^2$
$B^- \to D^{*+}\pi^-\pi^-$	< 0.4	$0.23 \pm 0.15 \pm 0.07$	$0.7 \pm 0.3 \pm 0.4$	
$B^- \to \psi K^-$	$0.08 \pm 0.02 \pm 0.02$	$0.10 \pm 0.07 \pm 0.2$	0.08 ± 0.04	$1.01 a_2^2$
$B^- \to \psi K^{*-}$	$0.13 \pm 0.09 \pm 0.03$			$4.33 a_2^2$
$B^- \to \psi K^- \pi^+ \pi^-$	$0.12 \pm 0.06 \pm 0.03$		0.12 ± 0.08	
$B^- \to \psi' K^-$	< 0.05		0.24 ± 0.19	
$B^- \to \psi' K^{*-}$	< 0.35			
$B^- \to D^0 D_s^-$	3.7 ± 1.9			$0.73 a_1^2$
$\bar{B}^0 \to D^+ \pi^-$	$0.27 \pm 0.08 \pm 0.05$	$0.51 \pm 0.27 \pm 0.14$	$0.28 \pm 0.12 \pm 0.09$	$0.48 a_1^2$
$\bar{B}^0 \to D^{*+} \pi^-$	$0.33 \pm 0.09 \pm 0.06$	$0.27 \pm 0.13 \pm 0.08$	$0.32 \pm 0.16 \pm 0.12$	$0.37 a_1^2$
$\bar{B}^0 \to D^{*+} \rho^-$	$1.9 \pm 0.9 \pm 1.3$			$1.18 a_1^2$
$\bar{B}^0 \to D^{*+} \pi^- \pi^+ \pi^-$	$1.5 \pm 0.4 \pm 1.0$	< 4.0	$3.9 \pm 1.1 \pm 1.8$	
$\bar{B}^0 \to D^{*+} a_1^-$	$1.8 \pm 0.5 \pm 1.2$			$1.63 a_1^2$
$\bar{B}^0 \to \psi \bar{K}^0$	$0.06 \pm 0.03 \pm 0.02$			$1.02 a_2^2$
$\bar{B}^0 \to \psi \bar{K}^{*0}$	$0.11 \pm 0.05 \pm 0.03$	$0.35 \pm 0.16 \pm 0.03$	0.30 ± 0.16	$4.36 a_2^2$
$\bar{B}^0 \to \psi K^- \pi^+$	$0.10 \pm 0.04 \pm 0.03$			
$\bar{B}^0 \to \psi' \bar{K}^0$	< 0.15			
$\bar{B}^0 \to \psi' \bar{K}^{*0}$	$0.14 \pm 0.08 \pm 0.04$			
$\bar{B}^0 \to D^+ D_s^-$	2.1 ± 1.1			$0.67 a_1^2$
$\bar{B}^0 \to D^{*+} D_s^-$	3.2 ± 1.7			$0.30 a_1^2$

†The ARGUS and previous CLEO results have been renormalized for equal charged and neutral B production on the $\Upsilon(4S)$.

The modes consisting of a D and a D_s have been observed for the first time. The branching fractions listed in Tables II and III are derived assuming that $B(D_s \to \phi \pi)$ equals 1.5%.[6]
For most of the modes the measured branching fractions are consistent with the previous measurements of both CLEO and ARGUS,[3] which are also listed in Table III. The branching fraction for $\bar{B}^0 \to D^{*+} \pi^+ \pi^- \pi^-$ is lower than that of ARGUS in part because the two groups make different assumptions about the background shape.[13] We do not confirm the decay $B^- \to \psi' K^-$ reported by ARGUS. The branching fraction for $\bar{B}^0 \to \psi K^{*0}$ of (0.11 ± 0.05)% is statistically consistent with previous CLEO and ARGUS measurements of (0.36 ± 0.17)% and (0.30 ± 0.16)%, respectively, although it is significantly smaller.
Although we have not established that $D^{*+} \pi^- \pi^+ \pi^-$ production is, in fact, $D^{*+} a_1^-$, the $\pi^+\pi^-\pi^-$ and $\pi^+\pi^-$ mass distributions are consistent

with that hypothesis. We quote a branching ratio on the assumption that all events satisfying the 3π and 2π mass cuts outlined above are indeed a_1's. (We also correct for unseen decay $a_1^{*+} \to \rho^- \pi^o$.)

The $D^{*+}\rho^-$ mode has also not been observed before. The B mass plot is shown in Fig. 13. Since the requirement on ΔE for $\bar{B}^o \to D^{*+}\rho^-$ is not very stringent due to CLEO's poor photon energy resolution, this mode might have background from other B decays, such as $\bar{B}^o \to D^{*+}a_1^-$ where one of the a_1 daughters has been missed. A Monte Carlo simulation showed, however, that effect due to $\bar{B}^o \to D^{*+}a_1^-$ is small since the measured branching fraction for $\bar{B}^o \to D^{*+}a_1^-$ is comparable to that of $\bar{B}^o \to D^{*+}\rho^-$. The large error in the branching ratio results from the dependence of the efficiency on the unknown ρ polarization.

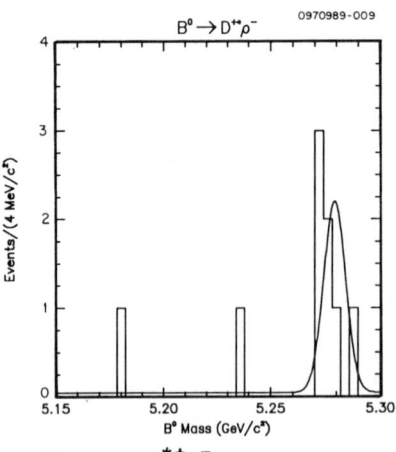
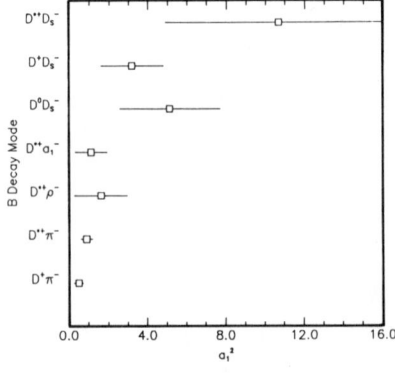

Fig. 13 $D^{*+}\rho^-$ mass   Fig. 14 a_1 from various decay modes

Also given in Table III are predictions by Bauer et al.[2] based on a factorization approach. The parameters a_1 and a_2 are related to external and internal W emission spectator diagrams. Most modes can be described with only one of these numbers, but the $D^o\pi^-$ requires both. ψ's can only be produced via the internal W emission diagram and therefore depend only on a_2. We show in Fig. 14 the a_1^2 values calculated from our data for the modes described only by a_1. The values for the modes involving a D^+ or D^{*+} and a light hadron are in good agreement. Averaging them gives $a_1^2=0.70\pm0.16$.[14] However, there is a large disagreement between these modes and the DD_s modes. This disagreement persists even if we change the $D_s \to \phi\pi$ branching ratio. We observed 12 events in the three modes involving a D_s. However, from the value of a_1 calculated from the other modes, only 1.5 events are expected. If the branching fraction $D_s \to \phi\pi$ is as large as 5%, the upper limit reported by the MARK III experiment,[15] the expected number of events is still only 4, much less than what is observed. We calculated a value of a_2 by averaging all four decay modes involving only a_2. We find that $a_2^2=0.035\pm0.010$. The ratio a_2/a_1 calculated from these branching fractions is 0.22 ± 0.04 and agrees with the expectation that the W internal emission diagram contributes much less than the external diagram.

In conclusion, we have measured the exclusive branching ratios of several decay modes. $\bar{B}^0 \to D^{*+}D_s^-$ and $D^+D_s^-$, and $B^- \to D^0 D_s^-$, as well as $\bar{B}^0 \to D^{*+}\rho^-$ have been observed for the first time. We find that the measured rates for the DD_s modes is somewhat inconsistent with the Bauer et al. prediction. The mass difference between $\bar{B}^0$ and B^- has been measured to be 0.4 ± 0.6 MeV.

We are deeply indebted to the member of CLEO. We also gratefully acknowledge the efforts of the CESR machine group who made this work possible.

REFERENCES

1. B. Gittelman and S. Stone, "B Meson Decay", in High Energy Electron Positron Physics, ed. by Ali and P. Söding, World Scientific, Singapore (1988), p. 275.
2. M. Bauer, B. Stech and M. Wirbel, Z. Phys. C34, 103 (1987); M.J. Savage and M.B. Wise, Phys. Rev. D39, 3346, (1989); D. Fakirov and B. Stech, Nucl. Phys. B133, 315 (1978); J.G. Korner, Proceedings of the International Symposium on Production and Decay of Heavy Hadrons, Heidelberg p. 279 (1987). The first two are applied to B decays.
3. Previous B meson mass and branching ratio measurements are reported in:
   CLEO: C. Bebek, et al., Phys. Rev. D36, 1289 (1987);
   ARGUS: H. Albrecht et al., Phys. Lett. 185B, 218 (1987); K. Schubert, "Review of B-Meson Decay Results", Talk presented at the Heavy Flavor Symposium, Ithaca, NY, USA, June 1989.
4. D. Andrews, et al., Nucl. Instr. and Methods 211, 47 (1983).
5. D. Cassel, et al., Nucl. Instr. and Methods A252, 325 (1986).
6. The derived value for $B(D_s^+ \to \phi\pi^+)$ is $1.5\pm0.8\%$. See S. Stone "Session Summary - Heavy Quark Decay", Cornell Preprint CLNS 89/925 (1989) to appear in Proceedings of XII Int. Workshop on Weak Interactions and Neutrinos, Ginosar, Israel, April 1989.
7. H. Albrecht, et al., Phys. Lett. 199B, 451 (1987).
8. M. S. Alam et al., Phys. Rev. Lett. 59, 22 (1987).
9. Particle Data Group, Phys. Lett. 204B, 1 (1988).
10. C. P. Singh et al., Phys. Rev. D24, 788 (1981); Lai-Him Chan, Phys. Rev. Lett. 51, 253 (1983); K. P. Tiwari et al., Phys. Rev. D31, 642 (1985); D. Y. Kim and S. N. Sinha, Ann. Phys. (N.Y.) 42, 47 (1985).
11. S. Ono, Acta Phys. Pol. B15, 201 (1984).
12. We use $B(D^{*+}\to\pi^+D^0)=57\pm4\pm4\%$ as reported in J. Adler et al., Phys. Lett. 208B, 153 (1988); and the D^0 and D^+ branching ratios reported in Phys. Rev. Lett. 60, 89 (1986). For ψ and ψ' branching ratios we use the values reported by the Particle Data Group, Phys. Lett. 204B, 1 (1988).
13. M.G.D. Gilchriese, Proceedings of XXIII International Conference on High Energy Physics, Berkeley, July 1986.
14. The branching ratio prediction assumes $|V_{cb}|=0.05$ and $\tau_B=1.2$psec.
15. R. Schindler as presented at XII Int. Workshop on Weak Interactions and Neutrinos, Ginosar, Israel, April 1989; to appear in proceedings.

EXPERIMENTAL RESULTS ON CHARM MESONS

REVIEW OF CHARM DECAY RESULTS

Gary Gladding
University of Illinois, Urbana, IL 61801

ABSTRACT. In this paper I review recent experimental results on the decays of the charmed mesons, D^0, D^+, and D_s.

Introduction

Extensive experimental studies of the decays of the D^0 and D^+ mesons have been carried out ever since their discovery[1] in 1976. Most of our theoretical knowledge of the weak decays of charmed mesons is derived from this work since much less is known about the weak decays of the D_s. In this paper I will not attempt to provide an exhaustive overview of charm decay; for that I refer the reader to the recent excellent review article by Morrison and Witherell.[2] Rather, I will provide here a survey of recent experimental results on the decays of the charmed mesons.

The weak decays of the charmed mesons can be subdivided into three distinct types, the leptonic, semileptonic and hadronic, as is displayed in Table I. Generally speaking, the theoretical interpretations of the results of measurements of the leptonic and semileptonic decays are much less ambiguous than those derived from hadronic decay measurements. However, the hadronic decays are usually easier to measure, and as a result we have a much larger sample of experimental measurements to work with to try to disentangle the theoretical complications arising from the strong interaction.

Table I			
Overview of D Decays			
Type	Prize	Theoretical Complications	Experimental Problems
Leptonic (e.g. $D^+ \to \mu^+\nu$)	f_D	none	low rate ($\approx 10^{-4}$) no mass peak need V_{cq}
Semileptonic (e.g. $D^0 \to K^-e^+\nu$)	V_{cq}	form factors	no mass peak quadratic ambiguity
Hadronic (e.g. $D^0 \to K^-\pi^+$)	dynamics (e.g. interference, annihihlation, FSI, etc.)	same as prize !	usual

© 1989 American Institute of Physics

D^0, D^+ decays

a) Leptonic decays

The decay constant f_D is a direct measure of the overlap of the wave functions of the heavy and light quarks in the D meson and can be determined from a measurement of the leptonic decay $D^+ \to \mu^+\nu$ from the expression:

$$B(D^+ \to \mu^+\nu) = \frac{G_F^2}{8\pi} f_D^2 \tau_D m_D m_\mu^2 |V_{cd}|^2 \left(1 - \frac{m_\mu^2}{m_D^2}\right)^2$$

where G_F is the Fermi constant, τ_D the lifetime of the D^+, m_D the D mass, m_μ the muon mass, and V_{cd} the Kobayashi-Maskawa matrix element. Currently, there is no measurement of this branching ratio. The best published limit comes from MarkIII[3] in which they search for the decay in the process:

$$e^+e^- \to \psi'' \to D^+ D^-$$
$$\quad\quad\quad\quad\quad\quad\; \hookrightarrow \mu^+\nu$$
$$\quad\quad\quad\quad\quad\quad\quad \hookrightarrow \text{hadronic tag}$$

The hadronic tag refers to any one of seven fully reconstructed hadronic decay modes of the D^-. A total of $2490 \pm 42 \pm 42$ hadronic tags are observed, but none of these events satisfies the requirements for a $D^+ \to \mu^+\nu$ decay opposite the hadronic tag. Using a likelihood ratio analysis, MarkIII sets a 90% confidence level upper limit on the branching ratio as:

$$B(D^+ \to \mu^+\nu) \leq 7.2 \times 10^{-4}$$

which, assuming $|V_{cd}|^2 = 0.0493$, corresponds to a limit on f_D of:

$$f_D < 290 MeV/c^2.$$

Current theoretical expectations[4] range from 100 to 250 MeV/c^2. It is very important to improve upon this limit to obtain a solid measurement in the coming years in order to provide what will be the benchmark test of lattice QCD calculations.

b) Semileptonic decays

The Kobayashi-Maskawa matrix elements V_{cd} and V_{cs} can most directly be obtained from measurements of the semileptonic decay rates of charmed mesons. For example, V_{cs} can be obtained from measurements of the decay $D^0 \to K^-e^+\nu$ from the expression:

$$B(D^0 \to K^-e^+\nu_e)\Gamma_{D^0} = (G_F^2 M_c^5/192\pi^3)|V_{cs}|^2 \int f_+^K(t) p^3 dt \quad .$$

The main theoretical complication in the extraction of V_{cs} from this expression lies in the evaluation of the form factor, $f_+^K(t)$. MarkIII attempts to minimize this complication by reporting a measurement[5] of the ratio of V_{cd} to V_{cs} derived from their determination of the ratio of the branching fractions $r = \frac{B(D^0 \to \pi^- e^+ \nu)}{B(D^0 \to K^- e^+ \nu)}$. That is, the uncertainty involved in the level of the SU(3) breaking of $f_+(t)$ is thought to be much larger than it is for the ratio of $\frac{f_+^\pi(t)}{f_+^K(t)}$. Figure 1 shows the U ≡ EMISS - PMISS distributions for a) $D^0 \to K^- e^+ \nu$ and b) $D^0 \to \pi^- e^+ \nu$ candidate events. Both plots show the expected peaking around 0, indicating a single missing neutrino. From the shapes of the expected backgrounds also shown in the plots, we see the background levels are low, approximately 1.5 events for $K^- e^+ \nu$ and 0.5 events for $\pi^- e^+ \nu$. The candidates shown (55 $K^- e^+ \nu$ and 7 $\pi^- e^+ \nu$) are found in a sample containing 3636 ± 54 ± 195 observed hadronic tags. From this sample, MarkIII reports the following measured values for the branching ratios:

$$B(D^0 \to K^- e^+ \nu) = 3.4 \pm 0.5 \pm 0.4\%$$
$$B(D^0 \to \pi^- e^+ \nu) = 0.39 ^{+0.23}_{-0.11} \pm 0.04\% \ .$$

It should be noted here that Fermilab experiment E-691[6], a photoproduction experiment featuring a silicon microstrip vertex detector, has also recently measured the $D^0 \to K^- e^+ \nu$ branching fraction and they obtain

$$B(D^0 \to K^- e^+ \nu) = 3.8 \pm 0.5 \pm 0.6\%$$

in good agreement with the MarkIII measurement.

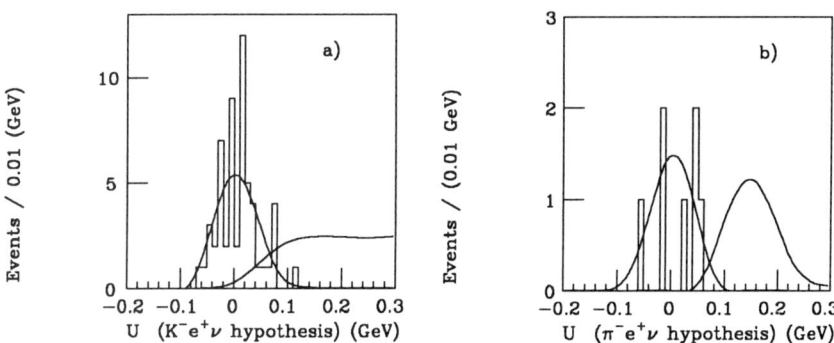

Figure 1. U ≡ EMISS - PMISS Distributions for
a) $K^- e^+ \nu$ and b) $\pi^- e^+ \nu$ events.
The curves to the right in each plot represent the
background shapes from a) $K\pi^0 e\nu$ and b) $Ke\nu$

To extract a determination of $\frac{V_{cd}}{V_{cs}}$ from the ratio of these branching fractions, MarkIII assumes the t dependence of the form factors can be described by a single pole at the D_s^* mass[6,7] and the ratio of the form factors at $t = 0$ to be unity. Specifically, they take:

$$\left|\frac{V_{cd}}{V_{cs}}\right|^2 = \left[\frac{f_+^K(0)}{f_+^\pi(0)}\right]^2 \left[\frac{\int [m_{D_s^*}^2/(m_{D_s^*}^2 - t)]^2 (E_K^2 - m_K^2)^{3/2} dt}{\int [m_{D^*}^2/(m_{D^*}^2 - t)]^2 (E_\pi^2 - m_\pi^2)^{3/2} dt}\right] \left[\frac{B(D^0 \to \pi e \nu)}{B(D^0 \to K e \nu)}\right]$$

$$= \quad (1) \quad \times \quad (0.51) \quad \times \quad \left[\frac{B(D^0 \to \pi e \nu)}{B(D^0 \to K e \nu)}\right]$$

$$= 0.057 \,^{+0.038}_{-0.015} \pm 0.005 \quad .$$

This determination is in good agreement with the theoretical expectation of 0.054, but increased statistics are certainly needed to make a significant test.

Fermilab experiment E-691 has also recently measured[8] the decay $D^+ \to \bar{K}^{*0} e^+ \nu$. Here the interest is not so much in determining V_{cs}, but rather in measuring some of the attendant strong interaction dynamics. In particular, there are now 3 form factors (2 axial vector and 1 vector) which are needed to describe the decay. Figure 2a shows the $K\pi$ mass distribution for their candidate events and demonstrates the dominance of the K^* signal. With "loose" cuts, they see a total of 318 events with the "right" sign electrons (i.e. with charge opposite to that of the K in the vertex) and a total of 66 with the "wrong" sign electrons. Requiring tighter cuts on the particle identification and on the vertex fit results in a reduction in the amount of background by a factor of five, while the signal is reduced by the expected factor of two. Normalizing their result to $B(D^+ \to K^- \pi^+ \pi^+)$ as measured by MarkIII[9], they obtain

$$B(D^+ \to \bar{K}^{*0} e^+ \nu) = 4.5 \pm 0.7 \pm 0.5\%$$

$$B(D^+ \to (K^- \pi^+)_{NR} e^+ \nu) = 0.3 \pm 0.2 \pm 0.2\% \quad .$$

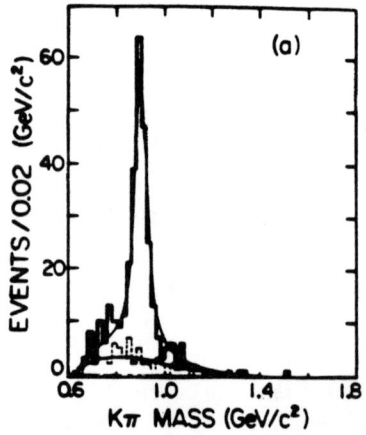
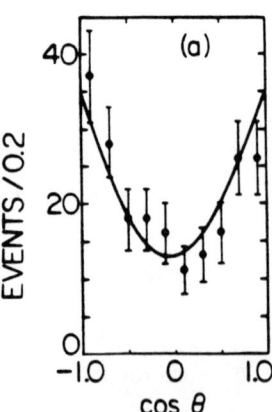

Figure 2. a) $K\pi$ invariant mass distribution b) $\cos\theta$ distribution from E-691: $D^+ \to K^- \pi^+ e^+ \nu$

This measurement can be combined with the previous measurement of $B(D^0 \to K^- e^+ \nu)$ and the measured lifetimes of the D^0 and the D^+ to obtain:

$$\frac{\Gamma(D^+ \to \bar{K}^{*0} e^+ \nu)}{\Gamma(D^0 \to K^- e^+ \nu)} = 0.45 \pm 0.09 \pm 0.07$$

which is about half of the theoretically expected value.

The polarization of the K^* can be determined from the $\cos\theta$ distribution, where θ is defined to be the angle between the π^+ and the D^+ direction in the K^* rest frame. In particular, they fit the angular distribution $W(\theta) = 1 + \alpha \cos^2(\theta)$ (shown in Figure 2b) for the parameter α, which then can be related to the ratio of the longitudinal to transverse polarization as

$$\frac{\Gamma_L}{\Gamma_T} = \frac{1+\alpha}{2} \quad .$$

They then obtain

$$\frac{\Gamma_L}{\Gamma_T} = 2.4 \, {}^{+1.7}_{-0.9} \pm 0.2 \quad .$$

This result is also difficult to accomodate within the usual models.

c) Hadronic decays

The simplest picture for the decays of the charmed mesons is the dominance of the spectator diagrams in which the valence c quark emits a virtual W which in turn produces either a $u\bar{d}$ or a $l\nu$ pair. In this picture, one expects the lifetimes of all charmed mesons to be equal. However, E-691 has measured[10]

$$\frac{\tau_{D^+}}{\tau_{D^0}} = 2.51 \pm 0.10 \quad .$$

Furthermore, MarkIII has measured[11]

$$\frac{B(D^+ \to e^+ X)}{B(D^0 \to e^+ X)} = 2.3 \, {}^{+0.5}_{-0.4} \pm 0.1$$

which demonstrates that this picture indeed is too simple and that the source of the lifetime difference lies in the hadronic decays. If we add to this picture "internal W emission" diagrams, where the virtual W produces a $u\bar{d}$ pair which then recombines with remaining quarks, we see that only in the case of the D^+ can this diagram lead to the same final state as for the spectator diagram and therefore give rise to possible interference. Indeed, this destructive interference is now generally held to be the dominant source of the difference in the D^+ and D^0 lifetimes. Unfortunately, the story doesn't stop here. Many decay modes have been measured which have prompted the introduction of other proposed players to this story such as final state interactions, W exchange diagrams, W annihilation diagrams and even penguins. The lesson one learns from this recent history is that "one experimental result cannot the theory determine". Consequently, the experimentalist can only continue to make accurate measurements of a comprehensive set of decays in order to establish a complete pattern which then must be fit into a unified, albeit complicated, theoretical picture.

There are several new measurements which have been reported at this conference. MarkIII has new results on $D^0 \to K^0\pi^0$, $D^0 \to K^0\eta$, and $D^0 \to K^0\eta'$ which are presented in detail by P. Kim in his contribution to these Proceedings[12] as well as a resonant substructure analysis of the $D \to K\pi\pi$ system which is detailed by F. DeJongh in his contribution to these Proceedings[13]. The main surprise in these results is that the Ka_1 component of the $K\pi\pi$ system is dominant, while the vector-vector component ($K^*\rho$) is found to be small.

The CLEO collaboration at CESR has submitted a paper[14] to this conference in which they report their observation of the decay $D^0 \to K^0\bar{K}^0$. The D^0's in this sample come from the data sets taken at the $\Upsilon(4S)$ and the $\Upsilon(5S)$. The K^0's were observed in the mode $K_S^0 \to \pi^+\pi^-$ where the $\pi^+\pi^-$ vertex was required to be well separated from the primary interaction point.

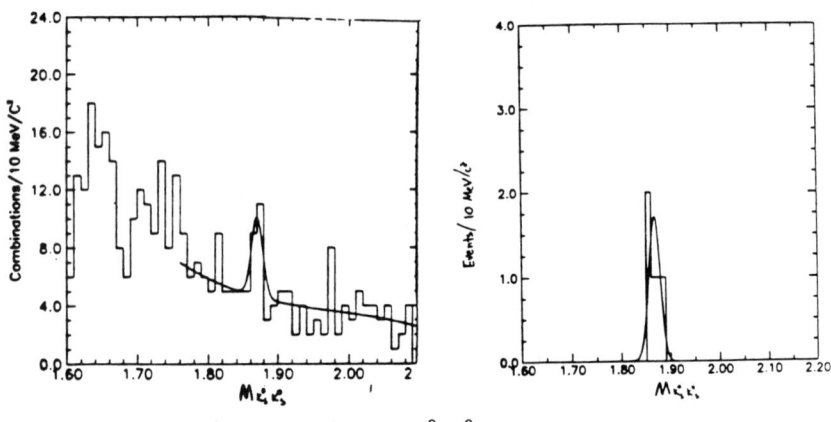

Figure 3. CLEO $K_S^0 K_S^0$ mass distributions
for: a) all events b) events with $D^{*+} \to \pi^+ K_S^0 K_S^0$

Figure 3 shows the $K_S^0 K_S^0$ mass distributions for a) all events and b) those events in which the $K_S^0 K_S^0$ pair can be combined with a π^+ in the event to form a D^{*+} candidate. A total of 12 ± 5 events are assigned to the D^0 signal in the inclusive case (with an estimated background of 0.3 events), while all 5 events in the D^* plot are consistent with the D^0 signal (with an estimated background of < 0.1 events). These signals can then be converted into an absolute branching fraction by first internally normalizing to their observation of $D^0 \to K_S^0\pi^+\pi^-$ and then multiplying by the MarkIII absolute measurement[9] of $D^0 \to \bar{K}^0\pi^+\pi^-$ as

$$B(D^0 \to K^0\bar{K}^0) = \frac{B(D^0 \to K_S^0 K_S^0)}{B(D^0 \to K_S^0\pi^+\pi^-)} B(D^0 \to \bar{K}^0\pi^+\pi^-) \ .$$

The CLEO results are then presented as:

$B(D^0 \to K^0 \bar{K}^0) = 0.06 \pm 0.03 \pm 0.01\%$ $(CLEO : inclusive)$

$B(D^0 \to K^0 \bar{K}^0) = 0.11 ^{+0.06}_{-0.04} \pm 0.02\%$ $(CLEO : requiring\ D^*)$.

These results then confirm the earlier observation by Fermilab experiment E-400[15] of :

$$B(D^0 \to K^0 \bar{K}^0) = 0.10 \pm 0.08\% \quad (E-400)$$

The significance of these results is somewhat unclear. In the absence of SU(3) flavor breaking, we can attribute this decay rate to final state interactions. It should be noted, however, that the decay $D^0 \to \pi^0\pi^0$ is the only KK or $\pi\pi$ mode which has not yet been observed. Once this decay is measured, we will have a complete set of KK and $\pi\pi$ modes which should allow the unraveling of the contributions from the non-spectator processes from those of the final state interactions.

D_s decays

The study of D_s decays is at an earlier stage of development than that of the D^0 and D^+ decays. However, since the D_s is distinguished from the D^0 and D^+ only by the flavor of the quark accompanying the charm quark, we will adopt the same framework for the discussion of decay mechanisms. Indeed, the D_s offers some new features, such as W annihilation diagrams which are Cabibbo-favored rather than Cabibbo-suppressed.

a) Leptonic decays

There is no published measurement or limit on the leptonic decays of the D_s. However, we can look forward to the interesting comparison that will be able to be made once measurements of both $D_s \to \tau\nu$ and $D_s \to \mu\nu$ are available.

b) Semileptonic decays

The only information to date concerning semileptonic decays of the D_s is the limit from MarkIII[16]

$$B(D_s^+ \to e^+ X) = 9 ^{+9}_{-7}\% \quad < 25\% \quad at\ 90\%CL.$$

If destructive interference is really the dominant cause of the D^+-D^0 lifetime difference, we would expect the semileptonic decay of the D_s to be more like that of the D^0 than that of the D^+, since destructive interference is impossible in D_s decays. In fact, since the lifetime of the D_s has been measured[10], we can convert the upper limit on the D_s inclusive semileptonic branching fraction to a lower limit on the hadronic width of the D_s with the result that it is almost twice as large as the hadronic width of the D^+[17], lending support to the hypothesis that the D^+-D^0 lifetime difference is largely due to destructive interference in D^+ decays.

b) Hadronic decays

Since the D_s has a valence s quark, we expect its decays to usually include s quark(s) in the final state if spectator processes are dominant. In fact, the large majority of currently observed D_s decay modes have kaons in the final state. In Table II, I list the current measurements of D_s decays to final states containing kaons[18].

Table II		
Branching fractions for D_s modes relative to $\phi\pi$		
[modes with kaons in the final state (including ϕ)]		
Decay Mode	Experiment	Result or Limit
$D_s \to \bar{K}^0 K^+$	MarkIII	$0.92 \pm 0.32 \pm 0.20$
	CLEO	$0.99 \pm 0.17 \pm 0.06$
$D_s \to \bar{K}^{*+} K^0$	CLEO	$1.2 \pm 0.21 \pm 0.13$
$D_s \to \bar{K}^0 \pi^+$	MarkIII	< 0.21 at 90% CL
$D_s \to \bar{K}^{*0} K^+$	E691	$0.87 \pm 0.13 \pm 0.05$
	ARGUS	1.44 ± 0.37
	MarkIII	$0.84 \pm 0.30 \pm 0.22$
	CLEO	$1.05 \pm 0.17 \pm 0.06$
$D_s \to \bar{K}^{*0} K^{*+}$	NA32	2.3 ± 1.2
$D_s \to \phi\pi^+\pi^0$	E691	$2.4 \pm 1.0 \pm 0.5$
	NA14	< 2.6 at 90% CL
$D_s \to (K^- K^+ \pi^+)_{NR}$	E691	$0.25 \pm .07 \pm .05$
	NA32	0.96 ± 0.32
$D_s \to \phi\pi^-\pi^+\pi^+$	E691	$0.42 \pm 0.13 \pm .07$
	NA32	0.39 ± 0.17
	Argus(a)	$1.11 \pm 0.37 \pm 0.28$
	Argus(b)	$0.41 \pm 0.13 \pm 0.11$
$D_s \to (K^- K^+ \pi^+ \pi^0)_{NR}$	E691	< 2.4 at 90% CL
$D_s \to (K^- K^+ \pi^- \pi^+ \pi^+)_{NR}$	E691	$< .32$ at 90% CL
	NA32	0.11 ± 0.07

The measurements shown in the table are in reasonable agreement with each other, but no mode has been measured particularly well yet. The most accurate measurements ($\approx 20\%$) come from CLEO in a paper[19] submitted to this conference. They measure 4 decay modes, including the first observation of

$D_s^+ \to K^{*+}K_S^0$. Also worthy of note from the table is the fact that E-691 observes $D_s \to \phi\pi^+\pi^0$ but sees no significant signal in the vector-vector component ($\phi\rho$), much as in the study of the $K\pi\pi\pi$ system in D decays.

The branching fractions in Table II are all given relative to that for $D_s^+ \to \phi\pi^+$ since there is currently no absolute measurement of D_s branching fractions. We expect $\approx 85\%$ of the D_s decays to be hadronic; the sum of all modes in Table II accounts for a total of $\approx 9\times$ BR($D_s \to \phi\pi$). To determine whether the modes listed in Table II contain most of the D_s hadronic modes we need to establish an absolute scale for $D_s \to \phi\pi$.

One indirect method for determining the $\phi\pi$ absolute branching fraction is to take the total D_s production at a fixed e^+e^- center of mass energy to be the difference between the theoretical prediction for the total charm production cross-section and the measured production of the known charmed particles (D^0, D^+, and Λ_c) at that energy. S. Stone[20] has recently performed such a calculation using the average of the CLEO and ARGUS measurements for $D^0 \to K\pi$, $D^+ \to K\pi\pi$ and $\Lambda_c \to pK\pi$ production at 10.5 GeV along with the MarkIII measurements of D^0 and D^+ branching fractions and preliminary determinations from CLEO and ARGUS of the Λ_c branching fraction to obtain:

$$B(D_s \to \phi\pi) = 1.5 \pm 0.8\% \ .$$

In principle, a direct determination of $B(D_s \to \phi\pi)$ can be made using fully reconstructed events at a fixed e^+e^- center of mass energy just as was done to obtain absolute D^0 and D^+ hadronic branching fractions[9]. MarkIII has recently performed such an analysis[21] by searching for fully reconstructed events from the reaction $e^+e^- \to D_s^{*+}D_s^-$, $D_s^{*+} \to \gamma D_s^+$ in a sample of $6.30 pb^{-1}$ taken at $\sqrt{s} =$ 4.14 GeV. A total of 28 final states are searched for containing the following D_s^+ modes: $\phi\pi^+, \bar{K}^0K^+, f_0(975)\pi^+, \bar{K}^*(892)^0K^+, \bar{K}^{*0}K^{*+}, \phi\pi^+\pi^+\pi^-$, and $\phi\pi^+\pi^0$. Figure 4 shows the mass distribution from this analysis. No events are found which lie in the signal region between 1.85 and 2.05 GeV/c^2.

An upper limit on $B(D_s \to \phi\pi)$ is obtained using a marginal likelihood technique which takes as input the measured relative branching fractions and correlations between measurements. This procedure results in an upper limit on $B(D_s \to \phi\pi)$ of 3.8%. This limit is then increased to account for systematic uncertainties to give the final result:

$$B(D_s \to \phi\pi) < 4.1\% \ at \ 90\% CL$$

Given these results on the absolute branching fraction of $D_s \to \phi\pi$, we are forced to conclude that the majority of the hadronic decay modes are not included in Table II. We therefore now turn to search for the missing D_s hadronic decays in modes not containing kaons in the final state. Table III gives a summary of the current experimental information on these modes.

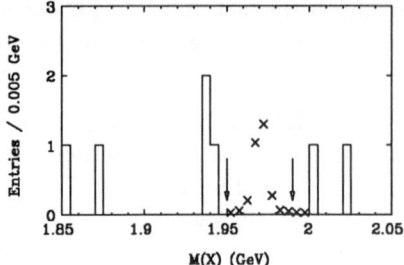

Figure 4. Mass distribution for candidates from the 28 final states. The points plotted with ×'s indicate the expected signal for $B(D_s \to \phi\pi) = 4.1\%$.

Table III		
Branching fractions for D_s modes relative to $\phi\pi$		
[modes with no kaons in the final state]		
Decay Mode	Experiment	Result or Limit
$D_s \to \rho\pi^+$	E691	< 0.08 at 90% C.L.
	Argus	< 0.22 at 90% C.L.
$D_s \to f_0(975)\pi^+$	E691	$0.28 \pm 0.1 \pm .03$
$D_s \to \omega\pi^+$	E691	< 0.5 at 90% CL
	E564	seen
$D_s \to (\pi^-\pi^+\pi^+)_{NR}$	E691	$0.29 \pm .09 \pm .03$
$D_s \to (\pi^-\pi^+\pi^+\pi^-\pi^+)_{NR}$	E691	< .29 at 90% CL
$D_s \to \eta\pi^+$	E691	< 1.5 at 90 % CL
	MarkII	3.0 ± 1.1*
	MarkIII	< 2.5 at 90% CL
$D_s \to \eta'\pi^+$	MarkII	4.8 ± 2.1*
	NA14	$6.9 \pm 2.4 \pm 1.4$
	MarkIII	< 1.9 at 90% CL
* assumes $B(D_s \to \phi\pi) = 4\%$		

D_s decays via the W annihilation diagram are Cabibbo-favored and should usually lead to final states containing no strange quarks. Searches for such decays have been performed for the $D_s^+ \to \pi^+\pi^-\pi^+$ and the $D_s^+ \to \pi^+\pi^-\pi^+\pi^0$ final states. In particular, the decay $D_s^+ \to \rho^0\pi^+$ has been studied by both ARGUS[22] and E-691[23]. Neither experiment sees a signal, and they quote limits relative to $D_s^+ \to \phi\pi^+$ of < 0.22 (ARGUS) and < 0.08 (E-691). E-691 has also searched for the decay $D_s \to \omega\pi$[24] and see no evidence for it at the level of $\frac{1}{2}$ the rate for $D_s \to \phi\pi$. Consequently, there is currently no evidence for D_s decay modes which can easily be associated with the W annihilation diagram.

Another possible source of the missing D_s decay modes has recently been suggested, namely modes involving η's or η''s. Unfortunately, there is currently real experimental disagreement in this area. We turn first to searches for the decay $D_s^+ \to \eta\pi^+$. Mark II has reported[25] seeing a signal in the reaction $e^+e^- \to D_s^+X, D_s^+ \to \pi^+\eta, \eta \to \gamma\gamma$. They make several requirements on the photons from the η, most notably that $\frac{E_{\gamma\gamma}}{E_{beam}} > 0.3$ and that neither η photon can be combined with other photons in the event to make a π^0, to obtain a total of 16 ± 6 $D_s^+ \to \eta\pi^+$ candidates. They quote a value for $\sigma_{D_s} B(D_s^+ \to \eta\pi^+) = 5.2 \pm 2.2$ pb. They cannot directly relate this measurement to the $\phi\pi$ branching fraction since they do not observe $D_s \to \phi\pi$. HRS[26] has reported a signal for $D_s \to \phi\pi$ at the same energy, but do not quote a value for $\sigma_{D_s} B(D_s \to \phi\pi)$ which could be used to provide the normalization of the MarkII result to the $\phi\pi$ rate. Therefore, MarkII chose to normalize to the $\phi\pi$ rate by making the standard theoretical assumption about the level of the D_s production which then translates to an absolute determination of $B(D_s^+ \to \eta\pi^+) \approx 12\%$. They then quote a number for $B(D_s^+ \to \eta\pi^+) \approx 3 \times B(D_s^+ \to \phi\pi^+)$, where they assumed the then current value of 4% for the $\phi\pi$ branching fraction. Note that current estimates[20] place the $\phi\pi$ branching fraction closer to 2%.

This result is, however, not confirmed by both E-691[24] and MarkIII[12]. E-691 searches for the $\eta\pi$ signal in which $\eta \to \pi^+\pi^-\pi^0$. They use two methods: (i) to fully reconstruct the $\pi^+\pi^-\pi^+\pi^0$ final state and (ii) to look at the shape of the $\pi^+\pi^-\pi^+$ mass distribution without requiring π^0 reconstruction. An order of magnitude increase in acceptance is gained in method (ii) at the cost of significant deterioration in the resolution of the signal. They quote the following results:

$$\frac{B(D_s^+ \to \eta\pi^+)}{B(D_s^+ \to \phi\pi^+)} = 1.5\ {}^{+1.1}_{-0.8} \pm 0.5\ < 3.2 \qquad \text{E}-691: \text{method (i)}$$

$$\frac{B(D_s^+ \to \eta\pi^+)}{B(D_s^+ \to \phi\pi^+)} = 0.0 \pm 0.5 \pm 0.5 \qquad \text{E}-691: \text{method (ii)}$$

They combine these measurements to give their final result as the limit

$$B(D_s^+ \to \eta\pi^+) < 1.5 \times B(D_s^+ \to \phi\pi^+)\ .$$

MarkIII reports at this conference searches for the process $D_s^+ \to \eta\pi^+$ in which $\eta \to \pi^+\pi^-\pi^0$ and in which $\eta \to \gamma\gamma$. Details of the analysis are given by P. Kim in his contribution to these Proceedings[12]. Figure 5 shows the $\eta\pi$ mass distributions for both methods.

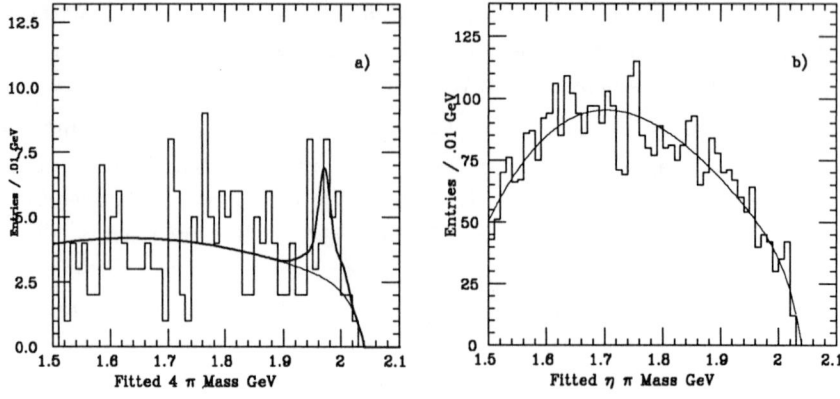

Figure 5. MarkIII $\eta\pi$ mass distributions
for: a) $\eta \to \pi^+\pi^-\pi^0$ b) $\eta \to \gamma\gamma$

No demonstrable signal is seen in either case with the results of fits given as:

$$\frac{B(D_s^+ \to \eta\pi^+)}{B(D_s^+ \to \phi\pi^+)} = 1.7 \pm 0.7 \pm 0.5 \qquad \text{MarkIII}: (\eta \to \pi^+\pi^-\pi^0)$$

$$\frac{B(D_s^+ \to \eta\pi^+)}{B(D_s^+ \to \phi\pi^+)} < 1.6 \qquad \text{MarkIII}: (\eta \to \gamma\gamma)$$

These measurements are then combined to give the final result as the limit

$$B(D_s^+ \to \eta\pi^+) < 2.5 \times B(D_s^+ \to \phi\pi^+) \ .$$

We now turn to searches for the decay $D_s^+ \to \eta'\pi^+$. Mark II has also reported a signal in this mode[25] from the reaction $e^+e^- \to D_s^+X, D_s^+ \to \pi^+\eta'$, $\eta' \to \pi^+\pi^-\eta, \eta \to \gamma\gamma$. After making similar cuts as in the $\eta\pi$ mode, they report $\sigma_{D_s}B(D_s^+ \to \eta'\pi^+) = 8.4 \pm 3.7$ pb. Normalizing this in the same way as they did for $\eta\pi$, they obtain $B(D_s^+ \to \eta'\pi^+) \approx 19\%$. Once again taking $B(D_s^+ \to \phi\pi^+) \approx 4\%$, they obtain $B(D_s^+ \to \eta'\pi^+) \approx 4.8 \times B(D_s^+ \to \phi\pi^+)$. This observation has recently been confirmed by the NA14$'$ photoproduction experiment[27] at CERN. They observe the signal in the decay mode $D_s^+ \to \pi^+\eta', \eta' \to \gamma\rho^0$. The $\gamma\rho\pi$ mass distribution for this experiment is shown in Figure 6.

They claim a total of 45 ± 11 events in the D_s mass peak which corresponds to :

$$\frac{B(D_s^+ \to \eta'\pi^+)}{B(D_s^+ \to \phi\pi^+)} = 6.9 \pm 2.4 \pm 1.4 \qquad \text{NA14}'$$

where the normalization is made to the observation of $D_s \to \phi\pi$ in the same experiment.

Once again, MarkIII, in a report to this conference[12] fails to confirm this observation of a D_s decay mode involving η's or η''s. They search for the $\eta'\pi$ decay

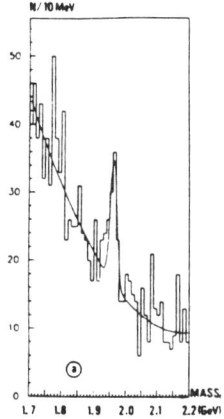

Figure 6. NA14' $\eta'\pi$ mass distribution for $D_s^+ \to \eta'\pi^+$ candidates. $\eta' \to \gamma\rho^0$.

mode in the reaction $e^+e^- \to D_s^{*-}D_s^+, D_s^+ \to \pi^+\eta', \eta \to \gamma\gamma$. A 2 constraint fit is performed in which the $\gamma\gamma$ pair is required to have the η mass, and the recoil mass is constrained to the D_s^* mass. Figure 7 shows the invariant mass distribution of the $\eta'\pi$ candidates.

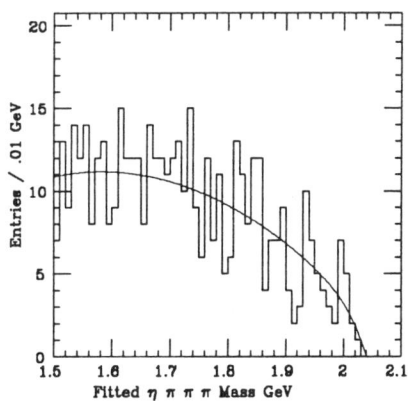

Figure 7. MarkIII $\eta'\pi$ mass distribution for $D_s^+ \to \eta'\pi^+$ candidates. $\eta' \to \pi^+\pi^-\eta$, $\eta \to \gamma\gamma$

No signal is seen; MarkIII then sets the limit:

$$\frac{B(D_s^+ \to \eta'\pi^+)}{B(D_s^+ \to \phi\pi^+)} < 1.9 \quad . \qquad \text{(MarkIII)}$$

Conclusions

The past year has seen several substantial contributions made to our understanding of the decays of the charmed mesons. In the semileptonic decays, MarkIII has made the first direct measurement of $|\frac{V_{cd}}{V_{cs}}|$ which agrees with the theoretical expectation, although improved accuracy is still desirable. E-691 has reported the surprising results that the rate for $D^+ \to K^*e\nu$ is about half of that expected and that the K^* itself is produced with a substantial longitudinal polarization. There are also indications from MarkIII that the semileptonic decay rate of the D_s is more similar to that of the D^0 than to that of the D^+, as expected. In the hadronic decays, the new results from CLEO on the decay $D^0 \to K^0 \bar{K}^0$ indicate that a quantitative understanding of these decays must include contributions from final state interactions. A measurement of $D^0 \to \pi^0\pi^0$ is still needed, however, to complete this picture of the $\pi\pi$ and KK decays of the D^0. From the new results from MarkIII that the $D \to K\pi\pi$ system is dominated by Ka_1 and from the E-691 upper limit on $D_s \to \phi\rho$, we can conclude that the vector-vector decays of charmed mesons are surprisingly small. The current state of understanding of the D_s hadronic decays can only be called confusing. An absolute measurement (e.g. for $B(D_s \to \phi\pi)$) is certainly needed to set the scale of these decays. Nonetheless, from the estimates from charm counting as well as from the direct limit from MarkIII, we can conclude that a large fraction of D_s hadronic decays are currently unaccounted for. There is no evidence at this time that the W annihilation diagram can provide the contributions necessary for the solution of this puzzle. Furthermore, there is a real experimental discrepancy concerning the $\eta\pi$ and the $\eta'\pi$ decay modes of the D_s, in that the large branching fractions reported for these modes by MarkII and NA14' have not been confirmed by MarkIII and E-691. It seems that only time (and continued work) will answer our question: "where have all the D_s decay modes gone?"

REFERENCES

1. G. Goldhaber et al., Phys. Rev. Lett. **37**, 255 (1976)
   I. Peruzzi et al., Phys. Rev. Lett. **37**, 569 (1976)

2. R.J. Morrison and M.S. Witherell, Ann. Rev. Nuc. Sci. **39**, 183 (1989)

3. J. Adler et al., Phys. Rev. Lett. **60**, 1375 (1988)

4. H. Krasemann, Phys. Lett **96B**, 397 (1980)
   E. Golowich, Phys. Lett. **91B**, 271 (1980)
   V. Mathur et al., Phys. Lett. **107B**, 127 (1981)
   T. Aliev et al., Sov. J. Nucl.Phys. **38**, 6 (1983)
   M. Suzuki, Phys.Lett. **142B**, 207 (1984)
   S. Godfrey et al., Phys. Rev. **D32**, 189 (1985)
   S. Godfrey, Phys. Rev. **D33**, 1391 (1986)
   C. Dominguez et al., Phys. Lett. **197B**, 423 (1987)
   L. Reinders, Phys. Rev. **D38**, 947 (1988)
   C. Bernard et al., Phys. Rev. **D38**, 3540 (1988)
   T. DeGrand et al., Phys. Rev. **D38**, 954 (1988)

5. J. Adler et al., Phys. Rev. Lett. **62**, 1821 (1989)

6. J. C. Anjos et al., Phys. Rev. Lett. **62**, 1587 (1989)

7. D. M. Coffman, Ph. D. thesis, California Institute of Technology, 1986 (unpublished)
8. J. C. Anjos *et al.*, Phys. Rev. Lett. **62**, 722 (1989)
9. J. Adler *et al.*, Phys. Rev. Lett. **60**, 89 (1988)
10. J.R. Raab *et al.*, Phys. Rev. **D37**, 2391 (1988)
11. R.M. Baltrusaitas *et al.*, Phys. Rev. Lett. **54**, 1976 (1985)
12. P. Kim in these Proceedings
13. F. DeJongh in these Proceedings
14. J. Alexander *et al.*, CLNS 89/940
15. J.P. Cumalat *et al.*, Phys. Lett. **210B**, 253 (1988)
16. D. Pitman, SLAC-PUB-4826 (1989)
17. R. Schindler, Proceedings of the Workshop on Weak Interactions and Neutrinos, Ginosar, Israel (1989), SLAC-PUB-4997
18. I take the experimentalist's prerogative here to include ϕ modes, since the ϕ is usually observed in its KK decay mode.
19. W.Y. Chen *et al.*, CLNS 89/920, submitted to Phys. Lett.
20. S. Stone, Proceedings of the Workshop on Weak Interactions and Neutrinos, Ginosar, Israel (1989), CLNS 89/925
21. J. Adler *et al.*, SLAC-PUB-5044 (1989), submitted to Phys. Rev. Lett.
22. H. Albrecht *et al.*, Phys. Lett. **195B**, 102(1987)
23. J.C. Anjos *et al.*, Phys. Rev. Lett. **62**, 125(1989)
24. J.C. Anjos *et al.*, Phys. Lett. **223B**, 267 (1989)
25. G. Wormser *et al.*, Phys. Rev. Lett. **61**, 1057 (1988)
26. M. Derrick *et al.*, Phys. Rev. Lett. **54**, 2568 (1985)
27. G. Wormser, Proceedings of XXIVth Rencontres de Moriond (1989), LAL89-10 (1989)

EXCLUSIVE SEMILEPTONIC DECAYS OF CHARMED MESONS

J.C. Anjos,[3] J.A. Appel,[5] A. Bean,[1] S.B. Bracker,[8]
T.E. Browder,[1,a] L.M. Cremaldi,[4,b] J.E. Duboscq,[1] J.R. Elliott,[4,c]
C.O. Escobar,[7] P. Estabrooks,[2] M.C. Gibney,[4] G.F. Hartner,[8]
P.E. Karchin,[9] B.R. Kumar,[8] M.J. Losty,[6] G.J. Luste,[8]
P.M. Mantsch,[5] J.F. Martin,[8] S. McHugh,[1] S.R. Menary,[8]
R.J. Morrison,[1] T. Nash,[5] P. Ong,[8] J. Pinfold,[2] G. Punkar,[1]
M.V. Purohit,[5,d] J.R. Raab,[1,e] A.F.S. Santoro,[3] J.S. Sidhu,[2,†]
K. Sliwa,[5] M.D. Sokoloff,[5,f] M.H.G. Souza,[3] W.J. Spalding,[5]
M.E. Streetman,[5] A.B. Stundžia,[8] M.S. Witherell,[1]

1. University of California, Santa Barbara, California, USA
2. Carleton University, Ottawa, Ontario, Canada
3. Centro Brasileiro de Pesquisas Fisicas, Rio de Janeiro, Brazil
4. University of Colorado, Boulder, Colorado, USA
5. Fermi National Accelerator Laboratory, Batavia, Illinois, USA
6. National Research Council, Ottawa, Ontario, Canada
7. Universidade de São Paulo, São Paulo, Brazil
8. University of Toronto, Toronto, Ontario, Canada
9. Yale University, New Haven, Connecticut, USA

a. Now at Stanford Linear Accelerator Center, Stanford, CA 94309
b. Now at University of Mississippi, Oxford, MS 38677
c. Now at Electromagnetic Applications, Inc., Denver, CO 80226
d. Now at Princeton University, Princeton, NJ 08544
e. Now at CERN, Division EP, CH-1211 Genève, Switzerland
f. Now at University of Cincinnati, Cincinnati, OH 45221
†Deceased

(The Tagged Photon Spectrometer Collaboration)

Presented by:
M.S. Witherell
Univ. of California, Santa Barbara, CA 93106

ABSTRACT

We present results on exclusive semileptonic decays of charmed mesons. Among these are unexpected results on $D \to K^* e^+ \nu_e$.

INTRODUCTION

Semileptonic decays of heavy quarks are of particular interest because they are the easiest decays to interpret. For this reason, they are used to determine the elements of the Kobayashi–Maskawa matrix. For the case of charm decay, the elements are constrained in the three-generation model to be $V_{cs} = 0.9743 \pm 0.0006$ and $V_{cd} = 0.220 \pm 0.0002$.[1] At the present level of theoretical and experimental knowledge, we can only check these values at a fairly crude level. For comparison, the effect of the third generation on V_{cs} is only 0.002, which is almost two orders of magnitude better than the direct experimental measurement.

Because the K–M matrix elements are so precisely known, in the three-generation scheme, we can use the measurements to tune up the form factor models. These models should then be even more accurate in B decays, and can be used for precise determinations of V_{cb} and V_{ub}, eventually. At the same time, the form factors are of interest themselves as a source of information about the wave functions of mesons containing heavy quarks.

Exclusive semileptonic decays of charmed mesons had been more difficult to measure than nonleptonic decays, because no narrow mass peaks are formed. We have been able to use good electron identification and vertex separation to achieve good background rejection and clean signals for these decays. The rates are well measured, and our attention is turning to more detailed measurements, such as angular distributions.

The description of semileptonic decay is straightforward. It is assumed that it proceeds by spectator decay only. In the weak interaction the charmed quark decays $c \to s(d)e^+\nu_e$ and the light antiquark $\bar{q}$ has no effect. All strong interaction effects are contained in the form factor, which describes the probability for the $s\bar{q}$ system to be bound as a K, K*, or other hadronic state. A simple isospin argument shows that $\Gamma(D^+ \to \overline{X^0}e^+\nu) = \Gamma(D^0 \to X^-e^+\nu)$ for each final state, in spite of the very different lifetimes. I will thus use partial rates rather than branching ratios in comparing results with theory.

INCLUSIVE DECAYS

We have had for some time good measurement from Mark III on the inclusive semileptonic decays.[2] They measure $B(D^0 \to e^+X) = (7.5 \pm 1.1 \pm 0.4)\%$ and $B(D^+ \to e^+X) = (17.0 \pm 1.9 \pm 0.7)\%$. These can be combined with the lifetime measurements from E691[3] to calculate the decay rate, $\Gamma(D^0 \to e^+X) = (17.4 \pm 2.5 \pm 0.9) \cdot 10^{10}$ s^{-1}, and $\Gamma(D^+ \to e^+X) = (15.7 \pm 1.8 \pm 0.6) \cdot 10^{10}$ s^{-1}. As expected, $\Gamma(D^0 \to e^+X) = \Gamma(D^+ \to e^+X)$, and the average value $\Gamma(D \to e^+X) = (16.3 \pm 1.5) \cdot 10^{10}$ s^{-1}.

In a simple quark decay picture,

$$\Gamma = \frac{G^3}{192\pi^3} m_c^5 \, f\left(\frac{m_s}{m_c}\right), \tag{1}$$

where the function f corrects for the finite mass of the strange quark and is about 0.5. This formula gives the right answer for $m_c = 1.6$ GeV, which is a reasonable value. Unfortunately, the dependence on the fifth power of the charmed quark mass means that the rate varies by a factor of 2 for a change of 0.2 GeV in the quark mass. To do better, we must use exclusive decays.

EXCLUSIVE DECAYS

The amplitude for Cabibbo-favored semileptonic decays can be written

$$A_{CF} = \frac{G}{\sqrt{2}} V_{cs} L^\mu H_\mu \tag{2}$$

where L^μ and H_μ are the leptonic and hadronic currents, respectively. For decay into a pseudoscalar X, D → Xeν, $H_\mu = (P_D + P_X)_\mu f_+^X(t)$, in the limit of small lepton mass. The momentum transfer variable $t = (P_D - P_X)^2$ is just equal to the square of the virtual W mass, $M_{e\nu}^2$. There is just one vector form factor, $f_+^X(t)$, needed to describe the decay. Often one assumes the t-dependence with a single pole form, $f_+^X(t) = f_+^X(0)\left[\frac{1}{1-t/M_y^2}\right]$, where M_y is the mass of the lowest $c\bar{s}$ resonance with $J^P = 1^-$, $M_{D_s^*} = 2.1$ GeV. More generally, this serves as a simple parameterization of the t-dependence. Assuming a mass of 2.1 GeV, the decay rate can be expressed

$$\Gamma(D^0 \to K^-e^+\nu) = |V_{cs}|^2 |f_+^X(0)|^2 \; 15.3 \cdot 10^{10} s^{-1}, \quad (3)$$

where we use the value of the form-factor at t=0 as the normalization parameter.

We looked at the decay chain $D^{*+} \to \pi^+ D^0$, $D^0 \to K^-e^+\nu_e$.[4] We require a K^-e^+ pair from the secondary vertex, in association with a π^+ from the primary. The electron identification efficiency is 72%, and the misidentification probability is 0.5%. Assuming that the Kπ comes from the decay $D^0 \to K^-e^+\nu$, one can calculate the Dπ mass. There is a clear excess in the right-sign spectrum at the D* mass, with a total of 250 events. The resulting branching ratio is $(3.8 \pm 0.5 \pm 0.6)\%$, corresponding to a decay rate of $(8.8 \pm 1.2 \pm 1.4) \times 10^{10}$ s^{-1}. The t-distribution is also measured, and a fit to the single-pole form gives a good fit with $M_y = 2.1 {}^{+0.4}_{-0.2}$ GeV/c^2.

The results agree with those from Mark III,[5] and a weighted average gives $B(D^0 \to K^-e^+\nu) = (3.5 \pm 0.5)\%$ and a decay rate of $B(D^0 \to K^-e^+\nu) = (8.2 \pm 1.2) \cdot 10^{10}$s^{-1}. Plugging this into the decay rate formula above gives the result $|f_+^K(0)|^2|V_{cs}|^2 = 0.54 \pm 0.08$. Assuming $|V_{cs}| = 0.975$, the best measurement of $f_+^K(0) = 0.75 \pm 0.05$. This agrees well with predictions of Wirbel, Stech, and Bauer[6] (0.76) and Dominguez and Paver[7] (0.75). Thus there is good agreement of the experimental result with the calculated form factor, assuming $|V_{cs}| = \cos\theta_c$.

The other dominant Cabibbo-favored decay is D → Kπeν. The first goal is to measure the decay rate for D → K*eν, and compare it to the calculations. The second is to measure the size of non-resonant Kπeν, and thus get a first measure of the importance of hadronic final states other than the K and K*. The general picture from form factor models is that[6,8] the K* to K ratio should be greater than 1.

We measured the decay $D^+ \to K^-\pi^+e^+\nu_e$.[9] Although the D* cut available for D^0 decay is not available, the vertex cuts are particularly effective because of the long D^+ lifetime. There is a signal of 250 events over 62 background as measured with the wrong sign ($K^-\pi^+e^-$) sample. With tighter cuts on electron-identification and vertex isolation, the numbers are 155 signal and 14 background.

The Kπ mass spectra for the two samples are shown in Figure 1. There is a clean K* peak, which clearly dominates the signal. The background contribution, averaged over the K* width, is about 10% with the standard cuts and 4% with the tight cuts. The results of the fit are shown in Table 1. Less than 20% of the decays are nonresonant, which corresponds to a very small fraction of the inclusive semileptonic decay rate. The decay rate for $D^+ \to \overline{K}^{*0} e^+ \nu$ combined with the E691 result for $D^0 \to K^- e^+ \nu$ leads to the ratio $\Gamma(D \to K^* e^+ \nu)/\Gamma(D \to K e^+ \nu) = 0.45 \pm 0.12$, which is significantly lower than expected. This calls for a re-examination of the form factors.

Table 1. E691 Results for $D^+ \to K^- \pi^+ e^+ \nu$

Mode	Branching Ratio (%)	Decay Rate $(10^{10} s^{-1})$
$D^+ \to \overline{K^{*0}} e^+ \nu$	$4.5 \pm 0.7 \pm 0.5$	$4.2 \pm 0.6 \pm 0.5$
$D^+ \to (K^- \pi^+)_{NR} e^+ \nu$	$0.3 \pm 0.2 \pm 0.2$	< 0.7

Do the exclusive decays to pseudoscalar and vector mesons saturate the measured inclusive rate? In Table 2 we list the known components of the semileptonic rate. For the nonresonant $K\pi e \nu$ decay, we multiply the $K^- \pi^+ e^+ \nu_e$ rate from E691 by the isospin factor of 1.5. The Cabibbo-suppressed decay can be estimated by taking the Cabibbo-favored decays and multiplying by, for example, the calculation for the ratio $\pi e\nu/K e\nu$ from reference 6. This takes into account the Cabibbo suppression, as well as differences in phase space and form factors, and should be quite accurate. The sum is $(14.1 \pm 1.5) \cdot 10^{10} s^{-1}$, compared to $(16.3 \pm 1.5) \cdot 10^{10} s^{-1}$ for the inclusive average. The missing rate is $(2.2 \pm 2.2) \cdot 10^{10} s^{-1}$, which is consistent with zero. There is not much room for $D \to K\pi\pi e \nu$, $K\eta e\nu$, etc.

Table 2. Semileptonic Decay Rate Summary

Mode	Source	Rate $(10^{10} s^{-1})$
$D^0 \to K^- e^+ \nu$	Mark III/E691	8.2 ± 1.2
$D^+ \to K^{*-} e^+ \nu$	E691	4.2 ± 0.8
$D^+ \to (K\pi)_{NR} e\nu$	E691	0.4 ± 0.4
$D \to (\pi, \rho) e\nu$	$\tan^2 \theta_c$	1.3 ± 0.2
Total		14.1 ± 1.5
Inclusive	Mark III	16.3 ± 1.5

ANGULAR DISTRIBUTIONS IN $D \to K^* e\nu$

Returning to the problem of the low $D \to K^* e\nu$ rate, the next step is to determine which of the form factors causes the discrepancy. There are three form factors in $K^* e\nu$ decay, one vector form factor $V(t)$ and two axial vector form factors $A_1(t)$ and $A_2(t)$. Fortunately, the extra information necessary to extract the form factors is available in the angular distributions, defined by three angles. The K^* decay angle θ_V is the angle between the π and the D momenta in the K^* center of mass. The leptonic decay angle θ_e is the angle between the electron and the D momenta in the rest frame of the virtual W. The angle between the decay planes is χ. The entire specification of an event is given by θ_e, θ_V, χ, and t.

The angular distribution for the general case has six terms:[10]

$$|A|^2 = G^2|V_{cs}|^2 t \times \Big\{ [(1+\cos\theta_e)^2|H_+|^2 + (1-\cos\theta_e)^2|H_-|^2]\sin^2\theta_V$$
$$+ [2(1-\cos^2\theta_e)|H_0|^2]\tfrac{3}{2}\cos^2\theta_V$$
$$- [2(1-\cos^2\theta_e)\mathrm{Re}(H_+H_-^*)]\tfrac{3}{4}\sin^2\theta_V\cos 2\chi - \tfrac{3}{2}\sin\theta_e[(1+\cos\theta_e)\mathrm{Re}(H_+H_0^*)$$
$$+ (1-\cos\theta_e)\mathrm{Re}(H_-H_0^*)]2\sin\theta_V\cos\theta_V\cos\chi \Big\}. \quad (4)$$

The functions H_+, H_0, and H_- correspond to helicity amplitudes of the virtual W^+. They are related to the form factors by the relations

$$H_\pm(t) = (M_D + M_{K^*})A_1(t) \pm 2\frac{M_D P_{K^*}}{M_D + M_{K^*}}V(t) \quad (5)$$

$$H_0(t) = \frac{1}{2M_{K^*}\sqrt{t}}[(M_D^2 - M_{K^*}^2 - t)(M_D + M_{K^*})A_1(t) - 4\frac{M_D^2 P_{K^*}^2}{M_D + M_{K^*}}A_2(t)]. \quad (6)$$

Integrating over the azimuthal angle χ, there are two terms: a transverse component with $\sin^2\theta_V$, and longitudinal with $\cos^2\theta_V$.

To get a first look at the relative amount of transverse and longitudinal decays, we have analyzed the $\cos\theta_V$ distribution, shown in Figure 2. A fit to the form $d\Gamma/d\cos\theta_V = 1 + \alpha\cos^2\theta_V$ gives the result $\alpha = 3.8^{+3.4}_{-1.8}$. This corresponds to a ratio of longitudinal to transverse of $\Gamma_L/\Gamma_T = (1+\alpha)/2 = 2.4^{+1.7}_{-0.9} \pm 0.2$. This ratio would be 0.5 for unpolarized K^*'s, and was expected to be 1.0 in the BSW model. The K^* is longitudinally polarized, somewhat more so than expected. The problem now is to find a convincing explanation for both the low K^* rate and the longitudinal polarization.

This fit uses only $\cos\theta_V$, and varies the relative size of Γ_L and Γ_T. We will next fit the same data sample using $\cos\theta_V$, $\cos\theta_e$, χ and t, varying the size of A_1, A_2, and V. It may be possible to learn the source of the discrepancy by measuring the individual form factors directly.

CONCLUSIONS

The average rate for $D^0 \to K^- e^+ \nu$ is $(8.2 \pm 1.2) \cdot 10^{10} sec^{-1}$, which agrees with expectations. We measure the decay rate for $D^+ \to \overline{K^{*0}} e^+ \nu$ to be $(4.2 \pm 0.6 \pm 0.5) \cdot 10^{10} sec^{-1}$, which is significantly smaller than the predicted value, and the K^* is longitudinally polarized, more than had been expected. The nonresonant decay $D \to K\pi e\nu$ is less than about 7% of the inclusive semileptonic rate, which confirms the picture that pseudoscalar and vector mesons final states are dominant.

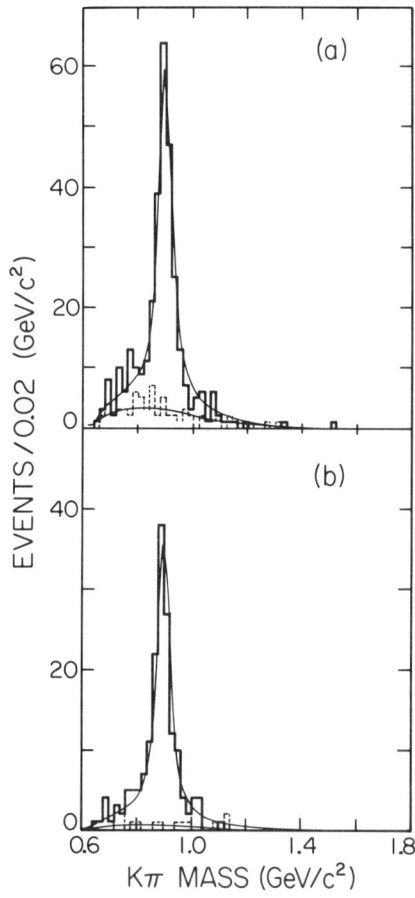

FIGURE 1

The $K\pi$ mass spectra for $D^+ \to K^-\pi^+e^+\bar{\nu}_e$ candidates from E691 with right-sign (solid) and wrong-sign (dashed) electrons: (a) loose cuts, (b) tight cuts. The curves are fits used to extract the K^* component.

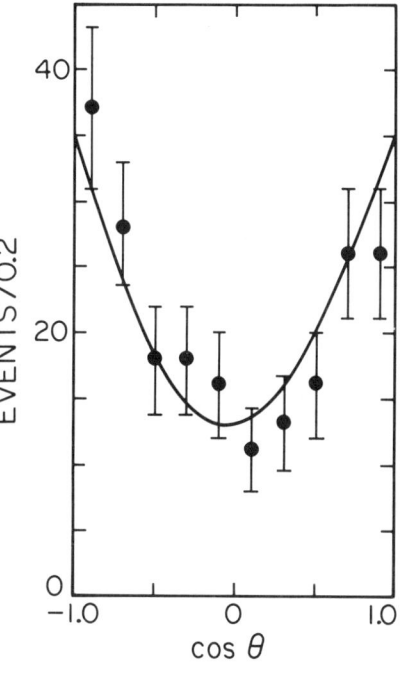

FIGURE 2

The $\cos\theta_V$ distribution for the $D^+ \to \overline{K^{*0}}e^+\nu_e$ candidate from E691. The curve is from a fit to background plus signal and corresponds to $\Gamma_L/\Gamma_T = 2.4$.

The major surprises are in the decay $D \to K^* e\nu$. There is a need to directly measure the form factors in this decay, and to compare with the many theoretical models that now exist. Future experiments will be able to do similar comparisons in $D \to \rho e\nu$ and $\pi e\nu$. With such information, a reliable model of semileptonic decays should be possible, which can then be applied to the problem of measuring V_{ub} and V_{cb} precisely.

REFERENCES

1. Particle Data Group, Review of Particle Properties, *Phys. Lett.* **B204**, 1 (1988).
2. R.M. Baltrusaitis et al., *Phys. Rev. Lett.* **54**, 1976 (1988).
3. J.R. Raab et al., *Phys. Rev. D* **37**, 2391 (1988).
4. J.C. Anjos et al., *Phys. Rev. Lett.* **62**, 1587 (1989).
5. J. Adler et al., *Phys. Rev. Lett.* **62**, 1824 (1989).
6. M. Wirbel, B. Stech, and M. Bauer, *Z. Phys.* **C29**, 637 (1985).
7. C.A. Dominguez and N. Paver, *Phys. Lett.* **207B**, 199 (1988).
8. N. Isgur et al., *Phys. Rev. D* **39**, 799 (1989).
9. J.C. Anjos et al., *Phys. Rev. Lett.* **62**, 722 (1989).
10. Adapted from J.C. Körner and G.A. Schuler, Mainz preprint MZ–TH/89-01 (1989).

CHARM FRAGMENTATION

Giancarlo Moneti
Syracuse University, Syracuse, NY 13244

ABSTRACT

The status of fragmentation models is surveyed, pointing out the physical meaning of the parameters used in the parton-shower plus string fragmentation models and the reliability of their determination by comparison of experimental data with full Monte Carlo simulation of the models. Results of charmed meson and baryon fragmentation distributions, as observed in e^+e^- annihilations, are then reviewed and the string model parameters obtained from them discussed. The excellent prospects for more accurate results in the near future are finally outlined.

STATUS OF FRAGMENTATION MODELS

The confinement of quarks and gluons (partons) and the mechanism through which they transform into groups of hadrons (fragmentation or hadronization), is still the least understood and mathematically most intractable problem of Quantum Chromo Dynamics (QCD).

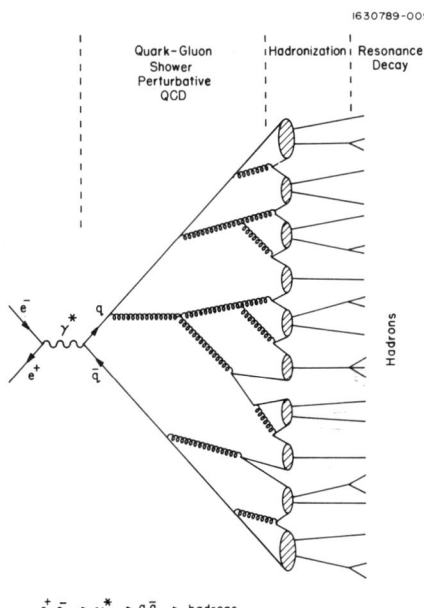

Fig. 1. The stages of e^+e^- annihilation into hadrons.

Experimentally the parton hadronization has been studied in a variety of high energy experiments, but e^+e^- annihilation into hadrons is a particularly suitable reaction, since the hadronic state is believed to evolve from a very simple initial condition: a high energy $q\bar{q}$ state. Starting from this simple state we can distinguish three stages in the evolution of the system (Fig. 1). In the first stage high energy quarks radiate, with relatively small momentum transfer, gluons that in turn generate other gluons and/or $q\bar{q}$ pairs. This is the so called **parton shower** that can be well described by perturbative QCD. Eventually, however, the second stage must take place: colourless $q\bar{q}$ pairs (or qqq and $\overline{qqq}$ triplets) must somehow form hadrons (stable or in excited states). This second stage is the one more properly called **hadronization**: it is at the heart of exploring the mistery of confinment, i.e. the still largely uncharted region of nonperturbative QCD. The third and last stage is the strong decay of excited hadrons. This is empirically known and parametrizable for the most common case. Let me remark, however, that we have no clear understanding of strong decays and that mastering confinement should also lead to their understanding.

© 1989 American Institute of Physics

In first, gross approximation, the second stage leads to an intermediate state made of the primitive $q\bar{q}$ pair plus a number of gluons (and possibly other quarks and antiquarks) with energies comparable to the initial center-of-mass energy. Empirically, each one of these partons hadronizes into a bunch of hadrons that have small transverse momentum with respect to the partons: a so called jet of hadrons.

Pioneering models of quark fragmentation[1] skillfully parametrized the hadronization of each parton separately (independent jet fragmentation). Various tricks were then used to enforce energy, momentum and flavor conservation on the entire event. Simple quantum-mechanical considerations were also applied[2] to successfully describe the essential features of heavy quark hadronization.

Although extremely useful in many other ways, these models' parameters were largely unrelated to the fundamental phenomenon of quark confinement and did shed little light on the nonperturbative phase of QCD.

Later progress in perturbative QCD allowed the construction of excellent models of the first (parton shower) stage of fragmentation, using the Leading Log Approximation (LLA)[3-4] resorting however, at least initially, to simple phase space considerations for the crucial second (hadronization) phase.

These models were soon shown to be clearly superior to the independent fragmentation models[5] and the Webber model better than Gottshalk's.[6]

The Lund group has incorporated in the most recent version of its Monte Carlo simulation (JETSET6.3),[7] a parton shower model that uses LLA plus $O(\alpha_s)$ exact matrix elements. As an alternative, Lund allows the use of parton shower generation based on QCD matrix elements up to $O(\alpha_s^2)$.

The Tasso collaboration has recently performed a detailed comparison of Webber's and Lund's LLA+$O(\alpha_s)$ models with their own data, as well as Mark II's and HRS's, of e^+e^- hadronic annihilation at c.m.e. between 12 and 41.5 GeV.[8] The distributions of 14 different global event-shape and individual charged track paramters, as well as the energy dependence of their averages, show a clear superiority of the Lund LLA+$O(\alpha_s)$ model over Lund's other option and Webber's model. There are hence good experimental reasons to concentrate on the comparison of experimental results with the Lund model, but I would also like to show that an even better reason is that it promises to tell us about the underlying fundamental hadronization process. This requires a brief description of the model.

THE LUND MODEL

What's the difference between Webber's and Lund's LLA+$O(\alpha_s)$ models? In Webber's model, at the end of the parton shower process all gluons are forced to split into $q\bar{q}$ pairs. One is then left with a number of colorless "clusters" that are let decay mainly into pairs of hadrons that conserve flavor, with probabilities proportional to the phase space available. Lund's model[9] always based the hadronization on the QCD-inspired relativistic string model.[10] In JETSET6.3 the strings extend from a quark to an antiquark with the intervening gluons represented by "kinks" in the string.

The tension of the string is assumed to have a constant value, estimated e.g. from the string model of Regge trajectories, $\kappa \equiv \frac{g^2}{4\pi} \sim 1~GeV/f = 0.2~GeV^2$ (where g is the strong coupling constant). With the $\overline{Q}$ and Q moving apart, their kinetic energy is transformed into string potential energy, making possible the breaking of the string, with probability Π per unit length and unit time, by creation of a $q\bar{q}$ pair and giving birth to a hadron (fig. 2). The probability dP of formation of a primary hadron is then given by the product of the probability $\Pi d\xi dt$ that the string breaks at space-time point ξ, t in $(d\xi, dt)$, times the probability e^{-A} that the string had not broken "earlier". Expressed in energy variables, this probability is:

$$dP = \Pi^2 \frac{dm^2}{2\kappa^2} dx^+ \frac{(1-x^+)^a}{x^+} \exp\left(-\frac{\Pi}{2\kappa^2} \frac{m_\perp^2}{x^+}\right), \quad (1)$$

where: $x^+ \equiv \frac{(E+p)_h}{(E+p)_Q}$ is a scaling light-cone energy and $m_\perp = \sqrt{m_h^2 + p_\perp^2}$ is the "transverse mass" of the hadron (to take the other two neglected space dimensions into account). $a = 1$ in the simplest version of the model. Eq. 1 is quite suitable for recursive use in a Monte Carlo simulation of $Q\overline{Q}$ fragmentation. Since the result of such an application is independent of whether the procedure starts from the quark or the antiquark, the function in eq. 1 is called the Lund Symmetric fragmentation function (LS FF). Slight modifications of eq. 1 allow to take account of the quark mass[11] and, through the diquark hypothesis to be discussed later, of baryon production.

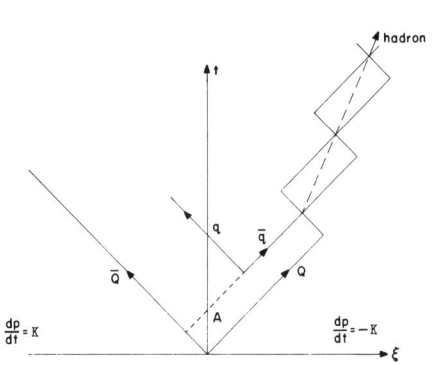

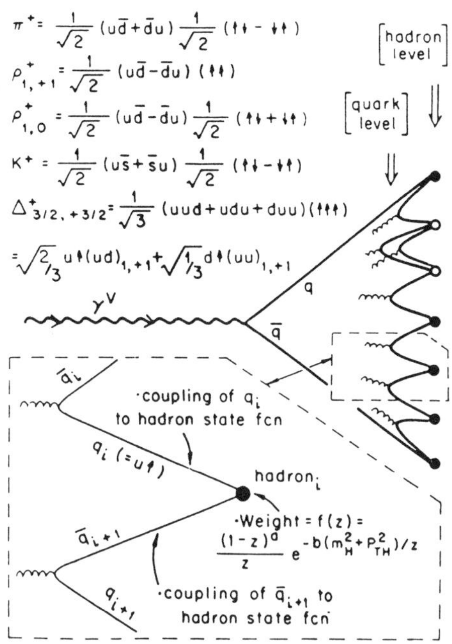

Fig. 2. The space-time evolution of the one-dimensional string bounded by massless quarks Q and $\overline{Q}$.

Fig. 3. Buchanan and Chun hadronization scheme.

The most common objection to the Lund Monte Carlo has been the large number of "free" parameters used to determine such things as the probability ratio s/d of breaking the string through creation of an $s\bar{s}$ pair (rather than $u\bar{u}$ or $d\bar{d}$), the ratio qq/q of a $qq\overline{qq}$ pair rather than a $q\bar{q}$ pair breaking the string, the probability V/(V+P) of generating a vector rather than a pseudoscalar meson, similar ratios to decide the spin of the baryons, etc.. Parameters like s/d and qq/q are not completely free, since they can be related to the tunnelling probability of creating quark or diquark pairs of different masses. Detailed analysis of the string breaking process leads also to predictions for V/(V+P) for different flavors.[12] The boldest step,[13] however, was to take the hadron mass dependence in eq. 1 literally, not only to obtain the x^+ distribution but also, in conjunction with the Clebsch-Gordan coefficients of the quark model spin-flavor hadron wave functions, to obtain the relative probability of generating different hadrons (fig. 3). This approach essentially leaves only the two LSFF parameters free, giving great predictive power to the model, displacing the problem back to the fundamental one of the hadron masses. The success, illustrated in fig. 4, in predicting meson and baryon frequencies over three orders of magnitude in e^+e^- annihilation is quite striking.

For the parameters Π and κ to have physical significance, we should find a way of measuring them and they should be "universal". What we can measure is the fragmentation distribution $D_q^h(x^+)dx^+$, the frequency with which a quark q generates the hadron h with x^+ in dx^+. For light quarks and hadrons the fragmentation distribution is dominated by secondary string breaking and the decay of resonances; thus it does not give us direct information on the LSFF parameters. On the contrary, we have good reasons to believe that heavy-flavored hadrons contain the primary heavy quark, since it is highly improbable that the string will break by creation of a heavy $q\bar{q}$ pair. For heavy-flavored hadrons we can then write the fragmentation function immediately from eq. 1:

$$D_Q^H(x^+)dx^+ = \Pi^2 \frac{dm^2}{2\kappa^2} dx^+ \frac{(1-x^+)^a}{x^+} \exp\left(-\frac{\Pi}{2\kappa^2}\frac{m_\perp^2}{x^+}\right) \qquad (2)$$

Except for the slight complication of decay of higher mass resonances, this fragmentation function is directly related to the string-breaking mechanism and to the measurement of its parameters. Hence the interest in charm (and bottom) fragmentation.

CHARM FRAGMENTATION

The study of charm fragmentation in e^+e^- annihilation is particularly attractive because the excited meson D^{*+} is most easily detected and is nearly background-free. Furthermore, at DORIS and CESR its production cross section is large and the relatively small c.m.e. minimizes the parton shower effects, thus making the results on hadronization less parton shower model dependent.

The D^{*+} fragmentation distribution measured by ARGUS[14] and CLEO[15] is shown in fig. 5. The leading role of the D^{*+} is evident and is in sharp contrast with what happens in a charm photoproduction experiment[16] where the charm quark is not the leading quark (fig. 6). The direct fit of the LSFF to the data in fig. 5 gives the following values for its parameters: $a = 1.03 \pm 0.12$ and $b = \frac{\Pi}{2\kappa^2} = 0.43 \pm 0.07$ GeV^2. These values, however, are not necessarily meaningful. In fact such a direct fit completely ignores parton shower effects, as well as those of QED radiation and forces the parameters to adjust for these effects, unrelated to hadronization. The only way to obtain correct hadronization parameters is to use the appopriate fragmentation function in a Monte Carlo program that also simulates QCD and QED radiation.

CLEO[15] has measured the fragmentation distribution for five different charm hadrons, D^{*+}, D^o, D^+, D_s, Λ_c, and performed a simultaneous fit of all of them using the LSFF in version JETSET5.2, then available, of the Lund Monte Carlo (fig. 7). The optimum values of the LSFF parameters:

$$a = 0.54 \pm 0.08 \pm 0.04 \quad \text{and} \quad b = \frac{\Pi}{2\kappa^2} = 0.53 \pm 0.03 \pm 0.03 \; GeV^2, \qquad (3)$$

substantially different from those of the direct fit mentioned above. CLEO has also used the fitted D^{*+} fragmentation distribution at 10.5 GeV c.m.e. thus obtained to predict such a fragmentation distribution at 30 GeV, using the QCD evolution equations.[17]

The result is nicely consistent with the D^{*+} fragmentation distribution obtained by TASSO, JADE, HRS and TPC at PETRA and PEP, thus showing that the same set of LSFF parameters describe the fragmentation into a variety of charmed hadrons at substantially different energies.

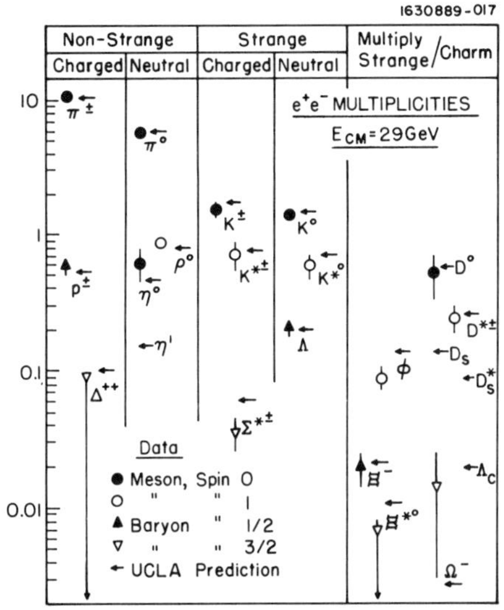

The simultaneous measuremnt of the D^{*+}, D^{*0}, D^o and D^+ production cross sections, allowed CLEO to measure another important hadronization parameter, the probability of generating a vector, rather than pseudoscalar, meson, $V/(V+P) = 0.85 \pm 0.11 \pm 0.17$ and also the D^{*+} decay branching ratio $B(D^{*+} \to D^o + \pi^+) = 0.52 \pm 0.07 \pm 0.11$. This parameters, or better their product, are also needed to get the correct relative normalization of the fragmentation distributions from the Lund Monte Carlo.

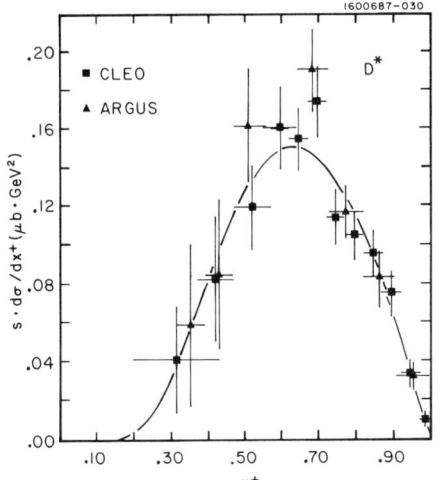

Fig. 4. Comparison of Buchanan and Chun's predictions with experimental hadron multiplicities in e^+e^- ennihilation at 29 GeV c.m.e..

Also in charm fragmentation we are however left with the residual problem of higher mass excited charm mesons, the so-called D^{**}'s. Four P-wave $c\bar{q}$ (q=d, u or s) states are expected to exist and at least three have been observed, the $D^{**o}(2428) \to D^{*+} + \pi^-$;[18] the $D^{**o}(2459) \to D^+ + \pi^-$[19] and the $D_s^{**+}(2535) \to D^{*+} + K^o$.[20] About 10% of the D^{*+} are decay products of the $D^{**o}(2428)$ and 3% of the $D^{**o}(2459)$ and 2% of the $D_s^{**+}(2535)$, while about 10% of the D^+ are decay products of the same $D^{**o}(2459)$. It would thus be nice to know accurately the production cross sections and decay branching

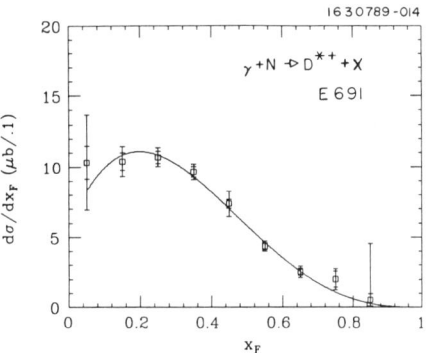

Fig. 5. The D^{*+} fragmentation distribution from the ARGUS[14] and CLEO[15] experiments. The curve is a direct fit of eq. 2.

Fig. 6. D^{*+} fragmentation distribution from the photoproduction experiment E691.[16]

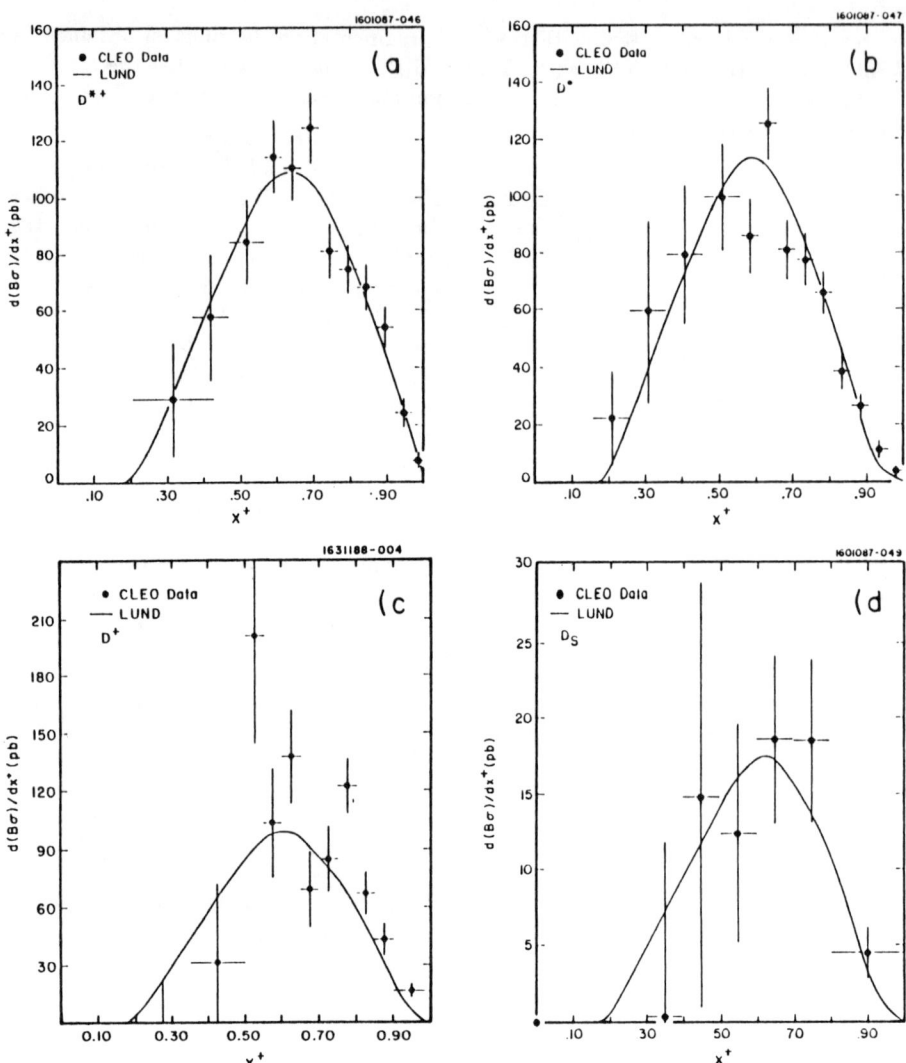

Fig. 7. Charm mesons fragmentation distribution for a) D^{*+}, b) D^o, c) D^+ and d) D_s from CLEO.[15] comapred with a fit of the Lund model.

ratios of these D^{**}'s in order to separate the effect of their decays from the direct charm hadronization into D^* and D; ignoring them may bias the values of parameters like a and b and $V/(V+P)$. The effect should not however be strong: because of its large mass, the daughter charm meson should retain most of the momentum of the parent excited state. These states should anyway be introduced in the fragmentation Monte Carlo's: everybody knows how to do it but, as far as I know, it has not been done yet.

The present knowledge of the fragmentation distribution of these D^{**o}'s is shown in fig. 8. It is rather primitive and will take time to improve appreciably.

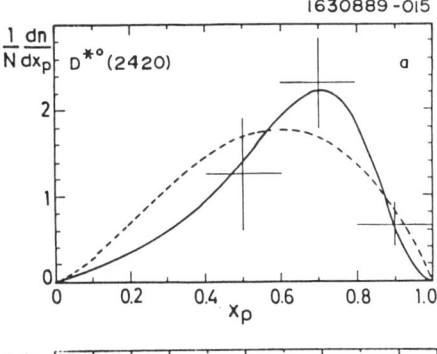

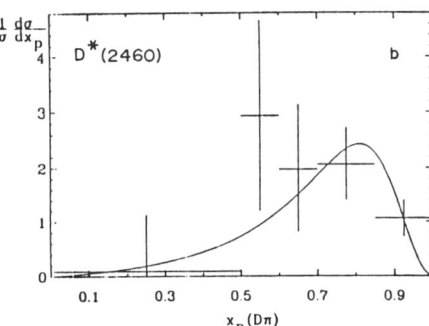

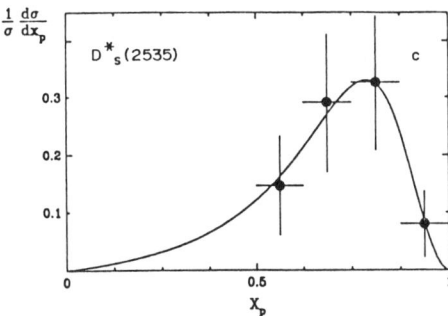

Fig. 8. D^{**} fragmentation distributions: a.) $D^{**o}(2420)$,[18] b) $D^{**o}(2459)$,[19] c) $D_s^{**+}(2535)$.[20]

D^{*+} POLARIZATION

Vector meson polarization (specifically D^{*+} polarization) can, in principle, provide additional information on the hadronization process. It turns out to be connected to the V/(V+P) ratio. This quantity has been generally treated as a, possibly flavor dependent, but energy independent parameter. Instead it turns out that the fragmentation function may well be of different shape for different spin and polarization states.

A simple way of parametrizing vector meson polarization is as follows. Of the four possible ways of combining the spins of the q and $\bar{q}$ in a meson, the two states with $S_z = 0$ can mix with probability f [$(1-f)$] to give a pseudoscalar [vector] meson. One can then easily get the vector meson spin density matrix and find that $\rho_{oo} = \frac{1-f}{2-f}$ and that $\frac{V}{(V+P)} = \frac{1-f}{2}$. The expected angular distribution of the D^o in the D^{*+} rest frame will be in the form $(1 + \alpha \cos^2 \theta)$ with $\alpha = \frac{3\rho_{oo}-1}{1-\rho_{oo}} = 1 - 2f$. Naïve spin counting leads to $f = 1/2$, $\frac{V}{(P+V)} = 0.75$ and $\alpha = 0$, i.e. no polarization.

Different models[0,4,21] lead to substantially different predictions for vector meson polarization and its energy dependence.

ARGUS[22] and CLEO[23] have measured D^{*+} polarization in continuum e^+e^- annihilation at 10.5 GeV c.m.e., and HRS,[24] with smaller statistical significance, at 29 GeV. The ARGUS and CLEO results, in the form of α vs x^+ are shown in fig. 9 (ARGUS errors are statistical only). They seem to hint some x^+ dependence, although

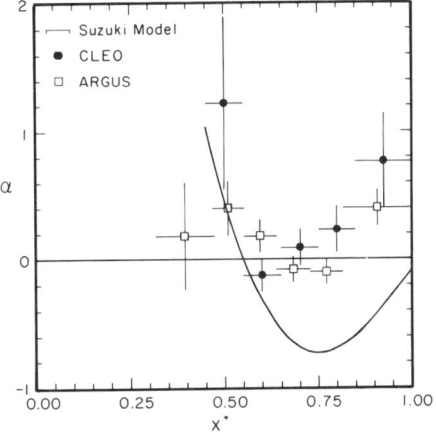

Fig. 9. Distribution of the D^{*+} polarization parameter α vs x^+.

the statistical model with $f = 1/2$ ($\chi^2/d.o.f. = 11/4$) cannot be excluded. The disagreement with the Suzuki model is stronger ($\chi^2/d.o.f. = 70/4$). Ignoring the x^+ dependence, the CLEO data give an acceptable fit ($\chi^2/d.o.f. = 11/9$) for $<\alpha> = 0.20 \pm 0.08 \pm 0.04$, implying $f = 0.40 \pm 0.04 \pm 0.02$ and $\frac{V}{(P+V)} = 0.80 \pm 0.02 \pm 0.01$, in agreement with the CLEO result from the total D^* and D production cross sections reported in the previous section.

Obviously we need statistically more accurate measurements to reach any firm conclusion.

CHARMED BARYONS

Baryon production in e^+e^- annihilation[25] is fairly well described, within the string model, by simply having the string broken by creation of a $\bar{q}\bar{q}qq$ pointlike "diquark" pair instead than the usual $q\bar{q}$ pair. Empirically, the probability of this happening seems to be close to 8%. This "diquark" mechanism was used in earlier versions of the Lund Monte Carlo. It implies strong rapidity correlations and transverse momentum anticorrelations between the baryon and the antibaryon thus produced. These and other angular correlations for protons and Λs have been clearly established.[26] They are well described by the Lund Monte Carlo that uses the LSFF and the diquark mechanism and strongly disgree with the cluster decay[4] model of hadronization. However the correlations between the baryon momentum components transverse to the jet axis favor the possibility that a meson be produced in-between the baryon and the antibaryon (the so called "popcorn" effect).

Although empirically successful, the "diquark" mechanism has several problems: (i) the existence of pointlke diquarks must be postulated and is not understood; (ii) the "popcorn" mechanism needs be added and is somewhat unnatural for pointlke diquarks; (iii) even with these ingredients, the string (Lund) model seems to predict proton and Λ fragmentation functions slightly harder than the experimental ones.[27]

Other models of baryon production ("color-fluctuations" plus random-recombination)[28-29] assume the popping from vacuum of $\bar{q}q$ pairs of a color different from that of the original string. One such pair will not break the string, but will change its color and make it possible for a second one to break the string, generating the baryon-antibaryon pair and naturally allowing the possibility of so called "popcorn" mesons in-between. These models are excluded by the experimental data unless strong space-time correlations between the $\bar{q}q$ pair creations are used.[29] A new way of introducing these correlations so as to allow use the LSFF for both meson and baryon production, with or without popcorn, was recently proposed.[27] Referring to fig. 10, they assume that the probability of getting a $q_R \bar{q}_R$ in the space-time region $[(0,0) - (\xi, \eta)]$ instead of being simply proportional to the area $\xi\eta$, be given by:

$$\int_0^\xi \int_0^\eta \frac{d\xi_R d\eta_R}{[(\xi_R - \xi)(\eta_R - \eta)]^\alpha}, \qquad (4)$$

that yields a fragmentation functionof the LSFF type:

$$f(x^+)dx^+ \propto \left(\frac{1-x^+}{x^+}\right)^{1-\alpha} \exp\left(-\frac{\Pi}{2\kappa^2} \frac{m_\perp^2}{x^+}\right) dx^+; \qquad (5)$$

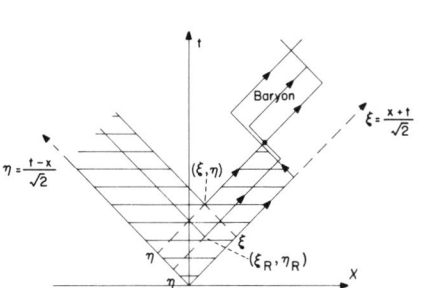

Fig. 10. Space-time diagram of color fluctuation correlations.

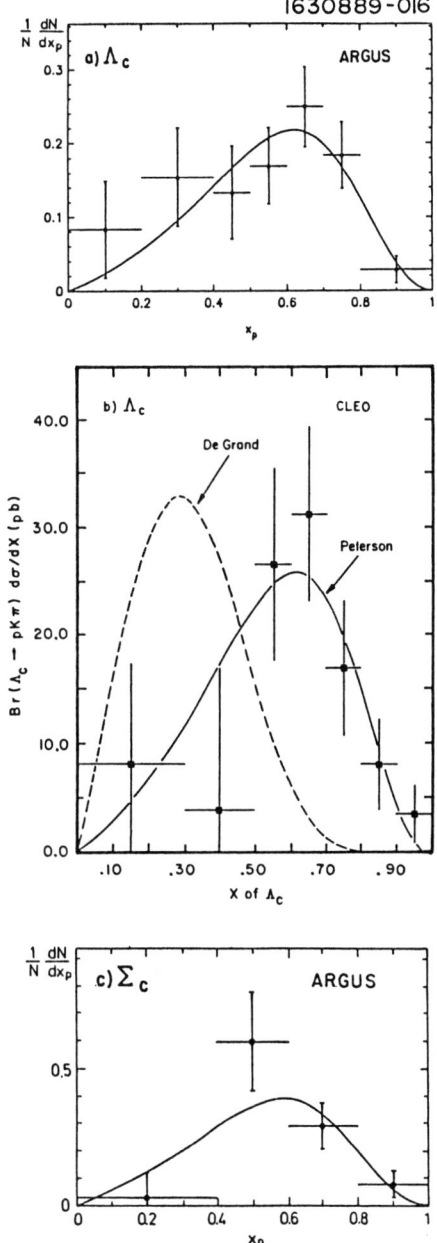

Fig. 11. a) Λ_c fragmentation distribution[30] with Peterson model; b) same[15] compared with Peterson and De Grand models; c) Σ_c fragmentation distribution[31] with Peterson model.

$\alpha = 0$ means no correlation, a case excluded by experimental results; $\alpha = 1$ means maximum correlation and results in the same fragmentation function given by the pointlike diquark hypothesis and no popcorn; the lower the value of α the higher is the probability of popcorn mesons. The high energy behaviour of baryon fragmentation distributions is sensitive to the value of α.

The study of charm baryon fragmentation distributions and correlations has many of the advantages already discussed for charm mesons. The Λ_c is the best detectable state but many higher mass charm baryons exist, have been detected and are known to decay to Λ_c, thus making the interpretation of its fragmentation distribution more difficult without knowing the ptoduction cross sections and decay branching ratios of these higher mass baryons. In fig. 11 we show recent results of Λ_c[15,30] and Σ_c.[31] We see that the Peterson model, using pointlike diquarks, fits the data quite well, while the De Grand model that uses uncorrelated $q\bar{q}$ creation can be clearly excluded. Fig. 12 shows the same data of figure 11.b) compared with the Lund model fitted to the charm mesons fragmentation distributions[15]; the Lund model gives a slightly too hard distribution; the model of ref. 27 with a value of $\alpha \simeq 0.5$ cures a similar problem for Λ and proton fragmentation distributions: it may be appropriate also in this case.

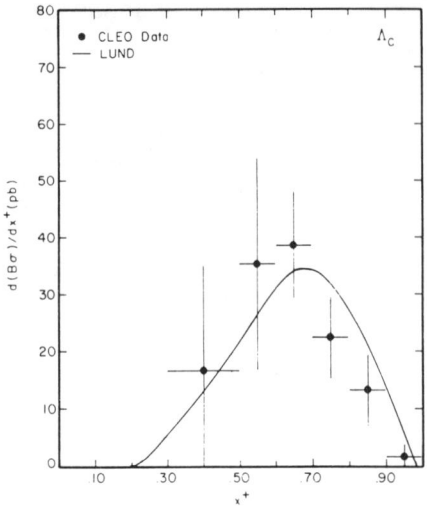

Fig. 12. Λ_c fragmentation distribution[15] compared with the Lund model.

FUTURE PROSPECTS

Much higher statistics is already available but not completely analized in the experiments at DORIS and CESR. Even larger data samples are expected in the next few years. As an example of what a factor of four in the data sample, together with substantially improved momentum and ionization resolution, can do, we show in fig. 13 preliminary CLEO results on D^+ and Λ_c fragmentation distributions from their 1987 data sample. They are to be compared to fig. 7.c) and 12. The relative errors on the fitted parameters are reduced by a factor of two relative to the previous results.

With these and future improvements we can hope to reach a level of experimental accuracy that will allow us to: (ii) clearly discriminate among models and (ii) reliably measure "universal" parameter of direct significance for the hadronization process.

Furthermore, with the advent of Z^o factories, we may get very important information on bottom fragmentation at 50 GeV.

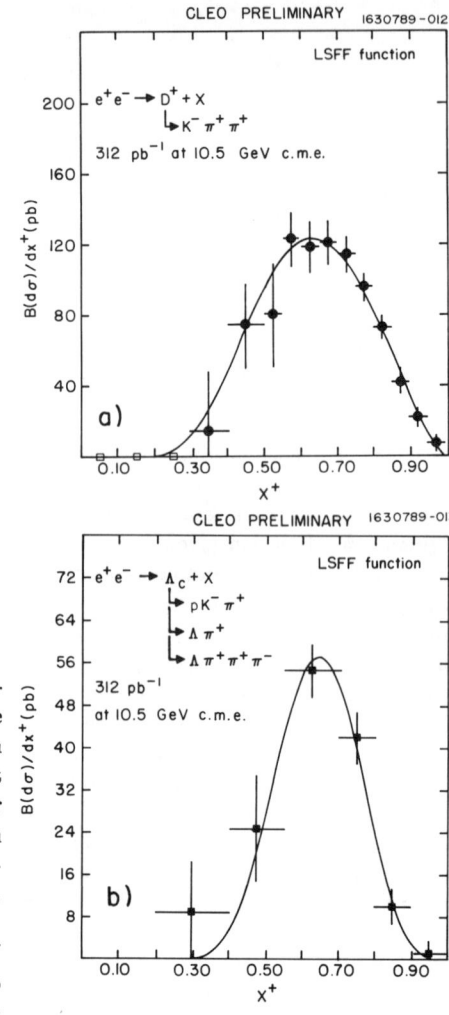

Fig. 13. Preliminary D^+ and Λ_c fragmentation distributions from the 1987 CLEO data sample.

On the theoretical side, we see considerable work beeing done[32] to put the hadronization models, and specifically the string model, on better theoretical foundation, hopefully allowing the transition from a stochastic model to one based on quantum mechanical amplitudes.

CONCLUSIONS AND RECOMMENDATIONS

With the combination of the parton shower model and the string model of hadronization we have a Monte Carlo procedure that successfully reproduce all the stages of the quark and gluon fragmentation and hadronization processes at all the observed energies. Most of the "free" parameters of the model are either eliminated or given fundamental physical significance.

Measurements of the fragmentation distributions promise to become accurate enough to be able to discriminate between models and to determine accurate values of the free parameters, to test their "universality".

In order to carry out this program, I recommend that all experimental results be compared with the Lund model of parton shower plus string hadronization. This comparison should be done with the results of the full Monte Carlo simulation contained in the Monte Carlo program JETSET6.3, specifying all the parameters used, so that fits of different experiments can be compared.

References

1) R. Field and R. P. Feynman, Nucl. Phys. **B161**, 1 (1978)

2) C. Peterson, D. Schlatter, I. Schmitt and P. M. Zerwas, Phys. Rev. **D27**, 105 (1983)

3) T. Gottschalk, Nucl. Phys. **B214**, 201 (1983), T. Gottschalk and D. Morris, Nucl. Phys. **B288**, 729 (1987)

4) G. Marchesini and B. R. Webber, Nucl. Phys. **B238**, 1 (1984), B. R. Webber, Nucl. Phys. **B238**, 492 (1984)

5) JADE Collab., W. Bartel et al., Phys. Lett. **101B**, 129 (1981); Phys. Lett. **134B**, 275 (1984); TPC Collaboration, H. Aihara et al., Phys. Rev. Lett. **54**, 270 (1985); Z. Phys. **C28**, 31 (1985)

6) MarkII Collaboration, A. Petersen et al., Phys. Rev. **D37**, 1, (1988)

7) T. Sjöstrand, M.Bengtsson, Comput. Phys. Commun. **43**, 367 (1987), M. Bengtsson and T. Sjöstrand, Phys. Lett. **185B**, 435 (1987)

8) W. Braunschweig et al., Z. Phys. **C41**, 359 (1988)

9) B. Andersson, G. Gustafson, G. Ingelman and T. Sjöstrand, Phys. Rep. **97**, 31 (1983)

10) X. Artru, G. Mennessier, Nucl. Phys. **B70**, 93 (1974); X. Artru, Phys. Rep. **97**, 33 (1983)

11) M.G.Bowler, Z. Phys. **C11**, 169 (1981)

12) M.G.Bowler, Z.Phys. **C22**, 155(1984)

13) C.D.Buchanan and S.B.Chun, Phys.Rev.Lett. **59**, 1997 (1987); preprint UCLA-HEP-89-003

14) ARGUS Collaboration, H. Albrecht et al., Phys. Lett. **150B**, 235 (1985)

15) CLEO collaboration, D.Bortoletto *et al.*, Phys. Rev. **D37**, 1719 (1988), Erratum **D39**, 1471 (1989)

16) J.C. Anjos *et al.*, FERMILAB-Pub-88/125-E, Sept. 1988

17) G. Altarelli, Phys. Rep. **81**, 1 (1982)

18) ARGUS collab., H.Albrecht *et al.*, Phys. Rev. Lett. **56**, 549 (1986)

19) ARGUS Collab., H.Albrecht *et al.*, Phys. Lett. **B221**, 422 (1989)

20) ARGUS Collab., R. Kutschke, these proceedings

21) R.Suaya, J.S.Townsend, Phys. Rev. **D19**, 1414 (1979); J.F.Donoghue, Phys. Rev. **D19**, 2806 (1979); M.Suzuki, Phys. Rev. **D33**, 676 (1986)

22) P.C.Kim, Ph.D. dissertation, Un. of Toronto, 1987 (unpublished)

23) CLEO collabor., D.Bortoletto *et al.*, Contribution to the DPF meeting of the APS, Storrs, CT, Aug. 1988

24) HRS Collab., S. Abachi *et al.*, Phys. Lett. **199B**, 585 (1987)

25) For a recent review and references to original papers, see D.H.Saxon, *Proc. Workshop on Diquarks*, Turin 1988 and RAL-88-102, Nov. 1988.

26) Mark II collab., C. de la Vaissiere *et al.*, Phys. Rev. Lett. **54**, 2071 (1985); TPC collab., H. Aihara *et al.*, Phys. Rev. Lett. **55**, 1047 (1985)

27) M.G.Bowler, P.N.Burrows and D.H.Saxon, Phys. Lett. **221B**, 415 (1989)

28) A. Casher, H. Neuberger and S. Nussinov, Phys. Rev. **D20**, 179 (1979)

29) B. Andersson *et al.* Physica Scripta **32** 574 (1985).

30) H.Albrecht *et al.*, Phys. Lett. **207B**, 109 (1988)

31) H.Albrecht *et al.*, Phys. Lett. **211B**, 489 (1988)

32) B. Andersson and W. Hofmann, Phys. Lett. **169B**, 364 (1986); X. Artru and M.G. Bowler, Z. Phys. **C37**, 293 (1988)

RECENT RESULTS ON P-WAVE CHARMED MESONS AT ARGUS

R. Kutschke
University of Toronto
Representing the ARGUS Collaboration

Abstract

Using the ARGUS detector at the DORIS II e^+e^- storage ring at DESY, we have obtained evidence for a new charmed-strange meson with a mass of $2535.9 \pm 0.6 \pm 2.0$ MeV/c^2. We have also measured the isospin mass splitting of the D_2^* system. Finally, we have separated the D_1^0 and D_2^{*0} contributions to the $D^{*+}\pi^-$ mass spectrum near 2420 MeV/c^2.

Over the past four years, two new charmed mesons, ascribed to P-wave bound states of a charmed quark and either a $\bar{u}$ or a $\bar{d}$ quark, have been observed [1-5]. A variety of models have been proposed which predict the masses and widths of these states and of their charmed-strange partners[6 – 8]. In this article, we report the observation of a new charmed-strange meson which decays into $D^{*+}K^0$. We also present the first measurement of the $M(D_2^{*+}) - M(D_2^{*0})$ isospin mass splitting. Finally, we present improved measurements of the mass and width of the $D_1(2420)^0$.

The data presented here were collected with the ARGUS detector at the DORIS II e^+e^- storage ring at DESY. A complete description of the detector, trigger conditions, multihadron selection criteria, luminosity determination and particle identification strategy can be found in reference [9].

The search for excited charmed-strange states[12] has been made using the decay chain $D_{sJ}^+ \to D^{*+}K^0$, [1] where the D^{*+} decays to $D^0\pi^+$, the K^0 is observed via $K_S^0 \to \pi^+\pi^-$ and where and the D^0 is reconstructed in the modes,

$$D^0 \to K^-\pi^+ \qquad (1)$$
$$\to K^-\pi^+\pi^+\pi^- \qquad (2)$$
$$\to K^-\pi^+\pi^0 . \qquad (3)$$

Each π^0, D^0, D^{*+} and K_S^0 candidate was kinematically fitted to its accepted mass[11]. In order to suppress combinatorial background, each $D^{*+}K_S^0$ combination was required to have $x_p > 0.6$, where $x_p = p/p_{max}$ and $p_{max}^2 = E^2(\text{Beam}) - m^2(D^{*+}K_S^0)$.

The resulting $D^{*+}K_S^0$ mass spectra are shown in Fig. 1a, for channels 1 and 2, and in Fig. 1b for channel 3. A prominent, narrow structure, at a mass of 2536 MeV/c^2 is observed in both spectra. In order to extract the parameters of the signal, each spectrum was fitted with the sum of a Gaussian, to parameterize the signal, and a first order polynomial multiplied by a square root threshold factor,

---

[1] References in this paper to a specific charged state are to be interpreted as implying the charge-conjugate state also.

to parameterize the background. The two spectra were fitted simultaneously, keeping the central values of the two Gaussians equal. This procedure yields a mass of $2535.9 \pm 0.6 \pm 2.0\,\text{MeV}/c^2$ and amplitudes of 8.5 ± 3 events, for the signal in channels 1 and 2 combined, and 7.5 ± 3 events for the signal in channel 3. Both fitted widths are consistent with the detector resolution. The fraction of the signal in each of the three channels is consistent with that expected from the acceptances and the known branching ratios[10][11]. The observed signal corresponds to a 6.7 standard deviation excess above the estimated background. If a more conservative, constant background parameterization is assumed, the significance is still 5.0 standard deviations. In the following, we shall refer to this state as the $D_{sJ}(2536)^+$.

To obtain an upper limit for the natural width, Γ, of the $D_{sJ}(2536)^+$, the signal was parameterized by a Breit-Wigner line shape convoluted with a Gaussian resolution. The widths of the two Gaussians were fixed to the detector resolution and the mass was fixed at $2535.9\,\text{MeV}/c^2$. This yields an upper limit of $\Gamma < 4.6\,\text{MeV}/c^2$, at the 90% confidence level.

In order to demonstrate that the signal is not an artifact of either the selection criteria, or of a kinematic reflection, a sideband study and a wrong charge study were performed. No significant signal was observed in either case.

The acceptance corrected fragmentation function of the $D_{sJ}(2536)^+$ is shown in Fig. 2. The overlayed curve is the result of fitting the fragmentation model of Peterson et al. [13], to the data. The Peterson parameter, ϵ[12], was measured to be $0.06^{+0.02}_{-0.01} \pm 0.02$. For comparison, the values of ϵ for the $D_1(2420)^0$, $D_2^*(2460)^0$ and $D_s^*(2113)^+$ are 0.07 ± 0.04[14], 0.06 ± 0.03[4], and $0.04^{+0.03}_{-0.01}$[15] respectively.

In order to determine the production cross section, the fragmentation function was extrapolated, using the Peterson model, to $x_p = 0$. The result was then divided by the integrated luminosity and known branching ratios[10][11] to obtain,

$$\sigma(e^+e^- \to D_{sJ}(2536)^+X) \cdot BR(D_{sJ}(2536)^+ \to D^{*+}K^0) = 16 \pm 5 \pm 3 \text{ pb},$$

at $E_{CM} = 10.30\,\text{GeV}$. If isospin invariance is assumed, one obtains,

$$\sigma(e^+e^- \to D_{sJ}(2536)^+X) \cdot BR(D_{sJ}(2536)^+ \to D^*K) = 32 \pm 9 \pm 6 \text{ pb}.$$

A search was also performed to determine whether the same state decays to $D^+K_S^0$, where $D^+ \to K^-\pi^+\pi^+$. No signal is observed near $2536\,\text{MeV}/c^2$ in the $D^+K_S^0$ invariant mass spectrum. From this we obtain,

$$BR(D_{sJ}(2536)^+ \to D^+K^0)/BR(D_{sJ}(2536)^+ \to D^{*+}K^0) \leq 0.43 \quad (90\% \text{ CL}).$$

Phase space considerations require that, for a state so close to the D^*K threshold, DK decays will, unless forbidden by some selection rule, dominate over D^*K decays. The suppression of the DK channel is, therefore, most easily understood if the state belongs to the unnatural spin-parity sequence. The lowest mass, excited, unnatural J^P, $c\bar{s}$ states, are predicted to be the two 1^+ members of the P-wave multiplet [6].

Although the $D_{sJ}(2536)^+$ is, at first sight, surprisingly narrow, a natural explanation is available. Unlike quarkonia and isovector mesons, the $c\bar{s}$ system has no conservation law which prevents the mixing of the $^3P_1$ and $^1P_1$ states. It can be demonstrated[7] that, under fairly general assumptions, triplet-singlet mixing will cause one of the observable states to broaden and the other to become narrow. One explicit calculation of the natural widths of P-wave $c\bar{s}$ mesons, including mixing of the $J^P = 1^+$ states, has been performed[8]. In that model the narrow $J^P = 1^+$ state is predicted to lie below the D*K threshold. When the observed masses of the D_{sJ}^+, D^{*+} and K_S^0 are used, however, the natural width is calculated to be about 3 MeV/c^2[16].

Another group has reported, as yet unconfirmed, evidence for two states which they interpret to be P-wave $c\bar{s}$ mesons[17]. One state, at a mass of 2537 ± 28 MeV/c^2, is reported to decay to $D_s^{*+}\gamma$ but not to D*K. Another, at a mass of 2564.3 ± 4.4 MeV/c^2, is claimed in the mode D*K. It is difficult to identify either of these states with the one discussed here.

A search for the isospin partner of the $D_2^*(2460)^0$ has been made in the decay mode $D^0\pi^+$, where $D^0 \to K^-\pi^+$. Again combinatorial background is reduced by requiring $x_p > 0.6$. An improvement of the signal to noise ratio was obtained by requiring $|\cos(\theta_\pi)| > 0.4$, where θ_π is defined as the angle between the π^+ and the $D^0\pi^+$ lab flight direction, both measured in the $D^0\pi^+$ rest frame. This cut was motivated by the observation of polarized production of the $D_2^*(2460)^0$[4]. Its isospin partner will have the same $\cos\theta_\pi$ distribution. The resulting $D^0\pi^+$ invariant mass spectrum is shown in Fig. 3. Two structures are observed. A Breit-Wigner was fitted to the higher mass structure, which was found to have a mass of 2471 ± 8 MeV/c^2, a natural width of 39 ± 20 MeV/c^2 and a statistical significance of 3.4 standard deviations. The errors are statistical only. We identify this as the isospin partner of the $D_2^*(2460)^0$ and will refer to it as the $D_2^*(2471)^+$. This corresponds to a mass splitting of,

$$M(D_2^{*+}) - M(D_2^{*0}) = 16 \pm 5, \text{ for the ARGUS } M(D_2^{*0})$$
$$= 12 \pm 5, \text{ for the world average } M(D_2^{*0})$$

The other structure in Fig. 3 is due to the process $(D_1^0, D_2^{*0}) \to D^{*+}\pi^-$ followed by $D^{*+} \to D^+(\pi^0, \gamma)$, where the neutrals are not detected.

The first P-wave charmed meson to be discovered was the $D_1(2420)$, which was observed in the process $D_1(2420)^0 \to D^{*+}\pi^-$, where $D^{*+} \to D^0\pi^+$. The $D^{*+}\pi^-$ mass spectrum also contains a contribution from $D_2^*(2460)^0$ decays, which appears as a shoulder on the high mass side of the D_1^0 peak. One tool with which to separate the two states is the angle, α, between the D^0 and the π^- directions, both measured in the D^{*+} rest frame. The angular distributions for the different

Table 1: Dependence of the mass and width of the $D_1(2420)^0$ on $\cos\alpha$.

$|\cos\alpha|$	Mass (MeV/c²)	Γ (MeV/c²)
> 0.75	2417.0 ± 2.2	25.3 ± 9.2
< 0.50	2429.6 ± 4.8	53.4 ± 15.5

Table 2: The mass and width of the $D_1(2420)^0$.

Group	Mass (MeV/c²)	Γ (MeV/c²)
ARGUS this work	2414 ± 2 ± 3	13 ± 6 ± 6
[14]	2419 ± 6	41^{+22}_{-14}
E691 [2]	2428 ± 8 ± 5	58 ± 14 ± 10
CLEO [3]	2428 ± 3 ± 3	30 ± 10 ± 10

allowed partial waves are,

$$\frac{dN}{d(\cos\alpha)} \propto \begin{cases} \sin^2\alpha & ; \ J^P = 2^+ \\ (1 + 3\cos^2\alpha) & ; \ J^P = 1^+, \text{ pure D} - \text{wave} \\ 1 & ; \ j^P = 1^+, \text{ pure S} - \text{wave} \end{cases}$$

Table 1 lists the results of a fitting a single Breit-Wigner to the $D^{*+}\pi^-$ mass spectrum for two different ranges of $\cos\alpha$. The inconsistency of the two results is evidence that the structure is indeed composed of more than one resonance. In order to make the best determination of the mass and width of the D_1^0, we require $\cos\alpha > 0.75$, to suppress $D_2^*(2460)^0$ decays, and $x_p(D^{*+}\pi^-) > 0.6$, to suppress the combinatorial background. The resulting $D^{*+}\pi^-$ mass spectrum is shown in Fig. 4. This spectrum was fitted with the sum of two Breit-Wigners. The mass and width of one Breit-Wigner were fixed to those of the $D_2^*(2460)^0$[4], while those of the other Breit-Wigner were left free, to parameterize the D_1^0. Table 2 lists the mass and width which result from this fit. For comparison, it also lists the previous measurements. From this we see that the D_1^0 is much narrower than initially believed.

In order to determine the fraction of the $D^{*+}\pi^-$ signal which comes from each of the two resonances, the $\cos\alpha$ cut was released and the $D^{*+}\pi^-$ invariant mass spectrum was fitted with the sum of two Breit-Wigners. The mass and width of both were fixed, one to those of the $D_2^*(2460)^0$ and the other to those determined above for the $D_1(2420)^0$. This procedure yields a preliminary result of,

$$\text{BR}(D_2^*(2460)^0 \to D^+\pi^-)/\text{BR}(D_2^*(2460)^0 \to D^{*+}\pi^-) \simeq 3 - 4$$

The theoretically expected value is model dependent and ranges from 1.5 to 4.

In summary we have observed a candidate for the D_{s1}^+ at a mass of 2535.9 ± 0.6 ± 2.0 MeV/c² with a width $\Gamma < 4.3$ MeV/c² at the 90% confidence level. We have also measured the isospin mass splitting of the D_2 system. Finally, we have used an angular analysis to separate the two components of the $D^{*+}\pi^-$ mass spectrum and, thereby, to obtain an improved measurement of the mass and width of D_1^0.

References

[1] H. Albrecht et al. (ARGUS), Phys. Rev. Lett **56** (1986) 549.

[2] J.C. Anjos et al. (E691), Phys. Rev. Lett **62** (1989) 1717.

[3] C. Bebek et al. (CLEO), Paper submitted to the 1987 Intern. Symp. on Lepton and Photon Interactions at High Energies (Hamburg, 1987); M.S. Alam (CLEO), presented at this symposium.

[4] H. Albrecht et al. (ARGUS), Phys. Lett. **221B** (1989) 422.

[5] J.A. Parsons, (ARGUS), to appear in the Proc. of the Twelfth Intern. Workshop on Weak Interactions and Neutrinos, Ginosar, Israel (1989); H. Albrecht et al., (ARGUS), "Resonance Decomposition of the $D^*(2420)^0$ through a Decay Angular Analysis", in preparation.

[6] A.De Rújula, H.Georgi and S.L. Glashow, Phys. Rev. Lett. **37** (1976) 785;
R. Barbieri et al., Nucl. Phys. **B105** (1976) 125;
D. Pignon and C.A. Piketty, Phys. Lett. **81B** (1979) 334;
E. Eichten et al., Phys. Rev. **D21** (1980) 203;
S. Godfrey and N. Isgur, Phys. Rev. **D32** (1985) 189;
J. Morishita et al., Phys. Rev. **D37** (1988) 159;
V. Gupta and R. Kögerler, Z. Phys. **C41** (1988) 277;
A.B. Kaidalov and A.V. Nogteva, Sov. J. Nucl. Phys **47** (1988) 321.

[7] J. Rosner, Comm. Nucl. Part. Phys. **16** (1986) 109.

[8] S. Godfrey and R. Kokoski, TRIUMF Preprint TRI-PP-86-51 (1986).

[9] H. Albrecht et al. (ARGUS), Nucl. Inst. Meth. **A275** (1989) 1.

[10] J. Adler et al. (MARK III), Phys. Lett. **208B** (1988), 152.

[11] Particle Data Group, Phys. Lett. **204B** (1988) 1.

[12] H. Albrecht et al. (ARGUS), "Observation of a New Charmed-Strange Meson", submitted to Phys. Lett. **B**.

[13] C. Peterson et al., Phys. Rev. **D 27** (1983) 105.

[14] J.C. Yun, Ph.D. thesis, Carleton University, Ottawa, Canada (1987), unpublished.

[15] J.A. McKenna, Ph.D. thesis, University of Toronto, Toronto, Canada (1987), unpublished.

[16] R. Kokoski, private communication.

[17] A.E. Asratyan et al., Z. Phys. **C40** (1988) 483.

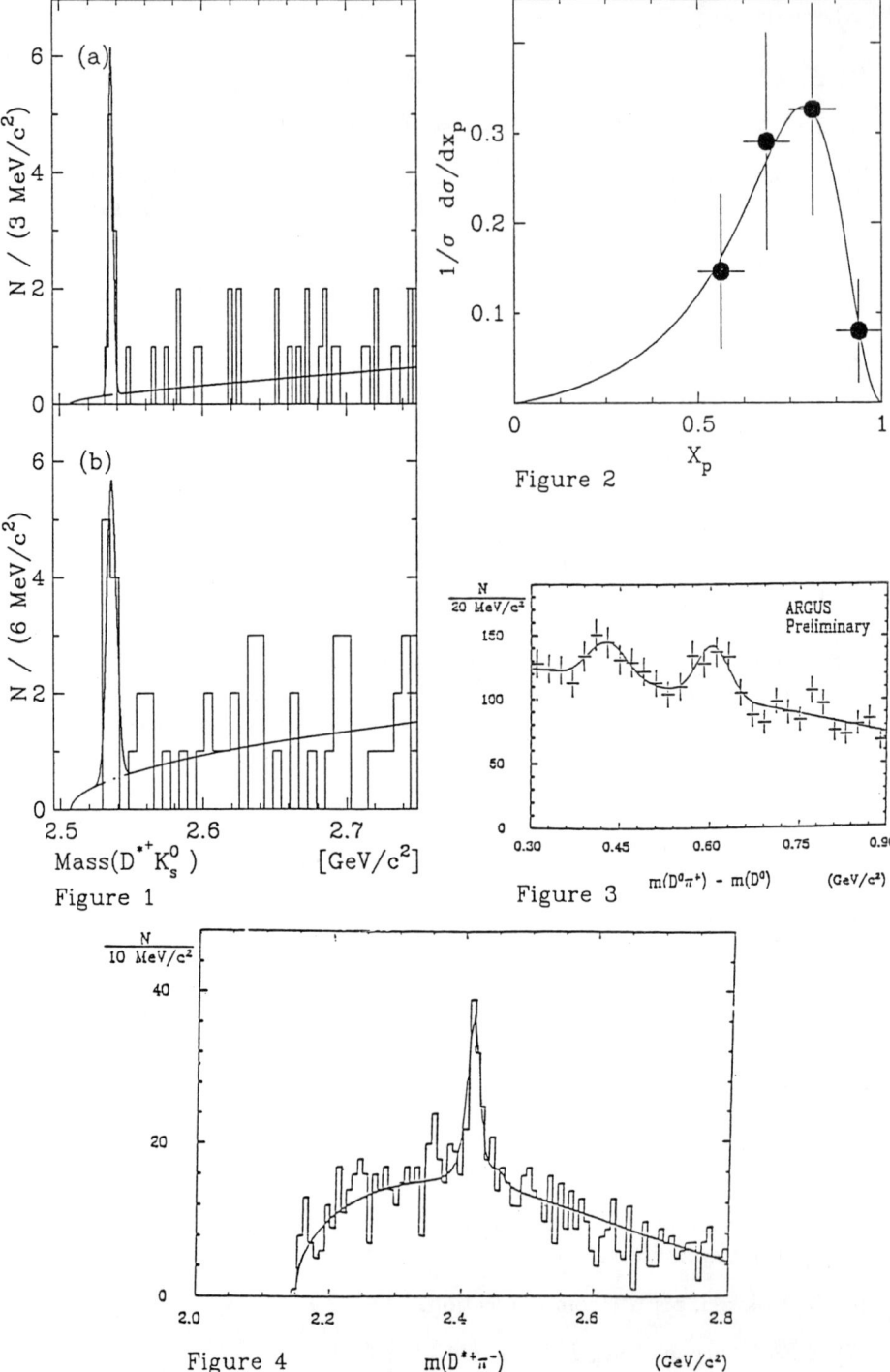

RESONANT SUBSTRUCTURE IN $K^-\pi^+\pi^+\pi^-$ AND $\bar{K}^0\pi^+\pi^+\pi^-$ DECAYS OF CHARMED D MESONS

F. DeJongh

California Institute of Technology, Pasadena, CA 91125

Representing the **Mark III** collaboration

ABSTRACT. We measure the resonant substructure of $D^0 \to K^-\pi^+\pi^+\pi^-$ and $D^+ \to \bar{K}^0\pi^+\pi^+\pi^-$ decays using a five-dimensional maximum likelihood technique to extract the relative fractions and phases of the amplitudes contributing to these final states. We obtain preliminary branching ratios of several decay modes including B($D^0 \to K^- a_1^+$) = .080 ± .008 ± .019, B($D^+ \to \bar{K}^0 a_1^+$) = .081 ± .020 ± .027, and B($D^0 \to \bar{K}^{*0}\rho^0$) = .023 ± .003 ± .007. The branching ratio for $D^0 \to \bar{K}^{*0}\rho^0$ is smaller than theoretically expected and the $\bar{K}^{*0}$ and ρ^0 are found to be polarized in the direction of their motion as seen from the D^0 frame.

Decays of D mesons to two pseudoscalar mesons (PP) and to a pseudoscalar and a vector meson (PV) have been studied in detail by numerous experiments.[1] The branching ratios for most of the PV decay modes were obtained by measuring the resonant substructure of the $D \to K\pi\pi$ final states.[2] Several models of hadronic decays of heavy mesons[3] have been able to describe these results successfully. The methods used in ref. 2 can be extended to study the resonant substructure of higher multiplicity final states and obtain information on new types of decay modes, such as decays to two vector mesons (VV), decays to a pseudoscalar and an axial vector meson (PA), and PP decays where one of the pseudoscalar mesons is a radial excitation of the pion or kaon. The Mark III data sample has a large number of reconstructed events in several $D \to K\pi\pi\pi$ final states, to which many of these decay modes are expected to contribute. We present herein an analysis of the resonant substructure of $D^0 \to K^-\pi^+\pi^+\pi^-$ and $D^+ \to \bar{K}^0\pi^+\pi^+\pi^-$ decays,[4] and results for branching ratios of VV, PA, and nonresonant decay modes.

The data were collected with the Mark III detector[5] at the SLAC e^+e^- storage ring SPEAR near the peak of the $\psi(3770)$, which decays predominantly to $D\bar{D}$. We select candidate $D^0 \to K^-\pi^+\pi^+\pi^-$ and $D^+ \to \bar{K}^0\pi^+\pi^+\pi^-$ events with the procedure used in an earlier analysis of D meson branching fractions.[6] Each $K\pi\pi\pi$ combination is kinematically constrained to the D mass, with the recoil mass allowed to vary. The signal can then be seen in the recoil mass plot as a peak at the D mass. With this type of constraint, all events have the same amount of phase space for the decay throughout the recoil mass plot. The recoil mass plots for the two final states discussed here are shown in fig. 1. We have 1281 ± 45 events in the D^0 final state and 184 ± 21 events in the D^+ final state.

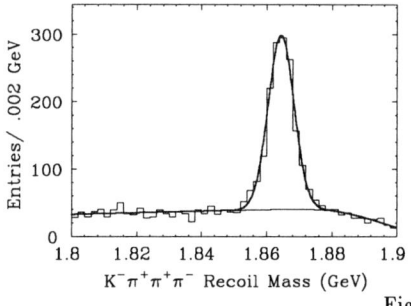
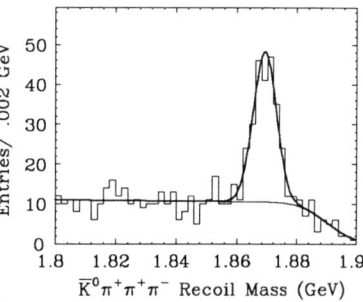

Figure 1

© 1989 American Institute of Physics

Table I Preliminary results for $D^0 \to K^-\pi^+\pi^+\pi^-$

Amplitude	Fraction	Phase	Branching Ratio[12]
4-Body Nonresonant	.233 ± .025 ± .10	-1.01 ± .08	.021 ± .003 ± .009
$\bar{K}^{*0}\rho^0$ Longitudinal	.014 ± .009 ± .01	-2.64 ± .28	Sum of L and T:
$\bar{K}^{*0}\rho^0$ Transverse	.152 ± .021 ± .05	-1.22 ± .11	.023 ± .003 ± .007
$K^-a_1^+$	.442 ± .021 ± .10	.0	.080 ± .008 ± .019
$K_1(1270)^-\pi^+$	.113 ± .028 ± .04	.44 ± .19	.031 ± .008 ± .011
$K_1(1400)^-\pi^+$	.011 ± .009 ± .03	.71 ± .43	< .012
$\bar{K}^{*0}\pi^+\pi^-$	.091 ± .018 ± .04	-3.31 ± .11	.012 ± .003 ± .005
$K^-\rho^0\pi^+$	.088 ± .023 ± .04	-.62 ± .09	.008 ± .002 ± .004

For a particular $K\pi\pi\pi$ final state, we determine the contribution of each decay mode using a maximum likelihood fit. The likelihood function $\mathcal{L}$ is a function in the five dimensional phase space defined by the four-momenta of the decay products of the D candidate, and consists of a signal term $\mathcal{L}_S$ and a background term $\mathcal{L}_B$:

$$\mathcal{L} = \frac{R_{S/B}\mathcal{L}_S + \mathcal{L}_B}{R_{S/B} + 1} \quad (1)$$

For each event we calculate the ratio of signal to background, $R_{S/B}$, as a function of recoil mass. We determine $\mathcal{L}_B$ with a fit to the events in the sideband regions of the recoil mass plot, and then fit $\mathcal{L}$ to the events in the signal region.

The function $\mathcal{L}_S$ is a coherent sum of complex amplitudes with the relative fractions f_i and phases α_i allowed to vary:

$$\mathcal{L}_S = \epsilon |\sum_{i=1}^{n} \sqrt{f_i}\, e^{i\alpha_i} B_i|^2 \phi \quad (2)$$

The efficiency ϵ is the probability of reconstructing an event as a function of position in the five-dimensional phase space, and ϕ is the four-body phase space function. We model each decay mode considered in the fit with a complex amplitude B_i. These amplitudes consist of a relativistic Breit-Wigner for each resonance in the decay chain, times a matrix element that depends on the spin and parity of the intermediate resonances and final decay products. These matrix elements are derived using either the Lorentz invariant amplitude formalism or the helicity formalism. The amplitudes are symmetrized with respect to the two identical pions.

The function $\mathcal{L}_B$ is an incoherent sum of functions G_i with the relative fractions allowed to vary:

$$\mathcal{L}_B = \epsilon \sum_{i=1}^{m} f_i G_i \phi \quad (3)$$

We model each source of background with a function G_i. For each final state, a set of functions including a $\bar{K}^*$ Breit-Wigner, a ρ Breit-Wigner, a product of a $\bar{K}^*$ and a ρ Breit-Wigner, and a constant term adequately describes the background.

Since ϵ and ϕ do not depend on the variables in the fit, they may be factored out when minimizing the sum over events of $-\ln(\mathcal{L})$. We evaluate ϵ and ϕ using Monte Carlo techniques when normalizing the B_i, $\mathcal{L}_S$, and $\mathcal{L}_B$, and when calculating one-dimensional projections of $\mathcal{L}$.

In previous analyses,[8-11] the resonant substructure of $D^0 \to K^-\pi^+\pi^+\pi^-$ decays was measured by fitting one-dimensional mass plots to obtain the fraction of $\bar{K}^{*0}\rho^0$, inclusive $\bar{K}^{*0}$, inclusive ρ^0, and nonresonant four-body. The advantage of the approach used here is that the amplitudes provide a complete description of the decay modes in the five-dimensional phase space and all the information available in the event is used in the fit, making it possible to fit to a general set of amplitudes, include interference, and obtain fractions for exclusive decay modes.

Table II Comparison between experiments

Channel	Mark I[8]	Accmor[9]	Argus[10]	E691[11]	Mark III
$\bar{K}^{*0} X$			$.39 \pm .03$	$.26 \pm .04 \pm .03$	$.201 \pm .019 \pm .03$
$\rho^0 X$			$.86 \pm .10$	$1.06 \pm .06 \pm .09$	$.825 \pm .029 \pm .06$
$\bar{K}^{*0} \rho^0$	$.1^{+.11}_{-.10}$	$.5 \pm .2$	$.35 \pm .06$		$.167 \pm .021 \pm .05$
$K^- \rho^0 \pi^+$	$.85^{+.11}_{-.22}$	$.2 \pm .2$	$.51 \pm .08$		$.088 \pm .023 \pm .04$
$K^- a_1^+$			$.51 \pm .08$†		$.442 \pm .021 \pm .10$
$\bar{K}^{*0} \pi^+ \pi^-$	$.0^{+.2}_{-.0}$	$< .18$	$.04 \pm .04$		$.091 \pm .023 \pm .04$
$K^- \pi^+ \pi^+ \pi^-$	$.05^{+.11}_{-.05}$		$.11^{+.11}_{-.05}$		$.233 \pm .025 \pm .10$

†In the Argus analysis, angular distributions of ρ^0 decays outside the $\bar{K}^{*0}$ bands were examined. The $K^- \rho^0 \pi^+$ component was found to be consistent with being entirely $K^- a_1^+$.

For each final state, a very large number of decay modes can potentially contribute. It is not practical to perform a fit that includes all possible decay modes simultaneously. Instead, we performed a large number of fits with different combinations of amplitudes. The fits with good likelihood gave similar results for the fractions of quasi-two-body amplitudes and the four-body nonresonant amplitude. Nonresonant $K^* \pi \pi$ and $\rho K \pi$ amplitudes also contributed. For these three-body amplitudes, there are several possibilities for the relative partial waves of the vector and two pseudoscalar mesons. The fits did not yield definitive results on which partial waves contribute. The systematic errors reflect the dependence of the fractions on which three-body amplitudes are included in the fit.

Since the a_1 meson is very broad, the $K a_1$ amplitudes are difficult to distinguish from nonresonant $K \rho \pi$ where the ρ and π are in a relative s-wave. Since the a_1 mass is very close to the maximum three-pion mass available in D decays, the $K a_1$ amplitude peaks about .1 GeV above the nonresonant amplitude. Fits in which the $K a_1$ amplitude was replaced by the three-body amplitude resulted in a significantly smaller likelihood, with a difference in $\ln(\mathcal{L})$ of at least 12 for $D^0 \to K^- \pi^+ \pi^+ \pi^-$ and 4 for $D^+ \to \bar{K}^0 \pi^+ \pi^+ \pi^-$. Therefore, it is assumed that this particular three-body amplitude does not contribute to the final states studied here.

The results of a representative fit to the $D^0 \to K^- \pi^+ \pi^+ \pi^-$ final state are in table I. The fractions have been scaled so that the likelihood function is properly normalized; because of interference, they do not sum to one. We included amplitudes for $K^- a_1^+$, $K_1(1270)^- \pi^+$, and $K_1(1400)^- \pi^+$ decays. For s-wave $\bar{K}^{*0} \rho^0$ decays, we considered two possibilities for the orientations of the spins of the $\bar{K}^{*0}$ and ρ^0 with respect to their direction of motion as seen from the D^0 rest frame. We included a transverse s-wave $\bar{K}^{*0} \rho^0$ amplitude, in which the spins have a parallel orientation, and a longitudinal s-wave $\bar{K}^{*0} \rho^0$ amplitude, in which the spins have a perpendicular orientation. We included a constant term for four-body nonresonant decays, in which all the final state mesons are in relative s-waves. We included one $\bar{K}^{*0} \pi^+ \pi^-$ amplitude in which the $\bar{K}^{*0} \pi^-$ system is in a pseudoscalar state, and one $K^- \rho^0 \pi^+$ amplitude in which the $K^- \rho^0$ system is in an axial vector state. The fractions obtained for the three-body modes do not depend strongly on the relative partial waves assumed for the vector and two pseudoscalar mesons.

One model based on a QCD factorization assumption has been used to make predictions for D meson decays to two vector mesons (VV) and to a pseudoscalar and an axial vector meson (PA).[13] In this model, the branching fractions are $B(D^0 \to \bar{K}^{*0} \rho^0) = 6.1\%$ and $B(D^0 \to K^- a_1^+) = 5.0\%$. We find $BR(D^0 \to K^- a_1^+) = 8.0\%$ which is very large as expected, and $BR(D^0 \to \bar{K}^{*0} \rho^0) = 2.3\%$ which is much smaller than predicted. Our results for $D^0 \to \bar{K}^{*0} \rho^0$ are consistent with complete transverse polarization.

Projections of $\mathcal{L}$ from this fit onto one-dimensional mass plots are shown in fig. 2. The $\pi^+ \pi^-$ combination with the higher mass is referred to as $(\pi^+ \pi^-)_{hi}$, and the $K^- \pi^+$ combination

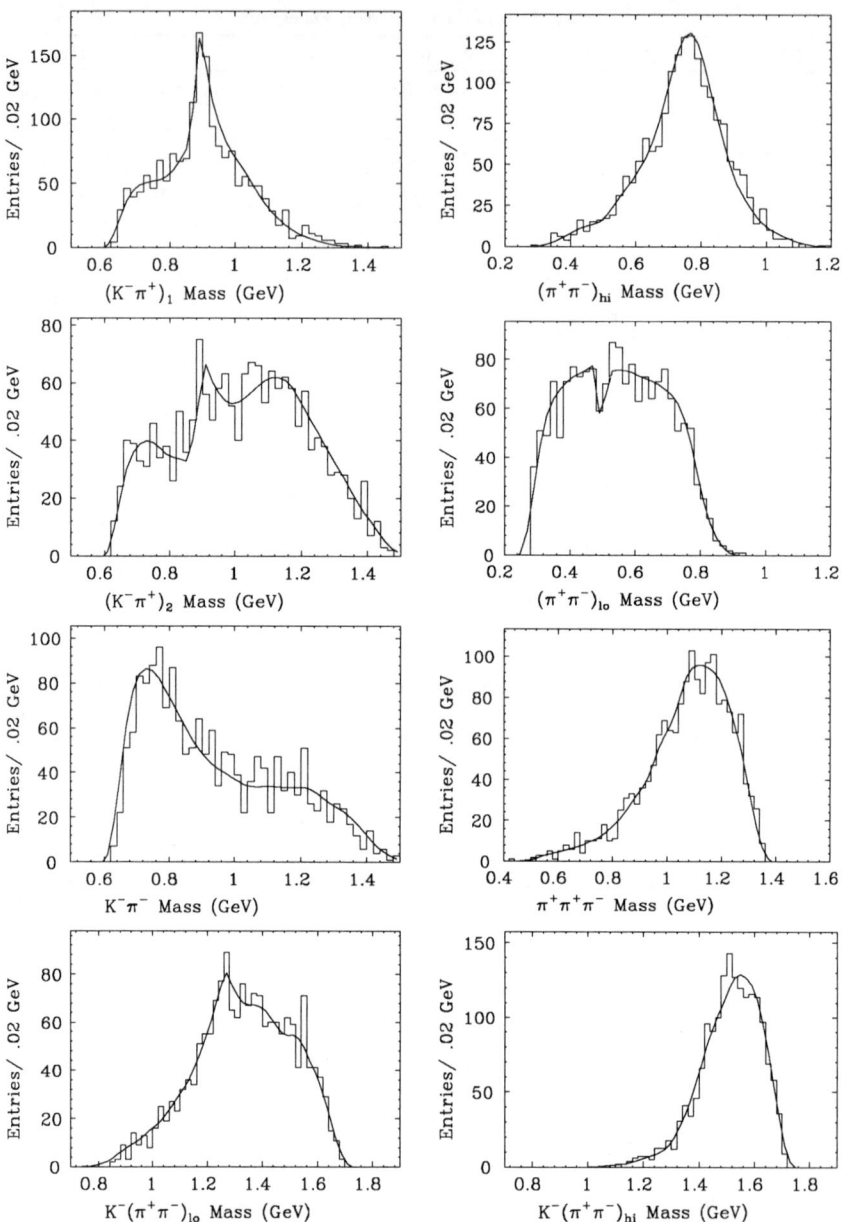

Figure 2. Projections of the likelihood function for $D^0 \to K^-\pi^+\pi^+\pi^-$. The solid lines, representing the projections of the likelihood function, are superimposed on histograms of the events in the signal region. The $\pi^+\pi^-$ combination with the higher mass is referred to as $(\pi^+\pi^-)_{hi}$, and the $K^-\pi^+$ combination formed with the π^+ not used in $(\pi^+\pi^-)_{hi}$ is referred to as $(K^-\pi^+)_1$. The deficit near .5 GeV in the $(\pi^+\pi^-)_{lo}$ mass plot is due to the rejection of $\pi^+\pi^-$ combinations which have a high probability of originating from a K_s decay.

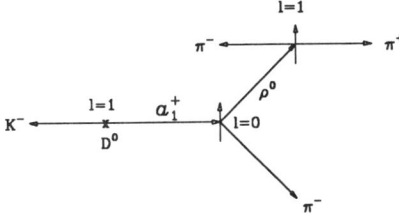

Figure 3. Illustration of the $K^-a_1^+$ amplitude. The small vertical arrows indicate the polarization of the a_1 and ρ. The relative orbital angular momentum at each vertex is shown. Because of the longitudinal polarization of the a_1 and subsequent polarization of the ρ, the π^- tends to be produced in a forward or backward direction with respect to the direction of the K^-, producing a distribution with an enhancement at low $K^-\pi^-$ mass.

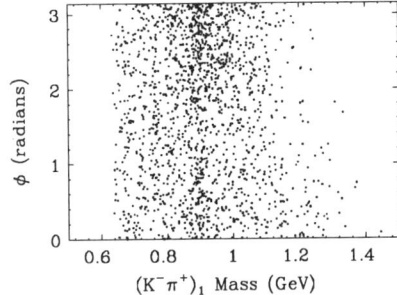

Figure 4. Scatter plot of $(K^-\pi^+)_1$ mass vs ϕ, where ϕ is the angle between the $\bar{K}^{*0}$ and ρ^0 decay planes as seen from the D^0 rest frame. In the $\bar{K}^{*0}$ band, an enhancement near $\phi=0$ and a larger enhancement near $\phi=\pi$ are visible. The transverse $\bar{K}^{*0}\rho^0$ amplitude is proportional to $\cos\phi$ and accounts for this distribution. Since the sign of this amplitude reverses from $\phi=0$ to $\phi=\pi$, there is more constructive interference near $\phi=\pi$.

Table III Preliminary results for $D^+ \to \bar{K}^0\pi^+\pi^+\pi^-$ [12]

Amplitude	Fraction	Phase	Branching Ratio
4-Body Nonresonant	.184 ± .052 ± .10	1.37 ± .17	.012 ± .004 ± .007
$K^-a_1^+$	.612 ± .053 ± .15	.0	.081 ± .020 ± .027
$\bar{K}_1(1270)^0\pi^+$	.010 ± .013 ± .02	1.30 ± .90	< .011
$\bar{K}_1(1400)^0\pi^+$	.163 ± .048 ± .08	.24 ± .26	.024 ± .009 ± .013

formed with the π^+ not used in $(\pi^+\pi^-)_{hi}$ is referred to as $(K^-\pi^+)_1$. Clear $\bar{K}^{*0}$ and ρ^0 peaks are visible, and are reproduced well by the fit. A peak at the $K_1(1270)$ mass is visible in the $K^-(\pi^+\pi^-)_{lo}$ mass plot, and is reproduced with an amplitude modeling the three largest decay modes of the $K_1(1270)$. The enhancement at low $K^-\pi^-$ mass is reproduced with the $K^-a_1^+$ amplitude, and is due to the longitudinal polarization of the a_1. We illustrate this effect in fig. 3. The presence of the transverse $\bar{K}^{*0}\rho^0$ amplitude leads to angular correlations between $\bar{K}^{*0}$ and ρ^0 decays. We show an example in fig. 4.

A comparison with results from other experiments is in table II. We calculate the total $\bar{K}^*$ and ρ content from our fits by summing the appropriate exclusive fractions and taking into account interference between the decay modes.

The results of a representative fit to the $D^+ \to \bar{K}^0\pi^+\pi^+\pi^-$ final state are in table III. No $\bar{K}^*\rho$ mode can contribute to this final state; to form a $\bar{K}^*$ we must have $D^+ \to K^{*-}\pi^+\pi^+$. No three-body mode was found to contribute at a statistically significant level to this final state.

Projections of $\mathcal{L}$ from this fit onto mass plots are shown in fig. 5. A clear K^{*-} peak is seen, and is reproduced well by the $\bar{K}_1(1400)^0\pi^+$ amplitude. As for the $D^0 \to K^-\pi^+\pi^+\pi^-$ final state, the $D^+ \to \bar{K}^0a_1^+$ decay mode is very large. In contrast to the $D^0 \to K^-\pi^+\pi^+\pi^-$ final state, the $D^+ \to \bar{K}_1(1400)^0\pi^+$ decay mode is present, and the $D^+ \to \bar{K}_1(1270)^0\pi^+$ decay mode does not contribute at a statistically significant level.

We gratefully acknowledge the dedicated efforts of the SPEAR staff. This work was

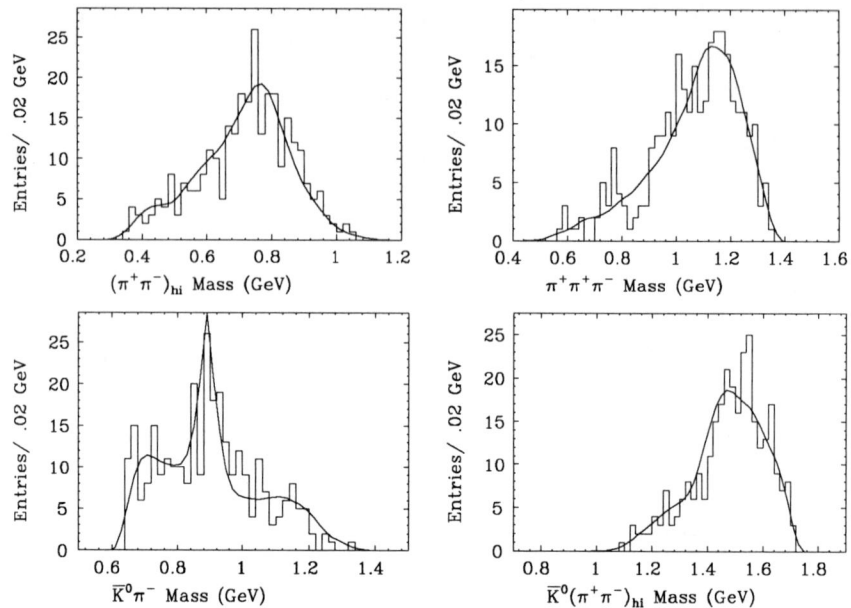

Figure 5. Projections of the likelihood function for $D^+ \to \bar{K}^0 \pi^+ \pi^+ \pi^-$.

supported by the Department of Energy, under contracts DE-AC03-76SF00515, DE-AC02-76ER01195, DE-AC03-81ER40050, DE-AC02-87ER40318, DE-AM03-76SF00034 and by the National Science Foundation.

REFERENCES

1. G.P. Yost et al. (Particle Data Group), Phys. Lett. **204B**, 1 (1988).
2. J. Adler et al., Phys. Lett. **196B**, 1 (1987).
3. For a review, see M.A. Shifman, in *International Symposium on Lepton and Photon Interactions*, edited by W. Bartel and R. Rückl (North-Holland, Amsterdam, 1987).
4. We adopt the convention that reference to a state implies reference to its charge conjugate.
5. D. Bernstein et al., Nucl. Instrum. Methods **226**, 301 (1984).
6. R.M. Baltrusaitis et al., Phys. Rev. Lett. **56**, 2140 (1986);
7. J. Adler et al., Phys. Rev. Lett. **60**, 89 (1988);
8. M. Piccolo et al., Phys. Lett. **70B**, 260 (1977).
9. R. Bailey et al., Phys. Lett. **132B**, 237 (1983).
10. P. Kim, Ph.D. thesis, University of Toronto, 1987 (unpublished).
11. J.C. Anjos et al., contributed paper, International Conference on Energy Physics, Munich, 1988
12. Branching ratios were calculated using $B(D^0 \to K^- \pi^+ \pi^+ \pi^-) = .091 \pm .008 \pm .008$ and $B(D^+ \to \bar{K}^0 \pi^+ \pi^+ \pi^-) = .066 \pm .015 \pm .015$ from ref 7, and the branching ratios of the intermediate resonances to the final state being studied.
13. M. Bauer, B. Stech and M. Wirbel, Z. Phys. C **34**, 103 (1987). Revised values of $a_1 = 1.2$ and $a_2 = -0.5$ are taken from B. Stech, preprint HD-THEP-87-18, 1987 (unpublished). The prediction for $B(D^0 \to K^- a_1^+)$ is from a private communication with B. Stech.

RESULTS ON HADRONIC DECAYS OF D AND Ds MESONS FROM MARK III

Peter C. Kim

Mark III Collaboration
Stanford Linear Accelerator Center
Stanford University, Stanford, CA

ABSTRACT

Recent results on hadronic weak decays of the charmed mesons, $D^0 \to \overline{K}^0\pi^0$, $\overline{K}^0\eta$, $\overline{K}^0\eta'$ and $D_s^+ \to \eta\pi^+$, $\eta'\pi^+$, from the Mark III at SPEAR are presented.

INTRODUCTION

Our understanding of the hadronic decays of the charmed mesons has been greatly improved by recent experimental measurements.[1] There now exist theoretical models[2,3] which fit the data relatively well. However, reconstruction of decay modes containing neutrals has been a diffcult task due to the large background levels and the modest photon energy resolutions of most detectors. We present here a study of the decays, $D^0 \to \overline{K}^0\pi^0$, $\overline{K}^0\eta$, and $\overline{K}^0\eta'$, which utilizes unique kinematic constraints possible at $D\overline{D}$ threshold. The above channels, presumed to be color-suppressed, could provide a test for the QCD color-matching hypothesis. One could in principle measure the mixing angle of the η and η' mesons. However, these channels may be greatly influenced by the final state interactions which could destroy any signatures at the quark diagram level.

In the D_s decay, only a few decay modes have been observed.[4] A recent estimate of the branching ratio for the decay, $D_s \to \phi\pi$ at the 1.5% level[5] indicates that the majority of D_s decays have not been observed. Therefore, recent reports of large rates for the decays $D_s \to \eta\pi$ and $D_s \to \eta'\pi$ from the MARK II[6] and NA14'[7] are of great interest. In a naive spectator model, these channels are expected to have similar rates compared to the $\phi\pi$ channel.[2,3,8]

© 1989 American Institute of Physics

ANALYSIS

Data samples of 9.3 pb^{-1} at $\sqrt{s} = 3.77$ GeV and 6.3 pb^{-1} at $\sqrt{s} = 4.14$ GeV were collected with the Mark III detector[9] at the e^+e^- storage ring SPEAR. Charge tracks are identified using the TOF information and neutral kaons by the decay $K_s^0 \to \pi^+\pi^-$.

The $\Psi(3770)$ resonance decays predominantly into $D\bar{D}$ pairs; each D meson carries away energy equal to the beam energy. In the study of the $D^0 \to \bar{K}^0\pi^0$, a 3-C kinematic fit is applied to $\pi^+\pi^-\gamma\gamma$ combinations with energy within 50 MeV of the beam energy. The three constraints are M_{K^0}, M_{π^0}, and that the missing recoil mass be equal to $M_{\pi^+\pi^-\gamma\gamma}$. Figure 1a shows the mass distribution of the $\pi^+\pi^-\gamma\gamma$ combinations where the fit probability is required to be greater than 10% and the fitted photon energy greater than 150 MeV. A fit with a Gaussian peak of a fixed width of 3.5 MeV plus a 3rd order polynomial background yields 74 ± 10 signal events. With a detection efficiency of 23% from a Monte Carlo study and $\sigma_{D^0} = 5.8 \pm 0.8$ nb,[10] we have $B(D^0 \to \bar{K}^0\pi^0) = 1.8 \pm 0.2 \pm 0.2\%$.

The same 3-C fit is performed on $\pi^+\pi^-\gamma\gamma$ combinations for the $\bar{K}^0\eta$ mode except the $M_{\gamma\gamma}$ constraint is changed to, $M_{\gamma\gamma} = M_\eta$. Combinations which have a fit probability greater than 5% and $E_\gamma > 120$ MeV are shown in Figure 1b. The curve shown is the result of a fit with a Gaussian peak ($\sigma = 3.0$ MeV fixed) plus a background of a shape: 3rd order polynomial $\times (M - M_{threshold})^\alpha$. There are 28 ± 10 events in the Gaussian signal; with a detection efficiency of 22%, it corresponds to $B(D^0 \to \bar{K}^0\eta) = 1.6 \pm 0.6 \pm 0.4\%$

A search for the decay $D^0 \to \bar{K}^0\eta'$ is made using the channel, $\eta' \to \rho\gamma$, $\rho \to \pi^+\pi^-$. Figure 1c shows the result of a 3-C fit of $\pi^+\pi^-\pi^+\pi^-\gamma$ combinations. Successful candidates must saitsfy the criteria that the invariant mass of the two pions from the ρ decay lie in the range, $600 < M_{\pi^+\pi^-} < 800$ MeV, $E_\gamma > 50$ MeV, and the fit probability be greater than 10%. A fit with a fixed width of 1.8 MeV Gaussian gives 1 ± 3 events. Detection efficiency from

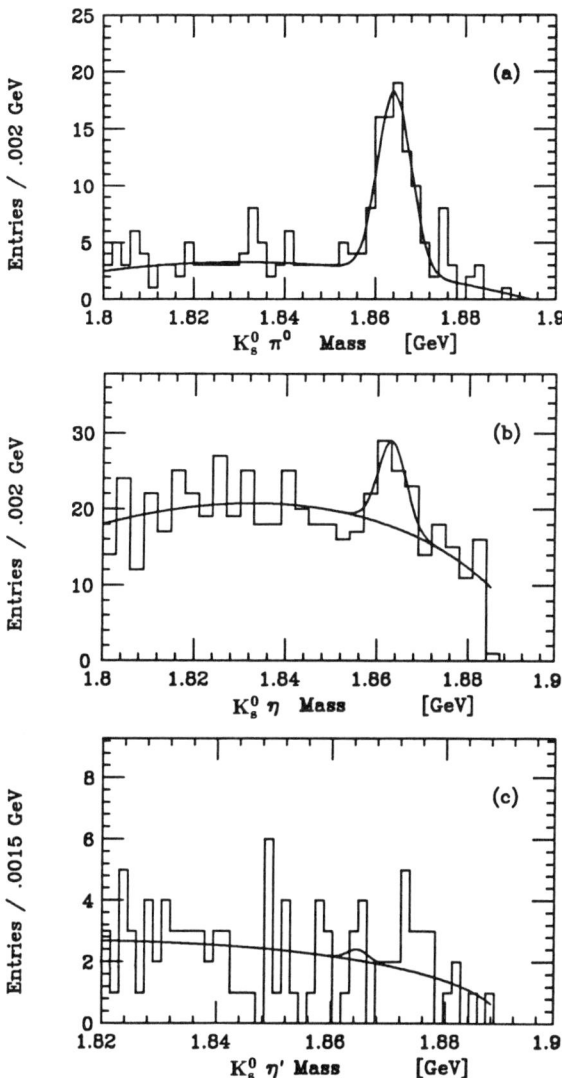

FIGURE 1. Fitted mass distributions; (a) $K_s^0\pi^0$, $\pi^0 \to \gamma\gamma$, (b) $K_s^0\eta$, $\eta \to \gamma\gamma$, and (c) $K_s^0\eta'$, $\eta' \to \rho\gamma$, $\rho \to \pi\pi$.

a Monte Carlo study is 5%. We set a 90% confidence level upper limit of $B(D^0 \to \overline{K}^0 \eta') < 2.7\%$.

At $\sqrt{s} = 4.14$ GeV, the D_s mesons are produced mainly in the reaction, $e^+e^- \to D_s^{\pm} D_s^{*\mp}, D_s^{*\mp} \to \gamma D_s^{\mp}$. We can apply a kinematic fit to the primary D_s with a missing D_s^* constraint. For the D_s from the D_s^*, this fit with a missing D_s^* results in a broadened signal shape. Therefore, to extract the number of signal events, we use a signal shape which is a mixture of the primary and secondary D_s mesons determined from a Monte Carlo simulation.

For the decay mode $D_s \to \eta\pi$, two different decay channels of the η are used, $\eta \to \pi^+\pi^-\pi^0$ and $\eta \to \gamma\gamma$. In the three-pion channel, candidate π^0's are selected by fitting two photons to the π^0 mass and taking the combinations which have a fit probability greater than 5%. A 2-C kinematic fit using the π^0 and missing D_s^* masses as constraints is performed to all $\pi^+\pi^-\pi^0\pi^+$ combinations. Combinations are retained if the fit probability is greater than 5%, $E_\gamma(\text{fitted}) > 70$ MeV, and $534 < M(\pi^+\pi^-\pi^0) < 564$ MeV. Figure 2a shows the fitted mass distribution of these combinations. A fit to a Monte Carlo determined signal shape and a background shape from the η side bands gives 16.6 ± 6.1 signal events. With a detection efficiency of 12.7%, we obtain $\sigma \cdot B(D_s^+ \to \eta\pi^+) = 44 \pm 16 \pm 12$ pb or $B(D_s \to \eta\pi)/B(D_s \to \phi\pi) = 1.7 \pm 0.7 \pm 0.5$. In the $\gamma\gamma$ channel, all two-photon combinations with a fit probabilty greater than 20% in a 1-C fit to the η mass, are selected as η candidates. Figure 2b shows the fitted mass plot of the $\gamma\gamma\pi^+$ combinations after a 2-C fit to M_η and a missing D_s^*, and requiring that each combination satisfy the following cuts: fit probabiblity $> 10\%$, $E_\gamma^{\text{hi}}(\text{fitted}) > 500$ MeV, and $E_\gamma^{\text{lo}}(\text{fitted}) > 200$ MeV. No signal is observed. Using a detection efficiency of 24% and including a 27% systematic error, we have $\sigma \cdot B(D_s^+ \to \eta\pi^+) < 42.5$ pb (90% C.L.). From these two measusments of the $\eta\pi$ mode, we proceed to calculate the 90% C.L. upper limit from a joint likelihood function and obtain $\sigma \cdot B(D_s^+ \to \eta\pi^+) < 66$ pb and $B(D_s^+ \to \eta\pi^+)/B(D_s^+ \to \phi\pi^+) < 2.5$.

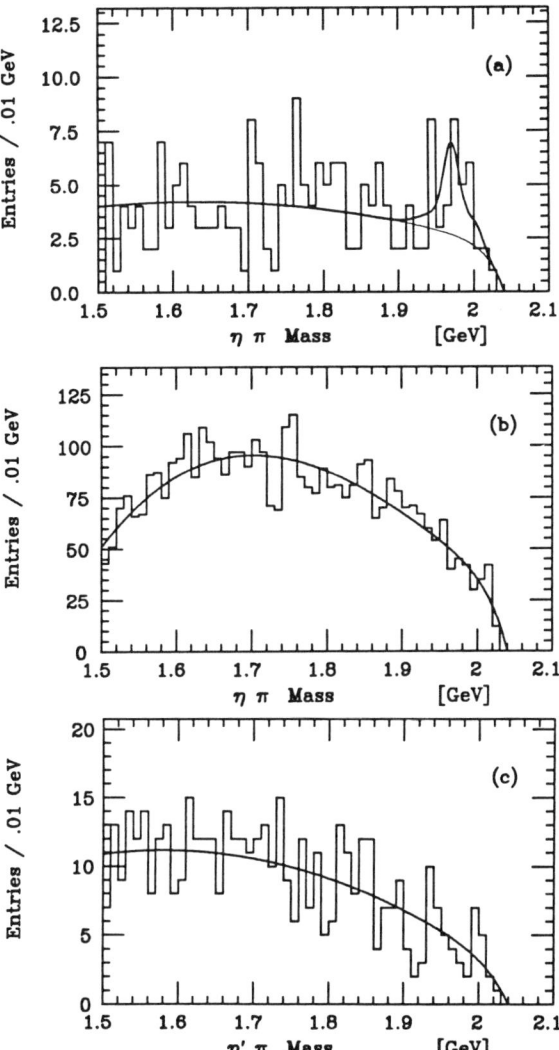

FIGURE 2. (a) $\eta\pi$ mass distribution, $\eta \to \pi^+\pi^-\pi^0$, (b) $\eta\pi$ mass distribution, $\eta \to \gamma\gamma$, and (c) $\eta'\pi$ mass distribution, $\eta' \to \eta\pi^+\pi^-$, $\eta \to \gamma\gamma$.

In the search for the $\eta'\pi$ decay mode of the D_s we look for the decay chain, $D_s^+ \to \eta'\pi^+$, $\eta' \to \eta\pi^+\pi^-$, and $\eta \to \gamma\gamma$. Photons from the η decay are fitted to M_η, with the fit probability required to be greater than 10%. A 2-C fit (M_η, and a missing D^*) to the $\gamma\gamma\pi^+\pi^-\pi^+$ combinations is then performed. Figure 2c shows the mass plot of the combinations with a fit probability $> 10\%$, E_γ(fitted) > 150 MeV, and $|M(\eta\pi^+\pi^-) - M(\eta')| < 15$ MeV. The background shape shown is from the η' side bands. We observe no signal events and set an upper limit of $B(D_s^+ \to \eta'\pi^+)/B(D_s^+ \to \phi\pi^+) < 1.9$ at the 90% C.L.

CONCLUSIONS

We have made preliminary measurements for branching fractions, $B(D^0 \to \overline{K}^0\pi^0) = 1.8 \pm 0.2 \pm 0.2\%$ and $B(D^0 \to \overline{K}^0\eta) = 1.6 \pm 0.6 \pm 0.4\%$, and set an upper limit of $B(D^0 \to \overline{K}^0\eta') < 2.7\%$ at the 90% C.L. For these channels, we conclude that there is either no color-suppression or that these channels are enhanced via the final state interactions. Searches for $\eta\pi$ and $\eta'\pi$ modes in D_s decays yield upper limits on relative branching fractions, $B(D_s \to \eta\pi)/B(D_s \to \phi\pi) < 2.5$ and $B(D_s \to \eta'\pi)/B(D_s \to \phi\pi) < 1.9$ at the 90% C.L., which are in disagreement with the MARK II and NA14' measurements.

This work was supported by the Department of Energy, under contracts DE-AC03-76SF00515, DE-AC02-76ER01195, DE-AC03-81ER40050, DE-AC02-87ER40318, DE-AM03-76F00034 and by the National Science Foundation.

REFERENCES

1. R. Morrison and M. Witherell, Ann. Rev. Nucl. Sci. **39**, 183 (1989).

2. M. Bauer, B. Stech and M. Wirbel, Z. Phys. **C34**, 103 (1987)

3. B. Blok and M. Shifman, Sov. J. Nucl. Phys. **45**, 103 (1987).

4. R. Schindler, Proceedings of the Worksop on Weak Interactions and Neutrinos, Ginosar, Isreal (1989), SLAC-PUB-4997.

5. S. Stone, Proceedings of the Worksop on Weak Interactions and Neutrinos, Ginosar, Isreal (1989), CLNS 89/925.

6. $B(D_s \to \eta\pi)/B(D_s \to \phi\pi) = 3.0 \pm 1.1$ and $B(D_s \to \eta'\pi)/B(D_s \to \phi\pi) = 4.8 \pm 2.1$, G. Wormser *et al.*, Phys. Rev. Lett. **61**, 1057 (1988)

7. $B(D_s \to \eta'\pi)/B(D_s \to \phi\pi) = 6.9 \pm 2.4 \pm 1.4$, G. Wormser, Proceedings of XXIVth Recontres de Moriond (1989), LAL89-10 (1989).

8. A. N. Kamal, N. Sinha and R. Sinha, Phys. Rev. D **38**, 1612 (1988).

9. D. Bernstein *et al.*,, Nucl. Inst. Meth. **226**, 301 (1984).

10. J. Adler *et al.*, Phys. Rev. Lett. **60**, 89 (1988).

Recent Results on D_s^+ Decays from E691

Greg Punkar
University of California, Santa Barbara, CA 93106
representing the Tagged Photon Collaboration[1]

ABSTRACT

Recent results on decays of the charmed strange pseudoscalar meson, the D_s^+, are presented. The data were obtained in Fermilab photoproduction experiment E691.[1] Topics discussed include: a probe of the weak annihilation mechanism by studying the decays $D_s^+ \to \rho\pi^+$ and $D_s^+ \to \omega\pi^+$; an observation of the decay $D_s^+ \to f_0(975)\pi^+$; two independent measurements of $D_s^+ \to \eta\pi^+$; comments on the vector-vector decays $D_s^+ \to \phi\rho^+$ and $D^+ \to \overline{K}^{*0}\rho^+$; and an accounting of measured D_s^+ decays.

I. WEAK ANNIHILATION

We begin with the search for $D_s^+ \to \rho\pi^+$, $\rho \to \pi^+\pi^-$. The $\pi^+\pi^-\pi^+$ mass spectrum from the E691 data sample is shown in figure 1. Peaks are clearly visible at the D^+ and D_s^+, as well as a structure below 1.8 GeV which is due to $D^+ \to K^-\pi^+\pi^+$ decays where the kaon is misidentified as a pion.

We search for resonant substructure by doing a Dalitz plot fit. Shown in figure 2(a) is the Dalitz plot distribution for $D_s^+ \to \rho\pi^+$ decays obtained from a Monte Carlo simulation. The corresponding distribution for $\pi^+\pi^-\pi^+$ combinations in the D_s^+ mass range from the data is shown in figure 2(b). No $\rho\pi^+$ signal is obvious in this plot.

The $D_s^+ \to \pi^+\pi^-\pi^+$ Dalitz plot analysis was done as follows: First the Dalitz plot was unfolded about its symmetry axis. Then events with $\pi^+\pi^-\pi^+$ invariant mass in the range 1.92-2.02 GeV were each mapped onto the Dalitz plot twice, once in the upper half and once in the lower half. In the fit function the $\rho\pi^+$ component was represented by two bands, one parallel to the x-axis and one parallel to the y-axis. Each of these bands was a Breit-Wigner distribution, with the $\cos^2\theta$ distribution characteristic of PV (pseudoscalar-vector) decays of pseudoscalar particles. Interference between the two bands resulted from adding the amplitudes for each band, then squaring the result. The $f_0(975)\pi^+$ component (subject of section II) was treated in an identical manner to the $\rho\pi^+$, except that a coupled-channel Breit-Wigner[2] was used (since the dikaon decay channel opens up in the high-mass tail of the f_o). We also allowed for non-resonant $D_s^+ \to \pi^+\pi^-\pi^+$ decays, and background $\pi^+\pi^-\pi^+$ combinations, in the fit. These latter two components were represented by constant-density functions on the Dalitz plot. Each component of the fit function was symmetric with respect to interchange of the x and y axes. Separation between background and signal components was achieved by fitting simultaneously to the Dalitz plot, and to the $\pi^+\pi^-\pi^+$ mass axis. On the $\pi^+\pi^-\pi^+$ axis, the background was represented by a linear function, and the D_s^+ signal components by a Gaussian of width 10.7 MeV (as determined by the Monte Carlo). A component was also allowed in the background for ρ's combining with random pions; the size of this component was determined by examining the $\pi^+\pi^-\pi^+$ mass regions above and below the D_s^+.

We also checked for interference between resonant and non-resonant components by allowing the non-resonant component to be an s-wave with a phase which floated in the fit. The fit was performed both with and without this interference.

Shown in figure 3(a) is the projection of data (two entries per event) and fit function for the case in which no interference between resonant and non-resonant components was allowed. The corresponding plot for the fit where such interference was allowed, is shown in figure 3(b). The bump at about 0.6 GeV^2 in the fit function in both plots is due to background $\rho\pi^+$ combinations. A better χ^2 was achieved for the fit of figure 3(a); this fit yielded -1.7 ± 7.7 $D_s^+ \to \rho\pi^+$ events. Using the E691 measurement of the decay mode $D_s^+ \to \phi\pi^+$ and correcting for detection efficiencies, this corresponds to

$$B(D_s^+ \to \rho\pi^+)/B(D_s^+ \to \phi\pi^+) \leq 0.08$$

at the 90% confidence level.

The decay mechanisms for $D_s^+ \to \rho\pi^+$ and $D_s^+ \to \phi\pi^+$ are shown in figures 4(a) and 4(b), respectively: the former must proceed by weak annihilation, and the latter must proceed by the spectator mechanism. Thus the ratio quoted above is evidence that annihilation amplitudes are small compared to spectator amplitudes.

However, a theoretical argument[3] states that $D_s^+ \to \rho\pi^+$ can be small, but annihilation can still be a large effect. This is based on the fact that the ρ wavefunction contains a relative minus sign between $u\bar{u}$ and $d\bar{d}$ components, whereas the wavefunction for the ω contains a relative plus sign. Thus these two components could interfere destructively for the ρ, but constructively for the ω. This argument, which is illustrated in figure 5, predicts that annihilation amplitudes may be comparable to spectator amplitudes (in spite of a very small $D_s^+ \to \rho\pi^+$ branching ratio), if $B(D_s^+ \to \omega\pi^+)$ is comparable to $B(D_s^+ \to \phi\pi^+)$. However, at this conference, Lipkin[4] has argued that the argument of reference 3 is incorrect. In any case, it is useful to independently measure $D_s^+ \to \omega\pi^+$ (which, like $D_s^+ \to \rho\pi^+$, must proceed by weak annihilation).

Hence we search for the decay $D_s^+ \to \omega\pi^+, \omega \to \pi^+\pi^-\pi^0$. The $\pi^+\pi^-\pi^+\pi^0$ mass spectrum for events with $\pi^+\pi^-\pi^0$ mass in the range

762-798 MeV is shown in figure 5(b). The $\pi^+\pi^-\pi^0$ mass spectrum for events with $\pi^+\pi^-\pi^+\pi^0$ mass in within 30 MeV of the D_s^+ mass is shown in figure 6(b). No signature for $D_s^+ \to \omega\pi^+$ is visible in either of these plots. A fit to figure 5(b) yields 0.0 ± 1.7 $D_s^+ \to \omega\pi^+$ events. This translates to

$$B(D_s^+ \to \omega\pi^+)/B(D_s^+ \to \phi\pi^+) \leq 0.5$$

at the 90% confidence level. This result, coupled with the stringent limit on $B(D_s^+ \to \rho\pi^+)$, provides strong evidence that weak annihilation is a small effect in two-body charm decays.

II. THE DECAY $D_s^+ \to f_0(975)\pi^+$

The spike near 1.0 GeV in figure 3(a) is consistent with the decay $D_s^+ \to f_0(975)\pi^+$, $f_0 \to \pi^+\pi^-$. This is an anomalous decay channel because, although there is no strangeness in the final state, the decay proceeds by the spectator mechanism. The fit discussed in section I (which included a coupled-channel Breit-Wigner for the $f_0(975)$ and is shown in figure 3(a)) yielded 24.2 ± 7.9 $D_s^+ \to f_0\pi^+$ decays. This corresponds to $B(D_s^+ \to f_0\pi^+)/B(D_s^+ \to \phi\pi^+) = 0.28 \pm 0.10 \pm 0.03$. We repeated the fit using the f_0 lineshape of the Mark III collaboration;[5] this lineshape is similar to the coupled-channel Breit-Wigner of reference 2, except it contains a long low-mass tail extending down to $\pi^+\pi^-$ mass threshold. When the Mark III lineshape is used for the f_0, we obtain 37.0 ± 16.4 $D_s^+ \to f_0\pi^+$ events, consistent with the number quoted above; but the χ^2 of the fit is ten units worse when the Mark III lineshape is used.

III. TWO INDEPENDENT MEASUREMENTS OF $D_s^+ \to \eta\pi^+$

In our study of $\pi^+\pi^-\pi^+\pi^0$ final states discussed in section I, we also searched for $D_s^+ \to \eta\pi^+, \eta \to \pi^+\pi^-\pi^0$. The $\pi^+\pi^-\pi^+\pi^0$ mass spectrum for events with $\pi^+\pi^-\pi^0$ mass in the range 536-560 MeV is shown in figure 5(a); a fit to this spectrum yields $2.8^{+2.1}_{-1.4}$ $D_s^+ \to \eta\pi^+$ events. The $\pi^+\pi^-\pi^+$ mass spectrum for events with $\pi^+\pi^-\pi^+\pi^0$ mass within 30 MeV of the D_s^+ mass is shown in figure 6(a). This spectrum is consistent with the number of $\eta\pi^+$ decays just quoted. This number translates into

$$B(D_s^+ \to \eta\pi^+)/B(D_s^+ \to \phi\pi^+) = 1.5^{+1.1}_{-0.8} \pm 0.5.$$

The kinematics of the $D_s^+ \to \eta\pi^+$ decay allows us to search for this decay without reconstucting the π^0. This is especially advantageous because our π^0 reconstruction efficiency is roughly an order of magnitude worse than our track reconstruction efficiency. In this case the $\pi^+\pi^-$ mass from the $\eta \to \pi^+\pi^-\pi^0$ decay must be in the range .28-.42 GeV; the $\pi^+\pi^-\pi^+$ mass from $D_s^+ \to \eta\pi^+, \eta \to \pi^+\pi^-\pi^0$ must be less than 1.833 GeV. A sharp edge at 1.833 in the $\pi^+\pi^-\pi^+$ spectrum would be a signature for $D_s^+ \to \eta\pi^+$ (a sharp edge at 1.733 would be signature for $D^+ \to \eta\pi^+$). We exclude a potentially large background from $D^+ \to K^-\pi^+\pi^+$ (with kaon misidentified as a pion) by vetoing $\pi^+\pi^-\pi^+$ combinations with $K^-\pi^+\pi^+$ mass in the range 1.85-1.89 GeV. Since we do not reconstruct the π^0, we must cut somewhat looser (by about 25%) on the miss distance of the $\pi^+\pi^-\pi^+$ reconstructed momentum vector from the primary event vertex, than we did in the search for $\pi^+\pi^-\pi^+$ decays.[6] The background was parametrized by using a high-statistics sample of $\pi^+\pi^-\pi^+$ combinations that had shorter lifetime and/or larger miss distance than the events that passed our analysis cuts. We also allowed for feedthroughs from other charm decays (this was found not to be a large effect).

The $\pi^+\pi^-\pi^+$ mass spectrum from a Monte Carlo simulation of $D_s^+ \to \eta\pi^+, \eta \to \pi^+\pi^-\pi^0$ decays is shown in figure 7(a). Figures 7(b) and 7(c) show the corresponding spectrum from our data sample, together with the best fits without and with (respectively) charm feedthroughs allowed for in the fit. Without the feedthroughs the fit gave -8.9 ± 15.2 $D_s^+ \to \eta\pi^+$ and 22.1 ± 18.9 $D^+ \to \eta\pi^+$ decays. With the feedthroughs included the fit gave -2.9 ± 14.6 $D_s^+ \to \eta\pi^+$ and 26.4 ± 18.1 $D^+ \to \eta\pi^+$ decays. (In each case the best number for $D^+ \to \eta\pi^+$ was obtained by setting the number of $D_s^+ \to \eta\pi^+$ to zero in the fit). The latter number for $D_s^+ \to \eta\pi^+$ translates to

$$B(D_s^+ \to \eta\pi^+)/B(D_s^+ \to \phi\pi^+) = -0.1 \pm 0.5 \pm 0.4.$$

To obtain our final result for $D_s^+ \to \eta\pi^+$ we take a weighted average of the two independent measurements quoted above, and obtain

$$B(D_s^+ \to \eta\pi^+)/B(D_s^+ \to \phi\pi^+) = 0.6 \pm 0.6 \pm 0.1 \leq 1.5$$

at the 90% confidence level. This measurement is 1.7 standard deviations from the recently published result of the Mark II collaboration[7] but in good agreement with the preliminary Mark III result.[8]

IV. VECTOR-VECTOR DECAYS

The model of Bauer, Stech, and Wirbel[9] is in astonishingly good agreement with experiment for charm decays to pseduscalar-vector (PV) and pseudoscalar-pseudoscalar (PP) states. Among the BSW predictions for vector-vector (VV) charm decays are $B(D_s^+ \to \phi\rho^+) / B(D_s^+ \to \phi\pi^+) = 6.3$ and $B(D^+ \to \overline{K}^{*o}\rho^+) = 17.0\,\%$. We examine $K^+K^-\pi^+\pi^0$ and $K^-\pi^+\pi^+\pi^0$ final states[6] which may contain the decays $D_s^+ \to \phi\rho^+$ and $D^+ \to \overline{K}^{*o}\rho^+$, respectively.

In final state $K^+K^-\pi^+\pi^0$ we demand that the K^+K^- invariant mass be consistent with a ϕ, and observe the spectrum shown in figure 8.

The fit yields 11.0 ± 3.6 $D_s^+ \to \phi\pi^+\pi^0$ events, which translates to $B(D_s^+ \to \phi\pi^+\pi^0)/B(D_s^+ \to \phi\pi^+) = 2.4 \pm 1.0 \pm 0.5$. Even if the entire $\phi\pi^+\pi^0$ decay is $\phi\rho^+$, this measurement is still far below the BSW prediction.

In the $K^-\pi^+\pi^+\pi^0$ final state we observe the spectrum of figure 9, for which the fit yields 91 ± 12 $D^+ \to K^-\pi^+\pi^+\pi^0$ decays, corresponding to $B(D^+ \to K^-\pi^+\pi^+\pi^0)/B(D^+ \to K^-\pi^+\pi^+) = 0.69 \pm 0.10 \pm 0.16$. If this entire final state is from $\overline{K}^{*0}\rho^+$, and we use the Mark III branching ratio $B(D^+ \to K^-\pi^+\pi^+) = (9.1\pm1.3\pm0.4)\%$[10], we get $B(D^+ \to \overline{K}^{*0}\rho^+) = (9.4\pm1.9\pm2.3)\%$, which again is far below the BSW prediction. Early indications are that, although the BSW model does a beautiful job for PP and PV charm decays, it may overestimate VV decays.

V. ACCOUNTING FOR D_s^+ DECAYS

There has been speculation that two-body decays may not dominate D_s^+ decays to the extent that they dominate D^0 and D^+ decays. To approach this question we can estimate the following ratio for D^0, D^+, and D_s^+:

$$R = \frac{\Sigma \; Cabibbo-favored\; PP, PV, VV\; branching\; ratios}{Total\; of\; Cabibbo-favored\; hadronic\; branching\; ratios}.$$

A more detailed calculation than the one presented here is given in reference 11. The denominator is estimated for D^+ and D^0 by using the accurate measurements of inclusive semileptonic branching ratios; for D_s^+ we assume that the total semileptonic branching ratio is the same for D_s^+ as for D^0, since their lifetimes are essentially equal. For all PP, PV, and VV modes we use experimental measurements when available, and otherwise we use the BSW model as a guide, assigning an *ad hoc* uncertainty of 50% to their predictions. The quantity R is in the range 40-45% for both the D^0 and D^+. For the D_s^+ there are large uncertainties in branching ratios for $\phi\pi^+$, $\eta'\pi^+$, $\eta\pi^+$, and VV final states. Nevertheless we can make some general observations:

If $B(D_s^+ \to \eta'\pi^+)/B(D_s^+ \to \phi\pi^+)$ is indeed in the range 5-6 (as reported at this conference,[12] and $B(D_s^+ \to \phi\pi^+)$ is in the range 3-4%, then

$$R_{D_s^+} \gtrsim R_{D^0} \cong R_{D^+}.$$

If $B(D_s^+ \to \eta'\pi^+) \cong B(D_s^+ \to \phi\pi^+)$ (as predicted by BSW), and $B(D_s^+ \to \phi\pi^+) \cong 3\text{-}4\%$, then

$$R_{D_s^+} \cong R_{D^0} \cong R_{D^+}.$$

Only if $B(D_s^+ \to \eta'\pi^+) \cong B(D_s^+ \to \phi\pi^+)$, and $B(D_s^+ \to \phi\pi^+) \gtrsim 1.5\%$, is there clearly a deficit in $R_{D_s^+}$.

REFERENCES

(1) For author list see M.Witherell, these proceedings. For a description of the experiment, see J.R.Raab et al., Phys. Rev. D **37**, 2391 (1988).

(2) G.Gidal et al., Phys. Lett. **107B**, 153 (1981).

(3) L.-L. Chau and H.-Y. Cheng, U.C. Davis preprint UCD-88-9 (1988).

(4) H.Lipkin, these proceedings.

(5) T.Browder, private communication.

(6) For more detailed discussion of the data analyses see: J.C.Anjos et al., Phys. Rev. Lett. **62**, 125 (1989); J.C.Anjos et al., Phys. Lett. **223B**, 267 (1989); and reference 11.

(7) G. Wormser et al., Phys. Rev. Lett. **61**, 1057 (1988). The Mark II collaboration has reported $\Gamma(D_s^+ \to \eta\pi^+)/\Gamma(D_s^+ \to \phi\pi^+) = 3.0 \pm 1.3$.

(8) P. Kim, these proceedings.

(9) M.Bauer, B.Stech, and M.Wirbel, Z. Phys C **34**, 103 (1987).

(10) J. Adler et al., Phys. Rev. Lett. **60**, 89 (1988).

(11) G.Punkar, Ph.D. thesis, University of California, Santa Barbara, preprint UCSB-HEP-89-02 (1989).

(12) G.Wormser, these proceedings.

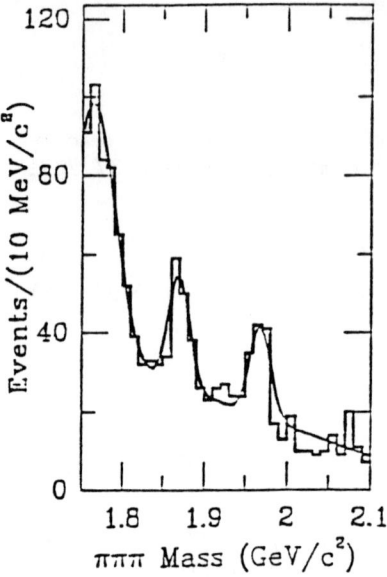

Figure 1. Inclusive $\pi^+\pi^-\pi^+$ mass spectrum used in search for $D_s^+ \to \rho\pi^+$.

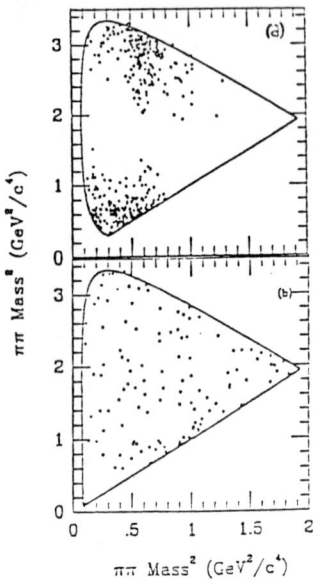

Figure 2. (a) Dalitz plot distribution from Monte Carlo simulation of $D_s^+ \to \rho\pi^+$ decay; (b) Dalitz plot distribution for data events with $\pi^+\pi^-\pi^+$ invariant mass in the range 1.953-1.983 GeV.

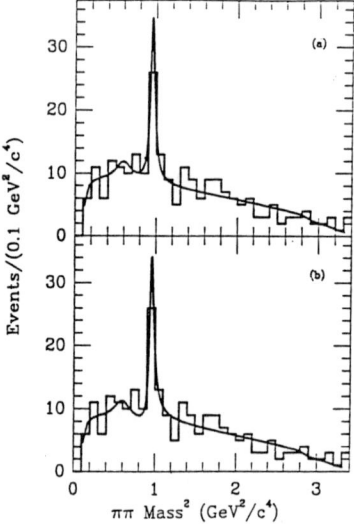

Figure 3. (a) Projection of data and fit function for fit where no interference was allowed between resonant and non-resonant amplitudes; (b) corresponding plot for case where such interference was allowed, assuming s-wave non-resonant amplitude.

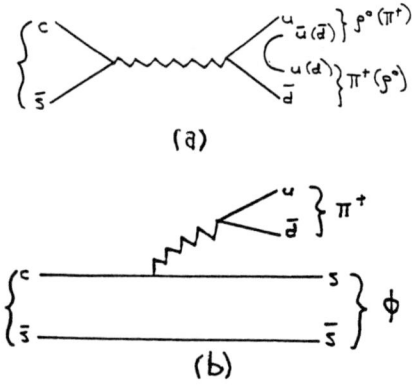

Figure 4. (a) Diagram for $D_s^+ \to \rho\pi^+$ decay (weak annihilation diagram); (b) diagram for $D_s^+ \to \phi\pi^+$ decay (spectator diagram).

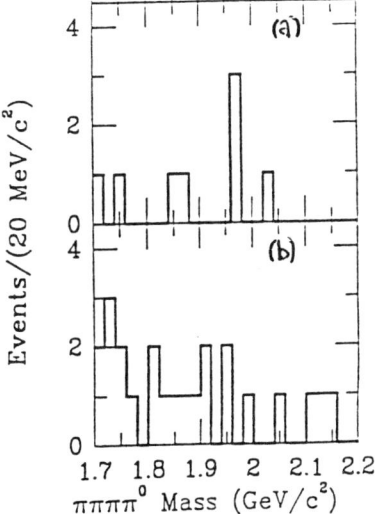

Figure 5. The $\pi^+\pi^-\pi^+\pi^0$ mass spectra for events with $\pi^+\pi^-\pi^0$ invariant mass in the range (a) 536-560 MeV, (b) 762-798 MeV.

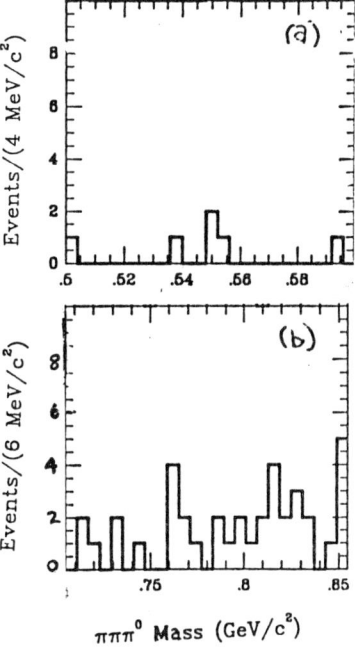

Figure 6. Two sections of $\pi^+\pi^-\pi^0$ mass spectrum for events with $\pi^+\pi^-\pi^+\pi^0$ mass in the range 1.938-1.998 MeV. The analysis cuts used for these plots are slightly looser than those used for figure 5.

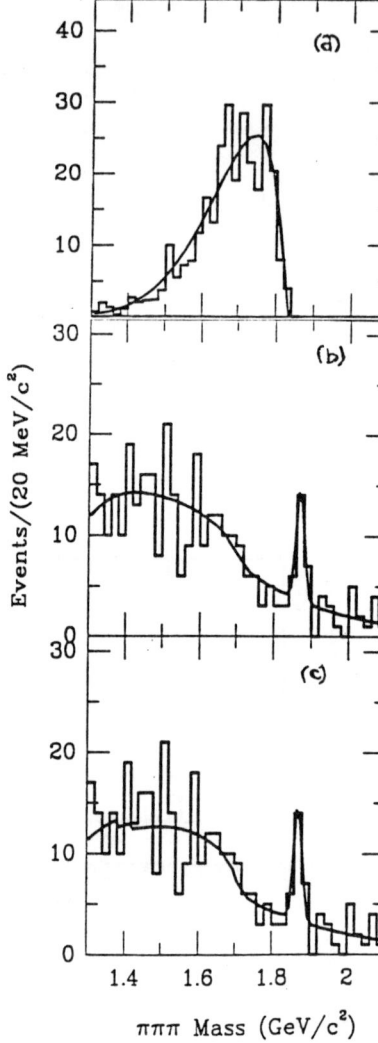

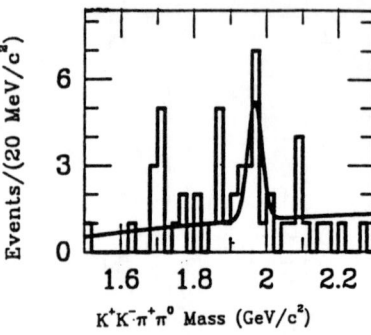

Figure 8. The $\phi\pi^+\pi^0$ mass spectrum which could contain events due to $D_s^+ \to \phi\rho^+$.

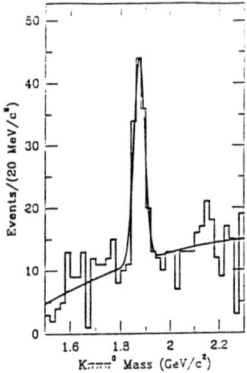

Figure 9. The $K^-\pi^+\pi^+\pi^0$ mass spectrum which could contain events due to $D^+ \to \overline{K}^{*0}\rho^+$.

Figure 7. (a) The $\pi^+\pi^-\pi^+$ mass spectrum from Monte Carlo simulation of the decay $D_s^+ \to \eta\pi^+, \eta \to \pi^+\pi^-\pi^0$; (b,c) the $\pi^+\pi^-\pi^+$ mass spectrum used in the search for $D_s^+ \to \eta\pi^+$. The solid curve in (b) is the best fit obtained when charm feedthroughs are not allowed in the fit; the curve in (c) is the best fit obtained when charm feedthroughs are allowed in the fit. The peak at 1.87 GeV is due to $D^+ \to \pi^+\pi^-\pi^+$ decays.

NA14' RESULTS ON D_S DECAYS

Guy Wormser
Laboratoire de l'Accélérateur Linéaire
I N2 P3, Université de Paris-Sud
Centre d'Orsay - Bâtiment 200
91405 ORSAY Cedex, FRANCE

ABSTRACT

NA14' is a charm photoproduction experiment, using a 100 GeV photon beam produced by the CERN SPS. Its set-up combines a high precision Silicium microvertex detector with a large acceptance spectrometer for both charged and neutral tracks.

Search for D_S decays in the mode $D_S \to \eta'\pi$ and $D_S \to \varphi\pi\pi^0$ are reported. A clear evidence is found for the former decay, with a preliminary branching ratio $Br(D_S \to \eta'\pi)/Br(D_S \to \varphi\pi) = 5 \pm 1.8 \pm 1.2$. An upper limit is given for the latter decay $Br(D_S \to \varphi\pi\pi^0)/Br(\varphi\pi) < 2.5$. Those 2 results are incompatible with theoretical expectations, where the first decay was supposed to be small and the second larger.

1. INTRODUCTION

All the major decay modes of the D^0 and D^+ mesons are now measured and can be successfully described with good success by models, either based on solid theoretical grounds like QCD sum rules [1], or using a more phenomenological approach [2]. It is striking to see how the situation differs for the D_s meson. Firstly, only decay modes with small branching ratios have been firmly established up to now. The pilot one, $D_S^{\pm} \to \varphi\pi^{\pm}$, is equal to ~ 2-3 %, according to various reports given at this conference [3]. Most other modes, K^+K^0, $\varphi 3\pi$, K^*K, are not larger than the pilot one. On the experimental side, the only modes with large branching ratios reported so far are $\eta\pi$ and $\eta'\pi$ by the Mark II collaboration [4], with branching ratios respectively 3 ± 1

© 1989 American Institute of Physics

and 5 ± 2.3 times the pilot branching ratios $Br(D_s \to \phi\pi)$. Concerning the $\eta\pi$ decay mode, it has been seen by 2 other experiments, Mark III and E 691 [5], and reports were given at this conference. Since these last 2 experiments give also upper limits from other analysis of the same decay channel, the preferred value for this branching ratio lies around 1.5 times $Br(D_s \to \phi\pi)$.

On the theoretical ground, a large branching ratio is predicted for the decay $D_s \to \phi\rho$, or more generally $\phi\pi\pi^0$, about 6 times $\phi\pi$ [2]. This pattern obviously motivated the NA14' collaboration to search for both decay modes $\phi\pi\pi^0$ and $\eta'\pi$.

2. THE NA14' SPECTROMETER

NA14' is a charm photoproduction experiment, using a 100 GeV photon beam produced by the CERN SPS. Its set-up combines a high precision Silicium microvertex detector with a large acceptance spectrometer for both charged and neutral tracks.

The NA14 spectrometer is described in greater detail elsewhere [6]. Of importance to the analysis described here, are the microvertex detector and the forward electromagnetic calorimeter. The microvertex detector comprises a Silicon active target, made of 32 300 µm thick Si planes and a Si microstrip telescope. The telescope is made of 10 planes of 50 µm pitch strips. This set-up enables a resolution of 100 µm and 300 µm along the flight direction for the primary and the secondary vertices.

The forward electromagnetic calorimeter is a large lead glass array, in association with a position detector made of 800 1.5 cm wide scintillators placed after 3 X^0 of lead glass. The energy and position resolution of the calorimeter are respectively 10 % $\sqrt{E}$ and 3 mm for a 2 GeV photon.

Our ability to reconstruct charm final involving neutrals is illustrated in Fig.1, where a clear signal of $D^0 \to K\pi\pi^0$ is visible (Fig.1a) and where the more delicate decay $D^{*0} \to D^0\gamma$ is seen (Fig.1b)

3. SEARCH FOR $D_S \to \phi\pi\pi^0$

A search for $D_S \to \phi\pi\pi^0$ was performed and no signal was observed, leading to the upper limit :

$Br(D_S \to \phi\pi\pi^0) / Br(D_S \to \phi\pi) < 2.5$ at 90% c.l.

Fig.2a shows the corresponding $\phi\pi\pi^0$ spectrum, where the normalizing signal $D_S^+ \to \phi^+\pi^+$ is displayed in Fig.2b. The π^0 efficiency was carefully monitored by studying the decay $D^0 \to K\pi\pi^0$, by comparing the rates of charged pions and π^0, and by a Monte Carlo program. This result is not incompatible with the positive evidence for the decay mode give by the E691 collaboration at this conference $Br(D_S \to \phi\pi\pi^0) / Br(D_S \to \phi\pi) = 2.4 \pm 1.2$ [5]. Both results clearly indicate an unexpectedly large suppression for this a priori avoured V-V decay mode.

4. SEARCH FOR $D_S \to \eta'\pi$

We search for this decay mode, using the $\eta' \to \rho\gamma$ decay mode, giving rise to a 3C1γ final state configuration. This is very favorable because the 3 charged tracks have to form a detached vertex, while only 1 photon is present in the final state, allowing the global efficiency to be sufficiently high. All 3 prong secondary vertices were selected, combined with photons found in the forward electromagnetic calorimeter ant the resulting track was required to point back to the primary vertex. A 3σ cut was applied for the separation between the 2 vertices. Furthermore, two opposite charge pions have to lie within the ρ mass region and the $\rho\gamma$ mass should lie within 15 MeV of the η' mass. Finally, all pions were asked not to be identified as K ou p in the Cerenkov counter. The resulting 3$\pi\gamma$ mass plot for the η' region and η' side bands is shown on Fig.3. A clear signal appears at the $D_S^\pm$ mass, with the expected width, while the side band distribution is completely smooth.

A clear η' peak is also visible when the $3\pi\gamma$ mass is selected in the D_S region (Fig.4). The D_S signal of 45 ± 11 events translate to :

$$\frac{Br(D_S \to \eta'\pi)}{Br(D_S \to \phi\pi)} = 5 \pm 1.8 \pm 1.2$$

where we have used the $\phi\pi$ signal observed in our data as reference. Both decay modes contribute about equally to the statistical error. This value updates the preliminary number given in [7].

The main source of systematic error is the uncertainly in the photon reconstruction efficiency. The lifetime of the D_S candidates has been measured and found to be compatible with our measurement in the $\phi\pi$ mode.

Mark III [3] has reported at this conference an upper limit of 1.9, in contradiction with our result.

Since no signal is observed at the D^+ mass, an upper limit can be set on the D^+ branching ratio to $\eta'\pi$:

$$Br(D^+ \to \eta'\pi) < 1\% \text{ at } 90\% \text{ c.l.}$$

5. CONCLUSION

The observed pattern of D_S decays is strikingly at odds with the expected one. The dominant D_S decay mode seems to be $D_S \to \eta'\pi$, while the a priori favoured one, $D_S \to \phi\pi\pi^0$ is found to be not very large.

Several authors [8] have considered the implications of such a pattern. Our upper limit in $D^+ \to \eta'\pi^+$ and the new Mark III limit presented at this conference on $D^0 \to K^0\eta'$ [3], are in contradicion with the prediction of Kamal et al.

The next important experimental clue lies, besides, of course, the measurement of the $\eta'\pi$ mode by other groups, in the relative yields $D_S \to \phi ev, \eta ev,$ and $\eta' ev$, D_S semileptonic decays, where only spectator decays contribute.

Concerning the $\varphi\pi\pi^0$ mode, there seems to be a unexplained trend to observe VV decay modes at lower rates than the prediction, in all D decays. Mark III [5] for instance, reported a very low branching ratio for $D^0 \to K^*\rho$.

FIGURE CAPTIONS

Fig.1a : $K\pi\pi^0$ mass spectrum

Fig.1b : $K\pi\gamma$ - $K\pi$ mass spectrum when $K\pi$ mass is in the D region

Fig.1c : $K\pi\gamma$ - $K\pi$ mass spectrum when $K\pi$ mass is outside D region

Fig.2a : $\phi\pi\pi^0$ mass spectrum

Fig.2b : $\phi\pi$ mass spectrum

Fig.3a : $\pi^+\pi^-\pi^\pm\gamma$ mass spectrum when $\pi^+\pi^-\gamma$ is in the η' region

Fig.3b : $\pi^+\pi^-\pi^\pm\gamma$ mass spectrum when $\pi^+\pi^-\gamma$ is outside the η' region

Fig.4 : $\pi^+\pi^-\gamma$ mass spectrum when $\pi^+\pi^-\pi^\pm\gamma$ mass is in the D_S region

REFERENCES

1. B. Blok and M.A. Shifman, ITEP preprint **37** (1986)

2. M. Bauer, B Stech, M. Wirbel, Z. Phys. C, **34** (1987) 103

3. M.S. Alam, Results from CLEO, These proceedings
   P. Kim Results from Mark III, These proceedings

4. G. Wormser et al., Phys. Rev. Lett., **61** (1988) 1057

5. G. Gladding, Results from Mark III, These proceedings
   G. Punkhar, Results from E691, These proceedings

6. R. Barate et al., Phys. Lett. **174B** (1986) 458
   R. Barate et al., Nucl. Instr. and Meth., **A235** (1985) 235
   R. Barate et al., Nucl. Instr. and Meth., **A253** (1987) 530

7. G. Wormser, LAL 89-10 (1989)

8. A. Kamal et al., Alberta Thy. 31.88
   I. Bigi, These proceedings

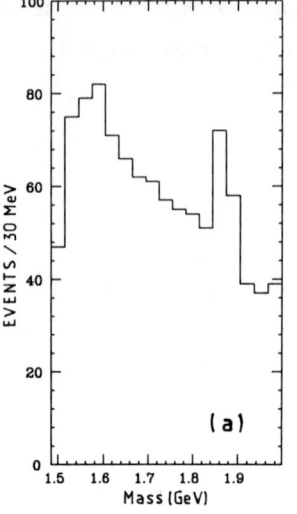
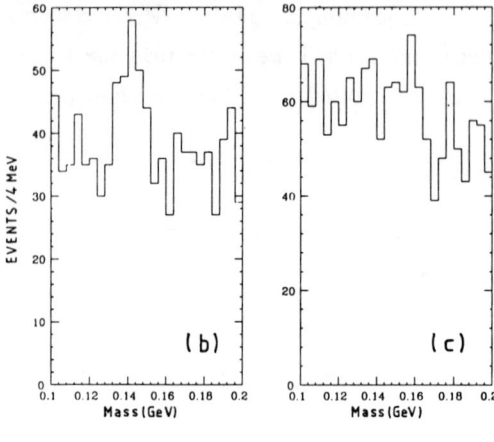

Fig.1: (a) Kππ° mass spectrum
(b) Kπγ-Kπ mass spectrum when Kπ mass is in the D region
(c) Kπγ-Kπ mass spectrum when Kπ mass is outside D region

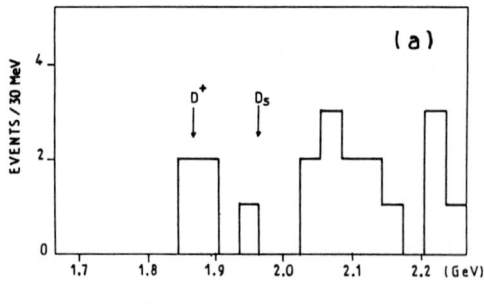
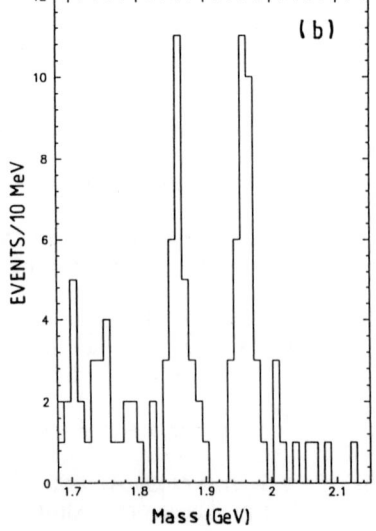

Fig.2: (a) ϕππ⁰ mass spectrum
(b) ϕπ mass spectrum

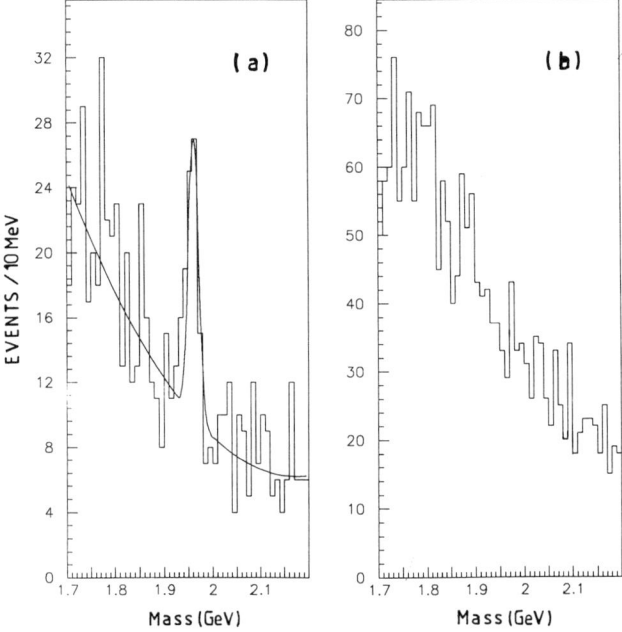

Fig.3: (a) $\pi^+\pi^-\pi^\pm\gamma$ mass spectrum when $\pi^+\pi^-\gamma$ is in the η' region
(b) $\pi^+\pi^-\pi^\pm\gamma$ mass spectrum when $\pi^+\pi^-\gamma$ is outside the η' region

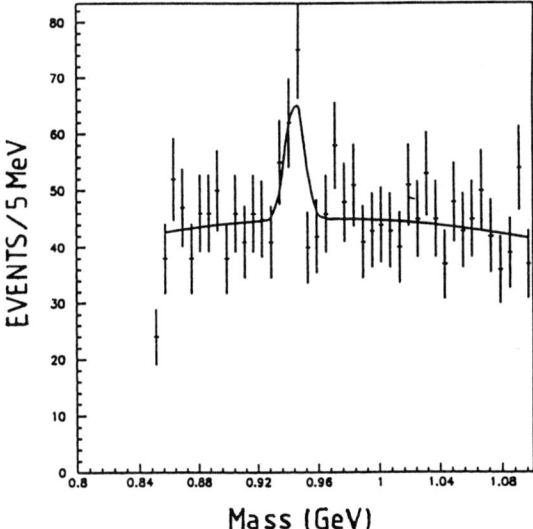

Fig.4: $\pi^+\pi^-\gamma$ mass spectrum when
$\pi^+\pi^-\pi^\pm\gamma$ mass is in the D_S region

Charmed Hadron Spectroscopy and Decay Results from CESR

by

M. S. Alam

State University of New York at Albany

and

The CLEO Collaboration

Abstract

We present recent results from CESR on the study of charmed baryon decays and excited charmed meson spectroscopy. We have measured the production of the charmed baryons Λ_c^+, Σ_c^0, Σ_c^+, Ξ_c^0 and Ξ_c^+ from e^+e^- annihilations in the continuum region below the $\Upsilon(4S)$ at center-of-mass energies around 10.5 GeV. Results are also presented on the continuum production of the excited charmed meson states $D^{**0}(2428)$, $D^{**0}(2461)$ and the $D_s^{**+}(2535)$.

1. Introduction

CESR operating at center-of-mass energies around 10 GeV has turned out to be a very useful place for the study of charmed particle spectroscopy and decays. With the discovery of the charmed quark, the SU(3) octet of $J^P = 1^-$ mesons and $J^P = 1/2^+$ baryons expanded to the SU(4) 16-plet and 20-plet, respectively. We have four new baryons with charm = 1, namely the singlet Λ_c^+(cud), the triplet Σ_c(cuu, cud, cdd), the doublet Ξ_c(csu, csd) and the singlet Ω_c^0(css), of which all except the Σ_c triplet decay weakly. Of these, the Λ_c^+ was discovered in 1979[1] and only recently have the Σ_c and the Ξ_c been reported with good statistics. The evidence for the Ω_c^0[2] is very weak. Study of charmonium ($c\bar{c}$) and bottomium ($b\bar{b}$) spectroscopy has added significantly to our knowlege of the strong quark-quark and hyperfine spin-spin interactions. The study of excited charm states with one heavy and one light quark will allow us to understand non-relativistic effects in quarkonium systems.

Charmed mesons and baryons are produced from the decays of B mesons as well as from e^+e^- annihilations to $c\bar{c}$ quark pairs on the continuum. Charmed particles produced from the decay of B mesons are kinematically limited to have their momenta less than 2.5 GeV/c or their fractional momenta x_p less than 0.5. Here $x_p = p/p_{max}$, where p and p_{max} are the momentum and its maximum kinematically allowed value, respectively. Further, $p_{max} = \sqrt{E_{beam}^2 - m_{\Lambda_c^+}^2}$. To avoid confusion from charmed particles produced from B decays, we only consider charmed particle production with x_p greater than 0.5 in all cases.

2. The Data Sample and Detector

The results presented in this paper are based on data collected with the CLEO detector at the Cornell Electron Storage Ring (CESR) during 1985 and 1987. The 1985 (1987) sample consists of 27 (117), 36 (101) and 78 (212) pb^{-1} of integrated luminosities at the $\Upsilon(3S)(\Upsilon(5S))$, at energies just below the $\Upsilon(4S)$ or continuum, and at the $\Upsilon(4S)$, respectively. The full data sample (571 pb^{-1}) has been used for the Σ_c analysis only; all other analyses are based on the 1987 sample only.

The CLEO detector and our hadronic event selection are described in detail elsewhere[3]. Charged tracks are measured in a central tracking system consisting of a 51-layer drift chamber, a ten-layer high precision drift chamber and a three layer straw-tube vertex detector inside a 1.5 Tesla magnetic field. Hadron identification is done using primarily dE/dx measurements from the 51-layer drift chamber or time-of-flight scintillation counters placed outside the solenoidal coil. A charged track is loosely (strongly) identified as a kaon or proton if its dE/dx (TOF) is consistent ($\pm 2\sigma$) with the expected value for that hypothesis but at least 1σ (2σ) away from that expected for the pion hypothesis[4]. K^0's are detected as K_S^0's. K_S^0's and Λ's are detected as secondary vertices decaying to $\pi^+\pi^-$ and $p\pi^-$, respectively. Ξ^-'s are detected as secondary vertices decaying to $\Lambda \pi^-$ [3].

3. Charmed Baryon Production and Decays

We report on the measurements of the production cross-section for the charmed baryons Λ_c^+, Σ_c^0, Σ_c^{++}, Ξ_c^0 and Ξ_c^+. The branching fractions of the charmed baryon Λ_c^+ into the decay modes $p\overline{K^0}$, $p\overline{K^0}\pi^+\pi^-$, $\Lambda\pi^+$, $\Lambda\pi^+\pi^-\pi^+$ and $\Xi^-K^+\pi^+$ relative to that into $pK^-\pi^+$ have been measured. Measurements of the mass differences $\Delta_{\Sigma_c} = M_{\Sigma_c^0} - M_{\Sigma_c^{++}}$ and $\Delta_{\Xi_c} = M_{\Xi_c^0} - M_{\Xi_c^+}$ are then presented.

3.1 Λ_c^+ Production and Decays The lack of data for charmed baryon decays has until now discouraged serious confrontation of theoretical models with data. Here, we present statistically significant measurements of the Λ_c^+ branching fractions into the decay modes $pK^-\pi^+$, $p\overline{K^0}$, $p\overline{K^0}\pi^+\pi^-$, $\Lambda\pi^+$, $\Lambda\pi^+\pi^-\pi^+$ and $\Xi^-K^+\pi^+$.

Figures 1 to 6 show the invariant mass distributions corresponding to particle combinations for the above decay modes with x_p greater than 0.5 from the 1987 data sample with 430 pb^{-1}. Since the particle identification and kinematic cuts are different for each decay mode, we discuss them separately. For the $pK^-\pi^+$ combinations, both the kaon and proton candidates are loosely identified. To reduce combinatorial background, only pions with momenta greater than 0.3 GeV/c are accepted. For the $p\overline{K^0}$ and $p\overline{K^0}\pi^+\pi^-$ decay modes, we use pK_S^0 and $pK_S^0\pi^+\pi^-$ combinations with loosely identified protons. We form all $\Lambda\pi^+$ and

$\Lambda \pi^+ \pi^- \pi^+$ combinations with opposite strangeness and charge. The corresponding combinations with the same strangeness and charge show no evidence for an enhancment. For the $\Xi^- K^+ \pi^+$ combinations, only positively identified kaons are used.

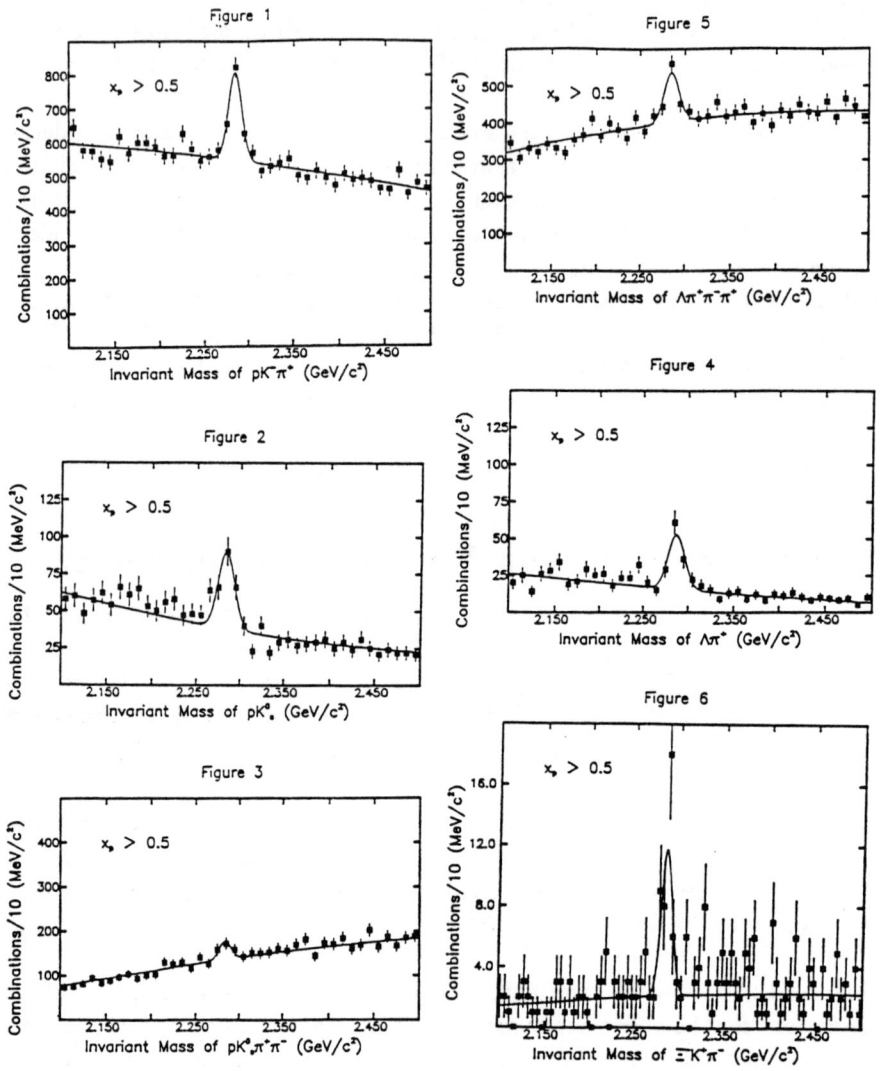

Figures 1 to 6. Invariant mass distributions with $x_p > 0.5$

We see clear enhancements around 2285 MeV/c^2 in all the figures. The signal areas and peak masses obtained from fits to the above distributions with a polynomial background and Gaussian signal function of width equal to the expected experimental resolution are presented in Table 1. To obtain the production cross-section times the branching fraction into a specific decay mode, the signal contribution for different x_p intervals is corrected by a momentum-dependent reconstruction efficiency and all contributions above x_p of 0.5 are added. The efficiency presented in the table is averaged for all x_p greater than 0.5. The first error is statistical while the second is systematic and reflects the uncertainty in our knowledge of the particle identification efficiencies. There is no model-independent method to calculate the absolute branching fractions, so we present all other branching fractions relative to that into $pK^-\pi^+$ with reduced systematic errors due to cancellations.

Table 1. Λ_c^+ Production Results

Decay Mode	Γ_{MC} MeV/c^2	Mass MeV/c^2	Raw Yield	ϵ_d %	$x_p > 0.5$ $\sigma \cdot Br$ (pb)	$B(\Lambda_c^+ \to$ Decay Mode) relative to $B(\Lambda_c^+ \to pK^-\pi^+)$
$pK^-\pi^+$	18	2284.5 ± 0.9	481 ± 54	18	$6.6 \pm 0.6 \pm 1.0$	$1.0 \pm 0.0 \pm 0.0$
$p\overline{K^0}$	25	2283.2 ± 1.8	131 ± 18	11	$2.9 \pm 0.4 \pm 0.3$	$0.44 \pm 0.07 \pm 0.08$
$p\overline{K^0}\pi^+\pi^-$	16	2285.2 ± 3.1	74 ± 23	7	$2.8 \pm 0.9 \pm 0.5$	$0.42 \pm 0.14 \pm 0.10$
$\Lambda\pi^+$	23	2287.3 ± 1.6	89 ± 13	17	$1.1 \pm 0.2 \pm 0.1$	$0.17 \pm 0.03 \pm 0.03$
$\Lambda\pi^+\pi^-\pi^+$	21	2285.0 ± 1.6	292 ± 42	16	$4.3 \pm 0.6 \pm 0.5$	$0.65 \pm 0.11 \pm 0.12$
$\Xi^-K^+\pi^+$	15	2285.5 ± 1.5	31 ± 7	8	$0.9 \pm 0.2 \pm 0.2$	$0.14 \pm 0.03 \pm 0.04$

Mass (weighted) = $2285.0 \pm 0.6 \pm 3.0$ MeV/c^2
Particle Data Book '88 = 2284.9 ± 1.5 MeV/c^2

3.2 Σ_c^{++} and Σ_c^0 Production and Mass Difference

The mass differences between the members of the same isospin multiplet receive contributions from the difference in the quark masses besides electromagnetic and spin-spin interaction effects. Theoretical estimates vary from -6 to 18 MeV/c^2 for charmed baryon multiplets[5]. Figure 7 shows the combined mass distribution for the Λ_c^+ in the decay modes Λ_c^+, $p\overline{K^0}$ and $\Lambda\pi^+\pi^-\pi^+$. Figure 8 displays the mass difference distributions $\Delta_m^{++} = M(\Lambda_c^+\pi^+) - M(\Lambda_c^+)$ and $\Delta_m^0 = M(\Lambda_c^+\pi^-) - M(\Lambda_c^+)$ corresponding to 1325 ± 95 Λ_c^+ candidates. We find 54 ± 11 and 48 ± 12 events correponding to Σ_c^{++} and Σ_c^0 production, respectively[6]. All measured values of Δ_m^{++}, Δ_m^0 are reported in Table 2. We find $\Delta_{\Sigma_c} = \Delta_m^0 - \Delta_m^{++}$ to be $+0.1 \pm 0.6 \pm 0.1$ MeV/c^2, where the systematic error is determined from our measurement of the mass difference $M(D^{*+}) - M(D^0)$. Our result is consistent with the ARGUS but not the E400 result[7,8]. Our result also indicates that the contribution from the quark mass difference must be equal and opposite to that from electromagnetic interactions.

Table 2. Σ_c^0 and Σ_c^{++} Mass Difference

Experiment	$M(\Sigma_c^0) - M(\Lambda_c^+)$ MeV/c^2	$M(\Sigma_c^{++}) - M(\Lambda_c^+)$ MeV/c^2	$M(\Sigma_c^0) - M(\Sigma_c^{++})$ MeV/c^2
E-400	$178.2 \pm 0.4 \pm 2.0$	$167.4 \pm 0.5 \pm 2.0$	$+10.8 \pm 2.9$
ARGUS	167.0 ± 0.5	168.2 ± 0.5	$-1.2 \pm 0.7 \pm 0.3$
CLEO	$167.9 \pm 0.5 \pm 0.3$	$167.8 \pm 0.4 \pm 0.3$	$+0.1 \pm 0.6 \pm 0.1$
E-691	$168.4 \pm 1.0 \pm 0.3$		
Theoretical Prediction: $M(\Sigma_c^0) - M(\Sigma_c^{++})$ = -6 to 18 MeV/c^2			

We have also measured that $(18 \pm 3 \pm 5)\%$ of the Λ_c^+'s are produced from the secondary decays of Σ_c with x_p greater than 0.5, to be compared with the ARGUS collaboration value of $(36 \pm 12 \pm 11)\%$.

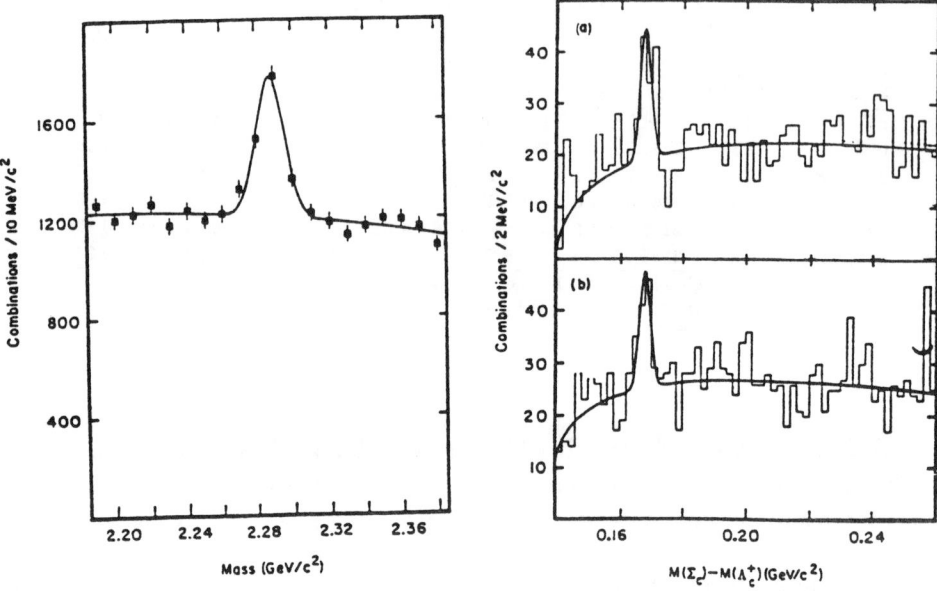

Fig. 7 Λ_c^+ mass from all decay modes.

Fig.8(a) $M(\Sigma_c^{++}) - M(\Lambda_c^+)$ and
Fig.8(b) $M(\Sigma_c^0) - M(\Lambda_c^+)$ mass-difference.

3.3 Isospin Mass Splittings of the Ξ_c
The Ξ_c^+ has been seen in hadro-production experiments. Evidence for the Ξ_c^+(csu) with a mass around 2460 MeV/c^2 in the decay mode $\Lambda K^- \pi^+ \pi^+$ has been presented from several hadron beam experiments [9,10,11]. CLEO has observed both Ξ_c^0 and Ξ_c^+ production from e^+e^- annihilations in the decay modes $\Xi^-\pi^+$ and $\Xi^-\pi^+\pi^+$, respectively [12,13]. This is the first observation of the Ξ_c^0.

We have 1006 ± 43 $\Xi^- \to \Lambda\pi^-$ candidates over a background of 563 random combinations. Figures 9 and 10 show the mass distributions corresponding to

$\Xi^-\pi^+$ and $\Xi^-\pi^+\pi^+$ combinations with x_p greater than 0.5. We see clear enhancements in these figures. Fits to the distribution with a Gaussian function over a smooth background yield 18.2 ± 5.0 Ξ_c^0 and 24.3 ± 6.3 Ξ_c^+ events at masses of 2472 ± 3 and 2467 ± 3 MeV/c^2, respectively. $\Delta_{\Xi_c} = M(\Xi_c^+) - M(\Xi_c^0)$ is measured to be $-5 \pm 4 \pm 2$ MeV/c^2, where the systematic error reflects both the uncertainties in the mass scale and the fitting procedure.

The production cross-section times branching fraction for the Ξ_c^0 and the Ξ_c^+ are respectively 0.38 ± 0.10 and 0.60 ± 0.16 pb for $x_p > 0.5$.

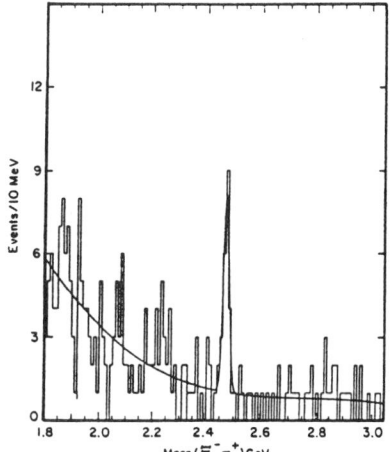
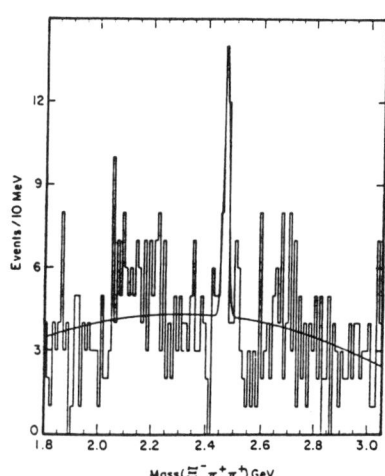

Figures 9 and 10. $\Xi^-\pi^+$ and $\Xi^-\pi^+\pi^+$ invariant mass distributions with $x_p > 0.5$

Two different kinds of strangely, charmed baryons can be formed from the (csd) quark combination, one being flavor SU(3) symmetric under s and d quark exchange and the other being symmetric. Examples of such pairs of states are Λ-Σ^0 and Λ_c^+-Σ_c^+. Kwong et al.[14] use a simple model based on QCD to predict the antisymmetric and symmetric states to have masses of 2505 and 2604 MeV/c^2. We expect the Ξ_c^0 to be the antisymmetric lighter state besides we could not see the higher mass symmetric state as it would decay to the lower mass state by the emission of a γ or a π^0.

4. P-Wave Charmed Meson Spectroscopy

The study of excited states with one heavy and one light quark has been limited by the lack of experimental data. For orbital angular momentum L = 1, there should be four states $^1P_1, ^3P_0, ^3P_1$ and $^3P_2$ with spin-parity $J^P = 1^+, 0^+, 1^+$ and 2^+, respectively. The 2^+ state can decay to both $D\pi$ and $D^{*+}(2010)\pi$ while, owing to parity conservation in strong decays, the 0^+ can decay only to $D\pi$ and the 1^+ state only to $D^{*+}\pi$.

Our study of the L = 1 $c\bar{u}$ D^{**0}(2420), reported earlier by ARGUS[15] and E691[16] reveals the presence of two different D^{**0} states[17]. We also report on the D^{**0}(2459) decaying into $D^+\pi^-$ and seen earlier by both the ARGUS[18] and the E691[19] collborations. We confirm the observation of the D_s^{**+}(2535), a $c\bar{s}$ excited meson decaying into $D^{*+}K_S^0$[20].

4.1 D^{**0}(2428) and D^{**0}(2460) Production

We have searched for $D^{**}(c\bar{u})$ states by first identifying D^{*+} and D^+ candidates. Figure 11 shows the mass difference $M(D^+\pi^-) - M(D^+)$ distribution with the x_p of both greater than 0.6, based on a sample of 16,925 $D^+ \to K^-\pi^+\pi^+$ candidates over a background of 180,000 random combinations. A fit to the distribution with a 2nd-order polynomial background and a Briet-Wigner function yielded a mass difference peak mean of $592\pm3\pm1$ MeV/c^2 which corresponds to a D^{**0} at mass $2461\pm3\pm1$ MeV/c^2. The width with 15 MeV/c^2 detector resolution subtracted is $20^{+9}_{-12}\pm9$ MeV/c^2. The production rate $\sigma(D^{**0}(2461) \to D^+\pi^-)/\sigma(D^+)$ for x_p greater than 0.6 is measured to be $(10\pm2^{+2}_{-1})\%$.

Figures 12(a) and 12(b) show the mass difference $M(D^{*+}\pi^-) - M(D^{*+})$ distribution plotted for the angular regions (a) $0.5 < |cos\phi_{D^{*+}}| < 1.0$ and (b) $0.0 < |cos\phi_{D^{*+}}| < 0.5$, where $\phi_{D^{*+}}$ is the D^{*+} polarization angle between the π^+ and the D^{**0} in the D^{*+} rest-frame. A Breit-Wigner fit to the two distributions yields signal masses and widths of 2426 ± 4 and 34 ± 10 MeV/c^2 for the first angular region and 2436 ± 5 and 73^{+15}_{-13} MeV/c^2 for the second. The inconsistency of the masses and widths in the two angular regions suggests that we are dealing with more than one state with overlapping masses.

If the D^{**0}(2461) is a 2^+ state, it is also expected to decay to $D^{*+}\pi^-$. A fit to the above mass difference distribution with two Breit-Wigner functions, one with mean and width fixed from the D^{**0}(2461) found in the $D^+\pi^-$ analysis, is carried out for the four $cos\phi_{D^{*+}}$ angular regions from -1.0 to 1.0. We see clear evidence for a second lower mass resonance with a weighted average mass difference of $418\pm2\pm1$ and width of $23^{+8}_{-6}{}^{+4}_{-3}$ MeV/c^2, where the experimental resolution of 12 MeV/c^2 has been subtracted in quadrature. The resulting D^{**0} mass is $2428\pm2\pm1$ MeV/c^2 and the production rate $\sigma(D^{**0} \to D^{*+}\pi^-)/\sigma(D^{*+})$ is measured to be $(8.9\pm1.1\pm0.5)\%$. The corresponding rate for the higher mass component is $(3.6\pm1.0^{+0.4}_{-0.8})\%$.

The polarization angle $\phi_{D^{*+}}$ distributions for the D^{**0}(2428) and D^{**0}(2461) are shown in Figures 13(a) and 13(b), respectively. We fit the two distributions to a $sin^2\phi_{D^{*+}}$ curve and a flat line. We find the lower mass component is inconsistent with a $sin^2\phi_{D^{*+}}$ (0.07% C.L.) distribution while the higher mass component is quite consistent (56.5% C.L.) with it, which is what is expected for the decay of a $J^P = 2^+$ state.

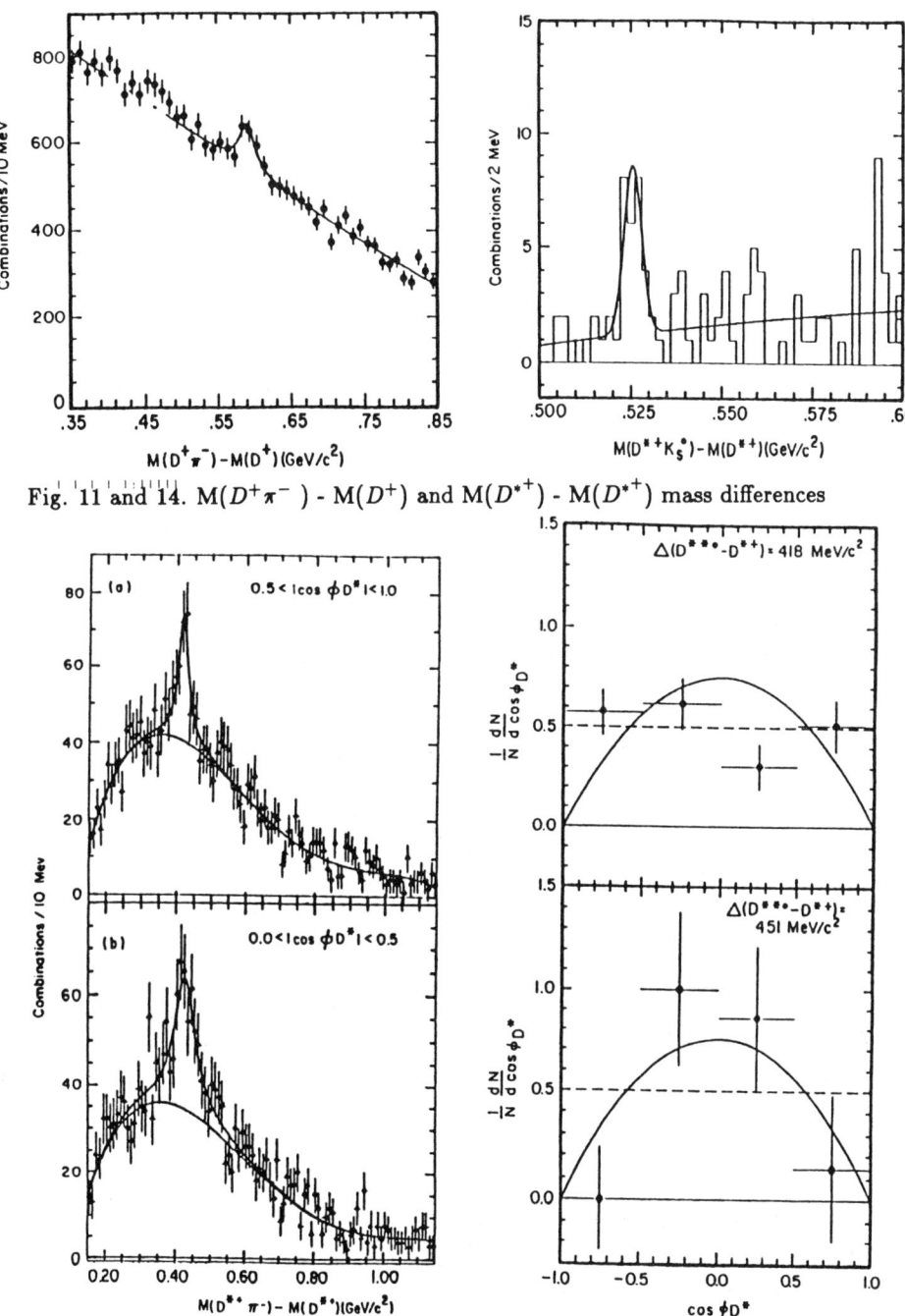

Fig. 11 and 14. $M(D^+\pi^-) - M(D^+)$ and $M(D^{*+}) - M(D^{*+})$ mass differences

Fig. 12(a) and 12(b) $M(D^{*+}\pi^-) - M(D^{*+})$ for two $\cos(\phi_{D^{*+}})$ regions. Fig. 13(a) and Fig. 13(b) are $\phi_{D^{*+}}$ distributions for the $D^{**0}(2428)$ and $D^{**0}(2461)$.

We therefore favor the assignment of $J^P = 1^+$ for the $D^{**0}(2428)$ state (if it is inconsistent with the 2^+ state, it must be a 1^+ state to decay into $D^{*+}\pi^-$), and $J^P = 2^+$ for the $D^{**0}(2461)$ state. Another way to confirm the above spin-parity assignment is to search for $D^{**0}(2428) \to D^+\pi^-$ which would only be possible if the $D^{**0}(2428)$ were a 2^+ state. We measure

$$B\left(D^{**0}(2428) \to D^+\pi^-\right)/B\left(D^{**0}(2428) \to D^{*+}\pi^-\right) < 0.24 (90\%\ C.L.)$$

and

$$B\left(D^{**0}(2461) \to D^+\pi^-\right)/B\left(D^{**0}(2461) \to D^{*+}\pi^-\right) = 2.3 \pm 0.8$$

The theoretical prediction[21] for this ratio for a 2^+ state is between 1.5 and 3.5, clearly supporting our assignment of $J^P = 2^+$ for the $D^{**0}(2461)$ state.

4.2 Production of $D_s^{**+}(2535)$

Figure 14 shows the $M(D^{*+}K_S^0)$ - $M(D^{*+})$ mass difference distribution for $x_p > 0.5$. A fit to the signal yields a mass difference peak at $525.5 \pm 0.7 \pm 0.4\ MeV/c^2$ with a $\sigma = 2.5 \pm 0.6 \pm 0.5\ MeV/c^2$ and signal area of 24 ± 6 events. This corresponds to a state mass of $2535.6 \pm 0.7 \pm 0.4\ MeV/c^2$. Since the observed width of the signal is consistent with the experimental resolution, we estimate the natural width $\Gamma < 5.4\ MeV/c^2$ ($90\% C.L.$). We measure the relative production rate $\sigma(D_s^{**+})/\sigma(D^{*+})$ to be $(1.9 \pm 0.5)\%$ for $x_p > 0.5$.

5. Summary and Conclusions

We now have statistically significant measurements for several decay modes of the Λ_c^+ relative to that into $pK^-\pi^+$. We find the mass splittings between the different members of the same isospin multiplet is consistent with zero for both the Σ_c and Ξ_c multiplets. The Ω_c^0 is still undiscovered. We have convincing evidence for two of the four $L = 1$ $D^{**0}(c\bar{u})$ states, with spin - parity assignments of $J^P = 1^+$ for the lower mass $D^{**0}(2428)$ and $J^P = 2^+$ for the higher mass $D^{**0}(2461)$. We have evidence for $D_s^{**+}(2535)$, the first $c\bar{s}$ state with $L = 1$. We have the very rich field of charmed baryon and excited meson spectroscopy, which is still very much uncharted nearly 15 years after the discovery of the charm quark.

References

1. G.S. Abrams et al., (MARK II) Phy. Rev. Lett. **44**, 10 (1980).
2. S.F. Biagi et al., (CERN) Z. Phy. **C28**, 175 (1985).
3. D. Andrews et al., (CLEO) NIM **47**, 211 (1983).
D. Cassel et al., (CLEO) NIM **A252**, 325 (1986).
4. T. Copie, Cornell University, internal memo CBX86-19, April, 1986. 5. D. Lichtenberg, Phy. Rev. **D16**, 231 (1977),

L. H. Chan, Phy. Rev. **D31**, 204 (1985),
W. Hwang and D. Lichtenberg, Phy. Rev. **D35**, 3526 91987).
6. T.J. Bowcock et al., (CLEO) PRL **62**, 1240 (1989).
7. M. Diesberg et al., (E400) PRL **59**, 2711 (1987).
8. H. Albrecht et al., (ARGUS) PL **B211**, 489 (1988).
9. S. Biagi et al., Phys. Lett. **B122**, 455 (1983).
Phys. Lett. **B150**, 230 (1985).
10. P. Coteus et al., Phys. Rev. Lett. **59**, 1530 (1987).
11. S. Barlag et al., XXII Int. Workshop on Weak Interactions and Neutrinos, (Ginosar, Israel, April 1989) and CERN Report-EP/88-106(1988).
12. P. Avery et al., Phy. Rev. Lett. **62**, 863 (1989).
13. M.S. Alam et al., Phy. Lett. **B226**, 401 (1989).
14. W. Kwong, J. Rosner and C. Quigg, Ann. Rev. Nucl. Par. Sci. **37**, 325 (1987).
15. H. Albrecht et al., Phy. Rev. Lett. **56**, 549 (1986).
16. J.C. Anjos et al., Fermilab preprint PUB88/155-E.
17. P. Avery et al., Cornell University preprint CLNS 89/839.
18. H. Albrecht et al., Phys. Lett. **B221**, 422 (1989).
19. J. C. Anjos et al., Phy. Rev. Lett. **62**, 1717 (1989).
20. ARGUS Collaboration, Int. Symp. on Heavy Quark Physics, (Cornell University, Ithaca, New York, June 13 - 17, 1989).
21. S. Godfrey and Isgur, Phy. Rev. **D32**, 189 (1985).

HADRO-PRODUCTION
OF HEAVY QUARKS

PHOTO- AND HADROPRODUCTION OF CHARM AND BEAUTY

Rollin J. Morrison
Univ. of California, Santa Barbara, CA 93106

ABSTRACT

Recent data on photo- and hadroproduction of charm and beauty are compared with predictions from perturbative QCD. Fragmentation properties in photo- and hadroproduction, and e^+e^- annihilation are compared.

INTRODUCTION

The lowest order QCD diagrams for the photo- and hadroproduction of heavy quarks are given in Figs. 1a and b, respectively. The single photon-gluon fusion diagram of Fig. 1a means that, in comparison with hadroproduction, photoproduction is a rather simple process and is more tractable theoretically.

A major breakthrough since the last heavy quark symposium is the higher order calculation of K. Ellis, P. Nason, and S. Dawson.[1,2] Photoproduction is now calculated through order $\alpha_{em}\alpha_s^2$ and hadroproduction through α_s^3. The results of these calculations indicate that the calculated cross sections are larger, by about a factor of 3 for hadroproduction, but that the shapes of the distribution in x_F and p_T^2 are not significantly affected. A rough guide to the theoretical confidence level for heavy quark production cross sections is given in Table I.

Table I
Relative Theoretical Confidence Level

Hadroproduction	Charm	Beauty	Top	
Photoproduction		Charm	Beauty	Top

0%   100%

The confidence level improves as the quark mass increases and is roughly one level better for photoproduction as compared with hadroproduction. In the following sections the emphasis will be on a comparison of the data with QCD predictions. In addition, fragmentation properties found in the different production processes and in e^+e^- experiments are compared.

PHOTOPRODUCTION OF CHARM

With photoproduction of charm we can learn about QCD from four experimental quantities; the magnitude of the cross section, the energy dependence of the cross section, and the p_T^2 and Feynman x distributions. The quark production cross section for photo-nucleon collisions is related to the fundamental photon-gluon cross section, $\hat{E}\frac{d^3\hat{\sigma}}{d\hat{p}^3}$, by

$$E\frac{d^3\sigma}{dp^3} = \int dx \, \frac{G(x)}{x} \hat{E}\frac{d^3\hat{\sigma}}{d\hat{p}^3}, \tag{1}$$

© 1989 American Institute of Physics

where x is the fraction of nucleon momentum carried by the gluon, and $\frac{G(x)}{x}$ is the gluon number density. The gluon distribution is parameterized by

$$G(x) = (1-x)^{n_g}. \qquad (2)$$

The experimentally determined quantities are dependent upon m_c and n_g, as indicated in Table II.

Table II
Dependences of Experimental Observables on m_c and n_g

Quantity	Sensitivity
$\sigma_{(c\bar{c})}^{tot}$	Mainly m_c
$\frac{d\sigma}{dE_\gamma}$	Steeper for large n_g, m_c
$\langle p_T^2 \rangle$	Larger for small n_g, large m_c sensitive to intrinsic K_T
$F(x_F)$	Sensitive to Fragmentation, n_g and m_c

A. THE TOTAL $c\bar{c}$ CROSS SECTION

The data[3] on the $c\bar{c}$ cross section is shown in Fig. 2. The procedure used in E691 is to measure $\sigma(D^0) + \sigma(\overline{D^0}) + \sigma(D^+) + \sigma(D^-)$ on Be for $x_F > 0.2$. This number is divided by two to correct for the separate counting of c and $\bar{c}$ and multiplied by $1/(1-.21)$ to correct for the 21% contribution given by the Lund Model for Λ_c^+ and D_s^+ production.* Finally we get the nuclear cross section from the Be measurements assuming $\sigma_A = \sigma_N A^\alpha$ with $\alpha = 0.94$. The value $\alpha = .94$ is chosen since this is the observation for inelastic ψ photoproduction[4]; this is also essentially the A dependence of the photon total cross section. The result is $\sigma_{\gamma N}^{c\bar{c}}(145\text{ GeV}) = 0.58 \pm 0.01 \pm 0.06$ μb at the average energy of 145 GeV.

Figure 2 also shows data from other experiments. The NA14 measurements were made at CERN in an experiment very similar to that of E691. The EMC and BPF measurements were made some years ago with virtual photoproduction using muons. There are no serious discrepancies within the measurements.

The theoretical curves are from Ellis and Nason. The central curve indicates the best value of the prediction for $m_c = 1.2$, 1.5, and 1.8 GeV/c for Figures 2a, b, and c, respectively. The outside curves indicate the estimate of the errors in the prediction due to uncertainties in the factorization/renormalization scale. It is clear that the magnitude of the cross section is inconsistent with low values of the charm quark mass.

---

* The use of the Lund values for the fraction of D_s^+ and Λ_c^+ production is motivated by the lack of measurements of the $D_s^+ \to \phi\pi$ and $\Lambda_c^+ \to PK^-\pi^+$ branching ratios. If we assume $B(D_s^+ \to \phi\pi) = 3\%$ and use the recently measured[5] $B(\Lambda_c^+ \to PK^-\pi^+) = 4.1\%$ we find that $\sigma^{c\bar{c}}$ is larger by about 5%.

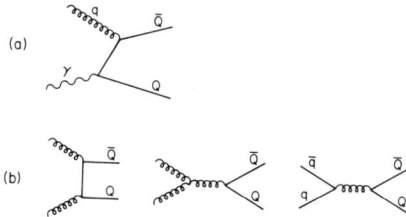

Fig. 1. Lowest order diagrams for photoproduction (a) and hadroproduction (b) of heavy quarks.

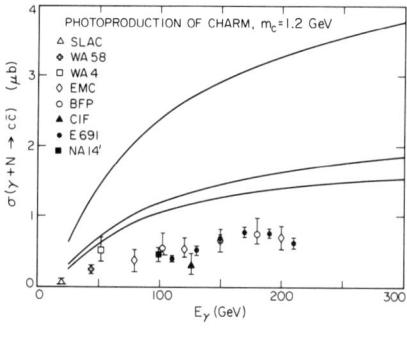

Fig. 2a

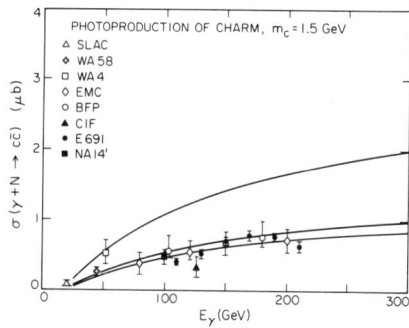

Fig. 2b

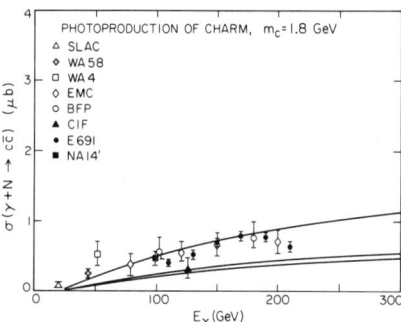

Fig. 2c

Fig. 2. Total charm photoproduction cross section. The data are from reference 3. The curves are order $\alpha_{em}\alpha_s^2$ predictions from Ellis and Nason, ref. 1, for charmed quark masses of 1.2 (a), 1.5 (b), and 1.8 (c) GeV/c^2. The central prediction is the calculation. The outer curves indicate the range of theoretical uncertainty. The data are from ref. 3.

B. THE DEPENDENCE OF $\sigma^{c\bar{c}}$ ON ENERGY, x_F AND p_T^2

As indicated in Table II the energy dependence of the total cross section is steeper either from larger values of m_c or n_g. An indication of the sensitivity to n_g is shown in Fig. 3. On the other hand the mean value of p_T^2 has the opposite dependence as n_g and m_c. This makes possible a unique determination of m_c and n_g from the energy and p_T^2 dependence. The E691 p_T^2 distribution is shown in Fig. 4 with predictions for D meson production for several values of n_g to indicate the sensitivity to this quantity. It should be noted that these curves do not include the effects of the intrinsic transverse momentum of the gluon.

The predicted x_F dependence of charmed quark production is shown in Fig. 5. This distribution is greatly modified, as shown in Fig. 6, when fragmentation into D mesons is taken into account. The x_F distribution is sensitive to the value of n_g and m_c as well as the details of the fragmentation function used. The E691 data are also shown in Fig. 6.

The x_F dependences of the NA14' and E691 data are compared in Fig. 7, where the data are normalized to the same value at $x_F = 0.25$. The data from the two experiments are in reasonable agreement. The 100 GeV NA14 data indicate a somewhat higher cross section at larger x_F than do the 145 GeV E691 data. It is not clear whether this is a significant effect.

It appears that the photoproduction data are consistent with QCD prediction and Lund Fragmentation using reasonable values of n_g and m_c. The fit to determine the best values of these parameters is not yet complete. The present indications are that larger values of both parameters are preferred.

C. FRAGMENTATION STUDIES — HIGH STATISTICS MODES

Due to the presence of the target we might expect fragmentation to be somewhat different in photoproduction as compared with e^+e^- experiments. The Lund model includes string connections to the target diquark. We will also compare with hadroproduction.

Table III
Production of charm particle types
in E691—high statistics modes

Particle	Signal	R $\frac{\sigma(\bar{c}x)}{\sigma(cx)}$	$\sigma^{Be}(c+\bar{c})$ μbarns $x_F > 0.2$
D^0	4252 ± 92	1.08 ± 0.03	$2.42 \pm 0.05 \pm 0.39$
$D^{*+} \to \pi^+ D^0$	988 ± 34	1.15 ± 0.07	$1.37 \pm 0.05 \pm 0.25$
$D^{*+} \to \pi^+ D^0$	1267 ± 47	1.23 ± 0.07	$1.60 \pm 0.06 \pm 0.31$
D^+	4864 ± 109	1.04 ± 0.03	$1.34 \pm 0.03 \pm 0.23$
$\langle R \rangle = 1.075 \pm 0.021$			

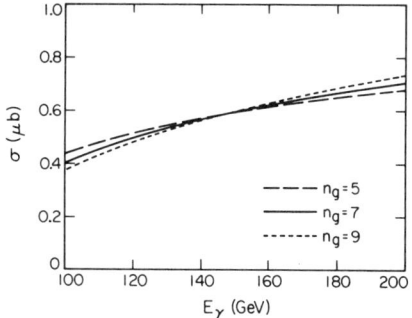

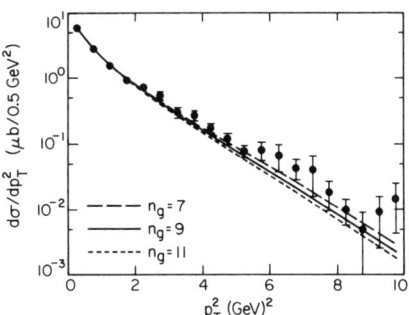

Fig. 3. Simulation of the energy dependence of the total charm cross section vs. energy for different values of n_g.

Fig. 4. The E691 P_T^2 dependence for the photoproduction of D mesons. The curves are the photon-gluon simulation with fragmentation for the case of a charm quark mass of 1.8 GeV/c^2 and for three values of n_g. The curves assume no intrinsic K_T for the gluon.

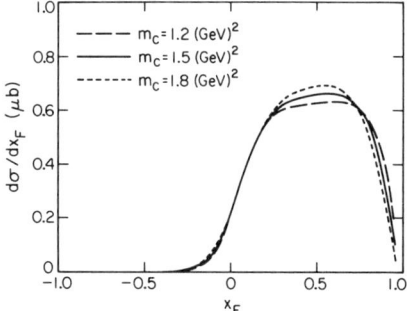

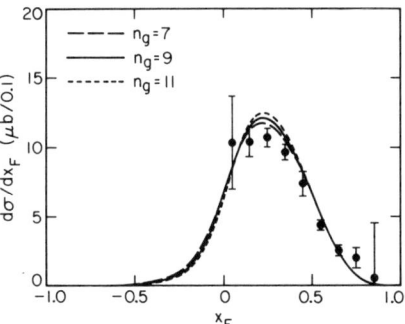

Fig. 5. The photon-gluon fusion x_F dependence for charmed quarks.

Fig. 6. The E691 x_F dependence. The curves are simulations with $m_c = 1.8$ GeV/c^2 and three values of n_g. The error in the simulation due to uncertainties in the fragmentation process is comparable to the difference in the curves for $n_g = 7$ and $n_g = 11$.

Some characteristics of the high statistics signals from E691 are shown in Table III. The D cross sections include D's from the strong decays of higher mass D*'s. Both D and D* cross sections include particles from the decays of higher mass resonances such as D**'s.

The particle ratios are consistent with production ratios, $\frac{\sigma(D^*)}{\sigma(D)} = 3/1$, expected from simple statistics arguments (and seen in e^+e^- events). There is a small but clear tendency for more $\overline{D}$'s than D's to be observed in the forward acceptance $x_F > 0.2$ of E691. This implies that $\frac{\sigma(cN)}{\sigma(\bar{c}N)} > 1$. This is consistent with the notion of the string connections of the c quark to a target diquark. It is also consistent with a small remnant of the $\overline{D}\Lambda_c^+$ associated production seen at low energies.

D. FRAGMENTATION STUDIES — LOW STATISTIC MODES

The values of the production cross section times branching ratio of particle types with low statistics in E691 are given in Table IV. An example of such a mode is the $D^{**}(2459)^6$ seen in Fig. 8. Two charmed baryons, D_s^+ states and three D^{**}'s have been observed in E691.

Table IV.
Production of Charm Particle Types—low statistics modes

Particle	Mode	R	$\sigma^{Be}(c+\bar{c}) \times B$ μbarns $x_F > .2$
D_s^+	$\phi\pi^+ + \overline{K^{*0}}K^+$	0.92 ± 0.14	$.0250 \pm .0022 \pm .0019$
Λ_c^+	$PK^-\pi^+$	0.79 ± 0.17	$.0404 \pm .0054 \pm 00.51$
Σ_c^0	$\pi^-\Lambda_c^+$		~ 0.005
$D^{**0}(2420)$	$D^{*+}(2010)\pi^-$		~ 0.2
$D^{**+}(2420)$	$D^{*0}(2010)\pi^+$		~ 0.2
$D^{**0}(2459)$	$D^+\pi^-$		~ 0.09

In Table V a comparison with particle ratios from continuum e^+e^- annihilation[7] at 10.5 GeV is given. From the ratio of $\sigma \cdot B$ for D_s^+ and nonstrange D's it appears that the probability of popping strange quarks out of the sea for photoproduction is the same as for e^+e^- annihilation. On the other hand charmed baryon production appears to be enhanced in photoproduction. In Section IV we will compare with hadroproduction. If we assume a branching ratio of 3% for $D_s^+ \to \phi\pi^+$ then $\sigma_{D_s^+}/\sigma_{D^0} + \sigma_{D^+} \sim 0.11$. Both CLEO[5] and ARGUS[5] find $B(D_s^+ \to PK^-\pi^+) \sim 4.1\%$. This gives a charmed baryon production ratio

$\sigma_{\Lambda_c^+}/(\sigma_{D^0} + \sigma_{D^+}) \sim .26.$

Table V.
Fragmentation Ratios for e^+e^- and Photoproduction

Experimental Quantity	Valence Quark Ratios	e^+e^- 10.5 GeV	Photoproduction E691
$\frac{\sigma(D^+)}{\sigma(D^0)}$	$\frac{c\bar{d}}{c\bar{u}}$	0.47 ± 0.11	0.55 ± 0.13
$\frac{\sigma \cdot B(D_s^+ \to \phi\pi)}{\sigma_{D^0}+\sigma_{D^+}}$	$\propto \frac{c\bar{s}}{c\bar{u}+c\bar{d}}$	$(4.4 \pm 1.1)10^{-3}$	$(3.4 \pm 0.9)10^{-3}$
$\frac{\sigma \cdot B(\Lambda_c^+ \to PK^-\pi^+)}{\sigma_{D^0}+\sigma_{D^+}}$	$\propto \frac{cud}{c\bar{u}+c\bar{d}}$	$(5.4 \pm 1.3)10^{-3}$	$(11 \pm 2)10^{-3}$

E. THE FUTURE — CONCLUSIONS

We can expect much better photoproduction data in the near future. Experiment E687 at the Fermilab has a much more intense photon beam. The first run for this experiment was severely compromised by a fire in the apparatus, but still appears to have been quite successful. The next run starting in January 1990 should have ~ 10 times the E691 statistics and should be at higher energy. From the point of view of production studies, it will be interesting to extend the range of the energy dependence measurements, to see the negative x_F behavior of the cross section, and to study charmed particle pair correlations. Much better statistics will be available for the study of semileptonic and hadronic decays, and for charmed meson and baryon spectroscopy.

A summary of the current state of the photoproduction of charm is that there are no serious discrepancies between experiments or between theory and experiment. The probability of fragmentation into baryons is larger than observed in e^+e^- annihilation.

PHOTOPRODUCTION OF BEAUTY

There is no data on the photoproduction of beauty. The reason for this can be seen from the predictions of Ellis and Nason shown in Fig. 9. The cross section at the mean E691 energy is only 2×10^{-4} that of charm. With their higher energy and luminosity E687 should see a few events. This will be extremely useful since the predictions for the photoproduction of beauty are thought to be more reliable than for charm. At UNK energies the cross section will be $\sim 10^{-4}$ of the total cross section.

The virtual photoproduction of beauty at the HERA, with $\sqrt{s} = 314$ GeV, will be a relatively copious process. The cross section for electron-proton production of beauty at this energy is predicted to be about 6 nbarns. Production via virtual photons dominates over virtual Z or W production.

HADROPRODUCTION OF CHARM

The theory of the hadroproduction of charm is difficult, as indicated by the

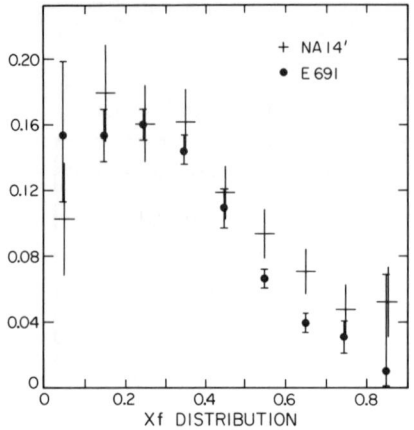

Fig. 7. A comparison of the E691 x_F dependence with that of NA14' using mean photon energies of 145 GeV and 100 GeV, respectively. The scale is arbitrary.

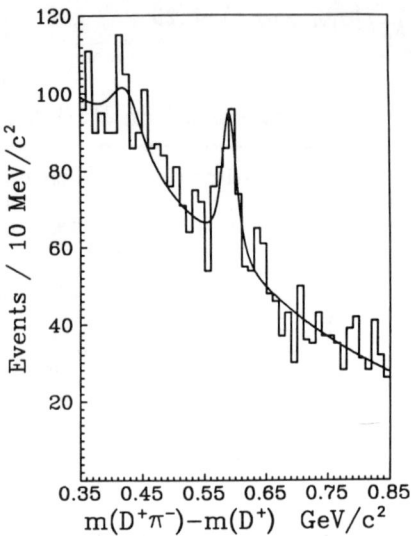

Fig. 8. The E691 $m(D^+\pi^-) - m(D^+)$ plot showing the fit for $D^{**0}(2459)$.

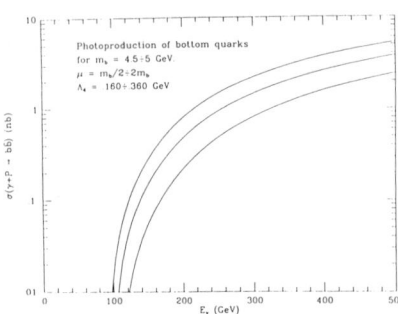

Fig. 9. The total cross section for the photoproduction of $b\bar{b}$ states vs. lab energy. The predictions shown are to order $\alpha_s^2 \alpha_{em}$ as computed by Ellis and Nason (Ref. 1). This figure is taken directly from Ref. 1.

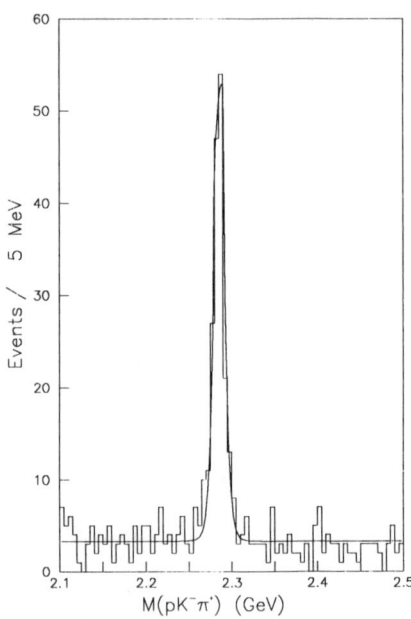

Fig. 10. The NA32' signal for $\Lambda_c^+ \to PK^-\pi^+$.

large K factor,
$$K = \frac{\sigma(\alpha_s^2+\alpha_s^3)}{\sigma(\alpha_s^2)} \sim 3. \qquad (3)$$

Compared with photoproduction, the experimental study of charm hadroproduction is also much more difficult. We now know from the experiments with the best signals that the fraction of charm events in hadroproduction is less than 1/5 the fraction of charm events in photoproduction. In addition, the multiplicity in hadroproduction is greater, leading to a further reduction of signal-to-noise. To a first approximation, the study of the photoproduction of charm is not possible without precision vertex detection for significantly reducing combinatorial background. It seems unlikely therefore that hadroproduction studies would be possible without the use of this technique.

The history of the hadroproduction of charm is, however, seriously affected by a large number of reported charm signals obtained without the help of precision vertex detection. These are all at about the level of the background, which is large, and they therefore correspond to large cross sections. In attempts to explain these confusing and often contradictory results more degrees of freedom were introduced. These include:
1. Dependence on incident particle type.
2. Leading particle effects.
3. Ultra steep energy dependences.
4. Differing dependences on A.

By leading particle effect we mean very forward production of charm particles which have one or more valence quarks in common with those of the projectile. By incident particle type dependence we mean any production cross section dependence on incident particle type not included in the leading particle category.

In order to make sense out of the data on the hadroproduction of charm I consider just those experiments which have clean signals in a number of channels. There are two experiments which provide most of our information on the hadroproduction of charm: NA27/E743[8] and NA32'.[9] In addition there are two very good newer experiments, WA82[10] and E653,[11] which are just producing results.

Experiment NA27 used a high resolution Lexan hydrogen bubble chamber, in conjunction with a magnetic multiparticle spectrometer, to identify charmed particles. At CERN 360 GeV/c π^-p and 400 GeV/c pp interactions were studied. The Lexan bubble chamber was then moved to the Fermilab where 800 GeV/c pp interactions were studied in experiment E743. The importance of NA27/E743 stems from the clean charm signals seen with pp interactions at two different energies and with π^-p and pp interactions at essentially the same energy. Since the data are for the proton, no assumptions must be made about the A dependence in order to obtain nucleon cross sections. The numbers of events obtained are not large enough to permit fine binning of production distributions.

The second major experiment is NA32'. The key to this experiment was precision vertexing using a combination of silicon microstrip and CCD detectors. The experiment used a 230 GeV/c π^- beam (with some K^-) on a copper target. The quality of the experiment can be seen from the spectacular $\Lambda_c^+ \to pK^-\pi^+$ signal shown in Fig. 10. This experiment has approximately 1000 very clean reconstructed charm decays in a large number of channels.

First we consider what we learn from NA27/E743. The value of $\sigma_{c\bar{c}}$ for pp

collisions at the two energies are shown in Fig. 11. Also shown are the prediction of QCD to order α_s^3 from Nason, Dawson, and Ellis,[2] as evaluated by Berger[12] for the two different predictions of structure functions of Duke and Owens[13] and Martin, Roberts, and Stirling.[13] The data are consistent with the theory for a reasonable value of m_c although the theoretical range for the size of the cross section is quite large. More importantly the data are in agreement with the energy dependence from QCD. *There is no anomalously large increase of the cross section with energy.*

From NA27 we can learn about incident particle dependences. The cross section, $\sigma(D) + \sigma(\bar{D}) \equiv \sigma(D/\bar{D})$, for 400 GeV/c pp collisions for all x_F is 30.2 ± 2.2 μb. For 360 GeV/c $\pi^- p$ collisions for $x_F > 0$ the value is 15.8 ± 2.7 μb. From this we learn that

$$\sigma^{c\bar{c}}(\pi p) \approx \sigma^{c\bar{c}}(pp). \qquad (4)$$

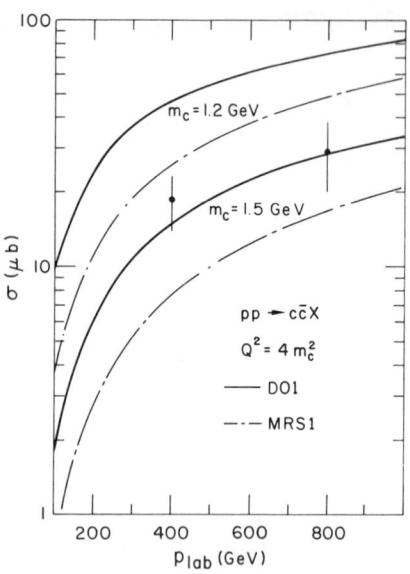

Fig. 11. Cross sections for $pp \to c\bar{c}X$ as a function of laboratory momentum for two choices of the charm quark mass m_c and two sets of parton densities (reference 13). The data are from NA27/E743. The figure is by E. Berger.

We can learn about leading diquarks from NA27 through their Λ_c^+ results in pp collisions. The Λ_c^+ could have a ud diquark in common with that of the incident proton. One might then expect more Λ_c production than $\bar{\Lambda}_c$ and one might also expect forward production with a rather large $\langle x_F \rangle$. In fact, NA27 finds $\sigma(\Lambda_c)/\sigma(\bar{\Lambda}_c) \sim 1$ and $\langle x_F \rangle_{\Lambda_c} \simeq \langle x_F \rangle_{\bar{\Lambda}_c} = 0.22 \pm 0.10$. *There is no strong leading diquark effect in NA27.* The cross section times branching ratio for NA27 for all x_F is, $\sigma \cdot B(\Lambda_c/\bar{\Lambda}_c \to 3\mathrm{prongs}) = 3.1^{+2.4}_{-1.8}\mu\mathrm{b}$.

What do we learn from NA32'? First, we compare NA32' pion cross sections with those from NA27. For 230 GeV/c π^- Cu collision NA32' measures $\sigma(D/\bar{D}) = 8.82 \pm 0.78$ μb for $x_f > 0$ and assuming A^1. To scale to the 360 GeV/c NA27 energy requires a factor of about 1.35. At 360 GeV/c the NA32' cross section would be $\sim 12 \pm 2\mu$barns. This is a bit smaller, but consistent with the NA27 value of 15.8 ± 2.7 μb. The two experiments are in perfect agreement with an A dependence of $A^{0.9}$, but clearly exclude $A^{0.75}$. We note that the new experiment, WA82[10] observes $A^{0.89 \pm 0.05 \pm 0.05}$.

We can compare Λ_c^+ production with pions and protons, again looking for evidence of leading diquarks. The NA32' result with 230 GeV/c π^- Cu interaction is

$$\sigma \cdot B(\Lambda_c/\bar{\Lambda}_c \to PK\pi) = 0.17 \pm 0.02 \ \mu\mathrm{b/nucleus} \qquad (5)$$

for $x_F > 0$ and assuming A^1. To compare with the NA27 result we correct by a factor of 1/2 for the missing x_F region, a factor of 1/1.35 for the different energies and a factor of 0.26 which is the fraction of three prongs observed by NA27 to be $PK\pi$. The equivalent value from NA27 is then $0.3^{+0.3}_{-0.2}$ to be compared with the NA32' value above. The conclusion is that Λ_c production by pions is about the same as with incident protons. *Again there is no dramatic leading diquark effect.*

The NA32' negative beam contained some K^- as well as π^-. This allowed a comparison of D_s^+ production by strange and nonstrange incident particles. In contrast to earlier results based on very low statistics, *NA32' now finds the production cross section with incident K's and pions to be the same.*

NA32' has performed a more detailed search for leading particle effects by comparing D production for the cases where the D contains a quark in common with the π (for example, D^0 contains a d quark) with those which do not. Parameterizing the cross section

$$d\alpha \propto (1-x_F)^n e^{-bp_T^2} \tag{6}$$

NA32' finds $n = 2.93 \pm 0.33$ and $n = 4.34 \pm 0.44$ for leading and nonleading cases, respectively. There appears to be a small leading particle effect at the 2σ level. This effect is small, however. For the leading particle case only 2.6% of the $x_F > 0$ part of the cross section has $x_F > 0.6$. *There is no large leading particle effect.*

From the NA32' Λ_c production measurement above and the NA32' value $\sigma \cdot B(D_s/\bar{D}_s \to KK\pi) = 0.08 \pm 0.01$ μb, with the E691 ratio $B(D_s \to KK\pi)/B(D_s \to \phi\pi) = 1.33$, we obtain the fragmentation values of Table VI. A striking feature of the comparison with other processes is that the fragmentation into D_s seems to be roughly the same for all processes, while the fragmentation into Λ_c is roughly four times as likely in pion production as compared with e^+e^- annihilation. Photoproduction is intermediate between these cases.

Table VI.
Fragmentation Ratios for e^+e^-, Photo- and Hadroproduction

Experimental Quantity	e^+e^- 10.5 GeV	Photoproduction E691	π^- Production NA32'
$\frac{\sigma \cdot B(D_s^+ \to \phi\pi^+)}{\sigma(D^0+D^+)}$	$(4.4 \pm 1.1)10^{-3}$	$(3.4 \pm 0.9)10^{-3}$	$(6.8 \pm 1.3)10^{-3}$
$\frac{\sigma \cdot B(\Lambda_c^+ \to PK^-\pi^+)}{\sigma(D^0+D^+)}$	$(5.4 \pm 1.3)10^{-3}$	$(11 \pm 2)10^{-3}$	$(19 \pm 2)10^{-3}$

A dramatic illustration of the conflicts in the hadroproduction data occurs for the case of the Ξ_c^+. The full power of the high resolution vertex technique is clear from the six clean $\Xi_c^+ \to \Xi^-\pi^+\pi^+$ and $\Xi_c^+ \to \Sigma^+K^-\pi^+$ decays observed by NA32'. As seen in Fig. 12, a total of four separated vertices are observed in these events. There is essentially no background. The value $\sigma \cdot B(\Xi_c^+ \to \Xi^-\pi^+\pi^+) = 0.04 \pm 0.02 \pm 0.02$ μb is quite reasonable. The suppression due to strangeness is roughly,

$$\frac{\sigma \cdot B(\Xi_c^+ \to \Xi^-\pi^+\pi^+)}{\sigma \cdot B(\Lambda_c^+ \to PK\pi)} \sim 0.2. \tag{7}$$

This is approximately the same value as $\sigma_{D_s}/\sigma(D^+ + D^0) = 0.2$ if we assume that $B(D_s \to \phi\pi) = 3\%$. The NA32' $\sigma \cdot B$ is about what one expects.

If we compare with the previously published Ξ_c^+ results of Table VII we find an irreconcilable conflict. The first observation with incident hyperons by WA62[14] leads to a $\sigma \cdot B$ of 0.63 ± 0.3 μbarns for $x_F > 0.6$. Even with the stiffest conceivable x_F dependence, $\sim (1 - x_F)^2$, only 6% of the cross section is for $x_F > 0.6$. Since we would not expect the branching ratios of the different observed modes to vary by large factors the implication is that hyperon production of Ξ_c is ~ 100 times that of pion production. But we have just seen that both NA32' and NA27 demonstrate that there are no large leading particle or incident particle type effects. If the WA62 measurement is correct hyperon production of charm is completely different from production by strange or nonstrange mesons or by nonstrange baryons. The second observation[15] of Table VII also indicates a cross section ~ 100 times that of NA32', in this case with incident neutrons. This is again totally in conflict with the picture of hadroproduction of charm which is clear from the NA27/E743 and NA32' results.

Table VII. Production of Ξ_c

Exp.	Beam/Target	Modes	Mass (MeV/c^2)	$\sigma \cdot B$ μb/nucleon	Comments
WA62	Σ^-	$\Lambda K^-\pi^+\pi^+$	2460 ± 15	0.63 ± 0.3	$\tau = 0.48^{+0.29}_{-0.18}$ ps
	135 GeV			$x_F > 0.6$	
E400	Neutrons	$\Lambda K^-\pi^+\pi^+$	$2459 \pm 5 \pm 30$	$7.5^{+4.5+1.9}_{-26-1.9}$	$\tau = 0.40^{+0.18+0.12}_{-0.10-0.10}$ ps
	600 GeV	$\Sigma^0 K^-\pi^+\pi^+$		$0 < x_F < 0.6$	
NA32'	π^-	$\Xi^-\pi^+\pi^+$	$2465.4 \pm 4.0 \pm 1.5$	$0.04 \pm 0.02 \pm 0.02$	$\tau = 0.34^{+0.21}_{-0.12}$ ps
	230 GeV	$\Sigma^+ K^-\pi^+$			$\frac{\Sigma^+ K^-\pi^+}{\Xi^-\pi^+\pi^+} = 0.17^{+0.24+0.03}_{-0.11-0.07}$

In the near future there will be a real step in the statistical precision in the hadroproduction of charm. WA82 has a high precision silicon vertex detector and a vertex based trigger. E653 has a silicon and emulsion definition of the charm vertex. Both have presented results at this symposium. E769 is reconstructing a sample which should result in about 5000 clean charm decays with different incident particles and on different nuclear targets. The CERN hyperon experiment has been described by Dr. S. Paul at this symposium. E781 is a new hyperon experiment being built at the Fermilab. E791 is an evolution from E769 (and previously E691) which is aiming for clean charmed samples of $\sim 100,000$ events. The charm rate in both of the hyperon experiments will be quite dramatic if the earlier measurement of WA62 is correct.

The conclusions on the hadroproduction of charm are that the results of NA27/E743 and NA32' are consistent with each other and with QCD. There are no strong leading particle effects, no strong incident particle type effects, and no ultra-steep energy dependences. The data favor an A^α dependence with $\alpha \sim 0.9$ as seen by WA82. Strange quark fragmentation is fractionally similar to that observed in e^+e^- and in photoproduction. Fragmentation into charmed baryons seems to be significantly enhanced as compared with e^+e^- and photoproduction.

The Ξ_c^+ is observed with a reasonable cross section by NA32'. This cross section is two orders of magnitude less than that of the two previous observations. By selecting just those experiments with clean, strong signals in numerous decay modes a picture of hadroproduction of charm emerges which agrees with QCD and which has no surprises. Since this ignores numerous conflicting results it is important that this picture be confirmed by future measurements.

HADROPRODUCTION OF BEAUTY

The cross section for the hadroproduction of beauty is small and is a strong function of energy at current fixed target energies. Predictions for $\sigma^{b\bar{b}}(\pi^- p)$ by Berger[12] using the order α_s^3 calculations of Ellis, Nason, Dawson are shown in Fig. 13.

One event has been seen in π^- emulsion interactions by WA75.[16] The only cross section measurement in a fixed target experiment is by WA78[17] using 320 GeV π^- uranium interactions and observing trimuons. Assuming an A^1 atomic number dependence WA78 measures a cross section of $3.27 \pm 0.24 \pm 0.9$ μb/nucleon. The $B\bar{B}$ production cross section is $\sim 10^{-7}$ of the total. From Fig. 13 it can be seen that this value is in reasonable agreement with QCD predictions, although the predictions still have a sizable range.

The cross sections are much larger at $p\bar{p}$ collider energies. UA1[18] has detected B's by means of single muons, low mass dimuons, high mass dimuons, and dimuons from J/ψ decays. CDF[19] has presented evidence at this symposium of B observations by means of a lepton correlated with a charm $K\pi$ mass peak. A problem in $p\bar{p}$ collider detection of B's, which have a mean p_T of about 5 GeV/c, is the necessity of avoiding the minimum bias small p_T background region. The data of UA1 are given in Fig. 14 integrated over p_T as a function of $p_T^{\min}$. The energy is $\sqrt{s} = 0.63$ TeV and $|y^b| < 1.5$. The single muons are from semileptonic B decay. The low mass dimuons are from sequential semileptonic decays of B and C. The high mass dimuons are from semileptonic decay of $B\bar{B}$ pairs. The data for J/ψ decays have been corrected for the large background of J/ψ from sources other than B decays. The data are in good agreement with $0(\alpha_s^3)$ calculation for $p_T^{\min} < 15$ GeV/c, as seen in Fig. 14, but there are no measurements for small p_T. The integrated cross section is $\sigma_{B\bar{B}} = 10.2 \pm 3.3$ μbarns.

As discussed by M. Witherell[20] at this symposium, hadroproduction is a prolific source of beauty at very high energies. The projections are illustrated in Table VIII. At SSC energies the fraction of the cross section associated with beauty is expected to be about the same as currently seen with charm in photoproduction.

CONCLUSIONS

The data on photoproduction of charm and hadroproduction of beauty are in good agreement with higher order QCD and standard fragmentation models. The predictions for charm hadroproduction are less reliable but here also the agreement is good for those experiments with clean signals in numerous channels.

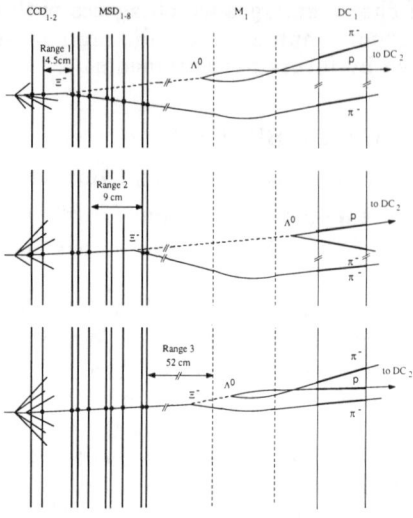

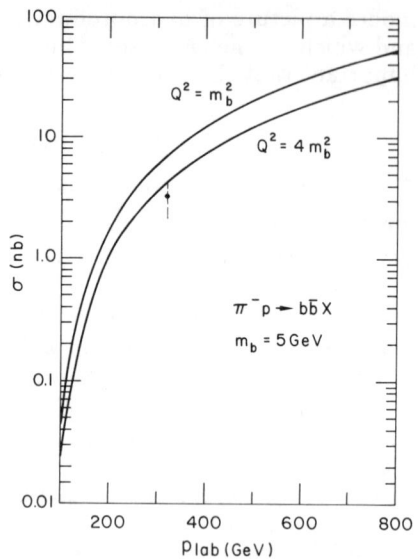

Fig. 12. Schematic event displays for Ξ_c^+ decays in NA32′ indicating the observation of four separated vertices in the events.

Fig. 13. Cross section through order α_s^3 for bottom quark production in $\pi^- p$ interactions as a function of lab momentum. The curves are from the calculation of Nason, Dawson, and Ellis as evaluated by E. Berger (reference 12). The measurement is by WA78 (reference 17). The figure is by E. Berger.

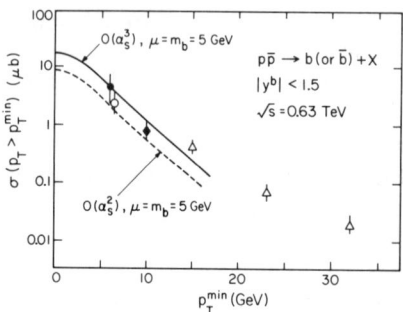

Fig. 14. The UA1 measurements of $p\bar{p} \to b$ (or $\bar{b}$) + X. The open circle is for high mass dimuons; the solid circle is for $B \to J/\psi$; the diamond is for low mass dimuons; and the open triangles are for single muons. The $0(\alpha_s^2)$ and $0(\alpha_s^3)$ are from reference 2.

Table VIII.
Hadroproduction as a Future Source of Beauty

Experiment	$\sqrt{s}$ GeV	$\sigma_{B\bar{B}}$ μb	$\frac{\sigma_{B\bar{B}}}{\sigma^{\text{tot}}}$
Tevatron Fixed	40	$\sim 3 \times 10^{-2}$	$\sim 10^{-6}$
UNK Fixed	70	$\sim 10^{-1}$	$\sim 3 \times 10^6$
RHIC	450	~ 10	10^{-4}
Tevatron $\bar{P}P$	1,800	~ 40	4×10^{-4}
SSC	40,000	~ 500	5×10^{-3}

REFERENCES

1. Predictions for photoproduction are from R.K. Ellis and P. Nason, Fermilab-Pub-88/54-T (1988).

2. Predictions for hadroproduction are from P. Nason, S. Dawson, and R.K. Ellis, *Nucl. Phys.* **B303**, 607 (1988); P. Nason, S. Dawson, and R.K. Ellis, *Particle Physics* **303**, 607 (1988).

3. J.J. Aubert et al. (EMC), *Nucl. Phys.* **B213**, 31 (1983); A.R. Clark et al. (BFD), *Phys. Rev. Lett.* **45**, 682 (1980); R.W. Forty (NA14), Proceedings of the XXIV Int'l. Conf. on High Energy Physics, Munich 1988, Springer-Verlag (Berlin, 1989), p. 668; J.C. Anjos et al. (E691), *Phys. Rev. Lett.* **62**, 513 (1989); M.I. Adamovich et al. (WA58), *Phys. Lett.* **B187**, 37 (1987); M.S. Atiya et al. (CIF), *Phys. Rev. Lett.* **43**, 414 (1979); this number has been corrected for the newest value of the $D^0 \to K^-\pi^+$ branching ratio; D. Aston et al. (WA4), *Phys. Lett.* **B94**, 113 (1980); K. Abe et al. (SLAC), *Phys. Rev. D* **33**, 1 (1986).

4. M.D. Sokoloff, et al., *Phys. Rev. Lett.* **57**, 3003 (1986).

5. CLEO reports a value $4.1 \pm 1.1^{+1.6}_{-1.3}\%$, talk by M.S. Alam, this symposium. ARGUS reports a value $4.2 \pm 1.5\%$, K. Schubert, this symposium.

6. J.C. Anjos, et al., *Phys. Rev. Lett.* **62**, 1717 (1989).

7. S. Stone, private communication.

8. M. Aguilar-Benitez, et al., *Phys. Lett.* **B189**, 476 (1987); R. Ammar, et al., *Phys. Lett.* **B183**, 110 (1987); M. Aguilar-Benitez, et al., *Z. Phys.* **C31**, 491 (1986); M. Aguilar-Benitez, et al., *Phys. Lett.* **B199**, 462 (1987).

9. S. Barlag, et al., contribution to XXIV Int. Conf. on High Energy Physics, Munich 1988. Note that the cross section scale has recently been corrected (private communication S. Kwan); S. Kwan, contribution to this symposium; S. Barlag, et al., contribution to the 12th Int. Workshop on Weak Interactions and Neutrinos, Ginosar, 1989.

10. D. Barberis, contribution to this symposium.

11. R. Lipton, contribution to this symposium.

12. E. Berger, Proc. of the XXIV Int. Conf. on High Energy Physics, ed. R. Koffhaus and J. Kühn, Springer-Verlag, 1989, p. 987.
13. D.W. Duke and J.F. Owens, Phys. Rev. D **30**, 49 (1984); A.D. Martin, R.G. Roberts, and W.J. Stirling, Phys. Rev. D **37**, 1161 (1988).
14. S.F. Biazi, et al., Z. Phys. C**28**, 175 (1985).
15. P. Coteus, et al., Phys. Rev. Lett. **59**, 1530 (1987).
16. J.P. Albanese, et al., Phys. Lett. **B158**, 186 (1985).
17. The published value of this cross section is $4.5 \pm 1.4 \pm 1.4$ μb per nucleon, M.G. Catanesi, et al., Phys. Lett. **B187**, 431 (1987). This measurement has been revised downward to $3.27 \pm 0.24 \pm 0.9$ μb, J. Conborg, "The WA78 Spectrometer", poster session, XXIV Int. Conf. on High Energy Physics, Munich, 1988 (see Ref. 12).
18. I.R. Kenyon, et al., Proc. of the XXIV Int. Conf. on High Energy Physics, ed. R. Kotthaus and J. Kühn, Springer-Verlag, 1989, p. 981.
19. G.W. Foster, talk at this symposium.
20. M.S. Witherell, talk at this symposium.

HADRONIC PRODUCTION OF HEAVY QUARKS IN A HYBRID EMULSION EXPERIMENT

Ronald J. Lipton
(Representing the E653 Collaboration)[1]
Carnegie Mellon University, Pittsburgh, Pa. 15217

ABSTRACT

Preliminary results are presented for hadronic production of charm in a hybrid emulsion experiment in an 800 GeV/c proton beam at Fermilab. The Feynman X distribution found for all produced charm is $(1-x)^{11.2\pm.9}$. No strong evidence was found for leading particle effects in proton emulsion interactions.

INTRODUCTION

In spite of important advances in experimental technique and detector technology nuclear emulsion remains a unique tool for some aspects of experimental particle physics. The unparalleled resolution and pattern recognition ability of emulsion allows for high efficiency detection of short decays and excellent efficiency for the reconstruction of charm pairs. Emulsion is uniquely suited for the investigation of complex topologies such as B pair production and decay, as well as simple topologies difficult to detect by counter techniques alone, such as the cascade decay $D_s \to \tau \nu_\tau \to \mu \nu_\mu \nu_\tau$.

In this paper we discuss preliminary results of Fermilab experiment 653, a muon triggered hybrid emulsion experiment using a high quality forward spectrometer. E653[2], shown in in figure 1, was designed to measure charm and beauty production and decay physics using pion and proton beams. Limitations placed on the muon trigger by π decay forced us to build a short spectrometer, which in turn imposed limits on particle identification (in our case only time of flight is used).

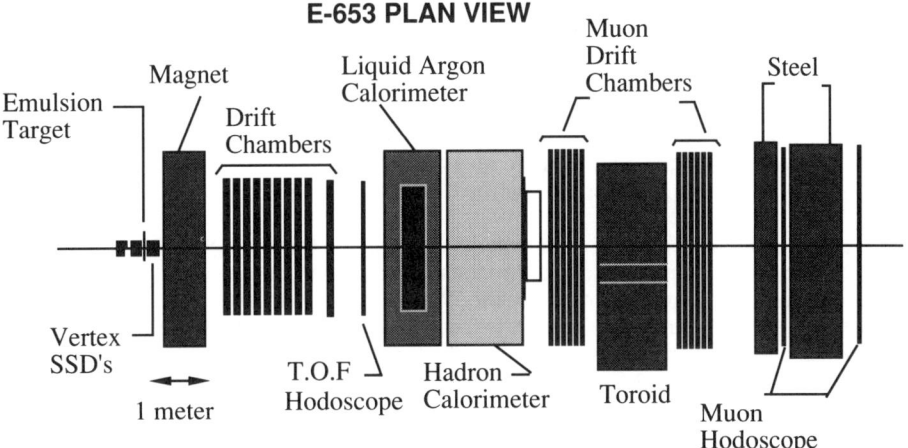

Figure 1. Plan view of the E653 spectrometer.

The primary constraints on a hybrid experiment of this type are limits imposed by emulsion scanning and measurement. Scanning is limited to about 20,000 events per year. In terms of our experiment this means that fewer than 1 interaction in 2000 must be selected for emulsion scanning. One event in 20 is selected online by the muon trigger. Offline event selection was based on secondary vertex reconstruction and impact parameter analysis of the muon. Reliable, high resolution vertex detection is crucial for this analysis. E653 employed 18 planes of 50μm pitch silicon strip detectors. The offline requirement that the muon have either a large impact parameter with respect to the primary vertex or be part of a reconstructed secondary vertex reduced the sample by a factor of 100.

E653 had two data runs, a 800 Gev proton run in 1985, and a run with 600 Gev pions in 1987. The data discussed in this paper is based on 2/3 of the proton exposure. Emulsion efficiencies are not yet included for this data, but they are not expected to be a strong function of X_F, and should not qualitatively change these preliminary results. Only statistical errors are quoted.

EMULSION ANALYSIS

Stacks with emulsion parallel ("horizontal") and perpendicular ("vertical") to the beam direction were exposed. The emulsion stacks were placed on a mechanical target mover assembly and moved continuously during the beam spill. The position of the emulsion was recorded for each trigger.

Extensive use is made of spectrometer information in the emulsion analysis. The analysis procedure for the vertical emulsion consisted of:

- Location of the primary vertex.
- Spectrometer track slopes are matched to those of tracks emerging from the primary vertex. If the muon is found at the primary the event is rejected.
- The event is scanned graphically (by computer display) to remove obvious interactions and K decays.
- Tracks at the primary which are not matched to spectrometer tracks are followed down through the emulsion to search for decays. Spectrometer tracks which do not come from the primary are "scanned back" from the emulsion exit. This procedure is shown schematically in figure 2.
- If a decay candidate is found the tracks slopes and vertex positions in the emulsion are measured.
- A search is made for the "partner" decay.

Decay candidate events are then reanalyzed using the track and vertex measurements as an aid to counter reconstruction. We can now analyze the events in a much simpler environment. Rather than trying to reconstruct three vertices (primary and two secondaries) per event using 18 tracks, we typically are searching for one decay vertex outside of the emulsion block using an average of 2.5 unmatched emulsion tracks and 5 unmatched spectrometer tracks per event.

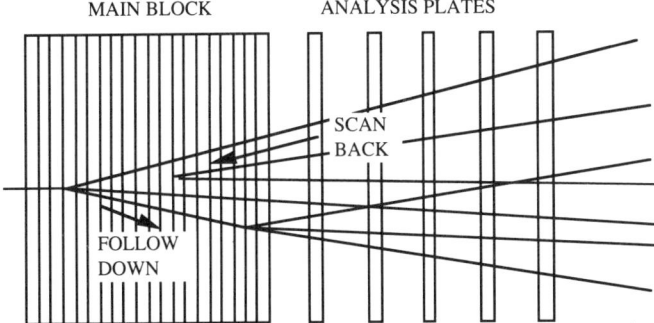

Figure 2. Vertical emulsion structure and schematic of scanning procedure.

Constrained fits are attempted for all events with approximate transverse momentum balance about the decay vertex using all reasonable parents and daughters. Zero constraint fits are used for events which appear to have a missing neutral. Only Cabibbo favored modes are used for these decays. The momentum was also estimated using decay kinematics. In this analysis the 0C fit closest to the estimated momentum was used. Monte Carlo studies show that the results are insensitive to the technique used for momentum estimation.

PHYSICS RESULTS

Our data sample consists of 699 candidate events identified solely on the basis of a single vertex with a decay topology. Approximately 25% of these events are low transverse momentum kinks or strange particle decays. There are 229 fit single decays and 127 pairs in the uncut sample. For pair candidates we require one muonic vertex, at least one emulsion vertex that passes weak cuts on transverse momentum balance and mass, and at least one vertex which passes tighter quality cuts. After these cuts 57 total pairs and 34 pairs with both vertices with kinematic fits remain. Single decay candidates must also appear in events with one good emulsion vertex. The candidate decay must also have an acceptable fit to a charm particle hypothesis.

Charm particle momenta are conventionally parameterized as:

$$\frac{dN}{dx_F} = (1-X_F)^n \, , \frac{dN}{dP_t} = e^{-bP_t^2} \, .$$

Figure 3 shows the X_F distribution for charm singles. The distribution is quite steep, falling as $(1-X_F)^{11.2\pm.9}$. Figure 4 shows data on charm production from ACCMOR(200 GeV/c π), LEBC (360 GeV/c π, 400 GeV/c P, and 800 GeV/c P), and E653[3]. There is a substantial increase in X_F slope as a function of beam energy. QCD predicts an effect of this nature but the predicted change in n between $\sqrt{s}$ of 23-38 is only about 2.5.[4]

If we separate the data into leading (D^0, D^-) and non-leading(D^0, D^+) components we find a small (~ 1 σ) difference in n. Table 1 compares our results with data from LEBC on pion and proton production of charm. While our data shows a slightly larger value of n for non-leading particles the NA27 PP data indicates a larger n for <u>leading</u> particles, which they ascribe to leading diquarks. We see no evidence of such an effect.

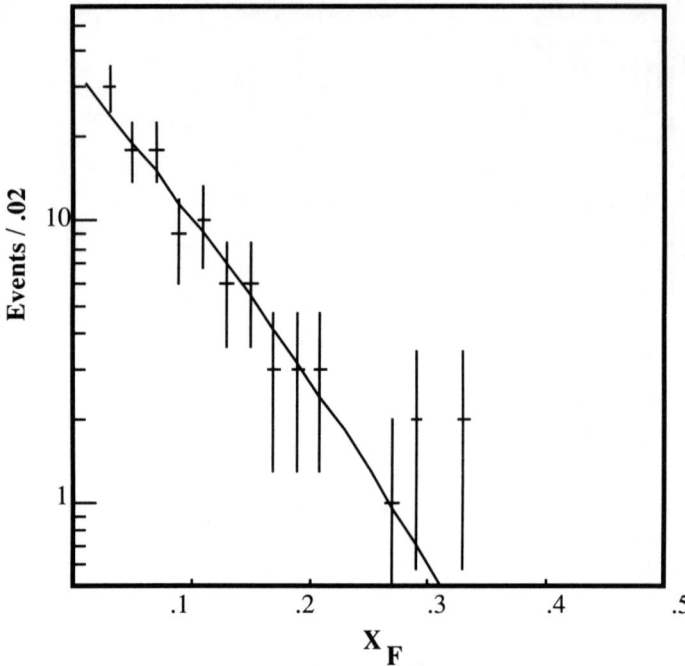

Figure 3. Xf distribution for all charm candidates. The solid curve is the fit to $(1-X_F)^n$.

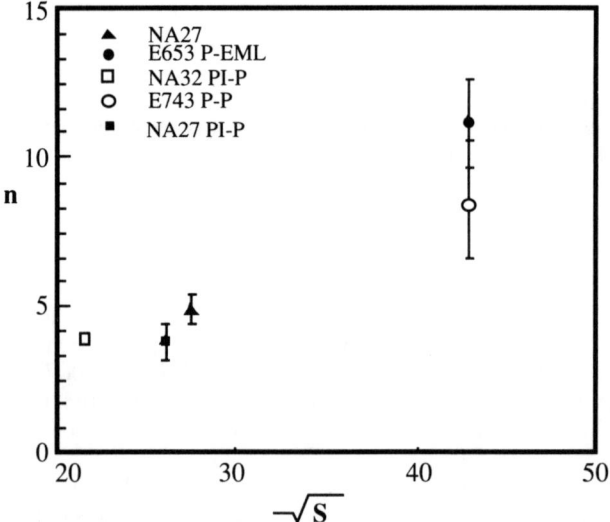

Figure 4. Data on the X_F slope, n, as a function of cm energy from E653, LEBC and NA32.

	E653 P EMUL 800 GeV	NA27 PP 400 GeV	NA27 πP 360 GeV
ALL	11.2±.9	4.9±.05	3.8±.63
LEADING	10.0±1.4	6.6±1.1	1.9±.8
NON LEADING	13.± 2.0	4.2±.08	7.9±1.5

Table 1. Results for leading and non-leading charm production.

In the minimal QCD model of hadronic charm production charm pairs are expected to be dominated by the fusion of a pair of gluons typically having small intrinsic transverse momentum. As a result of this the charm pair should be produced approximately back to back in the center of mass. In the laboratory this will show up as an anticorrelation of the charm particle direction in the plane transverse to the beam. We define the angle between the charm particles projected onto the transverse plane as Φ_T. In this simple model the distribution in Φ_T should peak near 180°. Other processes, such as gluon splitting or effects due to higher intrinsic transverse momenta will tend to wash out this peak. Our data is shown in figure 5. There is clear evidence of peaking near 180°. There is also a substantial flat region, larger than that expected in the Lund simulations. With improved statistics we can begin to plot these distributions as a function of other parameters such as pair transverse momentum or rapidity gap to look for more subtle effects.

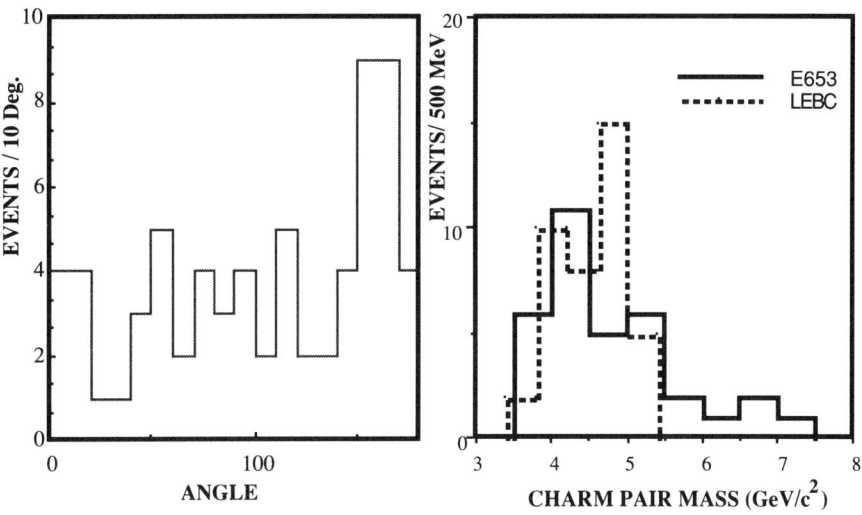

Figure 5. Pair Φ angle.

Figure 6. Charm pair invariant mass.

Figure 6 shows the pair invariant mass distribution along with published data from NA27. Care should be taken when comparing these distributions. The NA27 data consists of 17 events weighted by their efficiencies. The 34 E653 events have no acceptance of efficiency corrections applied.

CONCLUSIONS

Preliminary results from E653 have shown some interesting features. The production momentum distribution for charm is clearly very soft. This, taken with results from E743 at Fermilab, is consistent with a trend toward larger values of n (in $(1-X_F)^n$) as a function of $\sqrt{s}$. Nuclear effects may also contribute to this large value of n. We see no evidence for correlations between charm momentum and the number of heavy tracks (in effect this is a limit on A). Experiment 769 at Fermilab should clarify this issue. While we do see some hints of a difference in leading and non leading charm production, we do not see the dramatic leading effects seen in the LEBC pion data.

Our charm pair analysis is just beginning. The Φ_T distribution shows the expected peaking at 180°. Other distributions require more careful understanding of acceptance and efficiency. An understanding of charm pair production is crucial to a full understanding of the physics of heavy quark production. Are pairs correlated, anticorrelated, or uncorrelated in momentum? How accurate is the Lund color string picture of fragmentation? When does gluon splitting become an important part of heavy quark production? These issues need a large sample of pairs to be addressed fully.

Our analysis of the proton exposure should be complete by the end of this year. We expect an increase in statistics by a factor of ~1.4. Because E653 is a muon triggered experiment we have a large sample of semi-muonic charm decays. This should enable us to study the physics of semi-leptonic decays such as $D \rightarrow \pi \mu \nu$, $D \rightarrow K^* \mu \nu$, as well as search for the cascade $D_s \rightarrow \tau \nu$, $\tau \rightarrow \mu \nu \nu$. Analysis of the $D_s \rightarrow \tau$ cascade is challenging due to the small angle kink in $D_s \rightarrow \tau$ decay and will require special efforts on the part of the emulsion scanners.

Scanning is now beginning on data from the second run. This data includes substantial improvements in the beam, spectrometer, and the emulsion system. We expect a factor of ten increase in our sensitivity for charm decays and a factor of 25 increase in our beauty sensitivity.

---

1 The E653 collaboration consists of: N. Ushida, (Aichi University of Education); R.L. Lander, A. Mokhtarani, V.S. Paolone, J.T. Volk, J.O. Wilcox, P.M. Yager (University of California, Davis); R.M. Edelstein, A.P. Freyberger, D.B. Gibaut, R.J. Lipton, W.R. Nichols, D.M. Potter, J.R. Russ, Y. Zhang, (Carnegie-Mellon University); H.I. Jang, J.Y. Kim, M.Y. Pac, (Chonnam National University); B.R. Baller, R.J. Stefanski, (Fermi National Accelerator Laboratory); K. Nakazawa, S. Tasaka, (Gifu University); Y,S. Choi, K.H. Chung, D.C. Kim, I.G. Park, J.S. Song, C.S. Yoon, (Gyeongsang National University); M. Chikawa, (Kinki University); T. Abe, T. Fujii, G. Fujioka, K. Fujiwara, H. Fukushima, T. Hara, Y. Takahashi, K. Taruma, Y. Tsuzuki, C.Yokoyama, (Kobe University); S.D. Chang, B.G. Cheon, J.H. Cho, J.S. Kang, C.O. Kim, K.Y. Kim, T.Y. Kim, J.C. Lee, S.B. Lee, G.Y. Lim, I.T. Lim, S.W. Nam, T.S. Shin, K.S. Sim, J.K. Woo, (Korea University); Y. Isokane, Y. Tsuneoka, (Nagoya Institute of Technology); S. Aoki, A. Gauthier, K.

Hoshino, H. Kitamura, M. Kobayashi, K. Kodama, M. Miyanishi, K. Nakamura, M. Nakamura, Y. Nakamura, S. Nakanishi, K. Niu, K. Niwa, H. Tajima, (Nagoya University); J.M. Dunlea, S.G. Frederiksen, S. Kuramata, B.G. Lundberg, G.A. Oleynik, N.W. Reay, K. Reibel, C.J. Rush, R.A. Sidwell, N.R. Stanton, (The Ohio State University); K. Moriyama, H. Shibata, (Okayama University); T.S. Jaffery, G.R. Kalbfleisch, P.L. Skubic, J.M. Snow, S.E. Willis, W.Y. Yuan, (University of Oklahoma); O. Kusumoto, T. Okusawa, M. Teranaka, T. Watanabe, J. Yamato, (Osaka City University); H. Okabe, J. Yokota, (Science Education Institute of Osaka Prefecture); T. Ishigami, M. Kazuno, M. Kobayashi, F. Minakawa, H. Shibuya, S. Watanabe, (Toho University); Y. Sata, I. Tezuka, (Utsunomiya University); S.Y. Bahk, S.K. Kim, (Wonkwang University).

2 N. Ushuda et. al., "Hybrid Emulsion Experiment for the Detection of Hadronically Produced Heavy Flavor States", in preparation.

3 M. Aguilar-Benitez et. al., Phys. Lett., 189B (1987).
   M. Aguilar-Benitez et. al., Z. Phys. C, 31 (1986).
   R. Bailey et. al., Z. Phys. C, 30 (1986).
   R. Ammar et. al., Phys. Rev. Lett. 61, 2185 (1988).

4 R.K. Ellis and C. Quigg, FNAL Report No. FN445, 1987 and studies using the LUND Monte Carlo.

NEW EXPERIMENTAL RESULTS IN HADRONIC CHARM PRODUCTION

S. Kwan

EP Division, CERN, Geneva 23, Switzerland

Abstract: Recent results on hadroproduction of charm particles from the NA32 experiment are reported.

1. Introduction

In 1985–1986, the ACCMOR collaboration performed an experiment at the CERN SPS to study the production and decay properties of charmed particles produced in hadronic interactions. Its main aim was to study final states containing a pair of opposite charge kaons and/or protons, e.g. $\Lambda_c^+ \to pK^-\pi^+$. It was a continuation of the NA32 experiment [1] with a Cu target, an improved vertex detector and a dedicated trigger. For the first time, charge–coupled devices, CCD's [2] were used as vertex detectors. They provide high resolution space points ($\sim 5\mu m$) for tracks close to the target. A modified version of the FAMP trigger [3] was used to trigger on events containing a pair of opposite sign kaons and/or protons. A total of 17 million triggers have been recorded. Very clean samples of Λ_c^+, D_s^+, D^0 and D^+ are obtained and their production properties are reported.

Charm strange baryons have also been looked for in this data. The baryon character of a charmed baryon can be better ascertained in decays involving Σ^+ or Ξ^- hyperons through a clean reconstruction of their subsequent decay. Decays with a Λ^0 in the final state suffer more from combinatorial background because the Λ^0 track cannot be seen in the vertex detector and it is generally impossible to decide which vertex the Λ^0 is coming from. We have therefore concentrated our studies of charmed strange baryons in decay modes such as $\Xi_c^+ \to \Xi^-\pi^+\pi^+$ or $\Xi_c^+ \to \Sigma^+ K^-\pi^+$.

2. The experimental set–up and data analysis

A negative beam with a momentum of 230 GeV/c was used. Two CEDAR Cerenkov counters served for tagging incident pions and kaons (96% and 4% respectively). Hadronic charm decays into charged particles were fully reconstructed with the vertex detector and a large acceptance spectrometer [4]. The vertex detector consisted of two parts – a beam telescope with 7 silicon microstrip detectors (MSDs) and a vertex telescope with 2 CCDs at 10 and 20 mm behind the target and 8 MSDs positioned from 65 to 180 mm behind the target. A short 2.5 mm Cu target was used so that decay vertices could be observed in vacuum close to the primary vertex. The spectrometer consisted of 2 magnets and 48 planes of drift chambers arranged in 4 groups. Three multicellular threshold Cerenkov counters were used to identify π, K, p in the momentum range 4–80 GeV/c.

The trigger had two levels, a hardware trigger and the FAMP trigger [4]. Compared to a simple interaction trigger, this two–level trigger increased by a factor 5 the sensitivity of the experiment for decays with a pair of opposite charge K/p in the final state.

© 1989 American Institute of Physics

For all events, tracks are reconstructed in the drift chambers and particle identification is performed. Then the beam track and the secondary tracks are reconstructed in the beam and vertex telescopes respectively. For the secondaries, guidance from the drift chamber track is used to resolve ambiguities. Finally the primary vertex is reconstructed. Subsequently, events with the primary vertex inside the Cu target and containing at least 2 tracks not originating from the primary vertex (1% C.L.) are selected for further analysis. A search for secondary vertices is then performed. The fit is required to have a good χ^2 (99% C.L.) and the corresponding vertices have to be outside the Cu target to remove possible secondary interactions. To select fully reconstructed charm decays with no missing neutral particles, the total momentum vector of the decay tracks has to point to the primary vertex (99% C.L.). Finally for each accepted secondary vertex, all possible charged channels of charmed particles are considered taking into account the particle identification.

3. Production properties for Λ_c and D − mesons

Fig. 1 shows clean Λ_c^+, D_s^+, D^0 and D^+ signals on very little background.

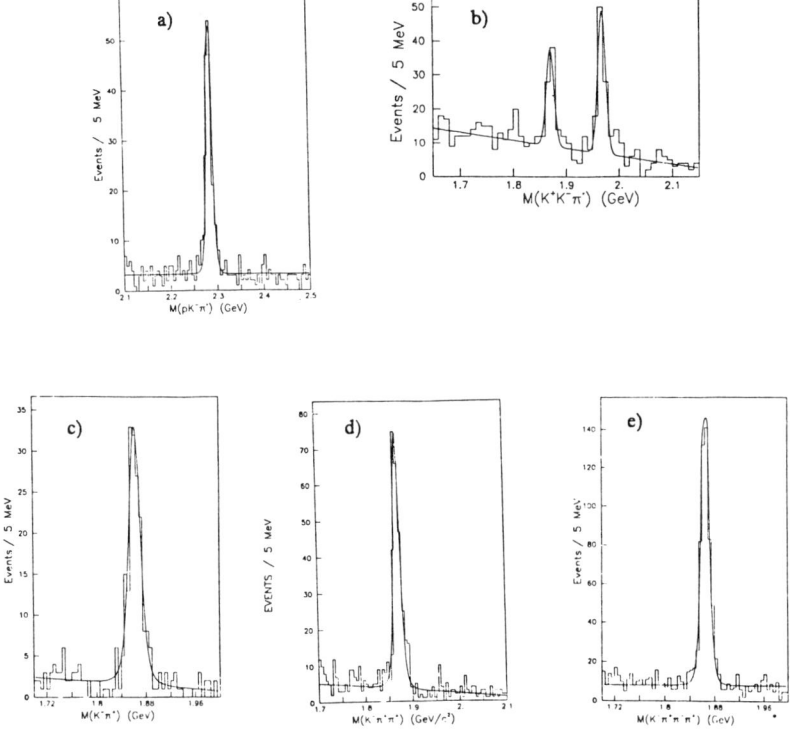

Fig. 1: Invariant mass distributions for a) $pK^-\pi^+$ b) $K^+K^-\pi^+$ c) $K^-\pi^+$ d) $K^-\pi^+\pi^+$ e) $K^-\pi^+\pi^-\pi^+$.

3.1 Acceptance calculation

The differential cross-section for charm particle production is usually parametrized by $(1-x_F)^n e^{-bp_T^2}$ where x_F is the longitudinal momentum of the charmed particle in the centre of mass system divided by the maximum possible value and p_T^2 is the transverse momentum of the charmed particle relative to the incident beam particle. In order to extract the x_F and p_T^2 distribution from the data, we have to correct for the acceptance of the selection criteria. Acceptance curves are determined by a Monte-carlo calculation which is described in [5]. Full simulation of the geometrical acceptance, reconstruction efficiency and measurement errors of the decay tracks, online trigger condition, particle identification and offline selection is done in the Monte-carlo. Fig. 2 shows the acceptance of the spectrometer for the decay modes under study as a function of x_F. The p_T^2 acceptance for all the decay modes are rather flat.

It has to be emphasized that our trigger selected preferentially any decay with a pair of kaons and/or protons of opposite charge in the final state. However, the kaon from the decay of the associated charmed partner or any accompanying pion, simulating kaons or protons at the trigger level, still permit decays with a single kaon in the decay products to fulfill the trigger condition with a sizeable efficiency. We have checked that the impact parameter and momentum distribution as produced in the Monte-carlo program agree well with our experimental disctributions.

3.2 Results

The parameters n and b are determined by a combined maximum likelihood fit to the invariant mass spectrum and the x_F and p_T^2 distribution for the charmed particles under study. The same procedure is also applied separately for leading and nonleading D-mesons. With an incident π^- beam, leading D-mesons are D^{*-}, D^-, D^0 and nonleading D-mesons are D^{*+}, D^+, $\bar{D}^0$. Table 1 summarizes the results of the fit.

The errors quoted are statistical only. The systematic error comes mainly from the event generation in the Monte-carlo program and some small differences in the geometry of the set-up of the vertex telescope between the 1985 and 1986 run. This gives a systematic error of about 10% in the value of n and about 7% in the value of b. All charmed particles are produced with similar production characteristics. There is only a small difference in the value of n between leading and nonleading particles.

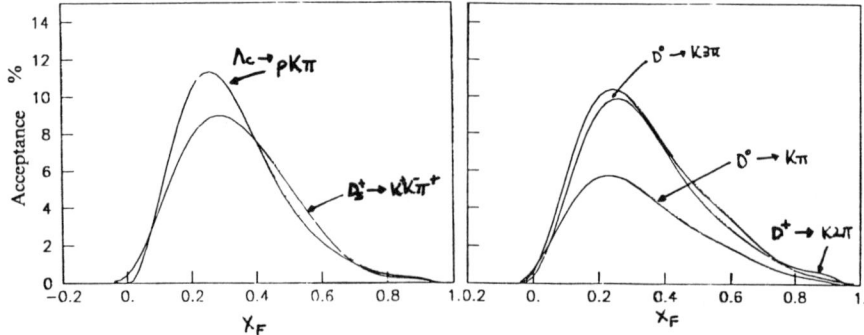

Fig. 2: Acceptance as a function of x_F for the various decay modes under study

CERN

Table 1: *Results for determining the production parameters n and b*

Charm particle	No. of events	n	b (GeV^{-2})
$\Lambda_c + \overline{\Lambda}_c$	146	$3.31^{+0.47}_{-0.45}$	$3.31^{+0.47}_{-0.45}$
Λ_c	72	$3.33^{+0.64}_{-0.60}$	$1.01^{+0.14}_{-0.13}$
$\overline{\Lambda}_c$	74	$3.28^{+0.63}_{-0.58}$	$0.74^{+0.11}_{-0.10}$
D_s	65	$3.35^{+0.83}_{-0.76}$	$0.70^{+0.10}_{-0.09}$
D^0	601	$3.57^{+0.27}_{-0.25}$	$0.82^{+0.04}_{-0.04}$
D^+	259	$3.50^{+0.56}_{-0.53}$	$0.86^{+0.06}_{-0.06}$
all D	858	$3.55^{+0.22}_{-0.20}$	0.83 ± 0.03
leading D	428	$2.93^{+0.34}_{-0.30}$	0.74 ± 0.04
non-leading D	430	$4.37^{+0.45}_{-0.40}$	0.94 ± 0.05

From the number of events, the acceptances and branching ratios of the different channels, we can determine the total inclusive D cross-section for $x_F > 0$. We have used the latest branching ratio as given by [6]. For the Λ_c^+ baryon and D_s^+ meson, we only calculate the inclusive production cross-section times the corresponding branching ratio since the branching ratios have not been accurately measured. The results are listed in Table 2 assuming an A^1 dependence of the charm cross-section.

Table 2: *Results of total cross-sections for the observed decay channels*

Charm particle	No. of events	$\sigma(x_F > 0)$ μb/N	$\sigma \cdot BR(x_F > 0)$ μb/N
Λ_c	120		0.17 ± 0.02
D_s^+	62		0.08 ± 0.01
D^0	524	6.03 ± 0.62	
D^+	227	2.79 ± 0.48	
all D	751	8.82 ± 0.78	
D^{*+}	150	3.32 ± 0.74	

CERN

The errors given are statistical only. The systematic error includes the uncertainties on the lifetimes of the charmed particles, on the values of n and b, on the Monte-carlo program (contribution of the associated charm partner to the acceptance and event generation) and in the case of D-mesons, on the branching ratio. This introduces a 16% incertitude in the cross section for Λ_c, 13% for D_s and 20% for the D-mesons.

A similar number of Λ_c and $\bar{\Lambda}_c$ are produced and we measure the ratio:

$$\sigma(\pi^- Cu \to \Lambda_c)/\sigma(\pi^- Cu \to \bar{\Lambda}_c) = 0.98 \pm 0.17.$$

4. Ξ_c search

For this study, the Ξ^- and Σ^+ hyperons are first reconstructed and details of the reconstruction procedures are described in [7]. Briefly, in both cases, a kink is looked for between a track seen only in the silicon detectors (the hyperon track) and a forward going track which may or may not be seen in the MSD's but is definitely reconstructed in the downstream spectrometer (the DC track). For Ξ^-, a Λ^0 is then looked for in the spectrometer which has the matching point between the hyperon track and the DC track as the production point. If the DC track is identified as that of a pion, the invariant mass of the Λ^0 and the pion is then calculated. In this way, clean Ξ and $\bar{\Xi}$ mass peaks are observed over a small background. For Σ^+, the DC track has to be identified by the Cerenkov counters as being that of a proton. Once a Σ^+ decay is recognized in this way, its momentum can be determined from the proton momentum and from the kink angle with the well-known twofold ambiguity.

For the $\Xi^-\pi^+\pi^+$ channel, we look for decay vertices having a χ^2 probability of better than 1% between the track of a reconstructed Ξ^- and two π^+ tracks. By requiring that 1) all the tracks to be incompatible with originating from the primary vertex; 2) the decay vertex to be outside the target; 3) the total momentum of the decay vertex to be compatible with pointing to the primary vertex with a χ^2 probability better than 1%, three events were selected from the total data sample. Their effective masses clustered around 2465 MeV/c^2. As a control sample, we have looked for $\Xi\pi\pi$ decay vertices with the two pions having opposite charge. No event was selected. We therefore interpret the three events as evidence for a new decay mode of the Ξ_c^+ baryon. All three events have been produced by incident pions. One decay is a Ξ_c^+ and the other two are $\bar{\Xi}_c^-$. We find the weighted mean mass of the three events to be 2465.4±4.0 MeV/c^2. The error given is statistical only.

For the $\Sigma^+K^-\pi^+$ channel, the analysis is similar except that this time, we look for vertices between the track of a reconstructed Σ^+ and two other tracks, which are compatible with being a K^- and π^+ respectively. The selection cuts applied are the same as specified for $\Xi^-\pi^+\pi^+$. Six events were selected out of the total sample. In 3 events, the kaon is ambiguous with a pion and one possible solution for the $\Sigma^-\pi^+\pi^-$ interpretation in all 3 cases is compatible with a Λ_c^+. The remaining 3 events have no K/π ambiguity and have a $\Sigma K\pi$ mass solution compatible with the Ξ_c^+ mass. To further investigate the background, we have looked for decay vertices with other charge states, viz. $\Sigma^+K^+\pi^-$ and $\Sigma^+K^-\pi^-$. Using the same selection criteria, five events were selected. All but one suffer from K/π ambiguity. Four are interpreted as Λ_c^+ decays. There is no conclusive interpretation for the fifth one. Considering the low level of background, we interpret the 3 events as evidence for another new decay mode of the Ξ_c^+. Two of them are produced by incident π^- and the third one is by K^-. The average mass of these three events is 2467.0±3.6 MeV/c^2. Combined with the three $\Xi^-\pi^+\pi^+$ events, we find the mean mass of the Ξ_c^+ to be 2466.3±2.7±1.5 MeV/c^2. The second error given here is systematic and comes from the uncertainties of the geometry and of the magnetic fields of our spectrometer.

We have determined the acceptance for the two decay modes with a Monte-carlo calculation assuming the Ξ_c^+'s are produced with similar characteristics as Λ_c^+. The lifetime of the Ξ_c was taken from [6]. Assuming linear A dependence, we deduce the product of production cross-section times branching fraction of $\Xi_c^+ \to \Xi^-\pi^+\pi^+$ to be 0.04±0.02±0.02 µb/nucleon for $x_F > 0.0$, where the first error is statistical and the second one is systematic. The latter comes from the large uncertainties in the lifetime, the values of n and b and in the event generation and trigger simulation in the Monte-carlo program. We also obtained the relative branching ratio: $B(\Xi_c^+ \to \Sigma^+ K^- \pi^+)/B(\Xi_c^+ \to \Xi^- \pi^+ \pi^+) = 0.25^{+0.35}_{-0.15}$.

5. Conclusion

By virtue of the precision of our vertex detector, clean samples of Λ_c^+, D_s^+, D^+ and D^0 have been obtained in 230 GeV/c π^-Cu interactions. Their production characteristics and cross-section have been measured. Because of the unique signature of a Ξ^- or Σ^+ in our set-up, we have observed 6 Ξ_c^+ events in two decay channels: $\Xi^- \pi^+ \pi^-$ and $\Sigma^+ K^- \pi^+$. We measure the mean mass of Ξ_c^+ to be $2466.3 \pm 2.7 \pm 1.5$ MeV/c^2 We have also determined the product of production cross-section times branching fraction $B(\Xi_c^+ \to \Xi^- \pi^+ \pi^+)$ to be $0.04 \pm 0.02 \pm 0.02$ μb/nucleon for $x_F > 0.0$.

References

[1]   H. Becker et. al., Phys. Lett. 184B (1987) 277.
[2]   R. Bailey et. al., Nucl. Instru. Meth. 213 (1983) 261.
[3]   C. Daum et. al., Nucl. Instru. Meth. 217 (1983) 361.
[4]   S. Barlag et. al., Phy. Lett. 184B (1987) 283.
[5]   S. Barlag et. al., CERN-EP/88-104.
[6]   Rev. of Particle Properties, Phys. Lett. 204B (1988).
[7]   S. Barlag et. al., CERN-EP/89-43.

PRODUCTION OF PROMPT ELECTRONS IN THE CHARM P_t - REGION AT $\sqrt{s}$ = 630 GeV.

O. Botner, L.O. Eek, T. Ekelöf, K. Fransson, A. Hallgren, B. Lund-Jensen
Dept of Radiation Sciences, Uppsala Univ., Box 535, S-751 21 Uppsala, Sweden

P. Kostarakis
Nuclear Research Centre Demokritos, Athens, Greece

G. Lenzen
University of Wuppertal, Federal Republic of Germany

ABSTRACT

The production of prompt electrons at $\sqrt{s}$ = 630 GeV has been measured in the p_t range 0.5 - 2.0 GeV/c using a Ring Imaging CHerenkov (RICH) counter installed in the UA2 experiment at the CERN p$\bar{\text{p}}$ collider. From our measurement of the electron/hadron ratio in the pt range 0.9 - 1.5 GeV/c we conclude that the charm production cross-section does not exceed 1.9 mb at 95% CL. As best estimate we obtain $\sigma(c\bar{c})$ = 684 ± 557(stat) ± 253(sys,exp) ± 205(sys,theory) μb. The RICH counter was operated successfully at the collider providing an electron/pion separation capability of better than $8 \cdot 10^{-5}$ at momenta below the pion threshold.

The measurements were made with a RICH [1], for electron identification, and a two-plane scintillator hodoscope, for triggering and track selection, installed in the UA2 detector [2]. This is the first experiment where a RICH has been used at a collider. Fig.1 shows the experimental lay-out, with the RICH installed in one of the 12 azimuthal end-cap sectors of UA2. The RICH covers the polar angle region 20° - 37.5° and an azimuthal range of 24°.

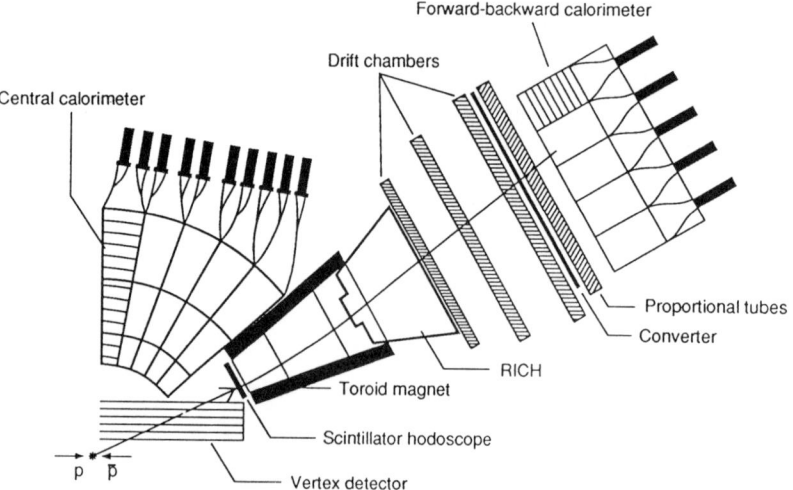

Fig.1 Cross-section of one quadrant of the UA2.

The π rejection power of the RICH was measured in a test beam at the CERN PS [3] with two 4m long threshold Cherenkov-counters in front of the RICH for electron identification. Fig.2 shows the probability of misidentifying a pion as an electron in the RICH, for a constant electron identification efficiency of 60% (closed circles). With the additional rejection from the electromagnetic calorimeter we get, below the π threshold at 3.5 GeV/c, a total rejection better than 10^{-5} (open circles). The π rejection grows worse for momenta above the threshold. We therefore imposed the cut p < 4 GeV/c.

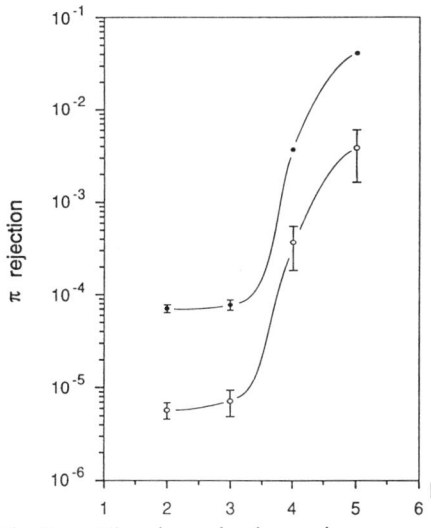

Data in the UA2 set-up were collected during the autumn of 1985. The trigger required signals in both scintillator-hodoscope planes in coincidence with signals obtained from "forward" scintillators close to the beam on both sides of the collision region and a transverse energy in the electromagnetic calorimeter above a certain threshold. Two thresholds were used, see table I. Minimum bias triggers, without any energy requirement, were recorded simultaneously for normalization.

Fig.2 The pion rejection, using :
RICH alone - closed circles,
RICH and calorimeter - open circles.

TABLE I   Integrated luminosity and collected number of triggers.

Nominal trigger threshold [GeV]	∫Ldt [μb⁻¹]	Number of triggers
–	7.83	$0.06·10^6$
0.8	1660	$0.6·10^6$
1.4	2720	$0.35·10^6$

To identify prompt electrons and reject triggers from trivial background sources, several requirements were imposed on the two data samples with thresholds. These requirements fall into two different categories :
1°. Event selection requirements:
- One track in the RICH sector, with segments in vertex chamber and drift chambers behind the toroidal magnet,
- pulse height in both scintillator-hodoscope planes consistent with a single particle and significantly lower than that expected for two such particles, and
- a good momentum fit with p < 4 GeV/c.

2°. Electron identification requirements:
- At least 3 photons detected within an annular region of width 20 mm centred at the position of the Cherenkov ring expected from the reconstructed track,
- the radius of the circle fitted to the detected photons consistent with 24 mm (24±2 mm), as expected for a $\beta = 1$ particle, and
- energy in the calorimeter consistent with that expected for a single electron with the measured momentum.

To further improve the rejection power of the RICH, all "noisy" events, defined as events containing more than three signals none of which could be assigned to either a traversing track or to a Cherenkov ring, were excluded.

The total efficiency of the electron identification requirements were estimated by Monte Carlo simulation to be 44%.

The inefficiency caused by the RICH is explained by the relatively low number of photons observed per Cherenkov ring (4.5 on the average), and by the strict condition on the fitted radius. The response of the RICH was well reproduced by the Monte Carlo.

Finally, a few events were removed by visual inspection of the detected image causing a negligible loss of real electrons.

The dominant background from trivial electron sources passing the above cuts comes from converting photons and Dalitz decays of $\pi°$ and η mesons and from the K_{e3} decays of charged and neutral kaons.

In the first case the conversion partner is close to the presumed single-electron candidate. Most of these events are rejected in the analysis by requiring single-tracks and by requiring that the pulse height in the scintillator hodoscope is inconsistent with that for two charged tracks. Events with hodoscope hits not associated to a track segment downstream of the magnet are also discarded. Conversions after the vertex-chamber region are removed by requiring, for the candidate track, hits in the two inner vertex chambers ("early" hits). Difficult cases are those where one of the electrons is lost before the scintillator hodoscope due to multiple scattering. Many of these events are anyway rejected when the second, non-converted photon deposits energy in the electromagnetic calorimeter.

The above methods eliminate most of the Dalitz decays as well, although here the probability for one of the partners escaping our solid angle before crossing the scintillator hodoscope is greater.

The weak decays $K^\pm \rightarrow e^\pm + \pi° + \nu$ and $K°_L \rightarrow e^\pm + \pi + \nu$ produce single electrons (positrons) which are difficult to distinguish from those originating from the processes $D^\pm \rightarrow e^\pm + X$ and $D° \rightarrow e^\pm + X$. The $K°_L$ - decays are removed if they occur outside the beam pipe by the requirement of "early" hits. The charged kaon decays are in most cases removed by requiring a good momentum fit for the candidate track, since this implies a track pointing at the vertex.

The requirement of "early" hits for the candidate track also removes background from Compton scattering occurring outside the beam pipe.

The residual background remaining after all cuts was calculated by Monte Carlo. For the shape of the $\pi°$ p_t distribution we have used the fit by UA2 [4] to the $p\bar{p} \rightarrow h+X$ data, since this extends to lower p_t values than the $p\bar{p} \rightarrow \pi°+X$ measurement. The $\eta/\pi°$ ratio was taken to be 0.6±0.16 [4]. Finally, the measured ratio $2\pi°/(h^+ + h^-)$ = 0.68±0.13 [4] was used to convert the residual background $e/\pi°$ ratio into an e/h ratio. The K_{e3} and the Compton contributions were estimated to be negligible compared to the $\pi°$ and η contributions. Fig.3 shows the dominant residual background.

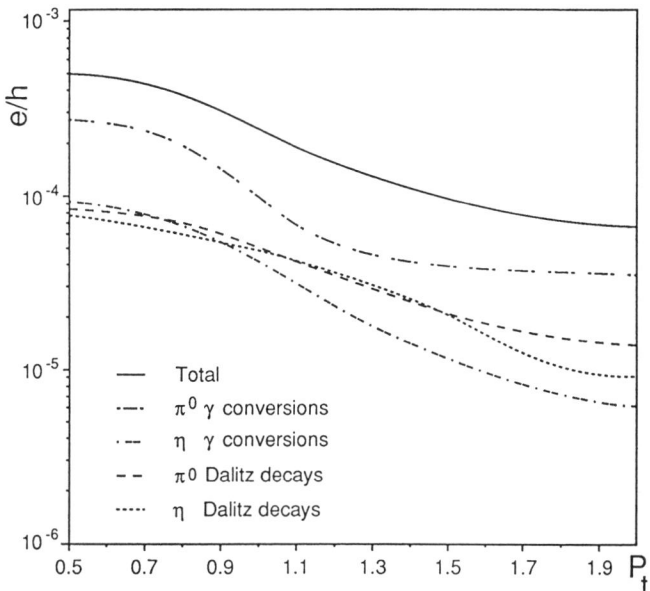

Fig.3 Residual background contributions to the e/h ratio as a function of p_t.

The sample remaining after application of the above selection criteria consists of 14 events with equal numbers of electrons and positrons.

A trigger acceptance correction was applied to get the yield of prompt electrons, which was then normalized to the measured yield of charged hadrons in the same p_t range, obtained by analyzing minimum bias events with the same geometrical acceptance and track-selecting requirements as in the electron case. Only events with a single track in the RICH sector were considered. The resulting sample consisted of 760 minimum bias events in the p_t range 0.5 - 2 GeV/c.

The cuts pertaining to track selection in general introduce losses of equal proportion for the electron (e) sample and for the sample of charged hadrons (h). Hence, their effects cancel out in the ratio e/h.

The measured e/h ratio is presented in table II for three momentum bins, the first two corresponding to the trigger with the lower nominal threshold and the last to the trigger with the higher threshold. The results have been corrected for trigger acceptance and electron identification efficiency. The errors quoted are statistical only. The systematic errors, mostly due to uncertainties in the electron identification efficiency and in the trigger acceptance correction, are estimated to be in total ±30%, ±13% and ±14%, respectively, for the three p_t bins.

The total residual background from π^0 and η conversions and Dalitz decays, expressed in terms of an e/h ratio, is also given in table II, where the quoted error is mainly systematic. We see that in the p_t range 0.5 - 0.9 GeV/c the measured e/h is dominated by background.

The systematic error of the measured value and that of the background have been added in quadrature.

TABLE II Resulting e/h values per p_t bin in units of 10^{-4}.

p_t [GeV/c]	$<p_t>$	measured	residual background	prompt e
0.5 - 0.9	0.6	3.42±1.71±1.03	4.45±0.93	-1.03±1.71±1.38
0.9 - 1.5	1.1	3.74±1.41±0.49	1.69±0.41	2.04±1.41±0.64
1.5 - 2.0	1.8	7.18±4.14±1.01	0.83±0.26	6.35±4.14±1.04

The background of misidentified hadrons is of importance only in the highest p_t bin where it is not larger than $1 \cdot 10^{-4}$. Below 1.5 GeV/c it is at most $9 \cdot 10^{-6}$ and can be neglected.

In the p_t range 0.9 - 1.5 GeV/c, the dominant process contributing to the prompt electron yield is expected to be semi-leptonic decay of charmed particles. It is thus possible to set an upper limit on the total charm production cross-section by assuming that all our prompt electrons have this origin.

To describe charmed-particle production and semi-leptonic decay characteristics in our kinematical region, a Monte Carlo programme with hadronization according to the Lund string-model [6] has been used, containing the quark-anti-quark and gluon-gluon fusion processes to order α_s^2 [5]. This Monte Carlo was used to extrapolate our results to the full phase space and to correct for the restriction in decay topology imposed by the requirement of a single track in the RICH sector. We obtain for the electron p_t range 0.9 - 1.5 GeV/c :

$\sigma(c\bar{c}) \leq 806 \pm 557(\text{stat}) \pm 253(\text{sys,exp}) \pm 242(\text{sys,theory})$ µb.

The second systematic error represents the uncertainty in the charm acceptance correction connected with the choice of Monte Carlo parameters.

Adding the different errors in quadrature the upper limit corresponding to a 95% confidence level is : $\sigma(c\bar{c}) < 1.9$ mb.

With $\sigma(b\bar{b}) \approx 10$ µb [7] the contribution from semi-leptonic b-decays for $0.9 < p_t < 1.5$ was estimated to be 5%. Using the Pythia Monte Carlo [8], decays of the vector mesons ρ, ω, ϕ into e^+e^- were estimated to contribute about 10 %. The absolute errors in these estimates are small in comparison to other errors. Neglecting them we obtain : $\sigma(c\bar{c}) = 684 \pm 557(\text{stat}) \pm 253(\text{sys,exp}) \pm 205(\text{sys,theory})$ µb.

P. Nason et al [9] have recently published results of a full calculation of the next-to leading order (α_s^3) QCD radiative corrections to the total cross-section for heavy quark-pair hadro-production. Based on this, G. Altarelli et al [10] have evaluated the total charm cross-section in pp-collisions up to $\sqrt{s} = 63$ GeV. Fig.4 shows the same type of evaluation extended to $\sqrt{s} = 630$ GeV [11]. Measurements of the total charm production cross-section in pp collisions at energies up to 63 GeV are plotted as well as our $p\bar{p}$ point at 630 GeV. The curves correspond to different values of the charmed quark mass and with different values of the Λ_5 parameter for the gluon structure function. Because of the high $\sqrt{s}$ and the relatively low quark mass in our case, the cross-section is sensitive to the behaviour of the structure functions in regions where no direct measurements exist so far. The value of the scale µ at which $\alpha_s(\mu)$ and the parton densities are calculated is set to 3 GeV. Using lower values for µ like $\mu \approx m_c = 1.5$ GeV decreases the $O(\alpha_s^2) + O(\alpha_s^3)$ cross-sections at $\sqrt{s} = 630$ GeV by about a factor of 2. The significant dependence on the value chosen for µ indicates that even higher order corrections are important.

As compared to the $O(\alpha_s^2)$ calculations the introduction of the $O(\alpha_s^3)$ corrections more than doubles the cross-section in the whole $\sqrt{s}$ range from 10 to 630 GeV. Our cross-section at 630 GeV is consistent with the theoretical expectation although it is clear that within the present theoretical and experimental uncertainties no distinction between the various calculations can be made.

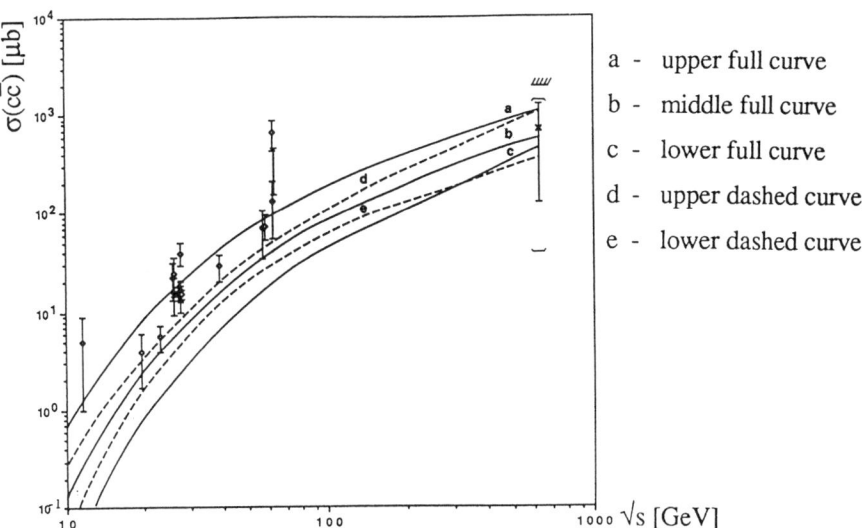

Fig.4 $\sigma(c\bar{c})$ in hadro-production. Open circles are pp measurements [12]. The cross is our p$\bar{\text{p}}$ result and the upper limit at 95% CL, assuming all electrons come from charm, is indicated. The curves correspond to next-to-leading order calculations [11] for pp collisions with the following parameters (GeV and GeV/c² respectively), using DFLM [13] structure-functions and $\mu = 3$ GeV:
a) $\Lambda_5 = 0.170$, $m_c = 1.2$ b) $\Lambda_5 = 0.170$, $m_c = 1.5$ c) $\Lambda_5 = 0.170$, $m_c = 1.8$
d) $\Lambda_5 = 0.250$, $m_c = 1.5$ e) $\Lambda_5 = 0.101$, $m_c = 1.5$.

REFERENCES

[1] O. Botner et al., Nucl. Instr. and Meth. A257 (1987) 580.
     B. Lund-Jensen, " Single-photon detectors for Cherenkov Ring Imaging ",
     Ph.D. Thesis, Uppsala University 1988.
[2] B. Mansoulié, Proc. 3rd Moriond Workshop on p$\bar{\text{p}}$ Physics and the
     W Discovery, La Plagne, 1983 (Ed. Frontieres, Gif-sur-Yvette 1983),
     p.609 and references therein.
[3] O. Botner et al., to be published.
[4] M. Banner et al., Z. Phys. C - Particles and Fields 27 (1985) 329.
[5] G. Ingelman, Computer Physics Communications 46 (1987) 217.
[6] B. Andersson, G. Gustafson, G. Ingelman, T. Sjöstrand,
     Phys. Rep. 97 (1983) 31.
     T. Sjöstrand, Computer Physics Communications 39 (1986) 347.
[7] C. Albajar et al., Phys. Lett. B213 (1988) 405.
[8] H. U. Bengtsson, T. Sjöstrand,
     Computer Physics Communications 46(1987)43.
[9] P. Nason et al., Nucl. Phys B303 (1988) 607.
[10] G. Altarelli et al., Nucl. Physics B308 (1988) 724.
[11] The calculations were made for us by P. Nason and G. Martinelli using the
     computer programmes referred to in [9] and [10].
[12] S. P. K. Tavernier, Rep. Prog. Phys. 50 (1987) 1439.
[13] M. Diemoz et al., Z. Phys. C - Particles and Fields 39 (1988) 21.
     J.V. Allaby et al., Phys. Lett. B197 (1987) 281.

SHADOWING AND NUCLEAR ABSORPTION IN J/Ψ HADROPRODUCTION

M. B. Gay Ducati

Instituto de Fisica, Universidade Federal do Rio Grande do Sul,
91500 Porto Alegre, Brasil

L. N. Epele, C. A. Garcia Canal

Laboratorio de Fisica Teorica, Departamento de Fisica,
Universidad Nacional de La Plata, 1900 La Plata, Argentina

ABSTRACT

The hadron-nucleus J/Ψ production is analysed in a phenomenological picture, considering the parton recombination model. There is an evidence for a nuclear suppression effect, besides shadowing and we obtain good agreement with experimental data.

INTRODUCTION

The suppression of J/Ψ in hadron-nucleus high energy interactions is repported in the literature and it increases with nuclear mass number A[1]. These results can be connected with data from EMC[2].

When treating these phenomena we are faced with the problem of evolution of quark (gluon) distribution functions, $q(x)$, x meaning Bjorken variable. The deep inelastic scattering (DIS) study have shown that dynamical effects must be taken into account, $q(x) \to q(x,Q^2)$, and also rescale x (quark mass threshold). Now, lepton-nucleus (ℓA) and hadron-nucleus (hA) results indicate an A dependence, so $q(x,Q^2) \to q(x,Q^2,A)$, which means that nuclear effects seem to be present at high energy regime.

NUCLEAR EFFECTS

It is well established that the cross section per nucleon for interactions between real photons and nuclei decreases with increasing A. This effect is called shadowing and similar effect is observed in hadron-nuclei scattering. Shadowing implies: $\sigma_{\gamma A} < A\sigma_{\gamma N}$, $\sigma_{hA} < A\sigma_{hN}$, $\sigma_{\gamma^* A} < A\sigma_{\gamma^* N}$. These facts are by no means obvious in the framework of the parton model, where, due to the intrinsic incoherence of partons, one should expect a cross section linearly growing with A, or target independent.

Shadowing effects have been clearly observed in muon-nucleus DIS[2] and have also been analysed. from a theoretical or phenomenological point of view[3,4].

Recent results on $\bar{p}A$ and π^-A at 125 GeV/c[1] show a J/Ψ resonance production dependent on the nuclear target. It is claimed by the authors that their results, the J/Ψ being suppressed in $p(\pi)W$ interactions relative to the rates obtained with lighter targets, are not concilable with EMC type effects previously repported. We pointed out that the kinematical region explored by this experiment, E537, is certainly coincident with the shadowing characteristic region of DIS, and does not correspond to the one where the standard EMC effect occurs[5].

An hA reaction can be described in terms of the momentum fraction of h and A carried by the corresponding partons, that we call x_1 and x_2 respectively. The invariant mass of the hard process (HP) is $M^2 = s x_1 x_2$, where s is the center of mass (CM) energy, and $x_F = x_1 - x_2$ measures the longitudinal fraction of momentum in the CM of the HP, and finally $\tau = x_1 x_2 = M^2/s$. Since in the experiment discussed τ is equal to 0.04, and $x_F > 0.5$, the range of x_2 (given by $x_2 = \frac{1}{2}[-x_F(1-\tau) + \sqrt{x_F^2(1-\tau)^2 + 4\tau}]$) lies between 0.05 and 0.18 as it is shown in figure 1.

The kinematical range explored in $\pi A \to J/\Psi$ is clearly coincident with the shadowing region in DIS, which permits a comparison of A-dependent effects in both reactions.

We study the A-dependent behaviour of the rate of cross sections

$$R_A = \frac{\sigma^{\pi A}}{A \sigma^{\pi D}}$$

first considering the factorisation proposed in the parton recombination model[3,6], assuming its validity in the gluon probe case. The rate is done by

$$R_A(x,Q_0^2,A) = R_S(x,Q_0^2,A)R_a(x,Q_0^2,A)$$

In this expression R_S is the shadowing factor which differs from one for small x, Q_0^2 is a fixed momentum transfer and R_a is the ratio of structure functions that includes all phenomenological aspects concerning the medium to $x \to 1$ region. Now, R_A should describe the A dependence behaviour in the whole x domain. A more detailed explanation is given in reference 7. Discussing briefly, in quark-parton language shadowing can be explained as due to a fusion of partons, most probably of gluons, from neighbour nucleons. Those ideas are not very new : since in the nuclear infinite frame the nucleus occupies the Lorentz contracted longitudinal size given by $\Delta Z_A \cong 2RM/P$, where R is the nucleus radius, M (=mA) the nucleus mass and P (=pA) the nucleus momentum, if the longitudinal size of the parton (quark or gluon) $\Delta Z > \Delta Z_A$, for a certain impact parameter, it can occur spatial superposition[8]. This effect can be present if $\Delta Z > \Delta Z_N$ for Z_N being the longitudinal size of the nucleon. The parton momentum is px, where p is the nucleon momentum and $\Delta Z \sim 1/px$, so x<1/2mR. This implies that at very small x, and same impact parameter, partons from different nucleons in a nucleus can correlate. The shadowing region begins, in principle, at $x \leq 1/2mr_0 \sim 0.1$. The recombination of partons will produce a negative contribution to the parton distribution functions[6]. In the case of J/Ψ production, which is dominated by the gluon-fusion mechanism it is necessary to extend the shadowing argument to gluon initiated processes.

Now we must recall that the J/Ψ hadroproduction cross section is generally suppressed, compared to the Drell-Yan continuum one due to absorption effects on the resonance when the reaction is in a nuclear target[9,10,11]. So the "shadowing" is not the only nuclear effect present in this process. The J/Ψ resonance can be absorbed if the production of the pair $c\bar{c}$ occurs at the interior of the target, if these interactions occur with the partonic constituents of J/Ψ with the nucleus, or the interaction of the J/Ψ constituents with other quarks or gluons[12], or scatterings of the J/Ψ itself with nucleus[13]. We consider all these combined effects in a phenomenological way by introducing an effective factor that we can fix from previous experiments. In order to introduce the strong screening, that should quantify the suppression due to J/Ψ absorption, we propose to factorise the cross section rate as

$$R_A = R_{SS} R_S R_a .$$

This new correction factor R_{SS} is supposed to be x_2 dependent. The experimental literature shows very clearly that the A dependence in J/Ψ production depends on the x_F (x_2) region[9,10], in the form

$$A^\alpha \begin{cases} x_2 \to 1 \to \alpha \cong 0.97 \\ x_2 \to 0 \to \alpha \cong 0.7 \end{cases}$$

The number of J/Ψ events decreases with respect to the number of dimuons as $A^{\alpha-1}$. This is precisely the factor R_{SS} we have introduced above.

We have studied the rate R_A for W using data from ref. 1 . Since their results are given as $R_{W/Be}$ we neglect possible shadowing in Be because $A_{Be} \ll A_W$. Hadron-nucleus J/Ψ data for W lies between $0.05 < x < 0.18$, and Drell-Yan data for W covers the range $0.15 < x < 0.45$[14]. In figure 2 we present our results obtained with the parton recombination model only, and combined with the strong screening effect, comparing R_W^{DY} and $R_W^{J/\Psi}$. The shadowing prediction for R_W^{DY} is in good accordance in the x_2 region, where data is available. This model is the same that is applied to DIS for Fe [3] and also agrees with recent results for Sn. So the result we obtained for R_W^{DY} is the prediction also for shadowing in $R_W^{J/\Psi}$, because we considered Be as D. However, hadron-nucleus data show a much stronger shadowing-like effect. We had already mentioned a fundamental difference between the two processes. The $\mu^+\mu^-$ pair produced from a γ^* in Drell-Yan cannot be absorbed in the nuclear matter as can occur with the $c\bar{c}$ pair in hadroproduction. We found that this strong screening effect can be parametrized in terms of an α parameter x_F (x_2) dependent, in other words, this effect depends on the kinematical domain. We obtain a good description of the J/Ψ suppression with a quadratic form

$$\alpha = 0.97 - 0.27 x_F^2 \ .$$

Our results are in accordance with the analysis of ref.13 which is based on calculations of cross-section, care taken of possible rescattering inside nuclear matter, and the α dependence on x_F being even, agrees with previous data[9,10].

CONCLUSION

The excellent result obtained indicates that both effects, gluon detected shadowing and nuclear absorption, are present at low x_2 J/Ψ hadroproduction.

The A dependent suppression of J/Ψ is due to two distinct factorisable nuclear effects. In order to continue a complete analysis of R_A^{DY} and $R_A^{J/\Psi}$, and a more rigorous comparison of shadowing and nuclear suppression dependence on x_2 , an extension of data to the whole x_2 domain is needed.

This work was supported by Centro Latino Americano de Fisica, CLAF, Fundaçao de Amparo a Pesquisa do Rio Grande do Sul, FAPERGS, Conselho Nacional de Desenvolvimento Cientifico e Tecnologico, CNPq, Brasil, and Consejo Nacional de Investigaciones Cientificas y Tecnicas, CONICET, Argentina.

REFERENCES

1. S. Katsanevas et al., Phys. Rev. Lett. <u>60</u>, 2121 (1988)
2. M. Arneodo et al., Phys. Lett. <u>B211</u>, 493 (1988), M. Arneodo, XXIV International Conference on High Energy Physics, Munich, 1988
3. E. L. Berger and J. Qiu, ANL-HEP-CP 88-42-1988
4. J. Kwiecinski and B. Badelek, preprint 1401/PH 1988, Warsaw University, Poland
5. L. Epele, S.H. Fanchiotti, C.A. Garcia Canal, M.B. Gay Ducati, preprint La Plata University 1988
6. A.H. Mueller and J. Qiu, Nucl. Phys. <u>B268</u>, 427 (1986), J. Qiu, Nucl. Phys. <u>B291</u>, 746 (1987)
7. L.N. Epele, C.A. Garcia Canal, M.B. Gay Ducati, preprint La Plata University nov. 1988, to be published
8. N.N. Nicolaev and V.I. Zakharov, Phys. Lett. <u>55B</u>, 397 (1975)
9. M.J. Corden et al., Phys. Lett. <u>B110</u>, 415 (1982)
10. J. Badier et al., Z. Phys. <u>C20</u>, 101 (1983), M.E. Duffy et al., Phys. Rev. Lett. <u>55</u>, 1916 (1985), H. Cobbaert et al., Phys. Lett. <u>B191</u>, 456 (1987)
11. S. Raha and B. Sinha, Phys. Lett. <u>B198</u>, 543 (1987)
12. S.J. Brodsky and A.H. Mueller, Phys. Lett. <u>B206</u>, 685 (1988)
13. A. Capella et al., Phys. Lett. <u>B206</u>, 354 (1988)

14 K. Freudenreich, XXIII International Conference on High Energy Physics, Berkeley, 1986.

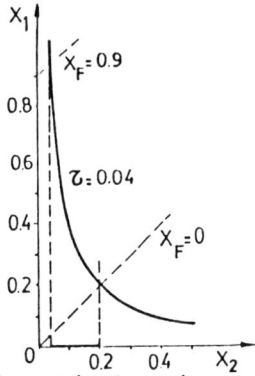

Fig. 1. Kinematical region for x_1 and x_2 in the E537 experiment. The x_2 variable runs in the shadowing characteristic region.

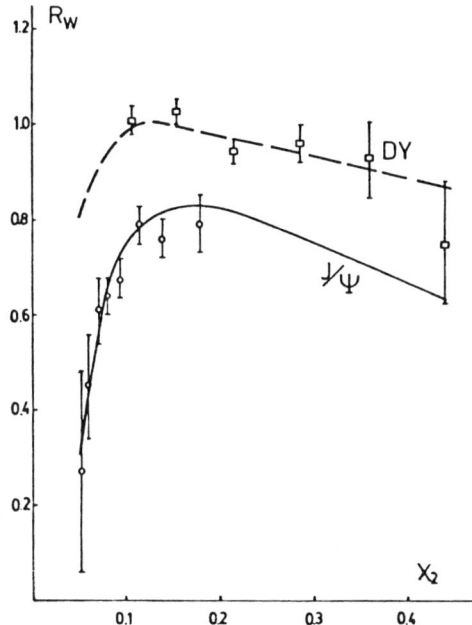

Fig. 2. Prediction of the shadowing model of ref. 3 for the ratio R_W^{DY} (dashed curve), and our prediction including absorption effects for the ratio $R_W^{J/\Psi}$ (solid curve). Data are from ref. 1 for J/Ψ (⌀) and ref. 14 for DY (⌀).

1. THE NEW HYPERON BEAM EXPERIMENT AT CERN

presented by S. Paul
Max−Planck Institut für Kernphysik, Heidelberg, W. Germany

for the
WA89−Collaboration

2. ABSTRACT

We are currently setting up an experiment aiming at the production of charmed strange baryons using a 350 GeV/c hyperon beam from the CERN SPS. This beam will be installed at the upgraded Ω−Spectrometer allowing to measure the forward production ($x_F > 0.5$) of charmed baryons, their lifetimes and decay modes. Other experimental interests will be the confirmation of the U(3100) and the production of the H−dibaryon by Ξ from the hyperon beam.

3. INTRODUCTION

Experiments with hyperon beams cover a broad spectrum of physics. The piece of physics most interesting for this conference is the production of charmed strange baryons. Since not much is known, the information on their mass, lifetime, the major decay modes and decay asymmetries is extremely important.

In a previous hyperon beam experiment [1] a new state called U(3100) was found. The decay into $\Lambda \bar{p} \pi^+ (\pi)$ lead to its interpretation as a diquonium state. Indications for its confirmation came from the BIS−2 collaboration in Serpukhov. A solid study of this object, if its existence can really be confirmed, should open an interesting field of physics.

Another challenging aspect of hyperon beams is the 'tagging' of the strangeness of the incoming hyperons ($\Sigma^-, \Xi^-, \Omega^-$). This feature could be used to study the production mechanisms of strange and charmed baryons, since projectiles with different quantum numbers could be compared. It also allows a search for exotic states with strangeness $>= 2$ as for example the double strange dibaryon (H) which has long been searched for as the best candidate for a multiquark ($n > 3$) system. Such a beam should also allow to do physics with incoming Ω, for instance the study of rare decays.

4. CHARMED STRANGE BARYONS

Fig. 1 shows the multiplett of the $1/2^+$ baryons, the vertical axis corresponds to the charm quantum number of the multiplett. The lower multiplett contains the well-known octett. From the 'first' floor, the $\Lambda_c, \Sigma_c^{\pm 0}$ are known which form the strangeness = 0 baryons. The first charmed strange baryon Ξ_c^+ (A^+) was found in a previous hyperon beam experiment WA62 [2] in the decay mode $\Xi_c^+ \to \Lambda K^- \pi^+ \pi^+$ (fig. 2) and later on confirmed by the neutron beam experiment E400 at Fermilab [3] (fig. 3). It should be noted, that WA62 did not have a microstrip detector and was purely relying on their driftchamber and MWPC reconstruction. They obtained a clear Ξ_c^+ signal at $x_F > 0.6$ only by Λ and K^- identification. The signal was clear enough to allow the determination of its lifetime to $\tau = 4.8 + 2.9 - 1.8$ 10^{-13} s. E400 however could not see the signal in the full data set but had to impose a lifetime cut to raise the signal above the background. The second (lower) peak in their spectrum corresponds to the same state decaying into Σ^0 which subsequently decays into $\gamma\Lambda$, the γ remaining undetected. This peak is not visible in the WA62 data. In 1989 the ACCMOR collaboration (NA32) published two new decay modes for the $\Xi_c^+ \to \Xi^- \pi^+ \pi^+$ and $\Xi^-_c \to \Sigma^+ K^- \pi^+$ [4]. Using a set of μ-strip detectors and a vertex telescope with CCD readout they were able to fully reconstruct five Ξ_c events. This year, the CLEO collaboration published the finding of the Ξ_c^0 decaying into $\Xi^- + \pi$ [5]. It was the first appearance of charmed strange baryons in e^+e^- colliders. During this conference, they also presented a signal for the Ξ_c^+, again decaying into $\Xi^- + \pi$ [6].

The apparently higher cross section for charmed strange baryon production in hyperon beams compared to other beams lead to the construction of a new hyperon beam with higher intensity and energy in conjunction with an existing large acceptance spectrometer of which the performance is well understood. In order to allow a wide range of physics to be studied simultaneously, the hard- and software of the spectrometer had to be upgraded to match the demand for fast triggering and high rate data taking.

5. THE APPARATUS

The new hyperon beam experiment consists of 2 parts, the beamline and the forward spectrometer.

5.1 The Beamline

The beamline comprises the proton beam hitting the hyperon production target followed by the 13 meter long back bend hyperon channel with 3 dipoles each providing a bending power of 8.4 Tm. During the first phase of the experiment this channel will be operated at 350 GeV/c with a momentum acceptance of $\pm 12\%$ and a divergence of ± 1.2 mrad in the non-bend and $\pm .6$ mrad in bending plane. A scintillating-fibre hodoscope will be placed between the second and third magnet. Together with 2 sets of μ-strip counters they will allow a measurement of the particle's trajectory and momentum. This information will be essential in the second phase of the experiment, when the hyperons should be tagged by means of a fast beam RICH, because of the large momentum byte accepted in this phase. The beam momentum will be reduced to 240 GeV/c where hyperons will be produced more copiously at the price of a much higher π background. A lower momentum also improves the feasability of hyperon tagging.

5.2 The Spectrometer

The experimental set up is shown in Fig. 3. Fig. 3b shows a blown up view of the target region. It consists of 2 targets (Be, Cu) being placed next to each other and surrounded by number of μ−strip detectors. These detectors (10 planes of 50μm pitch at first, with 5−10 planes of 25μm pitch added at a later stage) will allow the reconstruction of primary and secondary vertices and thereby help to further reduce the background especially in charm production.

The target region is followed by a 15 m long Λ decay region filled with 6 driftchamber sets to allow an easier reconstruction of 'longer' lived secondaries (Λ,Ξ). The spectrometer magnet (a superconducting coil with 1.5 meter vertical opening) is filled with MWPCs allowing a momentum determination of $\Delta p/p = p/p_0$ ($p_0 \approx 10000$ GeV/c). Behind the magnet are a number of scintillator hodoscopes for fast triggering and two large driftchambers. Charged decay particles will be identified by means of a wide angle RICH allowing a π/K separation up to 160 GeV/c with $\approx 3\ \sigma$. A leadglass calorimeter made of 650 blocks of 75·75 mm is used to detect photons from radiative decays and π^0 decays.

5.3 Data Acquisition

The trigger system consists of a number of decision levels starting from a fast interaction trigger, a requirement of a minimum multiplicity (M > 3), a crude momentum cut for the fastest particle up to a Λ reconstruction using the invariant mass of positive and negative particles. All the trigger decisions described above will be done using hardwired processors in the MBNIM standard. A VME processor finally will work on the data (< 1000/spill) stored in FASTBUS spillbuffers using the information coming from the RICH to detect events with a π as fastest particle. Those events can then be rejected.

6. CONCLUSION

The installation of a high energy hyperon beam at a modern high rate spectrometer opens a wide field of physics. In particular the study of the charmed baryon sector will help to understand the problem of masses and non perturbative QCD. After a first pilot run in november 1989 a first physics run next year should produce new and interesting information on this topic.

7. REFERENCES

[1] M. Bourquin et a;., Phys. Lett. 172B (1986) 113
[2] S.F. Biagi et al., Phys. Lett. 122B (1983) 455−460
   S.F. Biagi et al., Phys. Lett. 150B (1985) 230−234
   S.F. Biagi et al., Z. Phys. C28 (1985) 175−185
[3] P. Coteus et al., Phys. rev. Lett. 59 (1987) 1530
[4] S. Barlag et al., CERN−EP/89−43
[5] P. Avery et al., Phys. Rev. Lett 62 (1989) 863−865
[6] P. Avery et al., see these proceedings

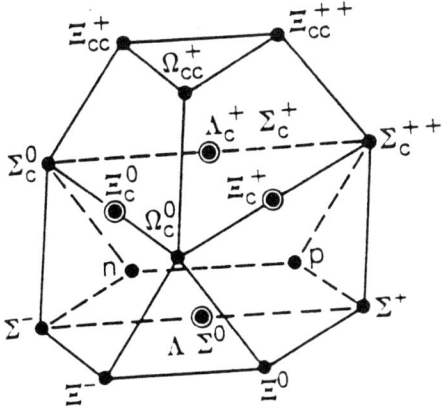

Fig. 1 : SU(4) multiplett of baryons with $J^P = 1/2^+$

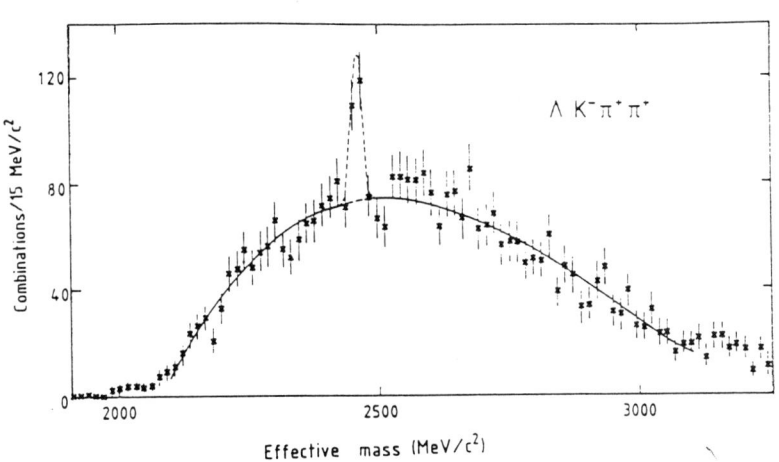

Fig. 2 : $\Lambda K^-\pi^+\pi^+$ invariant mass spectrum from WA62

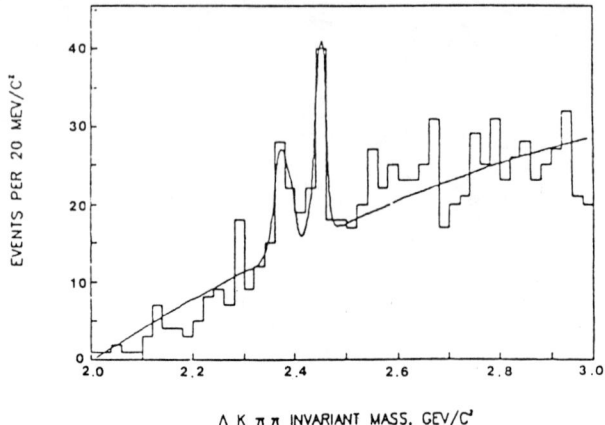

Fig. 3 : $\Lambda K^-\pi^+\pi^+$ invariant mass spectrum from E400 after a decay length cut.

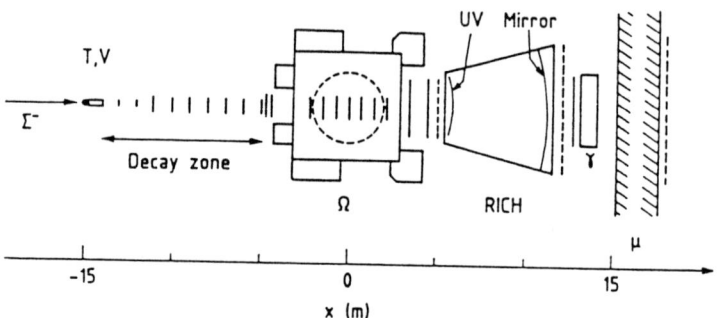

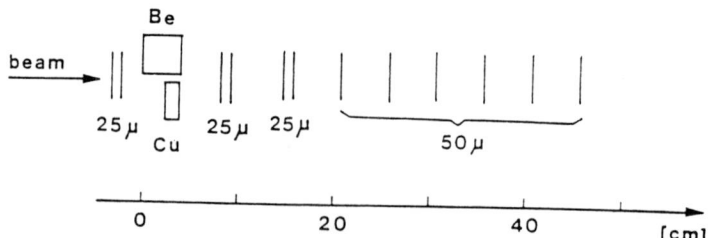

Fig. 4 : Scetch of the experimental set-up for WA89 at CERN.

Charm Hadroproduction with an Impact Parameter Trigger

A. Forino, R. Gessaroli, P. Mazzanti, A. Quareni, F. Viaggi
Univ. of Bologna/INFN

D. Barberis, W. Beusch, M. Davenport, J. P. Dufey, B. R. French,
A. Jacholkowski, K. Knudson, J. C. Lassalle, F. Muller
CERN

F. Antinori, M. Dameri, R. B. Hurst, B. Osculati, L. Rossi, G. Tomasini
Univ. of Genova/INFN

C. Meroni, N. Redaelli, D. Torretta
Univ. of Milano/INFN

J. L. Bailly, A. Buys, F. Grard, P. Legros
Univ. of Mons

M. I. Adamovich, Y. A. Alexandrov, S. G. Gerasimov,
S. P. Kharlamov, L. V. Malinina, M. V. Zavertyaev
Moscow Lebedev Phys. Inst.

(Presented by Dario Barberis)

ABSTRACT

We report results on hadroproduction of charm based on a sample of ~ 750 D mesons. Charmed particles, produced in a thin segmented target of Si and W by a 340 GeV/c π^- beam and selected by a novel trigger with an enrichment factor of ~ 15, are identified by the invariant mass of secondary vertices. Data were collected with the Ω' spectrometer at the CERN SPS supplemented by a silicon microstrip vertex detector. Assuming the parameterization $\sigma(\pi^- N \to DX) \sim A^\alpha$ where A is the atomic number, we find $\alpha = 0.89 \pm 0.05 \pm 0.05$ for a data sample with an average x_F of 0.2. The x_F and $p_\perp$ distributions of the charged D mesons are also described and the possibility of a leading effect is investigated.

© 1989 American Institute of Physics

Hadroproduction of charm is expected to provide a useful testing ground for perturbative QCD due to the large charm quark mass. The WA82 collaboration is studying charm hadroproduction using an impact parameter trigger which increases the charm content of recorded data by about a factor of 15 with respect to the 1/1000 signal-to-noise ratio of commonly used interaction triggers. The charmed particles are produced in a thin segmented target by a 340 GeV/c π^- beam and are identified by the invariant mass of secondary vertices. We present results, obtained from ~25% of our present data sample, on the nuclear dependence of the D production cross section and x_F and $p_\perp$ distributions of charged D mesons.

The study is conducted using the Ω' spectrometer [1] at CERN whose large acceptance, particle identification capability, and good effective mass resolution are well suited to the study of charm (Fig. 1). High precision charged particle tracking and vertex reconstruction are achieved by a silicon microstrip vertex detector (MVD, see Fig. 1) which supplements the standard Ω' tracking chambers. The MVD consists of an array of four 20μm and nine 50μm pitch microstrip planes arranged in a way which facilitates triggering. It is placed together with the target on an optical bench inside the Ω' magnetic field. A telescope of eight 20μm pitch detectors is used to measure the beam position. To select multivertex events the trigger logic, executed in ~350μs by the fast hardware processor MICE [2], identifies events with at least one track having an impact parameter (IP) between 0.1 and 1 mm, the range expected for charmed particles.[*]

Charm events have been selected from a data sample consisting of 8.8×10^6 triggers which corresponds to ~2×10^8 interactions in the target. This sample is ~$\frac{1}{4}$ of the total data currently on tape. A pre-selection of the data is done using the MVD information in the x-z plane (tracks are straight lines in this projection). Events are retained which have a reconstructed primary vertex in the target and two tracks crossing downstream of the target — one track with IP > 70μm and the

---

[*] IP is the distance in z of a track to the primary vertex. In the Ω' reference system the z-axis is vertical along the direction of the 18 kGauss magnetic field and the x-axis is in the beam direction.

other with IP $> 30\mu$m. The selected events are then reconstructed in space using the combined information of the MVD and the Ω' tracking chambers. Events are selected with reconstructed secondary vertices having the following characteristics:

1. the position of the secondary vertex is reconstructed in space outside and downstream of the target;

2. the separation between the primary and secondary vertices is $> 6\sigma$, where σ is the quadratic sum of the position measurement errors of the two vertices;

3. the total momentum vector of the secondary vertex tracks points to the primary vertex within 100μm;

4. the error on the secondary vertex invariant mass is < 12 MeV/c^2;

5. the lifetime of the decaying particle is > 0.2 ps.

For events satisfying these criteria invariant masses are calculated assuming the secondary vertices are produced by $D^+ \to K^-\pi^+\pi^+$, $D^0 \to K^-\pi^+$, and $D^0 \to K^-\pi^+\pi^+\pi^-$ (and charge conjugates) decays and taking all possible combinations of tracks (Fig. 2). Charm events are identified as those for which the invariant mass is within 3σ of the D meson mass. This sample consists of $\sim$750 events with an estimated background of 20%. The Ω' particle identification capabilities have not been used in the present analysis.

Reconciling the results of charm production experiments performed with different target materials has been so far complicated by a lack of knowledge of the nuclear dependence of the cross section. The cross section is commonly parameterized by $\sigma \sim A^\alpha$ where A is the nuclear mass number. In the QCD parton model the cross section is proportional to the number of quarks in the nucleus, so $\alpha \approx 1$. However, coherent effects may reduce the transparency of the nucleus and then α would be expected to approach the geometrical value of $\frac{2}{3}$. To evaluate the A dependence we have performed a measurement of the relative charm production cross sections on silicon and tungsten. Previous evaluations of the atomic number dependence of the charm hadroproduction cross section have been made either by measuring lepton yields in beam dump experiments [3, 4] or by comparing results from various experiments, each performed with a different target material [5, 6]. Both methods are model dependent and need large corrections; the results have shown conflicting indications. We use a segmented target which allows a direct measurement of charmed particle yields produced simultaneously on Si and W thereby minimizing systematic errors. The 1.25 mm thick target is divided along z into two equal sections. One section consists entirely of Si, the other is a sandwich consisting of a 800μm layer of W and a 450μm layer of Si. The beam is steered so that the two sections receive approximately equal intensity. The geometry of the target, reflected in the primary vertex z distribution (Fig. 3), permits a direct measurement of the ratio σ_W/σ_{Si}. Events within 50μm of the Si–W boundary are excluded. To correct for background contamination we assume that events outside the D peak are representative of the background under the peak. We find $\alpha_{bg} = 0.59 \pm 0.13$, a typical value for the production of light hadrons [7]. Correcting the observed numbers of events under the D peak for this background we find

$$\alpha_{charm} = 0.89 \pm 0.05 \pm 0.05.$$

The first error is statistical and the second is the systematic error arising from uncertainty in the relative beam flux on the two halves of the target.[*] The value of α for hadroproduction of light flavors is strongly dependent on x_F [7], defined as $\frac{p_\parallel}{p_\parallel^{max}}$ in the center-of-mass frame, and there is some indication of a similar dependence in charm production [8]. The average value of x_F for our charm sample is $\langle x_F \rangle = 0.2$; we have non-zero acceptance for all $x_F > 0$, as shown in Fig. 4. As a check, the same analysis procedure was applied to a sample of events with secondary vertices produced by $K_s^0 \to \pi^+\pi^-$. For a K_s^0 sample with $\langle x_F \rangle = 0.05$ we find $\alpha = 0.70 \pm 0.005 \pm 0.05$ in agreement with previous measurements [7].

---

[*] The relative beam flux is obtained from the ratio of the total numbers of reconstructed primary interactions in the two halves. Beam trigger runs have been used to measure the ratio of the effective interaction probabilities in the two halves.

The EHS collaboration performed a charm production experiment using a hydrogen target and reported the observation of a leading effect, *i.e.* higher values of x_F for charm particles which contain a quark identical to one of the beam particle valence quarks [9]. This finding has neither been confirmed nor disproved. Fig. 5 shows acceptance corrected x_F distributions for our D^+ and D^- samples. The curve is of the form $(1 - x_F)^n$ where $n = 3.40 \pm 0.45$ is determined from a fit to the combined $D^+ + D^-$ distribution. The line fits the data with $\chi^2 = 6.06$ for 7 d.o.f; however, we observe a systematic difference between the two distributions, so we apply a run test [10]. From a combination of the two statistical tests we find a 10% probability for the hypothesis that the observed D^+ and D^- distributions result from the same law.

The $p_\perp$ distribution of charm production data is often parameterized with an exponential of the form $e^{-ap_\perp^2}$. A single exponential of this form does not fit our data over the full range of $p_\perp$. However, the form $e^{-bp_\perp}$ does fit the data well (see Fig. 6) and we find $b = 2.00 \pm 0.12$ GeV^{-1} corresponding to a mean value of $p_\perp^2$ of 1.5 ± 0.2 GeV$^2/c^4$. Our $p_\perp$ acceptance is nearly flat as shown in Fig. 7. We note that our highest observed $p_\perp$ value (4 GeV/c) is well above those previously observed in hadroproduction.

To summarize, we have measured the relative charm production cross sections of π^- on silicon and tungsten and have found that with the parameterization $\sigma \sim A^\alpha$, $\alpha = 0.89 \pm 0.05 \pm 0.05$ for $\langle x_F \rangle = 0.2$. The differential cross section $d^2\sigma/(dx_F dp_\perp^2) \sim (1 - x_F)^n e^{-bp_\perp}$ fits the data well for $n = 3.40 \pm 0.45$ and $b = 2.00 \pm 0.12$ GeV^{-1}. There is a 10% probability that the slight excess of D^- over D^+ at $x_F \geq 0.44$ is due to a statistical fluctuation rather than to a leading effect.

REFERENCES

1. W. Beusch et al., CERN/SPSC 85-62 (1985) SPSC/P218.

2. J. Anthonioz-Blanc et al., CERN/DD 80-14 (1980).

3. M. E. Duffy et al., Phys. Rev. Lett. **55** (1985) 1816.

4. H. Cobbaert et al., Phys. Lett. **191B** (1987) 456.

5. H. Cobbaert et al., Z. Phys. **C36** (1987) 577.

6. M. MacDermott and S. Reucroft, Phys. Lett. **B184** (1987) 108.

7. C. S. Barton et al., Phys. Rev. **D27** (1983) 2580.

8. S. P. K. Tavernier, Rep. Prog. Phys. **50** (1987) 1439.

9. M. Aguilar-Benitez et al., Phys. Lett. **168B** (1986) 170, Z. Phys. **C31** (1986) 491, Phys. Lett. **169B** (1986) 106.

10. W. T. Eadie et al., *Statistical Methods in Experimental Physics*, North Holland Publishing Co., 1971.

FIGURE CAPTIONS

1. The Ω' spectrometer in the west hall of the CERN SPS and the silicon microstrip vertex detector.

2. Invariant masses of secondary vertex tracks for the assumptions (a) $K\pi\pi$, (b) $K\pi$, (c) $K\pi\pi\pi$, (d) sum of a–c.

3. Primary vertex position in z.

4. Global x_F acceptance for $D^\pm \to K\pi\pi$. Acceptance for D^0 is similar.

5. Acceptance corrected x_F distributions for D^+ and D^-.

6. Distribution of $p_\perp$ of $D^\pm$. The line is fit to the entire distribution.

7. Global $p_\perp$ acceptance for $D^\pm \to K\pi\pi$.

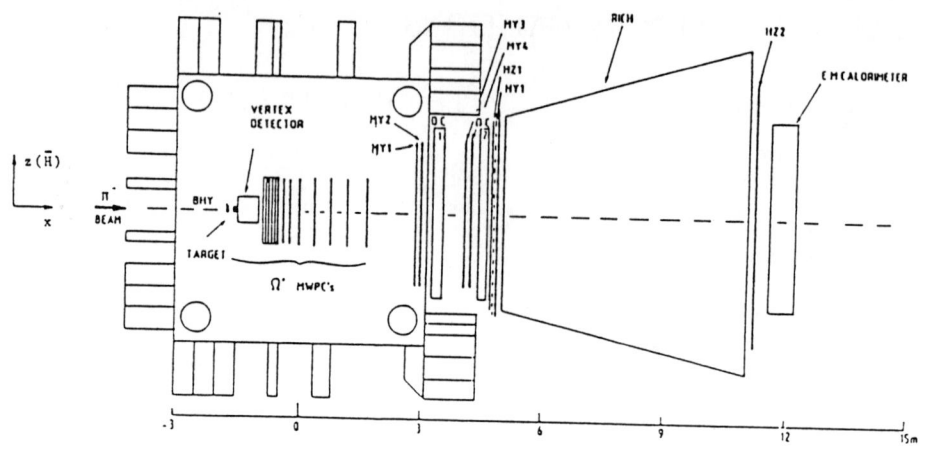

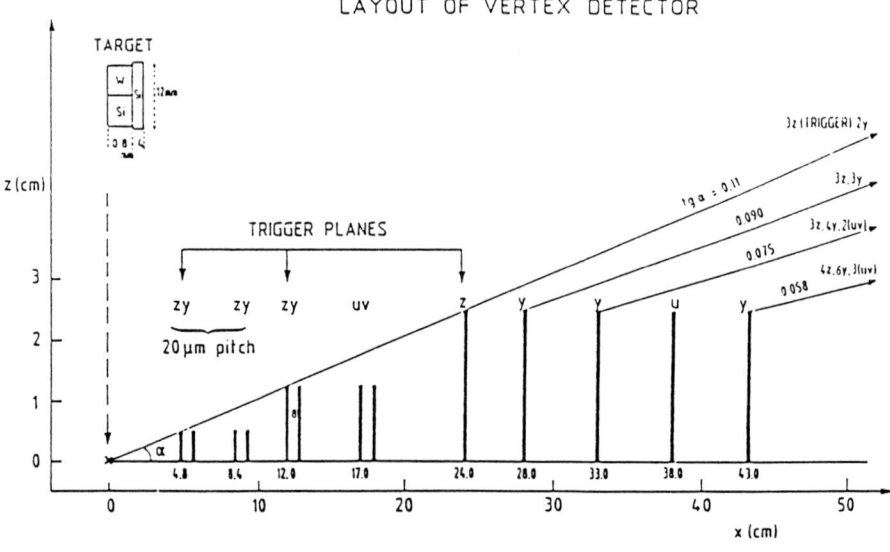

Fig. 1

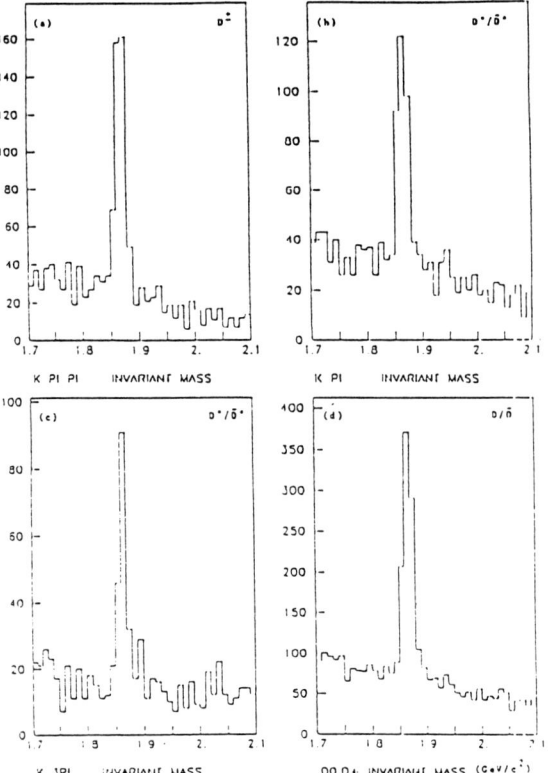

Fig. 2

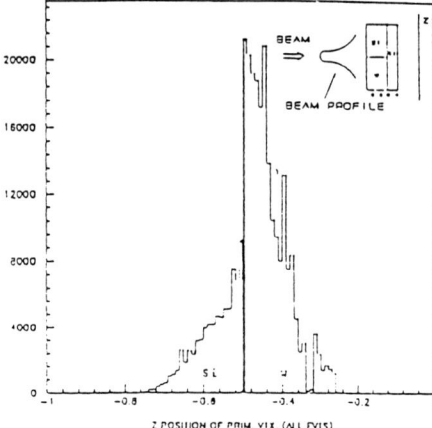

Fig. 3

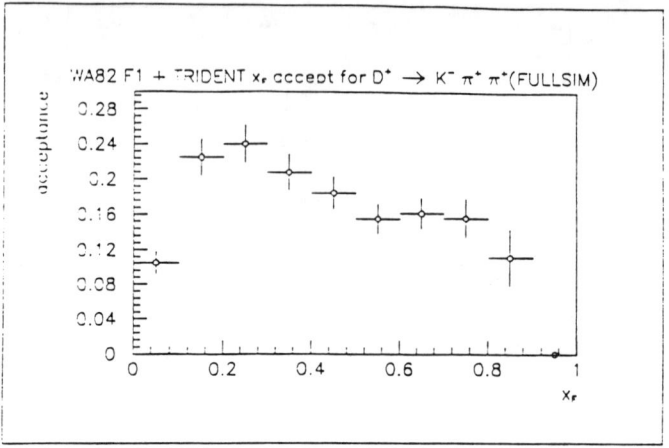

Fig. 4

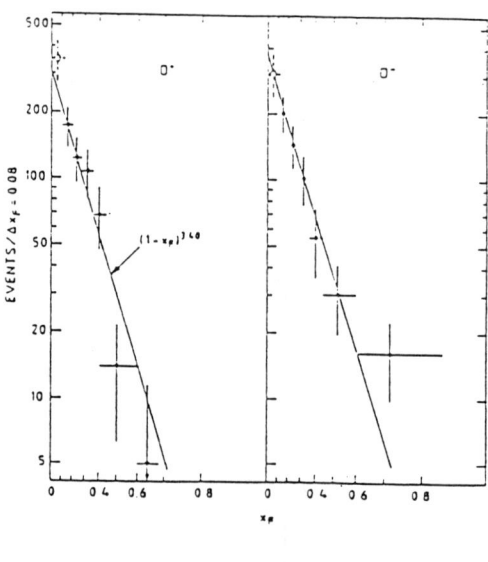

Fig. 3

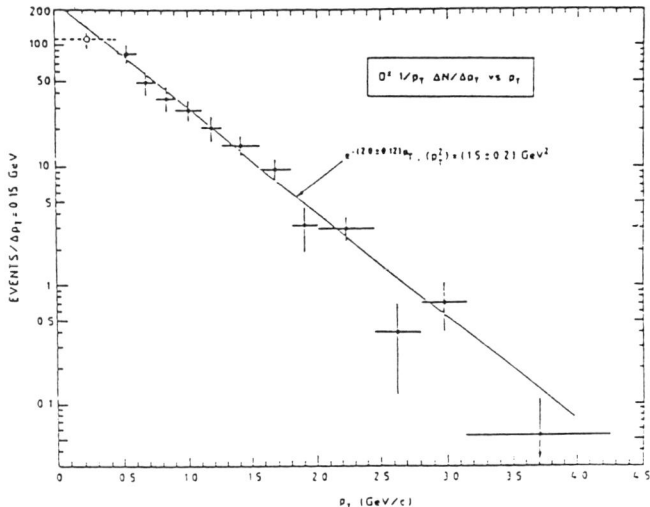

Fig. 6

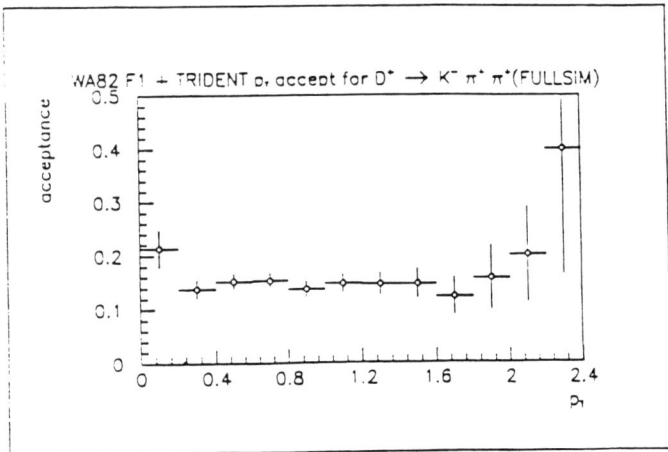

Fig. 7

QUARKONIUM

Υ SPECTROSCOPY: A REVIEW OF RECENT RESULTS*

P. Michael Tuts
Physics Department, Columbia University, New York, NY 10027

ABSTRACT

We review the most recent results on Υ spectroscopy from the CUSB, CLEO, ARGUS and Crystal Ball experiments.

INTRODUCTION

The Υ system has proved to be one of the richest systems in which to test QCD, measure α_s, test potential models (including the spin dependence), and measure hadronic transitions. In addition, the radiative decays are proving to be a sensitive hunting ground for light Higgs (as well as other exotica).

$\Upsilon(nS) \to \mu^+\mu^-$ BRANCHING FRACTIONS

The branching fractions for Υ decay into muon pairs, $B_{\mu\mu}$, can be used together with the e^+e^- leptonic widths and branching ratios to individual channels to determine the total and partial widths, Γ_{tot} and $\Gamma_{\mu\mu}$. Using those values, one can then extract the ratio $\Gamma_{ggg}/\Gamma_{\mu\mu}$ and hence the coupling constant of QCD, α_s, and the scale parameter of QCD, $\Lambda_{\overline{MS}}$. Significant new results for $B_{\mu\mu}$ have been obtained at CESR by the CUSB[1] and CLEO[2] experiments; in table I they are compared with the ARGUS[3] result.

Table I Recent $B_{\mu\mu}$ values

	$B_{\mu\mu}(1S)(\%)$	$B_{\mu\mu}(2S)(\%)$	$B_{\mu\mu}(3S)(\%)$
CUSB	$2.61 \pm 0.09 \pm 0.11$	$1.38 \pm 0.25 \pm 0.15$	$1.73 \pm 0.15 \pm 0.11$
CLEO	$2.52 \pm 0.07 \pm 0.07$		$2.02 \pm 0.19 \pm 0.33$
ARGUS	$2.30 \pm 0.25 \pm 0.13$		
Average	2.53 ± 0.08	1.38 ± 0.29	1.79 ± 0.17

We can use the known[4] values for $\Gamma_{had}\Gamma_{ee}/\Gamma_{tot}$ of 1.24 ± 0.05, 0.574 ± 0.034 and 0.415 ± 0.030 for the $\Upsilon(1S)$, $\Upsilon(2S)$, and $\Upsilon(3S)$ respectively to extract the total widths given in table II. If we also use the known branching ratios for all other transitions[6] we can extract the three gluon to muon pair partial widths,

---

* Work supported by the National Science Foundation and the Sloan Foundation

the QCD fine structure constant, and the QCD scale parameter, as shown in table II.

Table II Computed quantities from $B_{\mu\mu}$

	Γ_{tot} (keV)	$\Gamma_{ggg}/\Gamma_{\mu\mu}$	α_s(4.9 GeV)	$\Lambda_{\overline{MS}}$ (MeV)
$\Upsilon(1S)$	53.0 ± 2.6	32.1 ± 1.5	0.176 ± 0.003	164 ± 10
$\Upsilon(2S)$	43.4 ± 9.1	32.5 ± 8.3	0.176 ± 0.015	167 ± 54
$\Upsilon(3S)$	24.5 ± 2.8	29.3 ± 3.6	0.171 ± 0.007	146 ± 24

Where $\Gamma_{ggg}/\Gamma_{\mu\mu} = (1 - B_{\gamma gg} - B_{\pi\pi} - B_{E1})/B_{\mu\mu} - 3 - R$ and $R = 3.48 \pm 0.165$.[7] The QCD coupling constant is found [8,9] from the expression $\Gamma_{ggg}/\Gamma_{\mu\mu} = 5773\alpha_s^3[1 + c(\mu)(\alpha_s/\pi)]$ where we choose the scale, μ, to be $m_b \approx 4.9$ GeV as suggested by Kwong et al,[10] for which $c = 0.43$. The scale parameter is obtained in the usual way[11] for the $\overline{MS}$ renormalization scheme. It is interesting to note that since all annihilation widths scale together from any one Υ to another, then $B_{\mu\mu}(2S, 3S) = B_{\mu\mu}(1S) \times (1 - B_{NA}(2S, 3S))$, where B_{NA} is the branching ratio for Υ decays without $b\bar{b}$ annihilation. In this way we obtain $B_{\mu\mu}(2S) = (1.39 \pm 0.07)\%$ and $B_{\mu\mu}(3S) = (1.66 \pm 0.07)\%$, in excellent agreement with the measured values.

$\Upsilon(nS)$ TRANSITIONS

The hadronic transitions between Υ states have to date been successfully described in terms of a QCD multipole expansion of the gluon fields.[12] CLEO[5] has now obtained a large sample of $\Upsilon(3S)$ hadronic decays, with an improved detector, in which the observed dipion invariant mass spectrum (see Fig. 1) differs from that predicted by Ref. 12; theoretically, the situation remains unclear. The $\pi\pi$ recoil mass spectrum has been used to search for the singlet P state ($\equiv h_b$) in the decay $\Upsilon(3S) \to \pi^+\pi^- h_b$. No evidence is seen; the 90% CL upper limit branching ratio is less than 0.16%.

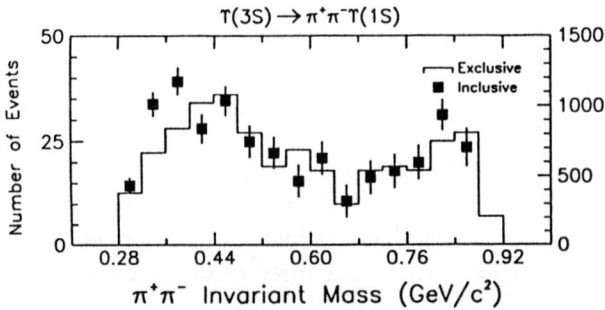

Fig. 1. Dipion mass spectrum for $\Upsilon(3S) \to \pi^+\pi^-\Upsilon(1S)$.

A detailed study of the fine structure of the Υ system and the properties of the triplet P states (χ_b, χ'_b) has been carried out by CUSB using photons from the electric dipole (E1) transitions $\Upsilon(3S) \to \gamma + (\chi_b, \chi'_b)$ in both inclusive and exclusive decay modes. In addition, we have used this large data sample on the $\Upsilon(3S)$ to search for other states such as the η_b and the D states. The observed fine structure can be compared to the generalized formulation of the spin dependence of the $Q\overline{Q}$ potential.[13] By inserting potentials of known Lorentz transformation properties in the Bethe-Salpeter equation and making a nonrelativistic reduction to order v^2/c^2, the fine structure of the χ_b states can be related to a vector potential, V, and a scalar potentials, S by: $M(^3P_2) = \overline{M} + a - 2b/5, M(^3P_1) = \overline{M} - a + 2b$, and $M(^3P_0) = \overline{M} - 2a - 4b$ where $a = (1/2M_Q^2)\langle 3V'/r - S/r \rangle$ and $b = (1/12M_Q^2)\langle V'/r - V'' \rangle$ are the spin-orbit and tensor contributions respectively, and $\overline{M}$ is the c.o.g. of the triplet P states.[14] Potential models (MR,[15] MB,[16] and GRR.[17]) differ in their choice of V and S.

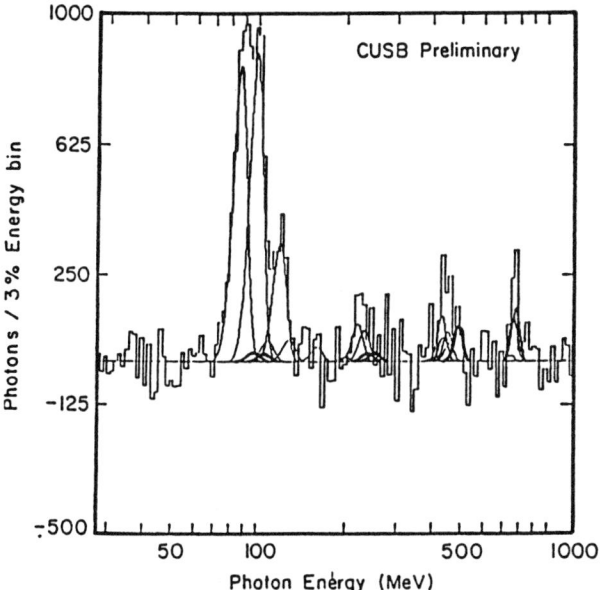

Fig. 2. Background subtracted photon spectrum from $\Upsilon(3S) \to \gamma X$.

CUSB has measured the fine structure using the background subtracted photon spectrum from the inclusive reaction $\Upsilon(3S) \to \gamma + X$ shown in Fig. 2. There are some 25 photon lines expected and the figure emphasizes the rich spectrum of states that are visible in the Υ system. Three prominent lines are visible at $87.6 \pm 0.5 \pm 1.5, 100.1 \pm 0.5 \pm 1.5$, and $121.0 \pm 1.6 \pm 1.5$ MeV, corresponding to the transitions $\Upsilon(3^3S_1) \to \gamma \Upsilon(2^3P_J)$ for $J = 2, 1, 0$ respectively. The corresponding electric dipole (E1) transition rates are measured to be $2.6 \pm 0.2 \pm 0.3, 2.7 \pm 0.2 \pm 0.3$ and $0.9 \pm 0.2 \pm 0.1$ keV respectively, which are in very good agreement with potential model expectations. Other photon transitions are visible in Fig. 2, corresponding to $\Upsilon(2P) \to \Upsilon(2S)$ ($E_\gamma \simeq 240$ MeV), $\Upsilon(3S) \to \Upsilon(1P)$ ($E_\gamma \simeq 450$ MeV), and $\Upsilon(2P) \to \Upsilon(1S)$ ($E_\gamma \simeq 750$ MeV) transitions which have now been observed for the first time in inclusive spectra.

We have also used this data to search for the η_b (via the $\Upsilon(3S) \to \pi\pi h_b \to \pi\pi\gamma\eta_b$ transition). Assuming that the observed excess of photons at $\simeq 496$ MeV is due to $h_b \to \gamma\eta_b$ transitions, we find $BR(3S \to \pi\pi h_b) \times BR(h_b \to \gamma\eta_b) = (6.4 \pm 3.9) \times 10^{-3}$. The search for D states is complicated by the fact that photons from $\chi'_b \to \Upsilon(1D)\gamma$ and $\Upsilon(1D) \to \gamma\chi_b$ transitions are swamped by the much stronger $\Upsilon(3S) \to \chi'_b\gamma \to \Upsilon(2S)\gamma\gamma$ transitions, and thus one can only show, at present, consistency with theoretical expectations.

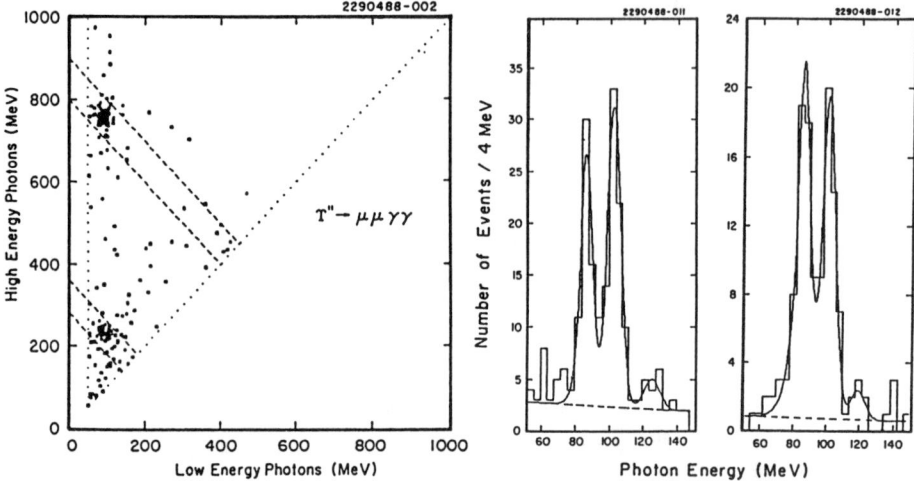

Fig. 3 $\Upsilon(3S) \to \mu\mu\gamma\gamma$ events. Fig. 4 Projection for lower (a), upper (b) cluster.

CUSB has also searched for $\Upsilon(3S) \to \gamma\gamma\ell\ell$ events from 6.59×10^5 $\Upsilon(3S)$ hadronic decays. Figure 3 shows a scatter plot of the higher energy photon ($E_{\gamma high}$) versus the lower energy photon ($E_{\gamma low}$) for $\mu\mu\gamma\gamma$ events at the $\Upsilon(3S)$ peak energy. The data cluster around 80–100 MeV for the lower energy γ and around 230 and 760 MeV for the higher energy photon, confirming their origin as being due to the cascade $\Upsilon(3S) \to \chi'_b\gamma \to (\Upsilon)\Upsilon(2S)\gamma\gamma$. The projections for all $\gamma\gamma\ell\ell$ events (Fig. 4) show two large resolved peaks and one smaller one. The third line is expected to be suppressed due to the large hadronic width (≈ 500–900 keV) of the χ'_{b0} state. The fitted photon energies, in MeV, are 86.1 ± 0.5, 101.5 ± 0.6, and 122.4 ± 2.6 for the $J = 2, 1,$ and 0 spin states respectively. Using the values from Ref. 9 we obtain the values in table III.

Table III Branching ratios and product branching ratios in %

χ'_{bJ}	$3S \to \chi'_{bJ}\gamma \to 2S\gamma\gamma$	$\chi'_{bJ} \to (2S)\gamma$	$3S \to \chi'_{bJ}\gamma \to 1S\gamma\gamma$	$\chi'_{bJ} \to 1S\gamma$
J=2	$1.94 \pm 0.30 \pm 0.37$	$19.5 \pm 3.0 \pm 3.7$	$0.86 \pm 0.15 \pm 0.10$	$8.6 \pm 1.5 \pm 1.0$
J=1	$2.18 \pm 0.29 \pm 0.39$	$24.3 \pm 3.2 \pm 4.3$	$0.76 \pm 0.12 \pm 0.07$	$8.4 \pm 1.4 \pm 0.8$
J=0	$0.32 \pm 0.16 \pm 0.06$	$5.3 \pm 5.6 \pm 1.0$	$0.10 \pm 0.06 \pm 0.01$	$1.7 \pm 1.0 \pm 0.2$

Although the hadronic widths of the χ'_b states are too small to be measured directly, we can combine the measured branching ratios together with the potential model predictions[18] for the E_1 rates for the χ'_b, and thus extract the hadronic widths shown in table IV, together with the predicted hadronic widths (including 1st order QCD corrections). Note that in the ratios of the hadronic widths there is agreement with the predictions, however we also note that there are large uncertainties in the calculation of the hadronic widths.

TableIV Hadronic widths (keV) of the χ'_b

State	$\Gamma^{expt}_{had}(\chi'_{bJ})$	$\Gamma^{theory}_{had}(\chi'_{bJ})$
$\chi'_{bJ=2}$	$74 \pm 11 \pm 11$	153
$\chi'_{bJ=1}$	$63 \pm 9 \pm 8$	51
$\chi'_{bJ=0}$	$360 \pm 157 \pm 39$	866

TableV Fine structure splitting

Transition	Energy (MeV)
$3^3S_1 \to 2^3P_2$	$86.8 \pm 0.3 \pm 1.5$
$3^3S_1 \to 2^3P_1$	$100.8 \pm 0.4 \pm 1.5$
$3^3S_1 \to 2^3P_0$	$121.5 \pm 1.2 \pm 1.5$

In the scatter plot (Fig. 3) near the reflection boundary in the region where both photons have ≈ 430 MeV, one sees events consistent with a transition from $\Upsilon(3S)$ to $\Upsilon(1S)$ due to the process $\Upsilon(3S) \to \chi_b\gamma \to \Upsilon(1S)\gamma\gamma$. Of the 8 events consistent with this transition, 2.4 are background, from which we obtain the product branching ratio $\sum_{J=2,1} BR(\Upsilon(3S) \to \chi_{bJ}\gamma) \times BR(\chi_{bJ} \to \Upsilon(1S)\gamma) = (9 \pm 5) \times 10^{-4}$. The EM transition rates ($\propto |\int \Psi_f \Psi_i r dv|^2$) between triplet S and triplet P $Q\overline{Q}$ states with different numbers of nodes in the wave functions are very sensitive to details of the potential. Three models, one with the nonrelativistic Cornell potential (EI),[19] one whose potential is reconstructed from the inverse scattering method (KR),[18] and a relativistic one (GRR)[17] predict product branching ratios ($\times 10^4$) of 100 ± 20, 4.5 ± 0.9, and 0.8 ± 0.2 respectively. Note that the range between predictions is about two orders of magnitude.

Using the combined "inclusive" and "exclusive" fine structure measurements combined, we find the values shown in table V. From these values we can calculate the relative spin-orbit, a, and tensor, b, interactions. In table VI we compare the values with potential model predictions. Another way of looking at the fine structure is by comparing the ratio $r = (M(^3P_2) - M(^3P_1))/(M(^3P_1) - M(^3P_0))$ to potential model predictions, as shown in table VII. The agreement of all the measured χ state data with the GRR[17] predictions is taken as proof that the confining potential is "scalar".

Table VI Spin-orbit vs tensor terms

	a (MeV)	b (MeV)
Expt	9.2 ± 0.3	1.9 ± 0.2
GRR	9.3	1.9
MR	6.5	2.1
MB	14.6	4.2

Table VII Fine structure ratio r

	χ_c	χ_b	χ'_b
Expt	0.48 ± 0.01	0.67 ± 0.06	0.67 ± 0.05
GRR	0.50	0.64	0.67
MR	0.42	0.42	0.42
MB	0.35	0.45	0.48

OTHER Υ DECAYS

CLEO[20] has the first, albeit weak, evidence of the decay $\Upsilon(1S) \to J/\psi + X$ for which they measure a branching ratio of $(0.11 \pm 0.04)\%$. Only speculations as to the mechanisms for J/ψ production exist at present.

Both CLEO[21] and the Crystal Ball[22] experiments have searched for two body radiative decays of the $\Upsilon(1S)$ into lighter mesons. Simple scaling arguments indicate a suppression of $\simeq 1/40 (= (e_b/e_c)^2 \times (m_c/m_b)^2)$ relative to the corresponding transitions in J/ψ radiative decays. With the present data samples, the experimental upper limits are beginning to reach those levels of sensitivity. In table VIII we list the present limits.

Table VIII The 90% CL upper limits on radiative $\Upsilon(1S)$ decays

Final State	$BR_{expt} \times 10^4$ (90% CL)	Scaling $\times 10^4$
$\gamma\eta(550)$	< 3.5 (CB)	0.2
$\gamma\eta'(957)$	< 13 (CB)	1
$\gamma f_2(1270)$	< 8.1 (CB)	0.3
$\gamma f_2(1270) \times f_2(1270) \to \pi^\pm \pi^\mp$	< 1.2 (CLEO)	
$\gamma f_2(1720) \times f_2(1720) \to K_S^0 K_S^0$	< 3.6 (CLEO)	
$\gamma f_2(1720) \times f_2(1720) \to \eta\eta$	< 4.3 (CB)	
$\gamma f_2(1720) \times f_2(1720) \to K^\pm K^\mp$	< 1.2 (CLEO)	.1
$\gamma\eta(1430) \times \eta(1430) \to K^\pm \pi^\mp K_S^0$	0.82 (CLEO)	1

LIGHT HIGGS AND GLUINO SEARCHES

A central problem in physics today is the detection of the Higgs; unfortunately there are very few restrictions on the Higgs mass. Light Higgs bosons have been searched for by CUSB in the radiative decays of Υ's. The decay $\Upsilon \to H + \gamma$ is due to the annihilation of the $b\bar{b}$ pair bound in the Υ into a photon and a Higgs. The predicted rate relative to the decay $\Upsilon \to \mu\mu$ is given by[23] $\Gamma(\Upsilon \to H + \gamma)/\Gamma_{\mu\mu} = G_F M_b^2/(\sqrt{2}\pi\alpha)(1 - M_H^2/M_\Upsilon^2)x^2$ where $x=1$ for a single Higgs doublet, and $x = v_1/v_2$ is the ratio of vev's for two Higgs doublets, however it was promptly pointed out[24] that QCD radiative corrections, to lowest order,[25] reduce the branching ratio for $\Upsilon \to H + \gamma$ by about a factor two. However, a more recent analysis[26] suggests that the corrections may not be as large.

CUSB has searched for such radiative decays using a sample of $400,000\Upsilon(1S)$ and $600,000\Upsilon(3S)$ events which have been added to their previous sample of $400,000\Upsilon(1S)$ events. The CUSB results are given both in terms of 90% CL upper limits on the $BR(\Upsilon \to \gamma + H)$ (see Fig. 5), and in terms of x vs Higgs mass (see Fig. 6) and are valid down to $2 \times m_\mu$ (at the low end, the continuum subtraction involves removing events from the reaction $e^+e^- \to \rho\gamma$ which is under revised study at the moment). The other curves shown are for the minimal standard model (dashed curve), including Ref. 26 corrections (dash-dot curve), and including Ref. 25 corrections (dotted curve). Thus, for the case of one Higgs doublet, the mass of the Higgs must be greater than ≈ 5 GeV, at the 90% confidence level.

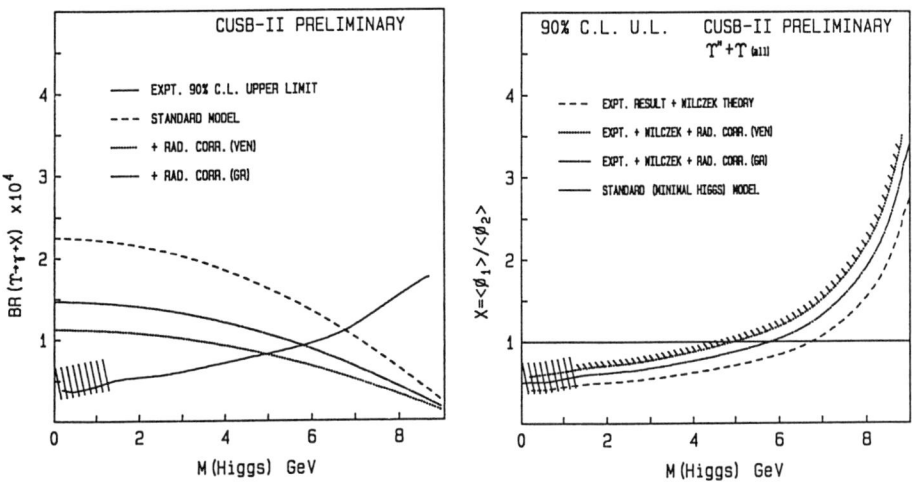

Fig. 5. 90% CL for $BR(\Upsilon \to \gamma + H)$.   Fig. 6. 90% CL for x.

This same data has been used to search for gluinos, since for gluino masses in the 1-5 GeV range, we expect gluinium bound states to exist with many of their properties very similar to those of η_c and η_b mesons. Comparing these results with theoretical calculations, one can exclude gluino masses between 0.1 and 3.6 GeV at the 90% CL.[27–29]

ARGUS[30] has searched for the exclusive decay $\Upsilon \to \gamma H$ Fig. 7) where $H \to \pi^+\pi^-$ is assumed to have a 45% branching ratio. In the $\pi\pi$ invariant mass region from about 290 to 570 MeV, the upper limit varies from $(3-4.5) \times 10^{-5}$, which lies below the radiatively corrected prediction of $\approx 5 \times 10^{-5}$.

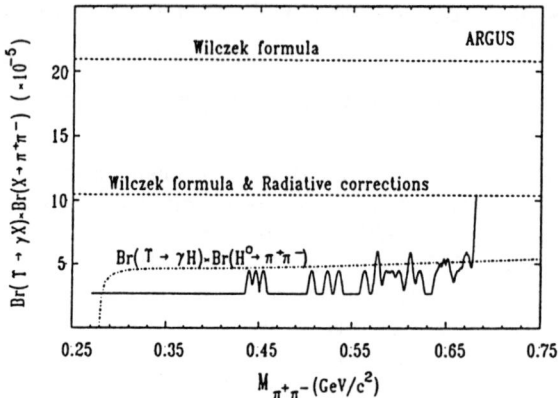

Fig. 7. ARGUS 90% CL upper limit on $BR(\Upsilon(1S) \to \gamma \pi^+ \pi^-)$.

EVIDENCE FOR B_s^* PRODUCTION AT THE $\Upsilon(5S)$

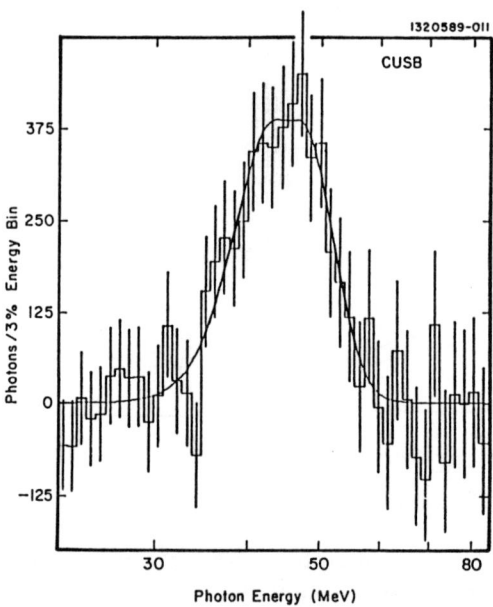

Fig.8. Background subtracted γ spectrum for $\Upsilon(5S) \to \gamma X$.

Information on the production of strange mesons is of interest to quark spectroscopy and to the questions of mixing and CP[31] violation in the B sector. A previous study[32] of $e^+ + e^- \to$ hadrons above the b–flavor threshold has shown a complicated structure for the cross section, which was well fit using a coupled channel[33] model calculation. From the same data we also obtained evidence for the existence of B^* mesons with $M(B^*) - M(B) = 52 \pm 2 \pm 4$ MeV. New data has been taken by CUSB-II, mostly at the $\Upsilon(5S)$, at a center of mass energy of 10.87 GeV. The much improved energy resolution of the CUSB-II detector allows to clearly observe a photon line from B^*decays and obtain the average mass difference. From an analysis of the background subtracted inclusive photon spectrum on the $\Upsilon(5S)$ shown in Fig. 8 we obtain $\langle E_\gamma \rangle = 47.5 \pm 0.4$ MeV, $N(\gamma)/n(\Upsilon(5S)) = 1.09 \pm 0.06$ and $\langle \beta \rangle = 0.156 \pm 0.01$. The observed line is however narrower than expected if due only to $B^*_{d,u}$ decays but quite consistent with large production of B^*_s, as predicted by the earlier model calculation. This is shown in Fig. 9 where we compare the observed photon spectrum (dashed line) with Monte Carlo calculations for the case of no Doppler broadening ($\beta = 0$) and maximal Doppler broadening ($\beta = .21$) from assuming only $B^*_{d,u}$ decays. Confirmation that the photons are from B^* decays is provided by tagging events that have high energy (1-3 GeV) electrons from B semileptonic decay and then looking at the inclusive photon spectrum (Fig. 10).

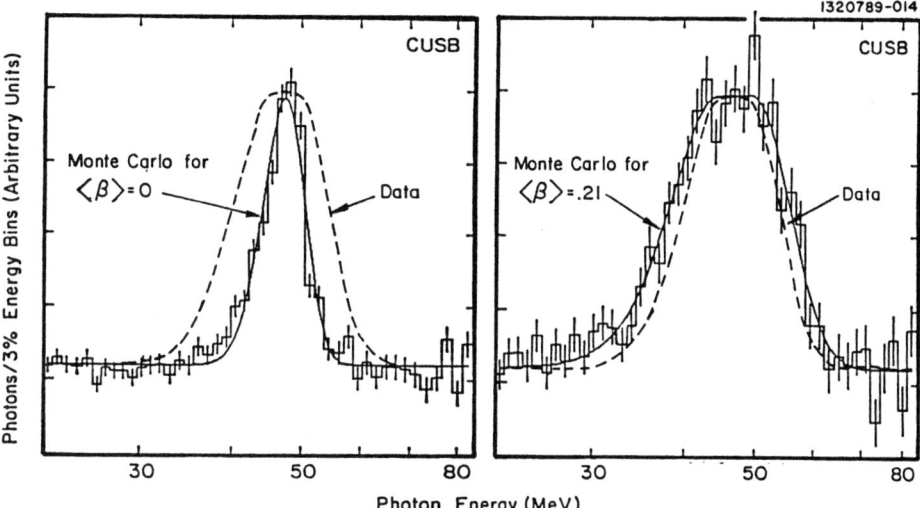

Fig. 9. Monte Carlo spectra for $\beta = 0$ and $\beta = .21$. Dashed line is the data.

The coupled channel analysis described above requires assumption about the masses of the B, B^*, B_s and B^*_s, in order to determine the position of the corresponding six thresholds for $B\overline{B}$, $B\overline{B}^*$+cc, $B^*\overline{B}^*$, $B_s\overline{B}_s$, $B_s\overline{B}^*_s$+cc and $B^*_s\overline{B}^*_s$ production. It also requires accurate calculations of the $\Upsilon(5S)$ wave function in order to obtain the decay amplitude into the six possible channels. It should be noted however that the uncertainty on the mass differences between vector and scalar B mesons can only reflect in a broader line, not a narrower one. If the mass difference between strange and non strange B's is changed from the values assumed in our previous fit, the result is to increase the fraction of strange B's. If the mass difference is reduced resulting in an increased β for B_s decays, a larger fractional production of B_s's is necessary to produce the observed narrow line. If the strange B's masses are increased, $B_s^* \overline{B}_s^*$ is strongly suppressed or no longer kinematically allowed, also requiring a larger fractional production of B_s's.

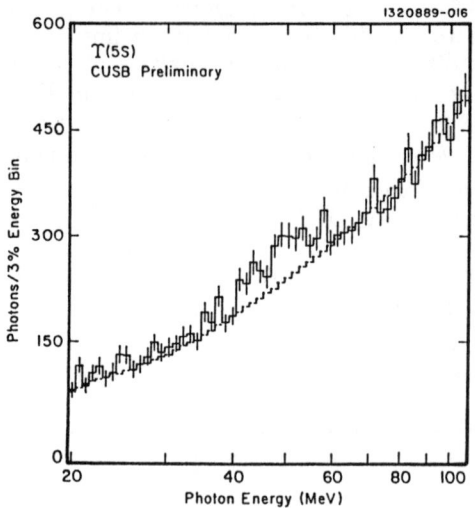

Fig. 10. High energy lepton tagged γ spectrum.

We know some of the factors contributing to the production ratios: $B\overline{B}:B\overline{B}^*+cc:B^*\overline{B}^*:B_s\overline{B}_s:B_s\overline{B}_s^*+cc:B_s^*\overline{B}_s^*$. The spin weights, W_s are in the ratio 1:4:7 for $BB : BB^* : B^*B^*$. P-wave production gives a factor p^3, including phase space. Production of nonstrange B's is suppressed because their high momenta and the large radius of the parent Υ's. Reasonable guesses for the ranges of the mass differences are possible: $80 < M(B_s) - M(B_d) < 104$ MeV, $25 < (M(B_d^*) - M(B_d)) < 75$ MeV, $0.7 < [M(B_s^*) - M(B_s)]/[M(B_d^*) - M(B_d)] < 1.0$

As a first approximation to the remaining factors we use (1) a factor $p^3 W_s$ for the relative production of the three $B\overline{B}$ and the strange pairs, but we allow for an arbitrary amount of the two species, and (2) a more sophisticated technique which includes the effects of the $\Upsilon(5S)$ decay amplitudes for various models (EI,[33] MN,[34] and BH[35]).

For all three cases we find that the 1σ limit never goes below 30% for the fraction of strange B meson production. Thus, given the successful coupled channel model description of the cross section above the b-flavor threshold, and given the good agreement with measurements of photons from excited B decays, we conclude that the fraction of B_s at the $\Upsilon(5S)$ is at least 30%, quite independently of assumptions on the strange B meson hyperfine splitting. Thus running at the $\Upsilon(5S)$ provides the highest yield of strange mesons, with a purity at least twice that of the continuum in e^+e^- annihilations.

Further evidence of B_s production is provided by measuring the shape of the semileptonic decay spectrum on the $\Upsilon(4S)$. In general, the more $B_s^* \overline{B}_s^*$

production, then the lower the endpoint of the spectrum, and the sharper the falling edge compared to pure (*i.e.* non-strange) $B\overline{B}$ production. A comparison of the spectrum with models for the two cases indicates that strange B production is favored. Confidence in the modelling is obtained by a measurement of the $\Upsilon(4S)$ semileptonic decay spectrum; the parameters obtained by CUSB ($BR = (11.1 \pm 0.3 \pm 0.5)\%$ and $|V_{ub}| = 0.047 \pm 0.004$ are in good agreement with the values presented at this conference.[36]

ACKNOWLEDGEMENTS

I would like to thank all the members of the CUSB group, particularly J. Lee–Franzini, as well as H. Schöder for discussions of ARGUS data.

REFERENCES

1. T. Kaarsberg et al. (CUSB), Phys. Rev.D **35**, 2265 (1987); T. Kaarsberg et al. (CUSB), Phys. Rev. Lett. **62**, 2077 (1989).

2. W.-Y. Chen et al. (CLEO), CLEO preprintCLEO 89-1 (1989).

3. H. Albrecht et al. (ARGUS), Z. Phys. C **35**, 283 (1987).

4. Particle Data Group, Phys. Lett. **204B**, 1 (1988).

5. I.C. Brock et al, CLEO preprint, CBX-88-22 (1988).

6. We use the PDG 88 values for the $\Upsilon(2S)$ and the new CUSB B_{E1} and CLEO $B_{\pi\pi}$ values from refs 1and 5for the $\Upsilon(3S)$.

7. Z. Jakubowski et al. (Crystal Ball), Z. Phys. C**40**, 49 (1988).

8. R. Barbieri, R. Gatto, R. Kögler and Z. Kunszt, Phys. Lett. **57B**, 455 (1975).

9. P.B. Mackenzie and G.P. Lepage, Phys. Rev. Lett. **47**, 1244 (1981).

10. W. Kwong, P.B. Mackenzie, R. Rosenfeld and J.L. Rosner, Phys. Rev. D**37**, 3210 (1988).

11. W.A.Bardeen, A.J. Buras, D.W. Duke and T. Mura, Phys. Rev. D**18**, 3998 (1978).

12. Y.P. Kuang and T.M. Yan, Phys. Rev. D**24**, 2874 (1981), and references therein.

13. E. Eichten and F. Feinberg, Phys. Rev. D**23**, 2724 (1981).

14. J.L. Rosner, in Experimental Meson Spectroscopy-1983, ed. by S.J. Lindenbaum. AIP, New York, (1984) 461.

15. P. Moxhay and J. L. Rosner, Phys. Rev. D**28**, 1132 (1983).

16. R. McClary and N. Byers, Phys. Rev. D**28**, 1692 (1983).

17. S.N. Gupta, S.F. Radford and W.W. Repko, Phys. Rev. D**30**, 2425 (1984).

18. W. Kwong and J.L. Rosner, Phys. Rev. D**38**, 279 (1988).

19. E. Eichten et al., Phys. Rev. **D17**, 3090 (1979); ibid. **21**, 203 (1980).
20. R. Fulton et al. (CLEO), CLEO preprint, CLEO 88-4 (1988).
21. R. Fulton et al. (CLEO), CLEO preprint,CLEO 89-7 (1989); D. Besson, these proceedings.
22. P. Schmitt et al. (Crystal Ball), DESY preprint, DESY 88-031 (1988).
23. F. Wilczek, Phys. Rev. Lett. **40**, 220 (1978); S. Weinberg, ibid. p. 223.
24. P. Franzini et al., Phys. Rev. **D35**, 2883 (1987), see in particular ref. 17 therein; M. Davier, in Proc. of the XXIII Int. Conf. on High Energy Physics, Berkeley, 1986, USA, Edited by S.C. Loken (World Scientific, Singapore,1987) p. 25.
25. M. I. Vysotsky, Phys. Lett. **97B**, 159 (1980); J. Ellis et al., ibid., **158B**, 417 (1985); P. Nason, ibid., **175B**, 223 (1986).
26. H. Goldberg and Z. Ryzak, preprint NUB-2954 (1988).
27. J.H. Kühn and S. Ono, Phys. Lett. **142B**, 436 (1984).
28. W.Y. Keung and A. Khare, Phys. Rev. **D29**, 2657 (1984).
29. T. Goldman and H. Haber, Physica **15D**, 181 (1985).
30. H. Albrecht et al. (ARGUS), contributed paper to the Munich Conference (1988).
31. J. Lee-Franzini, in Flavor mixing and CP violation, Proceedings of the Fifth Moriond Workshop, Jan 1985, edited by J. Tran Thanh Van, Editions Frontiere, Gif–sur–Yvette (1985); see also S. Ono, N. A. Törnqvist, J Lee-Franzini and A. Sanda, Phys. Rev. Lett. **55**, 2838 (1985).
32. D.M.J. Lovelock et al. (CUSB), Phys. Rev. Lett. **54**, 377 (1985).
33. Following E. Eichten et al., Phys. Rev. **D17**, 3090 (1979) and **D21**, 203 (1980); but assuming no S-D mixing, only P-waves and simple propagator form.
34. A. D. Martin and C. K. Ng, DPT Preprint, DPT/88/6.
35. N. Byers and D. S. Huang, FNAL Workshop on Beauty.
36. K. Schubert, these proceedings.

Radiative $\Upsilon(1S)$ Decays

Dave Besson

University of Florida

We report on a study of exclusive radiative decays of the $\Upsilon(1S)$ resonance. We have considered decays into final states of the form $\Upsilon \to \gamma n(\mathrm{h^+h^-})$, with $(\mathrm{h^+h^-})$ denoting a charged hadron pair (either $\pi^+\pi^-$, K^+K^-, or $p\bar{p}$), and n denoting the number of such pairs in the event. We measure branching ratios for $n=2$, 3 and 4, and search for structure in the recoiling hadronic system. The structure which we observe in the $n=1$ case is consistent with our expectations for the continuum process $e^+e^- \to \gamma$ X. No evidence is observed for two-body radiative decays of the $\Upsilon(1S)$ meson in any of the above channels.

The data sample under study consists of 15.5 pb^{-1} of $\Upsilon(1S)$ data collected with the CLEO detector as it existed prior to 1986, and 21.0 pb^{-1} of $\Upsilon(1S)$ data collected with the recently improved CLEO detector. The full data sample corresponds to 825,000 $\Upsilon(1S)$ decays and 150,000 hadronic continuum e^+e^- annihilation events under the resonance. We use a sample of 101 pb^{-1} collected on the continuum below the $\Upsilon(4S)$ for background studies (corresponding to 380,000 hadronic continuum events).

For $n > 1$ candidate events, we define p_h as the magnitude of the net momentum of the charged tracks. In real $\Upsilon \to \gamma n(\mathrm{h^+h^-})$ decays, the energy of the photon must equal p_h, and the angle θ between the photon direction and the direction of the net charged momentum must be 180°. Defining $\delta_p = (E_\gamma - p_h)/\sigma_E$ as the deviation between the energy of the observed photon (E_γ) and the net momentum of the charged tracks in units of the resolution σ_E of the electromagnetic calorimeter, we plot δ_p against the cosine of the angle (θ) between the γ and p_h. Event candidates are expected to populate the region near $\cos\theta = -1.0$ and $\delta_p = 0.0$. The plot for the case $n=2$, where we have assumed that all charged tracks in the event are pions (i.e., $\Upsilon \to \gamma \pi^+\pi^-\pi^+\pi^-$), is shown in Fig. 1.

By selecting the band with $|\delta_p| < 2.0$ and projecting onto the $\cos\theta$-axis, we obtain Fig. 2(a). Overplotted are the projections obtained using continuum data (open triangles), and the signal shape expected from $\Upsilon \to \gamma n(\mathrm{h^+h^-})$ Monte Carlo events (dashed line). Also shown are the projections obtained by carrying out this procedure for the cases $n=3$, 4 and 5. Extrapolating a flat background from the region $\cos\theta \geq -0.9995$ into the two leftmost bins on these plots (corresponding to the region $\cos\theta \leq -0.9995$), we observe evidence for an excess for the cases $n=2$, 3 and 4. In order to establish that the excesses observed for $n=2$, 3 and 4 are, in fact, due to radiative decays of the $\Upsilon(1S)$ meson, we have studied several possible background processes.

The level of continuum background in the region near $\cos\theta = -1.0$ (presumably due to the radiative process $e^+e^- \to q\bar{q}\,\gamma$, in which the photon is emitted from one of the incident leptons) has been verified as numerically consistent with expectations by running a Monte Carlo simulation with QED corrections.[1] We have addressed possible backgrounds from merged π^0 by determining the energy-weighted spread of each shower. We find that the average value of this parameter for Monte Carlo produced merged photons from π^0 decay is well separated from the values obtained for photons from true radiative Bhabha events and photons from our candidate exclusive $\Upsilon(1S)$ decay sample. The observed distribution in this parameter favors the single photon interpretation for our exclusive radiative decay event candidates. In addition, we have analyzed the data using $|q_{tot}|=1$ events; these are considered to be representative of the shape of the random background. We note that the total charge one events, shown in Fig. 2(b), give spectra which are flat in $\cos\theta$.

The net efficiency for observing our signal is obtained by using a detector simulation to determine geometric and tracking efficiencies. We use these efficiencies to calculate the branching ratios for the three ($n=2$, 3, or 4) modes. To determine the charged particle type recoiling against the photon, we use the particle identification capabilities of the main drift chamber in conjunction with an energy conservation criterion to differentiate between particle types. We calculate for each particle the deviation in units of the known dE/dx resolution under each particle-type hypothesis, and require that each particle have a dE/dx pulse height deposition in the inner drift chamber which is within $\pm 2.5\sigma_{dE/dx}$ of the assumed particle identity (either π, K, or p). We then calculate, for $n=2$ and 3 events, the difference between the known event energy and the sum of the measured photon energy and the charged particle energies under the particular combination of charged particle mass assumptions. We retain as the most likely particle mass assignments the hypothesis which best approximates energy conservation. Using this criterion, we find that, for $n=2$, all of our events are classified as either $\Upsilon \to \gamma\pi^+\pi^-\pi^+\pi^-$, $\Upsilon \to \gamma\pi^+\pi^- K^+K^-$, $\Upsilon \to \gamma\pi^+\pi^- p\bar{p}$, or $\Upsilon \to \gamma K^+K^- K^+K^-$. Since a true $\Upsilon \to \gamma\pi^+\pi^-\pi^+\pi^-$ event will sometimes be classified as something other than an $\Upsilon \to \gamma\pi^+\pi^-\pi^+\pi^-$ event, we unfold from our observed event-type distribution the true distribution as follows: equal numbers of Monte Carlo events of the varieties $\Upsilon \to \gamma\pi^+\pi^-\pi^+\pi^-$, $\Upsilon \to \gamma\pi^+\pi^- K^+K^-$, $\Upsilon \to \gamma\pi^+\pi^- p\bar{p}$ and $\Upsilon \to \gamma K^+K^- K^+K^-$ were used to determine a transformation matrix $\mathbf{T_n}$, which provides, for each type of event generated, the likelihood of it being classified as one of the four possible event-types. By inverting this matrix, and then applying it to the observed event-type distribution X, we unfold these observed distributions to obtain the true event-types.

We tabulate the signal size for each of the particle mass assumptions for the resonant data, the continuum data, and also the random background (using the $|q_{tot}|=1$ events) which populates the region under the signal. Subtracting the continuum background and the random background, we obtain the net signal size. We then apply the $\mathbf{T_n}^{-1}$ matrices to these observed excesses to determine the branching ratios. Results are given in Table I.

The $n=2$ data sample was also examined for the possibility that the final-state hadrons were being produced through some intermediate state (either ρ^0, K*, Δ or ϕ). No obvious sub-structure is observed in the $\pi^+\pi^-$, $p\pi^-$, or K^+K^- invariant mass plots; however, we notice an apparent clustering of events in the $\Upsilon \to \gamma\pi^+\pi^- K^+K^-$ event sample which suggests $\Upsilon(1S) \to \gamma K^{*0}\overline{K}^{*0}$. In particular, we find 7.4±3.3 events which fit this hypothesis in our $\Upsilon(1S)$ data sample. No such clustering is observed in the continuum dataset. The intermediate state $\Upsilon(1S) \to \gamma K^{*0}\overline{K}^{*0}$ therefore accounts for 26±11% of our observed $\Upsilon \to \gamma\pi^+\pi^- K^+K^-$ decay rate.

We subject $n=1$ candidate events to a kinematic fit, imposing overall energy-momentum conservation under the three possible hadron type assumptions. We retain events with

$\chi^2 < 30$ and calculate the invariant mass of the recoil hadrons. The scaled continuum data are shown in Fig. 3. with the data collected at the $\Upsilon(1S)$ resonance overplotted.

After performing a continuum subtraction, no obvious structure is observed in the recoil mass plot. Limits on branching ratios for exclusive radiative decays of the $\Upsilon(1S)$ into a photon plus another particle decaying into two charged tracks are presented in Fig. 4. for the three particle species.

In summary, we have observed exclusive radiative decays of the $\Upsilon(1S)$ meson in decay modes into a photon plus four, six, or eight charged tracks. Despite difficulties in ascertaining definite final states, we have observed positive evidence for the production of baryons in exclusive radiative decays. An analysis of the $\Upsilon \to \gamma \pi^+ \pi^- K^+ K^-$ mode indicates that roughly one-quarter of these decays are proceeding through the mode $\Upsilon(1S) \to \gamma K^{*0} \overline{K}^{*0}$. For the decay $\Upsilon \to \gamma(h^+ h^-)$, no structure is observed in the recoil hadron mass, and upper limits are set which are on the order of theoretical predictions.

Table I

Summary of $\Upsilon(1S) \to \gamma$ X analysis	
Decay mode:	Branching Ratio (10^{-4})
X=4π	2.5±0.7±0.5
X=2π2K	2.9±0.7±0.6
X=2π2p	1.5±0.5±0.3
Sum n=2	7.0±1.1±0.7
X=6π	2.5±0.9±0.8
X=4π2K	2.4±0.9±0.8
X=4π2p	0.4±0.4±0.4
Sum n=3	5.4±1.5±0.8
Sum n=4	7.4±2.5±2.5

References and Footnotes

1) F.A.Berends and R.Kleiss, *Nucl. Phys.* **B177**, 237 (1986).

FIGURE CAPTIONS

1. Number of observed $\Upsilon \to \gamma 2(h^+h^-)$ events as a function of the cosine of the opening angle between the charged particle net momentum and the photon direction '$\cos\theta$'(x-axis) vs. the net momentum difference between the photon and the charged particles 'δ_p'(y-axis). The peak bin contains 16 events.

2. a) Projections onto $\cos\theta$-axis for $n=2$, 3, 4 and 5; and b) projection obtained for events with total charge of magnitude 1.

3. $n=1$ invariant mass plots for continuum events satisfying kinematic fit criteria with $\Upsilon(1S)$ data overplotted.

4. Upper limits on $\Upsilon(1S) \to \gamma X$; $X \to h^+h^-$

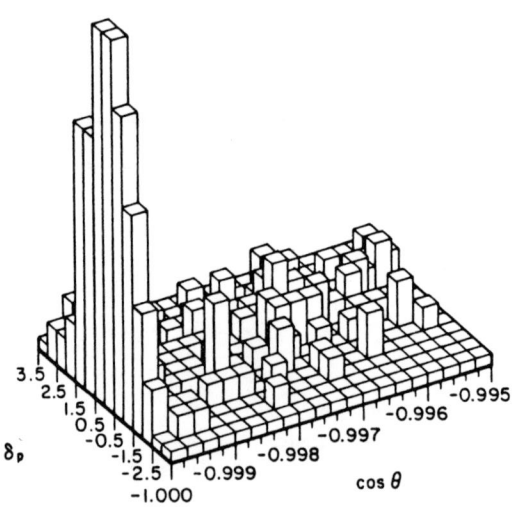

Figure 1

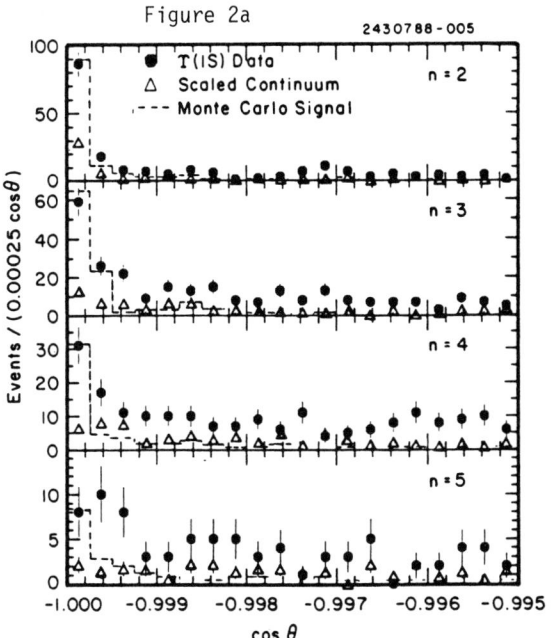

Figure 2a

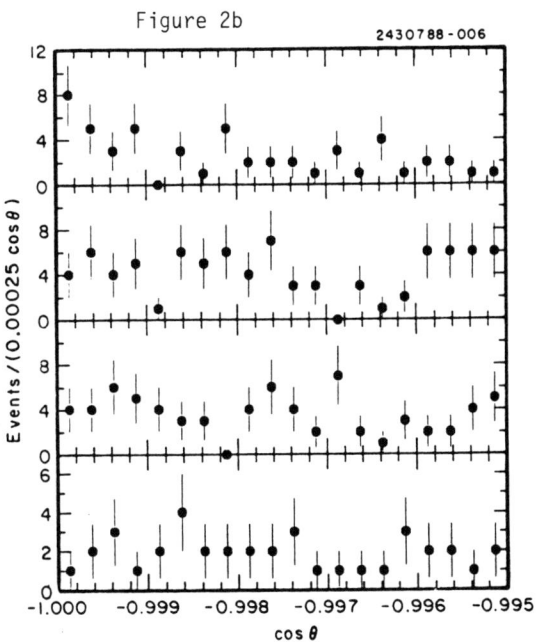

Figure 2b

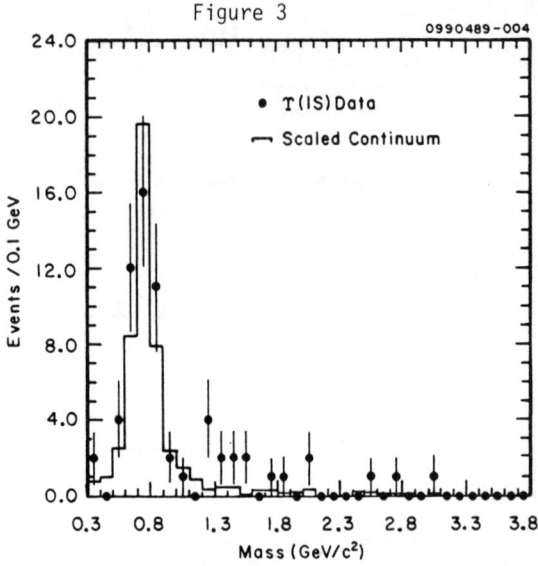

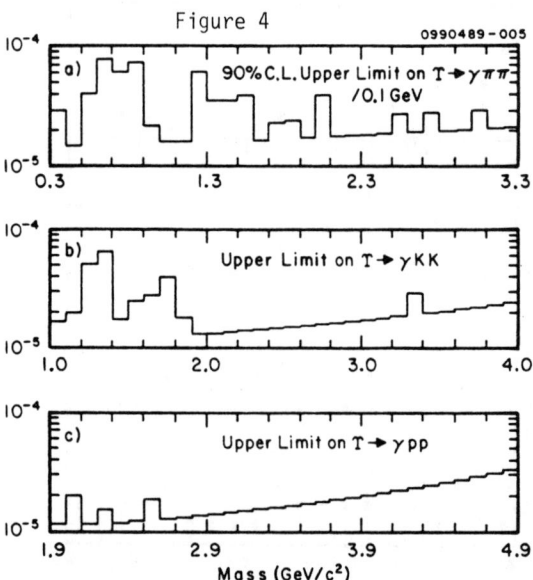

Study of ψ' Decays*

Walter H. Toki
representing the Mark III Collaboration
Stanford Linear Accelerator Center
Stanford University, Stanford, California 94309

Abstract

Hadronic decays of the ψ' are reviewed and a new preliminary upper limit of B(ψ'→ρπ)<7.0x10^{-5} at 90% C.L. from the Mark III is presented.

The ψ' is the 2^3S_1 $c\bar{c}$ quark-antiquark vector meson and is the only well established radial excitation. The hadronic decay rate of the ψ' relative to the J/ψ should scale as the ratio of the three gluon widths which are proportional to the leptonic widths divided by the full widths;

$$\frac{B(\psi' \to \text{hadrons})}{B(J/\psi \to \text{hadrons})} = \frac{B(\psi' \to ggg)}{B(J/\psi \to ggg)} = \frac{\Gamma(\psi' \to e^+e^-)\Gamma(J/\psi)}{\Gamma(J/\psi \to e^+e^-)\Gamma(\psi')} = (12.2 \pm 2.4)\%$$

This predicts that the absolute branching ratios of hadronic modes of the ψ' are 12% of the corresponding decay from the J/ψ or 8 times smaller. The Mark II group[1] measured several decays in 1982 and observed hadronic decays into several multibody decays but none of the vector-pseudoscalar pair combinations. More recently, the Crystal Ball group[2] observed the radiative decay to the tensor f(1270) but neither to the η nor the η'. The measured rates are shown in table 1. The second column contains the J/ψ branching rates and the third column the ratio of rates which should be ~12%. The results are very striking. Within the errors, the missing modes are vector pseudoscalar decays whereas the other modes are roughly consistent with the factor of 8 ratio. For some unknown reason the ψ' does not decay into this particular choice of mesons pairs. The largest J/ψ mode which is missing is the ρπ decay

Invited talk presented at the meeting of the 1989 International Symposium on Heavy Quark Physics at Cornell University, Ithaca, N.Y., June 13-17, 1989

---

*Work supported in part by the National Science Foundation and by the Department of Energy contracts DE-AC03-76SF00515,DE-AC02-76ER01195,DE-AC03-81ER40050, DE-AC02-87ER40318 and DE-AM03-76SF000324

Table 1. Comparison of ψ' and J/ψ decays

Mode	B(J/$\psi\to$mode)	Ratio($\psi' \div J/\psi$ in %)
$p\bar{p}$	0.22 %	8.6 ± 2.4 %
$p\bar{p}\pi^+\pi^-$	0.60	15.1 ± 4.1
$K^+K^-\pi^+\pi^-$	0.72	22.2 ± 9.0
$p\bar{p}\pi^0$	0.11	14.0 ± 6.3
5π	3.4	9.5 ± 2.7
7π	2.9	13.0 ± 7.0
$\gamma f(1270)$	0.14	9.0 ± 3.0
$\rho\pi$	1.28	<0.63
K^*K	0.75	<2.07
$\gamma\eta$	0.86	<1.8
$\gamma\eta'$	0.42	<2.6

and it appears to be suppressed by a factor 20. These results indicate we may not understand the simplest picture of how gluons form into hadrons. In this paper we present a search for this decay in a new sample of ψ' decays from the Mark III.

ψ' Data Sample

The analysis of the mode is based on a sample of $(236\pm35)\times10^3$ ψ' events taken with the Mark III detector at SPEAR. Approximately 40% of this data was taken in a five day run in 1982 and the remaining in a three month run in 1988. This corresponds to a total integrated luminosity of ≈0.5 inverse picobarns. The number of events were determined from a study of J/ψ production in ψ' decays in the topologies $\psi' \to \pi^+\pi^- + J/\psi$ where the J/ψ is observed from its direct decay into $\mu^+\mu^-$ and where it's production is inferred in the recoil from the $\pi^+\pi^-$ system. The Mark III experiment[3] is a solenoidal magnetic spectrometer optimized for SPEAR physics. This analysis utilized the tracking from the drift chamber[4] and the shower information from the barrel and endcap shower counters[5]. In addition, the 1988 data had the new Mark III vertex chamber[6] in operation which achieved an average position resolution of 60 microns on Bhabha events.

$\psi' \to \rho\pi$ Analysis

The decay $\psi' \to \rho\pi$ has the final state topology of two photons and two charged pions. The charged track selection requires:

1) two and only two oppositely charged tracks that each have two DC stereo hits.
2) shower counter energy < 0.5 GeV if the track entered the shower counter.
3) mass($\pi+\pi-$)< 2.9 GeV/c^2 to remove $\psi' \to J/\psi + \pi°\pi°$ events.

At least two photon candidates were required in the shower counters. The two charged tracks and the two photon candidates were subjected to a kinematic constrained fit (4-C) to the decay hypothesis $\psi' \to \gamma\gamma\pi^+\pi^-$. In order to reduce the feed down from a single photon and hadronic split off's, an angle cut is imposed to remove asymmetric decays by requiring, $\cos\theta(\pi°,\gamma)<0.98$, where the angle in the $\pi°$ frame is between the photon and the J/ψ. A conspicuous $\pi°$ signal in the $\gamma\gamma$ mass was observed as shown in Fig. 1 providing evidence for the decay $\psi' \to \pi^+\pi^-\pi°$. A hand scan is performed to remove events with extra photons. Twenty $\pi^+\pi^-\pi°$ candidates remain and of these events, two fall within the $\rho\pi$ bands (770±150 MeV/c^2). The first candidate had a mass, m($\pi^-\pi°$)=894 MeV/c^2 and the second had a mass, m($\pi^+\pi°$)=675 MeV/c^2. In Fig. 2 is the final Dalitz plot of the $\psi' \to \pi^+\pi^-\pi°$ events. There is no evidence for the triangular $\rho\pi$ bands in the Dalitz plot.

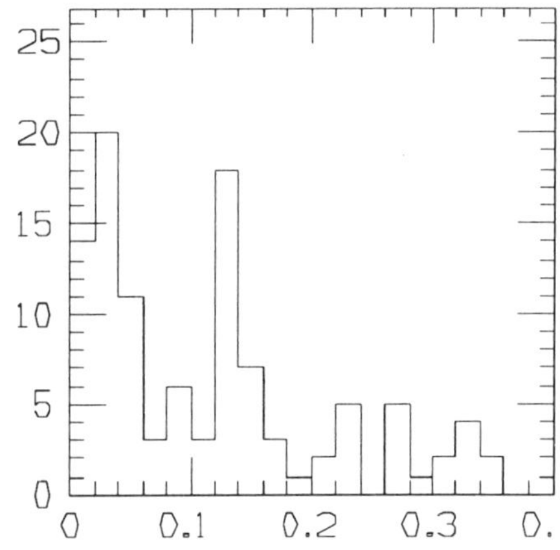

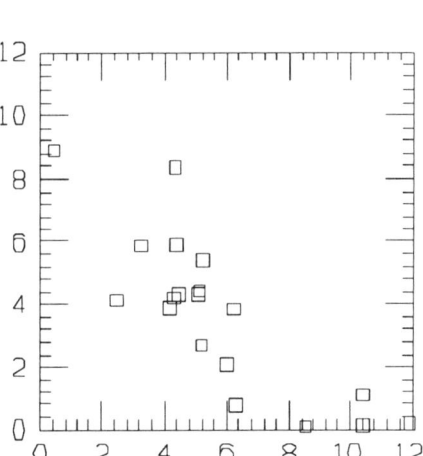

Fig. 1 $\gamma\gamma$ mass (GeV) from $\psi' \to \gamma\gamma\pi^+\pi^-$

Fig. 2 Dalitz plot of m^2($\pi^+\pi°$) vs m^2($\pi^-\pi°$) from $\psi' \to \pi^+\pi^-\pi°$

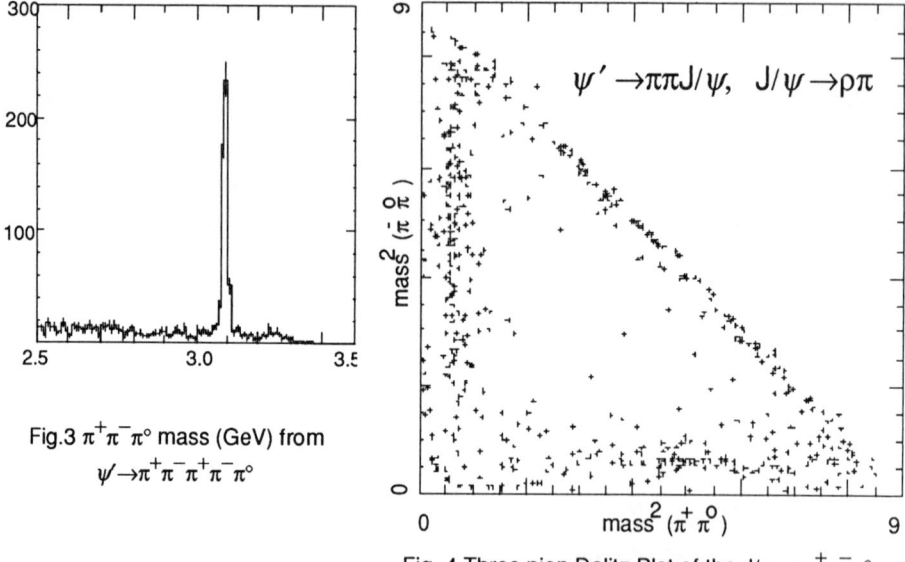

Fig.3 $\pi^+\pi^-\pi^0$ mass (GeV) from $\psi' \to \pi^+\pi^-\pi^+\pi^-\pi^0$

Fig. 4 Three pion Dalitz Plot of the $J/\psi \to \pi^+\pi^-\pi^0$ produced from the mode $\psi' \to \pi^+\pi^- J/\psi$

Several checks were performed. The same analysis program was used to analyze $J/\psi \to \pi^+\pi^-\pi^0$ in the J/ψ data except the e^+e^- center mass energy was changed to that of the J/ψ. A clear signal for $J/\psi \to \rho\pi$ was observed verifying that the program was correct. A second analysis was performed to detect the decay, $\psi' \to \pi^+\pi^- J/\psi$, $J/\psi \to \rho\pi$, $\rho\pi \to \pi^+\pi^-\pi^0$. In Fig. 3 is shown the $\pi^+\pi^-\pi^0$ mass and in Fig. 4 is the Dalitz plot of the three pions. An unmistakable J/ψ signal is evident and the Dalitz plot displays the three bands for the $\rho\pi$ decays. Hence, a $\rho\pi$ signal is definitely being produced in the ψ' data but not directly from the ψ'.

The events we observe are consistent with non-resonant 3 pion decays. If the two events we observe in the $\rho\pi$ bands are both from direct decay (this ignores the non-resonant background), we obtain;

$$B(\psi' \to \rho\pi) = \frac{2}{(.32)(236 \times 10^3)} = (2.6 \pm 1.7 \pm .5) \times 10^{-5}$$

where the Monte Carlo determined efficiency was $\varepsilon = .32$. If we set an upper limit at 90% C.L. for these two events, we obtain;

$$B(\psi' \to \rho\pi) < \frac{5.32}{(.32)(236 \times 10^3)} = 7.0 \times 10^{-5}$$

Assuming all the 3 pion events are non-resonant, we obtain the branching ratio of;

$$B(\psi' \to \pi^+\pi^-\pi^0) = (2.6 \pm 0.5 \pm 0.5) \times 10^{-4}$$

These results are comparable to the Mark II results[1] which were based on 1 million ψ' decays. They had one $\rho\pi$ candidate and set an upper limit of $B(\psi' \to \rho\pi) < 8.3 \times 10^{-5}$ at 90% C.L. and from four $\pi^+\pi^-\pi^0$ events they had measured a non-resonant signal of $B(\psi' \to \pi^+\pi^-\pi^0) = (8.5 \pm 4.6) \times 10^{-5}$.

Discussion of the Results

Our results confirm the Mark II measurements. The observed suppression of $\psi' \to \rho\pi$ may be due to the wave function differences between the J/ψ and the ψ' but this may not explain why in particular the vector-pseudoscalar combination is missing. The interference between the OZI and the electromagnetic decays could possibly create this suppression, however, the electromagnetic contribution is too small relative to the OZI rate to suppress selectively the vector pseudoscalar rate. It is not understood whether the ψ' decay is unexpectedly suppressed or the J/ψ decay is *enhanced*. Several models have been proposed;

- *Nambu & Freund* Model (1975)[7]

In this model the puzzle is not the suppression of the $\rho\pi$ rate in the ψ' decays, but the existence of the $\rho\pi$ rate from the J/ψ. The rate from the J/ψ should in fact be suppressed, but it is being produced by a constructive interference from a nearby resonance. This resonance is a Pomeron daughter called the "ϑ" meson which is an SU(4) singlet vector meson that can mix with the ω, ϕ and J/ψ. It is expected to decay into $\rho\pi$ and KK^* and very little into e^+e^- or $K\bar{K}$. The model also predicts the "ϑ" mass around 1.4-1.8 GeV with a width of 50-100 MeV.

• *Hou & Soni* Model (1982)[8]

As in the previous model, a vector gluonia, "ϑ" interferes with the J/ψ enabling vector-pseudoscalar decays. The mass of this resonance is predicted to lie around 2.4 GeV if the ratio of the $\rho\pi$ branching ratios is 1.25% instead of the 12%. The resonance mass will also increase higher if the suppression is larger. The resonance is expected to appear in the reactions J/ψ, $\psi' \to (\eta,\eta',\pi\pi) + \vartheta$, $\vartheta \to \rho\pi, K^*K$.

• *Brodsky, Lepage & Tuan* Model (1987)[9]

In this model, a vector gluonium resonance is proposed to lie within 100 MeV of the J/ψ and the interference produces the vector pseudoscalar decays of the J/ψ. This explains the puzzle as to why QCD hadron helicity conservation[10] fails to suppress the large $\rho\pi$ decay of the J/ψ. In addition this may explain why J/$\psi \to \phi S^*$ and not $\delta\pi$ is observed, since the ϑ mixes with the ϕ and enhances a mode that would otherwise be suppressed.

• *Slaugher & Oneda* Model (1988) [11]

In this model the vector glueball that mixes with the J/ψ is proposed also to explain why the decay J/$\psi \to \gamma\eta_c$ is so small (1.3%). In the charmonium model the M1 transition of J/ψ is expected to be a factor of three larger than observed.

• *Tornquist & Chaichian* Model (1988)[12]

In this model the puzzle is explained by introducing a hadronic form factor that exponentially decreases the two meson decays of the ψ' relative to the J/ψ by $\exp[-(m(\psi')^2 - m(J/\psi)^2/4K^2]$ where K is a parameter fitted to the data. This prediction explains the decay rate for $\phi \to \rho\pi$ and predicts a large suppression for many two meson modes.

• *Pinsky* Model (1989)[13]

In this paper it is pointed out that the radiative decay of the η from the ψ' is a hindered M1 transition. The radiative transitions to the η are predicted to scale as the η_c rates corrected for the phase space factors as;

$$\frac{B(\psi' \to \gamma\eta)}{B(J/\psi \to \gamma\eta)} = \left(\frac{p_\eta^{\psi'}}{p_\eta^{J/\psi}}\right)^3 \left(\frac{p_{\eta_c}^{J/\psi}}{p_{\eta_c}^{\psi'}}\right)^3 \frac{B(\psi' \to \gamma\eta_c)}{B(J/\psi \to \gamma\eta_c)} = 0.2\%$$

This predicts a small rate for the radiative η decay from the ψ'. Using a phenomenological model where the J/ψ has an OZI amplitude, F_v, to change into a light quark vector meson which subsequently decays into a vector pseudoscalar pair with a coupling, $G_{v \to vp}$, the ψ' is then expected to have an OZI amplitude, $F_{v'}$, to change into a radial light quark vector meson which subsequently decays into a vector pseudoscalar pair with a coupling, $G_{v' \to vp}$. This coupling, $G_{v' \to vp}$ for a radial vector meson to change into a vector pseudoscalar pair is then determined from measured transition rates for $\psi' \to \gamma \eta_c$ and found to be suppressed relative to $G_{v \to vp}$ which is evidence for a generalized hindered M1 transition. Using estimates for the OZI amplitudes, the final predicted rate for $\psi' \to \rho\pi$ is slightly less than 10^{-5}.

Summary

In conclusion, we have set an upper limit, B($\psi' \to \rho\pi$) <7.0×10^{-5} at 90% C.L. We observe, however, non-resonant three pion decays in the same data sample. This result confirms the previous search from Mark II.[1] These results are still theoreitcally puzzling and indicate the existence of more underlying complexity in the physics of hadronic decays of the J/ψ and ψ'. We thank the efforts of the SPEAR staff for the operation of the storage ring and the SLAC Technical staff for the operation of the LINAC that enabled the data runs for the ψ'

REFERENCES

[1] M. Franklin *et al.*, Phys. Rev. Letts. **51**, 963 (1983).
[2] R. Lee, Phd. thesis, Stanford University, SLAC Report-282, May 1985, unpublished.
[3] D. Bernstein *et al.*, Nucl. Instrum. Methods, **226**, 301 (1984).
[4] J. Roehrig *et al.*, Nucl. Instrum. Methods, **226**, 319 (1984).
[5] W. Toki *et al.*, Nucl. Instrum. Methods ,**219**, 479 (1984).
[6] J. Adler *et al.*, Nucl. Instrum. Methods, **276**, 42 (1989).
[7] P. Freund and Y. Nambu, Phys. Rev. Lett. **34**, 1645 (1975). see also J. Bolzan, W. Palmer, S. Pinsky, Phys. Rev. **14D**, 3202 (1976).
[8] W. Hou and A. Soni, Phys. Rev. Lett. **50**, 569 (1983).
[9] S. Brodsky, P. Lepage and S.F. Tuan, Phys. Rev. Lett. **59**, 621 (1987).
[10] S. Brodsky and G. Lepage, Phys. Rev. **24D**, 2848 (1981). Note that this rule is violated by the electromagnetic decay J/$\Psi \to \omega\pi$.
[11] S. Oneda and M. Slaughter, preprint LA-UR-88-617, February 1988.

[12] M. Chaichian and N. Tornqvist, preprint HU-TFT-88-11, March 1988.
[13] S. Pinsky, Ohio State University preprint, 1989.

TWO-PHOTON PRODUCTION OF THE η_c

T. Jensen
The Ohio State University, Columbus, OH 43210[*]
(Representing the CLEO collaboration)

ABSTRACT

A brief review of two-photon production of the η_c is given. Preliminary results from the CLEO experiment are presented and shown to be in good agreement with theoretical predictions.

INTRODUCTION

The Charmonium spectrum has been studied principally through e^+e^- production[1] of the triplet S states, J/ψ and ψ'. A complementary approach uses two-photon production of positive charge-conjugation states, η_c, η_c', χ_{c0}, and χ_{c2}. In what follows I will concentrate on two-photon production of the η_c, and describe preliminary studies of this process at the Cornell Electron Storage Ring (CESR) using the CLEO detector.

Radiative decays of resonant states provide a relatively clean probe of the underlying dynamics of these systems, and quark models have been quite successful in predicting rates for such processes. In these models[2], η_c and J/ψ are composed of a c and $\bar{c}$ quark in a relative S-wave state. The decay rate for $\eta_c \to \gamma\gamma$ or $\psi \to e^+e^-$ is then proportional to the square of the corresponding wave function at the origin.

$$\Gamma(\eta_c \to \gamma\gamma) = (4\pi\alpha^2 e_c^4 |\Psi(0)|^2/M_c^2)(1-3.4\alpha_s/\pi) \qquad (1)$$

$$\Gamma(J/\psi \to e^+e^-) = (16\pi/3)(\alpha^2 e_c^2 |\Psi(0)|^2/M_{J/\psi}^2)(1-16\alpha_s/3\pi) \qquad (2)$$

Perturbative corrections to the decay rates depend on the strong-coupling, which, for distances corresponding to charmonium states, has the value[2] $\alpha_s \sim 0.22$. Assuming the wave functions for the two states to be similar, and taking $M_{J/\psi} = 2M_c$, we obtain the relation,

$$\Gamma(\eta_c \to \gamma\gamma) = (4/3)\Gamma(\psi \to e^+e^-)(1+1.96\alpha_s/\pi). \qquad (3)$$

Using the measured value[3], $\Gamma(\psi \to e^+e^-) = 4.7\pm0.3$ keV, implies $\Gamma(\eta_c \to \gamma\gamma) \sim 7$ keV, which is a small fraction of the total decay width, $\Gamma_{tot}(\eta_c) = 10$ MeV. Furthermore, the observed decays of η_c have fairly small branching ratios, making high-statistics studies difficult.

Two-photon production of η_c was first observed by the PLUTO experiment[4], which measured a fairly large coupling, $\Gamma_{\gamma\gamma}(\eta_c) = 27\pm13$ keV. An experiment at the CERN ISR (R704), using a hydrogen gas-jet target in a circulating antiproton beam to measure the process $p\bar{p} \to \gamma\gamma$, found[5] $\Gamma_{\gamma\gamma}(\eta_c) = 4.3^{+3.4}_{-3.7}\pm2.4$ keV. The TPC/2γ collaboration searched for $\gamma\gamma \to \eta_c$ in a number of 4-track final states[6], and

*Supported by U.S. Dept. of Energy under contract DE-AC02-76ER01545

obtained the value $\Gamma_{\gamma\gamma}(\eta_c) = 6.4^{+5.0}_{-3.4}$ keV. TASSO has also studied several decay modes, and reports[7] $\Gamma_{\gamma\gamma}(\eta_c) = 19.9 \pm 6.1 \pm 8.6$ keV. The weighted average of these measurements is 7±3 keV which is in agreement with the theoretical estimate. However, given the wide range of values and large uncertaintities, the present measurements do little to restrict the theoretical models.

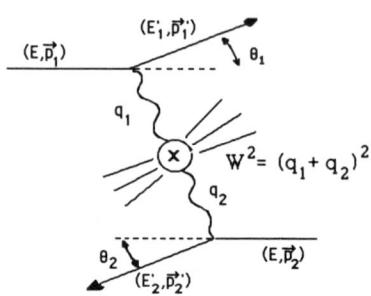

Fig. 1. Kinematics for the two-photon reaction $e^+e^- \rightarrow e^+e^-X$.

High-energy electron-positron colliders provide a useful source of virtual photons for studying photon-photon scattering[8]. This process is depicted in Fig. 1, where the incident electron and positron emit photons of 4-momenta q_1 and q_2 which combine to form the final state, X, characterized by invariant mass squared, $W^2 = (q_1+q_2)^2$. The photons are emitted predominantly at small angles, of order m_e/E, with respect to the beam, and follow roughly a bremsstrahlung energy spectrum, $\sim 1/E_\gamma$. The electron and positron generally deviate little from their incident direction, and thus are often not observed in the experiment.

When the final state is a narrow resonance, R, of mass M_R and spin J, the production cross section is,

$$\sigma(e^+e^- \rightarrow e^+e^-R) = (16\alpha^2/M_R^3)[\ln(E/m_e)]^2 F(M_R/2E)(2J+1)\Gamma(R \rightarrow \gamma\gamma), \quad (4)$$

where $F(x) = (2+x^2)^2 \ln(1/x) - (1-x^2)(3+x^2)$, E is the incident beam energy, and $\Gamma(R \rightarrow \gamma\gamma)$ is the partial decay width of the resonance to two photons. The selection rules for this process follow by considering the quantum numbers of the photon. Only neutral states with even charge conjugation (C=+) can be created. Allowed spin-parity combinations are, $J^P = 0^\pm, 2^\pm, 3^+, 4^\pm, \ldots$.

SELECTION OF 2-γ EVENTS AT CLEO

The CLEO I detector is a general purpose magnetic detector[9] which was not optimized for two-photon studies. There are no small-angle tagging capabilities. Furthermore, time-of-flight counters, required in the trigger, cover only the region $|\cos\theta|<0.65$ and are located outside the solenoid coil.

Nevertheless, this detector has some advantages over others. A major factor is the high luminosity of the CESR machine. The data sample used for our analysis consists of an integrated luminosity of 429 pb^{-1} accumulated at center-of-mass energies between 10.52 and 10.86 GeV.

The CLEO I detector has limited photon reconstruction capabilities, but has excellent charged-particle momentum and angular resolution. Thus we have restricted our analysis to final states having four charged tracks with net charge equal to zero. To aid in understanding the triggering efficiency, we required at least two tracks be within the range $|\cos\theta|<0.64$. As indicated in the previous section, the energy available to the two-photon system is typically much less than the center-of-mass energy. To separate 2-γ production from 1-γ annihilation processes, we required the total energy of the final state to be less than 9 GeV. To distinguish 2-γ events from those 1-γ events which have unobserved particles, we demanded that transverse momentum be balanced. The electron and positron from a 2-γ process tend to travel down the beampipe, so contribute little to the transverse momentum of the final state. If all particles from the 2-γ interaction are observed, we must find $|\Sigma \vec{p_t}|^2 \sim 0$. Owing to the excellent momentum resolution of the CLEO detector, we were able to place a stringent limit on the transverse momentum, requiring $|\Sigma \vec{p_t}|^2 < 0.006$ GeV^2.

Finally, to reduce combinatoric background, we used particle identification information. A K_s was identified through the reconstruction of a secondary vertex at least 2 mm from the primary vertex. The $\pi^+\pi^-$ invariant mass resolution is 4 MeV/c^2 (sigma). For particles which do not come from a secondary vertex we used dE/dx information from the 51-layer central drift chamber ($\sigma=6.5\%$), requiring that the measured ionization signal be within three standard deviations of the expected value for π, K, or p.

A candidate for 2-γ production of η_c followed by the decay $\eta_c \rightarrow K_s K^+\pi^-$ is shown in Fig. 2. In this event there is a net momentum of 1.75 GeV/c along the beam direction, a characteristic indication of two-photon production. The invariant mass distribution for all such candidates is shown in Fig. 3a. (It is implied that charge-conjugate decays are included in all distributions.) A clear signal is present at the mass of the η_c. The inset to this figure shows the distribution of the square of transverse momentum for these events, with the shaded region indicating the restriction imposed for selecting two-photon events. Figure 3b shows corresponding distributions for $K^+K^-\pi^+\pi^-$ final states where at least one of the $K^+\pi^-$ ($K^-\pi^+$) combinations is consistent with coming from a $K^*(892)$. Though by itself not statistically significant, there is an enhancement consistent with η_c production. For each decay mode the data was fit to a gaussian, plus an exponential for the background. The width of the gaussian was set according to Monte Carlo predictions (typically 25-30 MeV/c^2). A maximum-likelihood fit was performed, varying the area of each gaussian as well as the

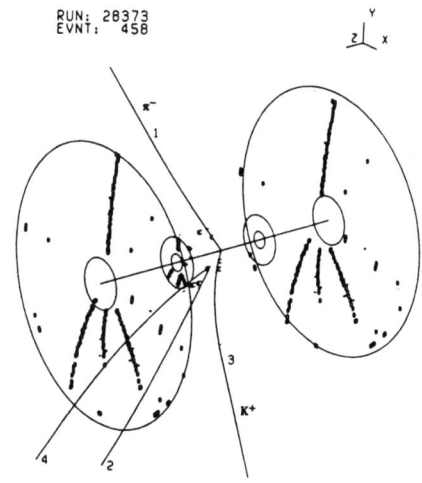

Fig. 2. Display of a candidate for the decay $\eta_c \rightarrow K_s K^+\pi^-$.

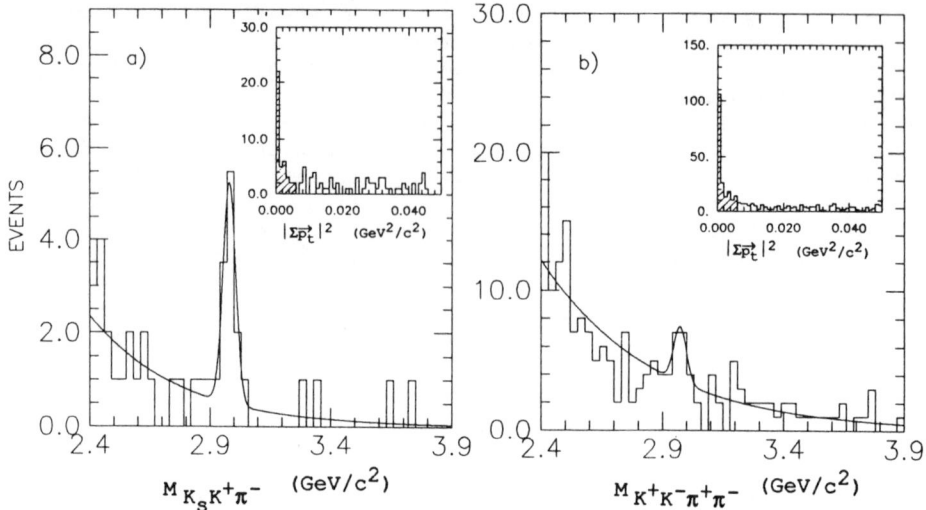

Fig. 3. Invariant mass distributions for two-photon production candidates for a) $\eta_c \to K_s K^+ \pi^-$, and b) $\eta_c \to K^{*0} K^- \pi^+$ (charge conjugate modes implied). The insets show the distribution of transverse momentum squared for these events.

background parameters. Our observations for the different decay modes are summarized in the second collumn of Table I. For decays other than $K_s K^+ \pi^-$ and $K^{*0} K^- \pi^+$ we quote 95% C.L. upper limits.

RESULTS

To convert the observed number of events into a measurement of the radiative decay width, we must account for the efficiency for observing the above processes. We have used a two-photon generator[10] which follows the prescription of Field[11] for calculating the QED factors for the process ee→eeX. The matrix element for the coupling $\eta_c \to \gamma\gamma$ takes the form appropriate for a pseudoscaler,

$$[(q_1 \cdot q_2)^2 - q_1^2 q_2^2]^{1/2} F(q_1^2, q_2^2), \qquad (5)$$

where we have assumed a J/ψ propagator for the form factor,

$$F(q_1^2, q_2^2) = \frac{1}{(1-q_1^2/m_{J/\psi}^2)} \frac{1}{(1-q_2^2/m_{J/\psi}^2)}. \qquad (6)$$

As we have restricted our analysis to very small values of transverse momentum, we are not very sensitive to the form factor.

η_c was allowed to decay to the final state particles according to phase space. The response of the CLEO detector to these generated events was simulated, and the resulting output processed by the standard data analysis programs. The efficiency for observing any of the 4-track decays of the η_c was found to be ~10%.

Table I. Preliminary CLEO results for radiative width of η_c

Decay mode [a]	events or 95% C.L.	$\Gamma_{\gamma\gamma} \cdot B$ (keV)	$\Gamma_{\gamma\gamma}(\eta_c)$ (keV)
$K_s K^+ \pi^-$	$10.4^{+4.1}_{-3.3}$	$0.17^{+0.07}_{-0.05}$	$9.4^{+3.7}_{-3.0} \pm 2.7$
$K^{*0} K^- \pi^+$	8.6	$0.17^{+0.09}_{-0.08}$	$8.5^{+4.6}_{-4.2} \pm 3.9$
$\rho^0 \rho^0$	<4.6	<0.07	<52
$K^{*0} \overline{K^{*0}}$	<4.7	<0.18	<40
$\phi\phi$	<3.0	<0.16	<47

[a] Charge-conjugate decay modes are implied.

Applying this correction to the observed number of events, we obtain the radiative width times branching ratio displayed in the third collumn of Table I. Using the 1988 Particle Data Group values[3] for η_c branching ratios yields radiative widths shown in the last collumn of this Table. The systematic errors account for uncertaintities in background shape used in fitting the distributions, as well as uncertaintities in efficiency for triggering and particle identification. In addition, the uncertaintity in the branching ratio for the given final state has been added in quadrature to the systematic error. As a preliminary result we quote the measurement from the $K_s K^+ \pi^-$ decay mode, $\Gamma_{\gamma\gamma}(\eta_c) = 9.4^{+3.7}_{-3.0} \pm 2.7$ keV. Measurements from other decay modes are consistent with this result.

A summary of the present status of the radiative width of η_c is presented in Fig. 4. Addition of the CLEO result reduces somewhat the uncertainty in the world average, but not enough to test perturbative corrections to quark models. Clearly more data are needed.

The potential for accumulating much more data is at hand. The high-luminosity B factories discussed at this symposium[12] offer unique opportunities for studying two-photon production of charmonium. The next generation of detectors being built for these machines will have enhanced capabilities making them well suited for two-photon studies. As an example, the CLEO II detector which will soon begin accumulating data, is vastly improved compared to the CLEO I detector. In this new detector, time-of flight counters are located inside the solenoid and cover 96% of the solid angle, allowing much more efficient triggering on two-photon events. In addition, photons will be observed in a high-resolution cesium-iodide calorimeter, which will allow many new decay modes to be studied.

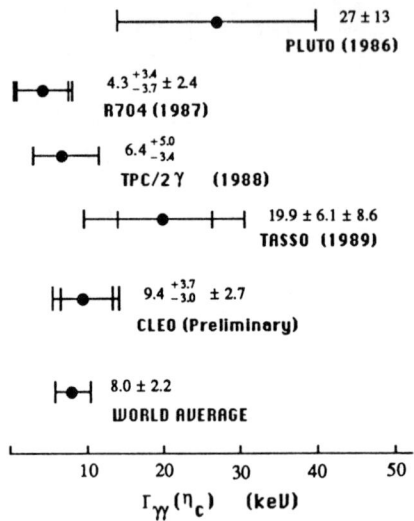

Fig. 4. Measurements of $\Gamma_{\gamma\gamma}(\eta_c)$. The inner error bars represent statistical errors, and outer error bars include systematic errors added in quadrature.

ACKNOWLEDGEMENTS

I thank my colleagues on the CLEO experiment for their helpful suggestions in the preparation of this report. I thank John R. Smith for several informative discussions regarding Monte Carlo generation of two-photon processes.

REFERENCES

1. See, for example, W. Toki, in Charm Physics, Proceedings of the CCAST Symposium/Workshop, Beijing, ed. Min-han Ye and Tao Huang (1987), p 89.
2. For a recent review see W. Kwong, J.L. Rosner, and C. Quigg, Ann. Rev. Nucl. Sci. 37, 325 (1987).
3. G.P. Yost et al. (Particle Data Group), Review of Particle Properties, Phys. Lett. B204 (1988).
4. Ch. Berger, et al. (PLUTO), Phys. Lett. 167B, 120 (1986).
5. C. Baglin, et al. (R704), Phys. Lett. 187B, 191 (1987).
6. H. Aihara, et al. (TPC/2γ), Phys. Rev. Lett. 60, 2355 (1988).
7. W. Braunschweig et al. (TASSO), Z. Phys. C41, 533 (1989).
8. For general reviews of the two-photon process see H. Kolonoski, Two-Photon Physics at e^+e^- Storage Rings, Springer Tracts in Modern Physics, Vol 105. Berlin/Heidelberg/New York (1984); M. Poppe, Int. J. Mod. Phys. A1, 545 (1986).
9. D. Andrews et al., Nucl. Instrum. Methods 211, 47 (1983); D.G. Cassel et al., Nucl. Instrum. Methods A252, 325 (1986).
10. J. R. Smith, "Two-Photon Production of the f^0 Meson", PhD Thesis, UC Davis (1982), unpublished.
11. J.H. Field, Nucl. Phys. B168, 477 (1980); G. Bonneau, M. Gourdin, and F. Martin, Nucl. Phys. B54, 573 (1973).
12. K. Berkelman, "B Factory Plans at Cornell"; J. Chauveau, "Detector for the B Factory at PSI"; D. Hitlin, "A High Luminosity, Asymmetric BB Factory", in these proceedings.

Inclusive Hadron Production at 10 GeV

R. Waldi
Institut für Experimentelle Kernphysik, University
7500 Karlsruhe, Fed. Rep. Germany

ABSTRACT

Recent results of the ARGUS collaboration on inclusive momentum and angular distributions of charged hadrons produced in direct $\Upsilon(1S)$ decays and nonresonant e^+e^- annihilation at 10 GeV are presented, which allow investigation of quark and gluon fragmentation. The data demonstrate some of the shortcomings of present fragmentation models.

INTRODUCTION

Multihadron final states in e^+e^- annihilation are produced via quark and antiquark fragmentation, those from direct $\Upsilon(1S)$ decays originate from the hadronization of three gluons. In our present understanding these are factorizing two-step processes of parton production and subsequent hadronization; the second step is modelled as a parton shower[1] or a breaking string[2]. Experimental data from both reactions can test these ideas.

Two data samples obtained with the ARGUS detector[3] at the DORIS II storage ring have been used for the analysis on inclusive stable hadrons: One comprising an integrated luminosity of 6.6 $events/pb$ at the $\Upsilon(1S)$ energy and 2.6 $events/pb$ in the nearby continuum has been used to determine inclusive momentum spectra[4], since the drift chamber performance, which plays an important role for charged particle identification, has been studied in great detail there. The second, which is used to investigate charged hadron production at very high momenta[5] corresponds to 220 $events/pb$ in integrated luminosity, and includes data from runs at the $\Upsilon(4S)$ resonance energy. The $\Upsilon(4S)$ decays into two B mesons, which in turn cannot have any decay product beyond an x_p value of 0.53, where $x_p = p/p_{\max}$. Since we examine a momentum region $x_p > 0.74$, we can include these data safely.

INCLUSIVE MOMENTUM SPECTRA

To investigate stable hadron production, multihadron events with at least four charged tracks were used. From these events, charged tracks were selected, which originate at the interaction region, and have $|\cos\theta| < 0.85$.

Charged particles can be identified using information from two detector components: the main drift chamber, providing dE/dx information, and the time-of-flight (TOF) counters, which surround the drift chamber at a radial distance of 95 cm from the interaction point. The dE/dx samples from all hit wires are corrected for space charge effects and averaged, discarding the highest 30% and lowest 10% of the values[3]. The track cuts ensure that at least 10 hits contribute to this truncated mean. The time of flight is corrected for a pulse height dependent time shift. Distributions $f_j(dE/dx|\vec{p})$ and $g_j(TOF|\vec{p})$ have been precisely parametrized for the selected data, using clean samples of electrons (from radiative Bhabha events), pions (from K_s^0 decays) and protons (from Λ decays), in the selected range of momentum p and polar angle θ. The spectra for charged pions, kaons and antiprotons are obtained from a maximum likelihood fit of

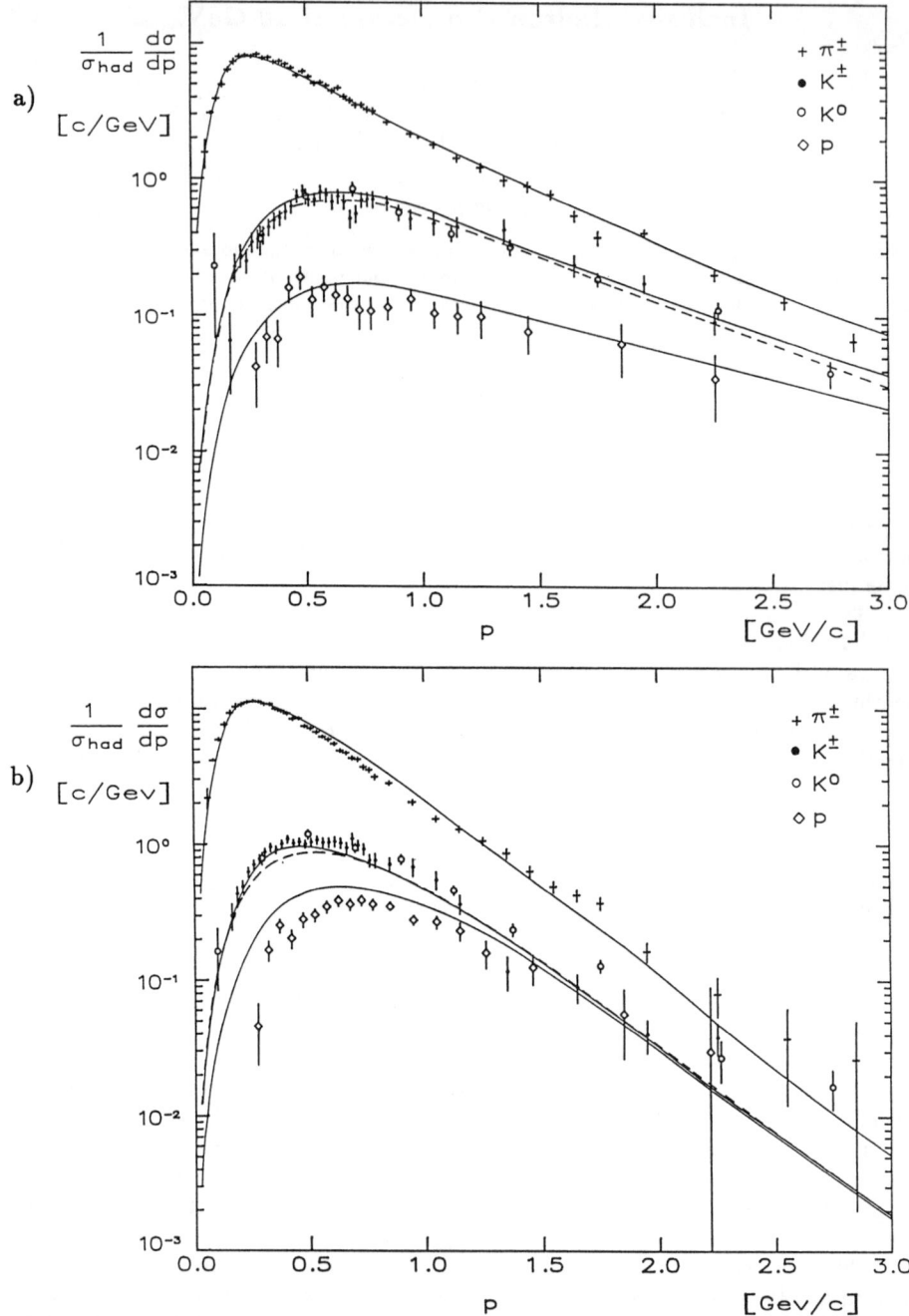

Fig. 1 Inclusive momentum distribution of $\pi^{\pm}$, $K^{\pm}$, $K^0/\overline{K}^0$ and $p/\overline{p}$ in e^+e^- annihilation at $10\,GeV$ (a) and in direct $\Upsilon(1S)$ decays (b). The curves are predictions of the Lund model.

Table 1 Mean multiplicity of stable hadrons in quark and gluon fragmentation (disjoint sets, i.e. contributions from K and Λ decays are not included in the π and p rates)[4,6,7]

hadron	$q\bar{q}$	$\Upsilon(1S)_{\text{dir}}$	ratio $\Upsilon : q\bar{q}$
$\pi^\pm$	$5.69 \pm 0.03 \pm 0.10$	$6.69 \pm 0.03 \pm 0.12$	1.175 ± 0.023
π^0	$3.04 \pm 0.07 \pm 0.31$	$3.73 \pm 0.23 \pm 0.38$	1.23 ± 0.17
$K^\pm$	$0.89 \pm 0.02 \pm 0.02$	$0.91 \pm 0.02 \pm 0.02$	1.02 ± 0.04
$K^0/\overline{K^0}$	$0.91 \pm 0.05 \pm 0.03$	$1.03 \pm 0.04 \pm 0.04$	1.14 ± 0.09
$p/\bar{p}$	$.212 \pm .012 \pm .012$	$.361 \pm .011 \pm .023$	1.70 ± 0.15
$\Lambda/\overline{\Lambda}$	$.092 \pm .003 \pm .008$	$.228 \pm .003 \pm .021$	2.48 ± 0.15

a sum of these distributions to all selected tracks, using as free parameters the numbers of $\pi^\pm$, $K^\pm$, $p+\bar{p}$, $e^\pm$ and $\mu^\pm$ in each momentum bin. The numbers of electrons and muons are further constrained to fit predetermined distributions $h_e(p)$ (from data) and $h_\mu(p)$ (from Monte Carlo).

K_s^0 mesons were selected requiring a secondary vertex with two oppositely charged tracks, separated by at least $2\,cm$ from the interaction point. The charged tracks had to be compatible with the pion hypothesis, with a normalized likelihood[3] exceeding 0.09. Furthermore, the angle ω between the momentum of the K_s^0 and its direction of flight, determined as the vector from the primary vertex to its decay vertex, had to be small, $\cos\omega > 0.9$, and the angle between the momenta of any of the two pion candidates and the K_s^0 had to fulfill $|\cos\theta_{K,\pi}| < 0.98$ to reject converted photons.

In fig. 1, momentum distributions of $\pi^\pm$, $K^\pm$, $K^0/\overline{K^0}$ and $p/\bar{p}$ in e^+e^- annihilation at $10\,GeV$ and direct $\Upsilon(1S)$ decays are presented. We consider $\pi^\pm$, π^0, $K^\pm$, $K^0/\overline{K^0}$, $p/\bar{p}$ and $\Lambda/\overline{\Lambda}$ as a disjoint set of "stable" hadrons; hence we subtract remaining contributions from K_s^0 and Λ decays from our pion and proton spectra. All spectra in the continuum are corrected for contributions from the process $e^+e^- \to \tau^+\tau^-(\gamma)$, which constitutes a 4.8% fraction of the data after the cuts, and radiative corrections have been applied. To obtain particle spectra from direct $\Upsilon(1S)$ decays, the underlying continuum, as well as electromagnetic decays of the $\Upsilon(1S)$ to $q\bar{q}$ have been subtracted.

The average multiplicities are obtained as integrals over the extrapolated momentum distributions. The results are listed in table 1, where also published ARGUS results on other stable hadrons are included.

The curves in fig. 1 are results of the Lund Monte Carlo program[8] (version 6.2). Although there is coarse agreement, the model predictions differ from the data in many details. Especially, the proton enhancement in $\Upsilon(1S) \to ggg$ decay is overdone by Lund, whereas the Λ ratio[7] is underestimated. However, Scheck has shown[9] that this can be improved by adjusting not fragmentation parameters, but Λ_c decay properties. In fact, the origin of the baryon enhancement is not mainly due to special properties of gluon fragmentation, but due to the fact, that about 40% of the continuum data are $c\bar{c}$-jets, with hard fragmentation function and hence reduced phase space in the remaining fragmentation string. To shed more light on gluon fragmentation properties, separate studies of heavy and light quark jets in the continuum will be required.

INCLUSIVE ANGULAR DISTRIBUTIONS

Common to all fragmentation models is the production of jets, where the hadrons follow more or the less the direction of the momentum of one of the primary partons: Their mean transverse momenta relative to the jet axis are—independent of both its own energy and the energy of this parton—of the order of a few hundred MeV. As a consequence, the $\cos\theta$ distribution of fast hadrons is expected to reflect the distribution of the quark antiquark pair, which is predicted from lowest order QED to be

$$\frac{d\sigma(e^+e^- \to q\bar{q})}{d\Omega} = \frac{\alpha^2}{4s} \cdot \beta \cdot [(2-\beta^2) + \beta^2 \cos^2\theta] \tag{1}$$

which is flat at threshold, where the quark velocity $\beta = \sqrt{1-4m^2/s}$ is small, and approaches rapidly the asymptotic distribution $(1+\cos^2\theta)$ as $\beta \to 1$.

Therefore we also looked at $\cos\theta$ distributions of charged hadrons in the continuum. The uncorrected data have been parametrized as

$$\frac{dn(e^+e^- \to h^\pm X)}{d\Omega} \propto [1 + A(x_p)\cos^2\theta] \tag{2}$$

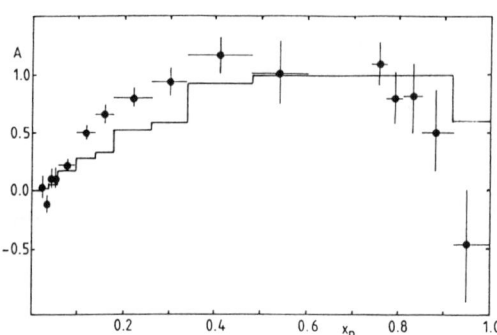

Fig. 2 Parameter $A(x_p)$ of the polar angle distribution for charged hadrons in e^+e^- annihilation events, as obtained from a fit of uncorrected data to (2). The solid histogram is the result of the Lund Monte Carlo program.

In fig. 2 we show $A(x_p)$ together with the Lund Monte Carlo prediction (passed through a detector simulation program). The expected behaviour is indeed observed—in qualitative agreement with the Lund histogram. Quantitatively, however, there are significant differences at low x_p. We investigated two possible sources of this discrepancy: Changing α_s within a reasonable range, 0.17 to 0.25, did not give any significant changes in the inclusive angular distribution. Therefore we exclude an overestimation of gluon radiation as a possible source. Changing the width of the transverse momentum distribution of the primary hadrons with respect to the parton axis does alter fig. 2. However, only a reduction of $\sigma_\perp$ to zero, i.e. producing no $p_\perp$ at all, could give approximate agreement.

Also included in fig. 2 are the results of a second analysis, using charged hadrons with $x_p > 0.74$ and $|\cos\theta| < 0.8$. Additional cuts are applied here[5] in order to reduce the larger amount of background reactions, while the cut on the charged multiplicity is released to $n_{ch} \geq 3$.

Leptons as well as hadrons from tau decays in $e^+e^- \to \tau^+\tau^-(\gamma)$ are subtracted. The 4000 remaining hadrons correspond to a visible cross section of $(18.2 \pm 0.6)\,pb$, which is in good agreement with the Lund model prediction of $(16 \pm 1)\,pb$ for such events. Also the momentum distribution of the fast tracks agrees with the Lund model. This gives us confidence that we have selected just a corner of phase space of inclusive hadron production from otherwise unbiased e^+e^- annihilation data. This is also confirmed by an investigation of the inclusive momentum distribution of charged tracks in the hemisphere opposite to a fast track with $x_p > 0.925$. This is compared in fig. 3 to the corresponding events from the Lund Monte Carlo as well as to "ordinary" multihadron events in the e^+e^- annihilation continuum, which have been selected without requiring a track with $x_p > 0.74$, but applying all other cuts appropriately. The agreement between both data samples is unexpectedly good, whereas Lund is a little bit at variance with our fast track sample.

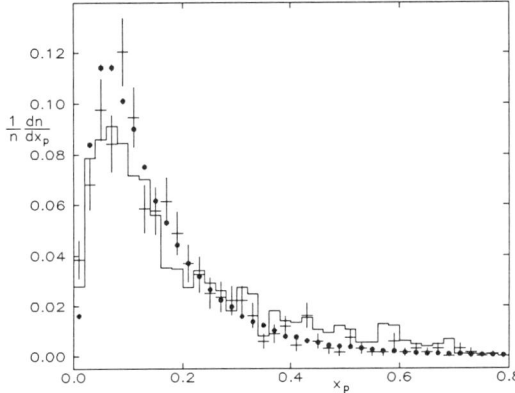

Fig. 3 Distribution of observed charged hadrons in x_p (crosses), which are opposite to a fast track with $x_p > 0.925$, compared to Lund model predictions for those events (solid histogram) and tracks from "ordinary" multihadron events (circles).

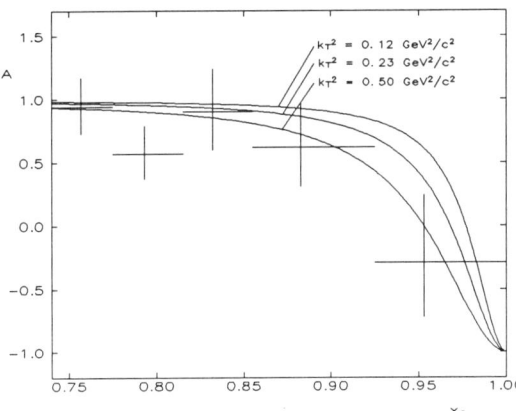

Fig. 4 Results of a fit of $d\sigma/d\Omega \propto 1 + A \cdot \cos^2\theta$ to the acceptance corrected polar angle distribution of charged hadrons, compared to predictions from a higher twist calculation[10].

The result of the fits of (2) to the $\cos\theta$-distribution of the fast tracks is shown in the right part of fig. 2. It deviates from the naive expectation at the highest momentum bin ($x_p > 0.925$). Part of this deviation is due to the fact, that the detector resolution in $|p|$ depends on θ, and therefore the population of tracks in this bin has a larger contribution from lower momentum tracks near $\cos\theta \simeq 0$ than at the ends of the $\cos\theta$ distribution. Since the Lund Monte Carlo events passed a full detector simulation, this effect lowers A also there in the highest bin.

To examine the remaining discrepancy further, we compare in fig. 4 the fit results of acceptance corrected angular distributions, which are expected to give $A = 1$, to a calculation of higher twist terms by Berger[10]. He finds for pions

$$\frac{d\sigma(e^+e^- \to \pi X)}{d\Omega} \propto$$
$$(1-z)^2(1+\cos^2\theta) + \frac{4}{9}\frac{p_\perp^2}{s}\sin^2\theta$$
(3)

where $z = E_\pi/E_{beam}$, and $p_\perp$ is the transverse momentum of the pion with respect to the primary quark. Equation 3 describes a rapid but smooth transition from the $(1+\cos^2\theta)$ behaviour at moderately high z to the exclusive limit $(1-\cos^2\theta)$ for $e^+e^- \to \pi^+\pi^-$. An identical distribution is expected for kaons, but protons, which contribute about 10–15% to the fast hadrons, may show a different behaviour. Obviously, the behaviour of the data is in qualitative agreement with this prediction, although it is only about 2σ distant from $A = 1$ at the highest x_p bin.

CONCLUSIONS

The precision of data on quark and gluon fragmentation is steadily improving, and enough details become visible to observe shortcomings of well-tuned fragmentation models. This demands new efforts in the theoretical description and modelling of hadronization processes. Coherence effects should be included, as non-factorizing amplitudes for parton production and hadronization (higher twist), but there are others not mentioned in this talk like Bose-Einstein correlations. Also, quantum number correlations beyond flavour, like isospin and spin, will certainly influence the details of hadronic final states. Finally, inclusion of more excited states (meson and baryon resonances) should make a fragmentation model more realistic.

On the experimental side, still the best laboratory to examine gluon fragmentation properties are direct $\Upsilon(1S)$ decays. However, these data should be compared to light quark fragmentation, and additional efforts of the experimentalists are required to separate those from the 40% $c\bar{c}$ jets in the e^+e^- annihilation continuum.

REFERENCES

1. R. D. Field, S. Wolfram, Nucl. Phys. **B213**, 65 (1983);
   G. Marchesini, B. R. Webber, Nucl. Phys. **B238**, 1 (1984);
   B. R. Webber, Nucl. Phys. **B238**, 492 (1984).
2. B. Andersson et al., Phys. Rep. **97**, 31 (1983) and references therein.
3. H. Albrecht et al. (ARGUS), Nucl. Instr. Meth. **A275**, 1 (1989).
4. H. Albrecht et al. (ARGUS), DESY 89–014, to appear in Z. Phys. C;
   W. Funk, Diplomarbeit, IHEP–HD/88–3, Heidelberg 1988;
   S. Werner, Diplomarbeit, IHEP–HD/87–1, Heidelberg 1987.
5. B. Schwingenheuer, Diplomarbeit, IHEP–HD/89–3, Heidelberg 1989.
6. H. Albrecht et al. (ARGUS), DESY 89–066.
7. H. Albrecht et al. (ARGUS), Z. Phys. **C39**, 177 (1988).
8. T. Sjöstrand, Comp. Phys. Comm. **27**, 243 (1982) and **39**, 347 (1986).
9. H. Scheck, Phys. Lett. **B224**, 343 (1989).
10. E. L. Berger, Z. Phys. **C4**, 289 (1980).

SEARCH FOR TOP

LATEST RESULTS OF THE SEARCH FOR TOP AT UA1

Martti Pimiä

(UA1 Collaboration, CERN, Geneva, Switzerland)
Helsinki University, Finland

ABSTRACT

UA1 is continuing its search for new heavy quarks in the muon channel, using 1.3 pb^{-1} of new data taken during the CERN p$\bar{\text{p}}$ Collider run in the autumn of 1988. The properties of isolated muons accompanied by jets are consistent with the predictions from the Standard Model but do not show a signal for a new heavy quark. Combining all the UA1 data (2 pb^{-1}), and using all the channels available, a new lower limit on the mass is obtained at 56 GeV/c^2 (95% CL) for the t-quark and 33 GeV/c^2 (95% CL) for a fourth-generation charge 1/3 quark (b'-quark).

1. INTRODUCTION

UA1 has made a systematic study of heavy-flavour production in proton-antiproton collisions using muons. Both the production mechanism and the topology of standard heavy-flavour (b- and c-quarks) events have been studied in detail and have led to a test of QCD [1], the first observation of B^0-$\bar{B}^0$ mixing [2] and the measurement of the b-quark total production cross section [3].

The knowledge of standard heavy-flavour properties allow us to predict their contribution to a new heavy-quark signal background. We present here an update on our searh for new heavy quarks [4], using a sample which includes recent 1988 data (1.3 pb^{-1}). A larger data sample (4 pb^{-1}) is expected from the 1989 SPS collider run.

2. SEARCH FOR THE t - QUARK IN THE 1988 DATA

UA1 is now investigating a t-quark mass range above 40 GeV/c^2, where the channel W → t$\bar{\text{b}}$ dominates over the strong production channel (t$\bar{\text{t}}$).

2.1 The UA1 detector in 1988

The new data allow for the t-quark search in the muon channel only, as the electromagnetic calorimeters were removed to prepare the installation of the new uranium-TMP calorimeter [5]. The calorimetry is still adequate for measuring missing transverse energy and jet topologies that are relevant to the search for new heavy quarks.

The only change in the muon identification is due to the fact that the decay background is slightly increased because of the longer decay path for hadrons. The muons are detected up to |η| < 2.3, and the inclusive muon trigger acceptance is good within |η| < 1.5, which includes most of the muons from the heavy t-quark semileptonic decays.

2.2 Comparison of 1988 data with Monte Carlo for b- and c-quark production

The b- and c-quarks are the known prototypes of heavy quarks, and they constitute one of the main backgrounds to new heavy quark signatures. The first step in the search for new heavy quarks is the study of b- and c-quarks in a control sample where b- and c-production dominates:
- one muon with $12 \text{ GeV/c} < p_t < 15 \text{ GeV/c}$, $|\eta| < 1.5$,
- at least one jet with $E_t > 12 \text{ GeV}$ and $|\eta| < 2.5$ and the jet axis outside a cone of $\Delta R = (\Delta\Phi^2 + \Delta\eta^2)^{1/2} = 1$ around the muon direction.

Monte Carlo (ISAJET) generated events corresponding to 10 pb^{-1} were used to simulate the raw data of the present detector configuration. They were then processed with our standard reconstruction program and compared with the experimental data. The new data are in good agreement with the Monte Carlo. The dominant contributions are $b\bar{b}$- and $c\bar{c}$-production and charged hadron (π,K) decay background.

The isolation properties of muons are related to their production mechanism. We describe the isolation by the variable $I = [(\Sigma E_t/3)^2 + (\Sigma P_t/2)^2]^{1/2}$. The sums extend to calorimeter cells (ΣE_t) and Central Detector tracks (Σp_t) in a cone $\Delta R = 0.7$ around the muon direction. As expected, most of the muons are produced inside jets as shown by the distribution of I, see figure 1.

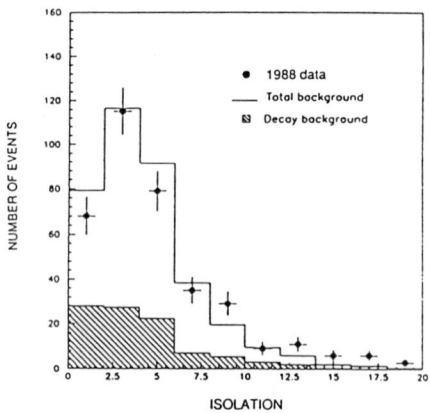

Fig. 1 Muon isolation I for the control sample

2.3 Search for the t-quark in the single muon + jets channel

This search is made by selecting isolated muons accompanied by at least two jets. The best sensitivity to a possible t-quark signal is obtained by requiring an isolated, large-p_t muon with $I < 2$, $p_t > 12$ GeV/c, and at least two jets with $E_t(\text{jet}_1) > 15$ GeV and $E_t(\text{jet}_2) > 7$ GeV. The cut on the transverse mass of the muon-neutrino system to reject W background is $m_t < 60$ GeV/c^2. With these cuts, we observe 19 events.

An improved rejection against $b\bar{b}$ and $c\bar{c}$ background is obtained by introducing two additional cuts:
(1) cut on the angle θ^* between the axis of jet$_2$ and the incoming antiproton in the centre of mass of the system (μ, jet$_1$, jet$_2$, ν_t): $|\cos\theta^*| < 0.8$,
(2) cut on the azimuthal angle $\Delta\Phi$ between the muon and jet$_1$: $|\Delta\Phi| < 150°$.

These cuts are used because in the $b\bar{b}$ and $c\bar{c}$ events the second jet generally comes from initial state

m_t (GeV/c)^2	40	50	60	70
Isolated μ + $\geq$ 2 jets	8.9	6.5	3.9	1.3
Additional cuts: 1) $|\cos\theta^*_{j2}| < 0.8$ 2) $\Delta\phi_{\mu,j_1} < 150°$	5.5	4.0	2.5	0.8

Table 1 Number of t-quark events expected for various t-quark masses

gluon bremsstrahlung and is produced with a large value of |cosθ*|, and in $b\bar{b}$ and $c\bar{c}$ events the muon and jet$_1$ tend to be coplanar with the beam line. After cuts 1 and 2 we observe 5 events.

Table 1 shows the number of t-quark events expected from the above selection in our 1988 data, and for various t-quark masses. Table 2 shows the details of the comparison between data and predictions for all processes except for the t-quark.

	K/π decays	W/Z	Drell-Yan J/ψ, Υ	$b\bar{b}$ $c\bar{c}$	Total MC	Data
μ + ≥ 2 jets	6.9	2.3	2.2	10.1	21.5	19
μ + ≥ 2 jets + cuts 1 and 2	3.2	1.1	0.1	1.2	5.6	5

Table 2 Comparison between 1988 data and simulation of standard processes

Figure 2 shows that the isolation distribution of muons for the 1988 data agrees well with the prediction from background processes. To reduce the systematic uncertainty from $b\bar{b}$ and $c\bar{c}$ predictions, the background distribution is normalized to the data in the control region I > 2. With this procedure, we estimate a background of 20.5 ± 2.3 (syst) events in the signal region I < 2, where we observed 19 events.

As in our previous work [4], we can obtain an upper limit on the t-quark cross section that is consistent with the number of events observed in the bin I < 2. Using the predicted cross section as a function of the t-quark mass m_t, a new limit on m_t can therefore be derived for the 1988 data alone. The result is m_t > 44 GeV/c^2 (95% CL). With the additional cuts on |cosθ*| and ΔΦ, although the signal-to-background ratio is improved, we obtain a similar limit: m_t > 42 GeV/c^2 (95% CL). In deriving these limits we combined in quadrature the systematic errors listed in Table 3.

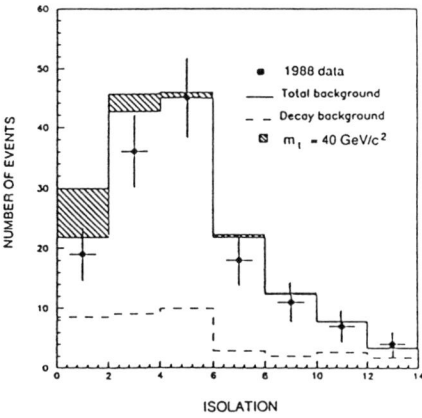

Fig. 2 Muon isolation I for the signal sample

Systematic error source	$t\bar{t}$ (%)	$t\bar{b}$ (%)
Theory	30	10
Efficiency of jet cuts	12	12
Integrated luminosity	15	-
Number of W → μν events observed	-	14
Efficiency of muon cuts	10	10

Table 3 Systematic errors on the number of expected t-quark events.

2.4 Search for the t-quark in the muon-pair channel

Since the W channel is dominant for the t-quark between 40 GeV/c^2 and 80 GeV/c^2, it is natural to search for events where the produced t-quark and b-quark decay semileptonically into a muon. The muon coming from the t-quark is expected to have a large transverse momentum and to be isolated, whilst the muon from the b-decay is not isolated. The two muons have the same electric charge. Here we use all the data collected by UA1, corresponding to 1.96 pb^{-1}.

We made the following selection:
- $p_t(\mu_1) > 8$ GeV/c, $|\eta(\mu_1)| < 1.6$, $p_t(\mu_2) > 3$ GeV/c,
- mass(μ_1,μ_2) > 4 GeV/c^2 (suppress J/ψ background and most of Drell-Yan),
- $I(\mu_1) < 6$ (loose isolation), $I(\mu_2) > 2$ (non-isolated muon, Drell-Yan suppression),
- at least one jet with $E_t > 10$ GeV.

With this selection, 43 events are found in the data, whilst we predict 48±4 (syst) from bb and cc, and less than 13 from decay background. The prediction for the t-quark contribution is 5.2, 3.0 and 1.4 events for m_t of 40, 50 and 60 GeV/c^2, respectively. This selection provides no sensitivity to the t-quark production. An increased sensitivity can be obtained by making use of all the event properties that can differentiate between t and non-t events.

We define a likelihood variable $L = \Pi P_{top}(X_i) / P_{bot}(X_i)$, where P_{top} and P_{bot} are the probability density functions for the variable X_i for t-quark and for $b\bar{b}$- and $c\bar{c}$-events, respectively. Π indicates a product for all values of the index i. The variables chosen are

(1) the isolation of the fastest muon in a cone $\Delta R < 0.4$,
(2) the transverse momentum of the fastest muon, $p_t(\mu_1)$, and
(3) the difference in the azimuthal angle between the two muons, $\Delta\Phi(\mu_1,\mu_2)$.

Fig. 3 shows the distribution of the logarithm of the likelihood for all the UA1 data, the prediction for backgrounds, and the prediction from $t\bar{b}$ and $t\bar{t}$ (multiplied by 10) The data do not show a signal for t-quark production.

If we select events with log(L) > 1, we find no event in the data, whilst we predict 2.6±0.3 (syst) events for backgrounds, and 3.2 or 2.1 t-quark events for m_t of 40 or 50 GeV/c^2, respectively. This result can be converted into a lower bound on the t-quark mass at $m_t > 40$ GeV/c^2 (95% CL).

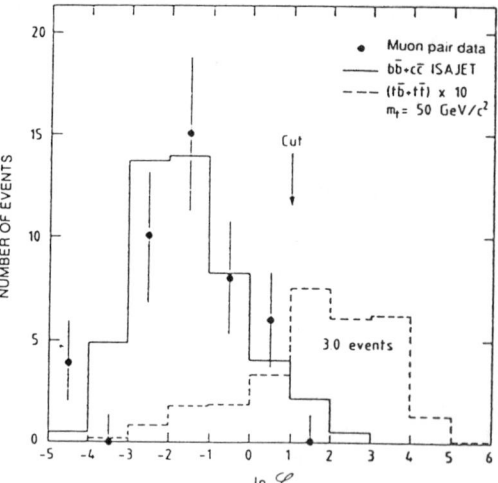

Fig. 3 Distribution of the logarithm of the likelihood function for muon pairs. Comparison of data with a simulation of standard processes and with the contribution from the t-quark multiplied by a factor of 10 ($m_t = 50$ GeV/c^2).

2.5 Search for the t-quark in the muon-electron pair channel

In the same way as t-quark events can produce muon pairs, they can also produce muon-electron pairs. The main difference is that electrons need to be isolated in order to be identified. The only data sample available comes from data taken up to 1985 and corresponds to 0.55 pb^{-1}.

The selection used requires one muon with $p_t > 3$ GeV/c and one electron with $E_t > 8$ GeV in $|\eta| < 1.5$. This selection gives 10 events, and background calculations give 12 events for $b\bar{b}$ and $c\bar{c}$. The number of events predicted from t-quark production is 3.8 or 1.9 events for $m_t = 40$ or 50 GeV/c^2, respectively. The signal-to-background ratio can be improved by requiring a neutrino with $E_t > 10$ GeV, since the neutrino momentum tends to be harder for t-quark events than for $b\bar{b}$ and $c\bar{c}$ events. With this E_t cut we do not find any event in the data, whilst we predict 1.6 events from $b\bar{b}$ and $c\bar{c}$, and 1.7 or 1.2 from t-quark with $m_t = 40$ or 50 GeV/c^2, respectively. From this result we obtain a lower limit for t-quark mass of $m_t > 25$ GeV/c^2 (95% CL).

3. t - QUARK MASS LIMIT

Table 4 summarizes all the UA1 results on the t-quark search for the different channels used. All these channels can be combined to obtain an overall mass limit for the t-quark. Most systematic errors are correlated between channels: for instance, the error in the integrated luminosity, the error in the theoretical t-quark production cross section. We take the most unfavourable case, where we assume that all the systematic errors are correlated. This will give us a conservative limit for m_t.

Channel	Data	Background ± syst. error	Signal $m_t = 40$ GeV/c^2	Signal $m_t = 50$ GeV/c^2	m_t limit (GeV/c^2)	$\int L\, dt$ (pb^{-1})
e + jets ($\leq$ 1985)	26	26.0 ± 2.8	11.0	8.5	41	0.69
μ + jets ($\leq$ 1985)	10	11.4 ± 1.2	7.4	4.7	40	0.55
e + μ (1985)	0	1.6 ± 0.1	1.7	1.2	25	0.55
μ + jets (1988)	19	20.5 ± 2.3	8.9	6.5	44	1.3
μ + jets (1988) plus cuts 1 and 2	5	4.3 ± 0.5	5.5	4.0	42	1.3
$\mu\mu$ (1982–88)	0	2.6 ± 0.3	3.2	2.1	43	1.96

Table 4 Summary of present UA1 results on t-quark mass limits (95% CL)

Figure 4 shows the 90% and 95% CL limits on the t-quark production cross section as a function of m_t. If the t-quark exists, as expected in the Standard Model, it must be heavy compared with other known quarks and leptons. The present UA1 limits are

$$m_t > 56 \text{ GeV/c}^2 \text{ (95\% CL)},$$
$$m_t > 60 \text{ GeV/c}^2 \text{ (90\% CL)}.$$

4. b' - QUARK MASS LIMIT

The search for the fourth-generation charge 1/3 quark (b'), is similar to the search for the t-quark. Using the same data samples as those in the search for the t-quark, we obtain a lower limit on the mass of the b'-quark: $m_{b'} > 33$ GeV/c^2 (95% CL).

5. CONCLUSION

The present UA1 detector has been used for the search of new heavy quarks, t-quark and b'-quark, in the muon channel. In the additional data taken in 1988, we do not find a signal for t-quark, nor for b'-quark production. Combining the new data with the data collected before 1988, the new UA1 limits on m_t and $m_{b'}$ are:

$m_t > 56$ GeV/c^2 (95% CL),
$m_{b'} > 33$ GeV/c^2 (95% CL).

UA1 is continuing to explore a t-quark mass region where $W \to t\bar{b}$ dominates. The 1989 Collider run will add about 4 pb^{-1} of integrated luminosity and extend our sensitivity to t-quark masses approaching the kinematical limit of the W channel.

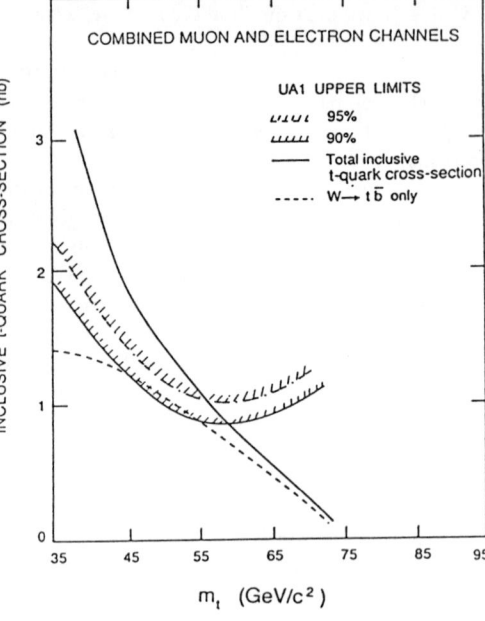

Fig. 4 Combined muon and electron channel limits on the t-quark production cross section as a function of its mass m_t. Comparison with theoretical prediction (full line).

The financial support of Valtion luonnontieteellinen toimikunta, Suomen Akatemia, Finland, is gratefully acknowledged.

REFERENCES

[1] C. Albajar et al. (UA1 Collab.), Phys. Lett. **200B** (1988) 380.
     C. Albajar et al. (UA1 Collab.), Z. Phys. **C37** (1988) 489.
     C. Albajar et al. (UA1 Collab.), Phys. Lett. **186B** (1987) 237.
[2] C. Albajar et al. (UA1 Collab.), Phys. Lett. **186B** (1987) 247.
[3] C. Albajar et al. (UA1 Collab.), Phys. Lett. **213B** (1988) 405.
[4] C. Albajar et al. (UA1 Collab.), Z. Phys. **C37** (1988) 505
[5] UA1 Collaboration, Design report of a uranium-TMP calorimeter for the UA1 experiment with ACOL, CERN UA1 TN/86-112 (1986)
     M. Albrow et al., Nucl. Instrum. Methods **A265** (1988) 303.
     A. Gonidec, C. Rubbia, D. Schinzel and W.F. Schmidt, preprint CERN-EP/88-36 (1988), submitted to Nucl. Instrum. Methods.
     UA1 Collaboration, Performance of a UA1 uranium-TMP calorimeter module, CERN/SPSC 89-23 (1989).

SEARCH FOR NEW HEAVY QUARKS AT TRISTAN

Hiroshi Sakamoto
National Laboratory for High Energy Physics, KEK
Oho, Tsukuba, Ibaraki 305, Japan

ABSTRACT

Study to search new heavy flavors has been considerably progressing at the TRISTAN e^+e^- collider as the beam energy has been pushed up. Three general purpose detectors, AMY, TOPAZ and VENUS have accumulated about $25pb^{-1}$ of luminosities at the beam energies from 25 GeV up to 30.4 GeV. R measurements, event shape analyses, isolated lepton and photon analyses were performed and new mass limits for top and b' quarks have been obtained. As for b', not only the charged current decay mode, but also the flavor changing neutral current decay were taken into account. Measured R values above $\sqrt{s} = 56\text{GeV}$ were slightly higher than the standard model prediction and this can be interpreted that the Z^0 mass may be somewhat lighter than as was used.

INTRODUCTION

The TRISTAN e^+e^- collider[1] has been in operation since Nov. 1986 at National Laboratory for High Energy Physics *KEK*. The beam energy of TRISTAN has been pushed up to 30.4 GeV after superconducting RF cavities had been installed. There are 3 large detectors, AMY[2], TOPAZ[3] and VENUS[4] which accumulated their integrated luminosities of about $25pb^{-1}$ and now they are increased at the rate of $300nb^{-1}$ per day.

Search for new heavy flavors has been eagerly continued at these detectors. A top quark, which is thought to exist but not found yet, and a b' quark, which is the fourth generation $-1/3e$ charged quark, are the subjects of the search.

TOTAL HADRONIC CROSS SECTION

If a new flavor is open at this energy region, the total cross section of hadron productions is significantly increased according to the charge of the new flavor. Usually an R value is employed to measure the increase. The R value is defined as the ratio of the total hadronic cross section to that of the lowest order QED cross section of the $e^+e^- \to \mu^+\mu^-$ process, and is obtained experimentally as follows;

$$R = \frac{N_{obs} - N_{bg}}{\mathcal{L}\epsilon(1+\delta)\sigma_0} \tag{1}$$

© 1989 American Institute of Physics

where N_{obs} and N_{bg} are the numbers of the observed hadronic events and the expected background events respectively. The value $\mathcal{L}$ is the integrated luminosity, ϵ is the detection efficiency and δ is the electroweak radiative correction factor. The lowest order QED cross section is noted as σ_0.

The selection criteria for hadronic events are summarized as follows.

- Number of charged tracks ≥ 5.

- Visible energy (E_{vis}) $\geq$ beam energy (E_{beam}).

- Sum of longitudinal momenta < 0.4 of E_{vis} (AMY, TOPAZ) or E_{beam} (VENUS).

- Enough energy deposit in the calorimeters (AMY, VENUS) or enough invariant mass in at least one hemisphere (TOPAZ).

As for the radiative correction, the contribution of Z^0 bosons is no longer negligible at these energies so that the full calculation up to the order α^3 is employed [5]. The correction factor and the detection efficiency were obtained using Monte Carlo simulations.

Systematic errors were estimated for luminosity measurements, radiative corrections, efficiency evaluations, event selection stabilities and background estimations and were finally found to be 4.2% (AMY), 5.4% (TOPAZ) and 4.2% (VENUS).

The combined R values of 3 groups are plotted in Fig. 1. In the figure, theoretical predictions of 5 flavors, with fully open top and b' quarks are shown. As seen from this figure, the existence of the top at this energy region can be rejected easily. As for the b' quark, at this level of statistics, we cannot draw a definit conclusion from R.

SEARCH FOR A TOP QUARK

In the next step, event shape analyses were performed on the hadronic data. Just above a new flavor threshold, a quark pair is created almost at rest so the event shape is expected to become spherical which is in contrast with 2 or 3 jet topologies of lighter quark pair productions. There introduced were several kinds of event shape variables, i.e. Q values and thrust (T), sphericity (S), aplanarity (A), acoplanarity (A_{cop}) and so. The Q values (Q_1, Q_2, Q_3) are the eigenvalues of the sphericity tensor $M_{\alpha\beta}$ which is written as

$$M_{\alpha\beta} = \frac{\sum_i p_{i\alpha} p_{i\beta}}{\sum_i |\vec{p}_i|^2} \qquad (\alpha, \beta = x, y, z) \qquad (2)$$

and other values are defined as

$$S = \frac{3}{2}(Q_1 + Q_2) \qquad (3)$$

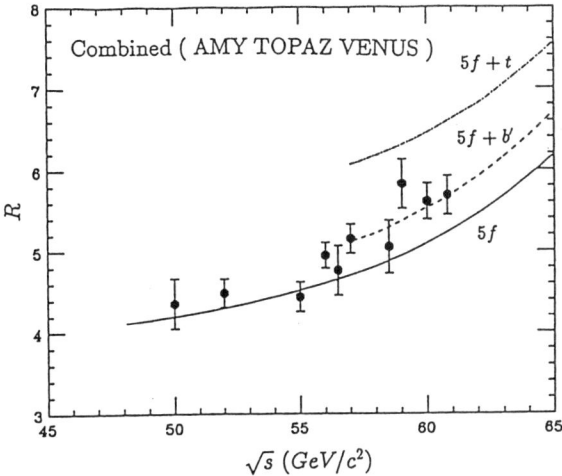

Fig. 1: Combined R values. The data points are obtained by combining the results from AMY, TOPAZ and VENUS. Theoretical predictions for 5 flavors only (solid), with a fully open b' quark (dashed) and with a top quark (dotdashed) are also drawn.

$$A = \frac{3}{2}Q_1 \qquad (4)$$

$$T = \text{Max}\frac{\sum_i |\vec{p_i} \cdot \vec{n}|}{\sum_i |\vec{p_i}|} \qquad (5)$$

$$A_{cop} = 4\text{Min}(\frac{\sum_i |\vec{p_i} \cdot \vec{n}|}{\sum_i |\vec{p_i}|})^2 \qquad (6)$$

Experimental data of these variables were compared to Monte Carlo predictions with/without a new flavor. The thrust and aplanarity distributions obtained by the TOPAZ group are given in Fig. 2 where the experimental result favours the prediction of 5 flavors only. From this result, together with an isolated muon analysis mentioned later, they placed a new lower limit of the top mass at 29.7 GeV/c^2. Similar analyses were also performed at AMY and VENUS and the VENUS group gave the limit of 29.0 GeV/c^2 from the Q plot analysis.

SEARCH FOR A b' QUARK

From the analyses above, the existence of top quarks at this energy region is rejected. How about b' quarks? As the mass of a top quark might be heavier than the mass of b', we have to consider two cases of a b' decay. One is a charged current (CC) decay to c or u quarks. In this case the decay takes place by changing the generation by two. From an analogy of the suppressed

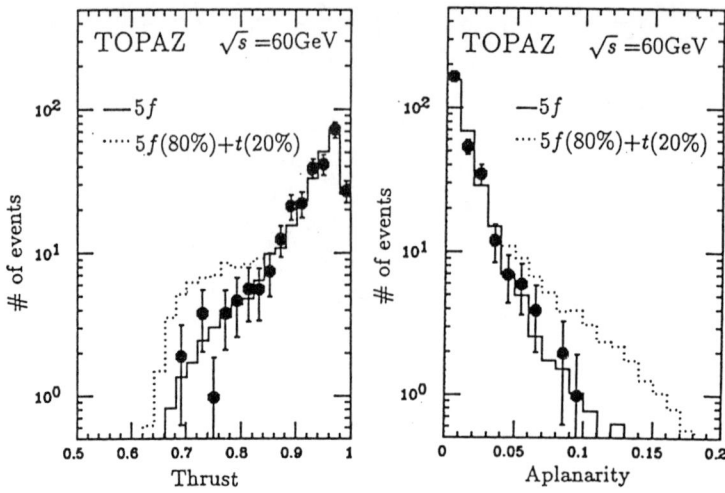

Fig. 2: Thrust and aplanarity distributions obtained by the TOPAZ group. The experimental data are plotted together with the Monte Carlo prediction by 5 flavors (solid line) and 5 flavors + top quark (dotted line).

non-charmed b decay, the charged current decay of b' to c or u may be also suppressed. So we have to take account of the other decay mode, a flavor changing neutral current (FCNC) decay including a loop diagram emitting a photon or a gluon at the intermediate state[6]. As the signals from these two decay modes are somewhat different, the appropriate analyses for each decay modes must be introduced.

I. Charged Current Decay

In the CC decay, a b' quark emits W to turn into c or u. The W boson then produces a $l\nu$ pair or a pair of up-type and down-type quarks. So the signal from the CC decay is, an isolated, energetic lepton, or a spherical multi-hadronic event shape.

In the isolated lepton analyses, in addition to the standard hadronic event selection mentioned above, the lepton identification and its isolation angle selection are introduced. In the VENUS case, a muon is identified by its track in the muon chambers and an electron is identified by an E/p method using a cluster in the electromagnetic calorimeters and the corresponding track in the tracker. The isolation angle is defined as the maximum opening angle of a cone around the lepton in which the energy flow except for the lepton itself is less than 1 GeV/c. As the isolated lepton selection, the threshold lepton momentum is set at 4GeV/c and the threshold isolation angle at 30°. Results of the isolated μ and e analyses from VENUS are plotted in Fig. 3 and Fig. 4 respectively. The experimental data of below and above 60 GeV are plotted in

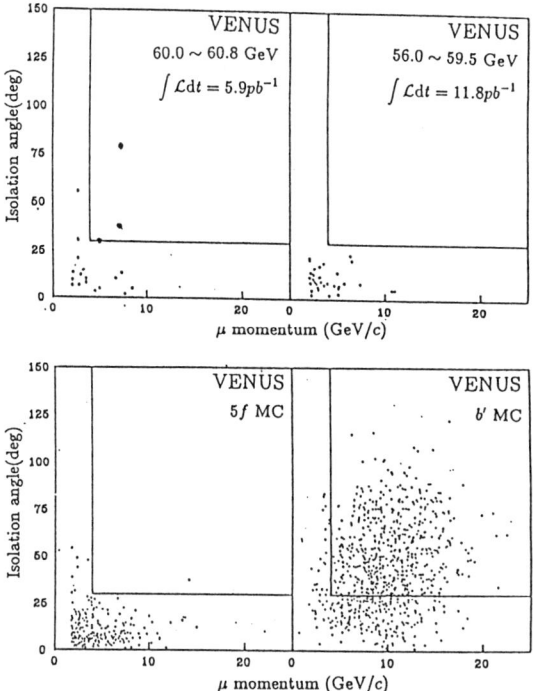

Fig. 3: Scatter plot of μ momenta and the isolation angles obtained at VENUS. Experimental data (upper) above (left) and below (right) 60 GeV and Monte Carlo prediction (lower) by 5 flavors (left) and b' (right) production are shown.

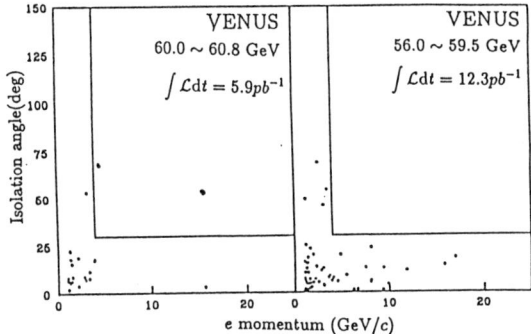

Fig. 4: Scatter plot between e momenta and the isolation angles obtained at VENUS. Experimental data above (left) and below (right) 60 GeV are shown.

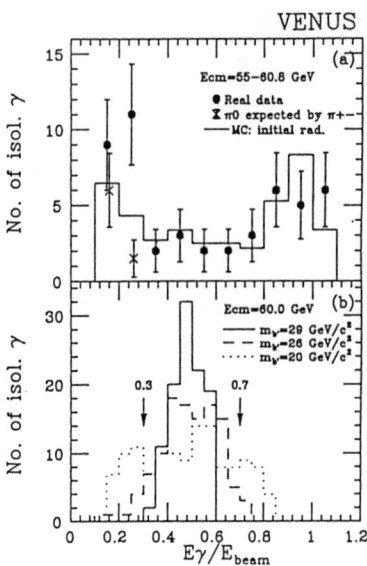

Fig. 5: Energy spectra of isolated photons in multihadronic events measured by VENUS. Experimental data (a) and Monte Carlo expectation from b' decay (b) are shown. In Fig (a), prediction by the initial radiation is drawn in a solid line and neutral pion background expected from charged pion data is plotted together.

separate frames in both Figs. In both cases, several isolated lepton events were observed above or equal 60 GeV but no events survived below 60 GeV. An event shape analysis for hadronic events using the thrust and the acoplanarity were also performed at VENUS and after cuts of thrust < 0.75 and acoplanarity > 0.15, 8 events were survived at the luminosity of $5.9pb^{-1}$ at 60.0 and 60.8 GeV while only 7 events were left at $12.3pb^{-1}$ below 60 GeV. From these VENUS results, b' around $28 \text{GeV}/c^2$ is not excluded at this statistics.

Similar analysis on the isolated μ events was also performed at TOPAZ and they obtained the b' mass limit of $28.5 \text{ GeV}/c^2$.

II. Neutral Current Decay

In the FCNC decay, a b' quark decays through a loop and a photon or a gluon is emitted from the intermediate state. So the signal from the FCNC decay is, an isolated energetic photon or a spherical multi-hadronic event shape. In the multi-hadronic case, as the initial decay takes a form of a two body decay kinematics, the event shape is thought to become planar around the threshold energy. An isolated photon analysis was performed as follows. Identification of γ was done by finding a cluster in electromagnetic calorimeters which had no

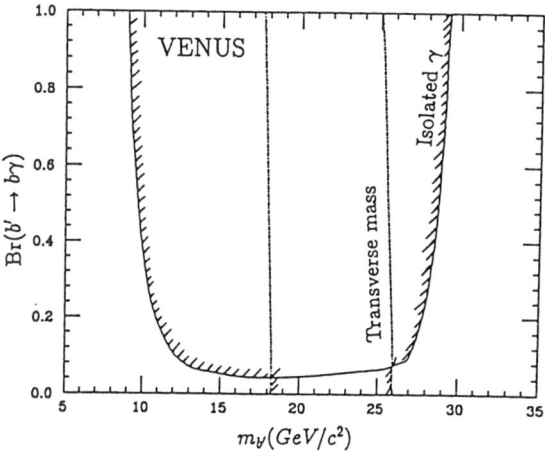

Fig. 6: Excluded region in the domain of b' mass and the branching ratio $Br(b' \to b\gamma)$ obtained by VENUS. The region drawn by a solid line is obtained by the isolated photon analysis and the band drawn by two dotdashed lines is from the transverse mass analysis.

cut	MC 100% 5f	MC 90% 5f 10% b'(CC)	MC 90% 5f 10% b'(FCNC)	Data $\sqrt{s}$ 56-60.8
$T < 0.82$	211	190+148=338	190+128=318	205
$A_{cop} > 0.14$	157	141+110=251	141+110=251	162
$LBW > 0.17$	109	98+96=194	98+133=231	111

Table 1: Result of the thrust (T), acoplanarily (A_{cop}) and lesser bi-width (LBW) cuts on the experimental data and Monte Carlo predictions obtained by AMY.

associated tracks in the tracker. As the isolation condition, events which had other clusters (≥ 1GeV) or any good tracks in a cone of 30° were rejected. The energy spectrum of isolated γ is shown in Fig. 5 where the experimental data are plotted together with an expected neutral pion contamination which is derived from the experimental energy spectra of charged pions. In order to compare the Monte Carlo expectation from b', a cut was set at the range from 0.3 to 0.7 E_γ/E_{beam} and no excess was found at this region.

From the kinematic characteristics, multihadronic events are thought to exhibit planar shapes near the mass threshold. So some different analysis was performed for the gluon emission of the FCNC b' decay. A transverse mass variable, defined as follows, is introduced.

$$X = \frac{\sum P_{t_1} \times \sum P_{t_2}}{E_{vis}^2} \quad (7)$$

First, the thrust axis is used to divide an event into two hemispheres and, next, an axis is searched to minumize the sum of the transverse momenta toward the axis in each hemisphere. This variable is expected to be more sensitive to such an event topology. By placing a cut that $X > 0.08$, and by comparing the yield with the Monte Carlo expectation, a limit was obtained in the plane between the b' mass and the branching ratio $Br(b' \to b\gamma)$. In Fig. 6, excluded region obtained from the isolated photon analysis and the transverse mass analysis are plotted.

Similar, but somewhat different analyses were done at AMY and TOPAZ. In the AMY analysis, a variable 'Lesser Bi-Width (LBW)' is introduced which is defined as the smaller transverse momentum in the two hemispheres introduced above, and normalized by the visible energy.

$$LBW = (\frac{\sum p_t}{E_{vis}})_{smallerside} \quad (8)$$

By this LBW analysis, they found a fairly well agreement between the experimental data and the 5 flavor Monte Carlo prediction. The result is shown in Table 1.

A back-to-back 4 jet analysis was performed at TOPAZ and including the result of the isolated γ analysis, they also obtained the excluded region in the space of the b' mass and the FCNC branching ratio.

SUMMARY

Before concluding this paper, the next question should be commented on. If the increase of R is not due to the b' production, how is this to be explained? One possible answer is that the Z^0 mass may be somewhat lower than that was used. By fitting the R distribution by putting the Z^0 mass as a free parameter, the 3 groups obtained the following values.

$$\text{AMY} \quad 88.1^{+1.7}_{-1.5} \text{GeV}/c^2$$
$$\text{TOPAZ} \quad 89.8 \pm 2.2 \text{GeV}/c^2$$
$$\text{VENUS} \quad 88.9^{+1.5+2.5+0.4}_{-1.3-2.2-0.4} \text{GeV}/c^2$$

Thus, all 3 groups claim that the Z^0 mass seems lighter. Anyway, this question will be settled soon.

Not so large but apparent excess from the 5-flavor standard model prediction is observed in R values above 56 GeV measured by the TRISTAN 3 groups. The R result does not favour the t production at these energies and by including the event shape analyses, a new lower limit of the t mass was obtained. As for the b' production, due to rather poor statistics, we can say nothing about that from the R result. So we had to make a detailed analysis on the b' production. Two decay modes, i.e. the charged current decay and the flavor changing neutral decay, were studied using several kinds of analyses. From the CC decay analysis made by VENUS, the existence of b' around 28 GeV is not excluded at this statistics, and much more luminosity is necessary to make a definit conclusion. From the FCNC analysis, an excluded region in the plane of the b' mass and the branching ratio $Br(b' \to b\gamma)$ was obtained.

REFERENCES

[1] F. Takasaki, Nucl. Phys. B(Proc. Suppl.) **3**(1988)39.

[2] H. Sagawa et al., Phys. Rev. Lett. **60**(1988)93.

[3] I. Adachi et al., Phys. Rev. Lett. **60**(1988)97.

[4] H. Yoshida et al., Phys. Lett. **B198**(1987)570.

[5] J. Fujimoto and Y. Shimizu, Mod. Phys. Lett. **A3** (1988)581.

[6] W. Hou and R. Stuart, Phys. Rev. Lett. **62**(1989)617.

STATUS OF THE TOP QUARK SEARCH AT CDF

G. William Foster for the CDF Collaboration:

F. Abe,[8] D. Amidei,[4] G. Apollinari,[11] M. Atac,[4] P. Auchincloss,[14] A. R. Baden,[6]
A. Bamberger,[19] A. Barbaro-Galtieri,[9] V. E. Barnes,[12] F. Bedeschi,[11]
S. Behrends,[12] S. Belforte,[11] G. Bellettini,[11] J. Bellinger,[18] J. Bensinger,[2]
A. Beretvas,[4] J. P. Berge,[4] S. Bertolucci,[5] S. Bhadra,[7] M. Binkley,[4] R. Blair,[1]
C. Blocker,[2] A. W. Booth,[4] G. Brandenburg,[6] D. Brown,[6] E. Buckley,[14]
A. Byon,[12] K. L. Byrum,[18] C. Campagnari,[3] M. Campbell,[3] R. Carey,[6]
W. Carithers,[9] D. Carlsmith,[18] J. T. Carroll,[4] R. Cashmore,[19] F. Cervelli,[11]
K. Chadwick,[4] G. Chiarelli,[5] W. Chinowsky,[9] S. Cihangir,[4] A. G. Clark,[4]
D. Connor,[10] M. Contreras,[2] J. Cooper,[4] M. Cordelli,[5] D. Crane,[4] M. Curatolo,[5]
C. Day,[4] S. Dell'Agnello,[11] M. Dell'Orso,[11] L. DeMortier,[2] P. F. Derwent,[3]
T. Devlin,[14] D. DiBitonto,[15] R. B. Drucker,[9] J. E. Elias,[4] R. Ely,[9] S. Errede,[7]
B. Esposito,[5] B. Flaugher,[14] G. W. Foster,[4] M. Franklin,[6] J. Freeman,[4]
H. Frisch,[3] Y. Fukui,[8] Y. Funayama,[16] A. F. Garfinkel,[12] A. Gauthier,[7] S. Geer,[6]
P. Giannetti,[11] N. Giokaris,[13] P. Giromini,[5] L. Gladney,[10] M. Gold,[9]
K. Goulianos,[13] H. Grassmann,[11] C. Grosso-Pilcher,[3] C. Haber,[9] S. R. Hahn,[4]
R. Handler,[18] K. Hara,[16] R. M. Harris,[9] J. Hauser,[3] T. Hessing,[15] R. Hollebeek,[10]
L. Holloway,[7] P. Hu,[14] B. Hubbard,[9] B. T. Huffman,[12] R. Hughes,[10] P. Hurst,[7]
J. Huth,[4] M. Incagli,[11] T. Ino,[16] H. Iso,[16] H. Jensen,[4] C. P. Jessop,[6]
R. P. Johnson,[4] U. Joshi,[4] R. W. Kadel,[4] T. Kamon,[15] S. Kanda,[16]
D. A. Kardelis,[7] I. Karliner,[7] E. Kearns,[6] R. Kephart,[4] P. Kesten,[2] R. M. Keup,[7]
H. Keutelian,[7] S. Kim,[16] L. Kirsch,[2] K. Kondo,[16] S. E. Kuhlmann,[1] E. Kuns,[14]
A. T. Laasanen,[12] J. I. Lamoureux,[18] W. Li,[1] T. M. Liss,[7] N. Lockyer,[10]
C. B. Luchini,[7] P. Maas,[4] M. Mangano[11] J. P. Marriner,[4] R. Markeloff,[18]
L. A. Markosky,[18] R. Mattingly,[2] P. McIntyre,[15] A. Menzione,[11] T. Meyer,[15]
S. Mikamo,[8] M. Miller,[3] T. Mimashi,[16] S. Miscetti,[5] M. Mishina,[8] S. Miyashita,[16]
Y. Morita,[16] S. Moulding,[2] A. Mukherjee,[4] Y. Muraki,[16] L. Nakae,[2] I. Nakano,[16]
C. Nelson,[4] C. Newman-Holmes,[4] J. S. T. Ng,[6] M. Ninomiya,[16] L. Nodulman,[1]
S. Ogawa,[16] R. Paoletti,[11] A. Para,[4] E. Pare,[6] J. Patrick,[4] T. J. Phillips,[6]
R. Plunkett,[4] L. Pondrom,[18] J. Proudfoot,[1] G. Punzi,[11] D. Quarrie,[4] K. Ragan,[10]
G. Redlinger,[3] J. Rhoades,[18] F. Rimondi,[19] L. Ristori,[11] T. Rohaly,[10]
A. Roodman,[3] A. Sansoni,[5] R. D. Sard,[7] A. Savoy-Navarro,[19] V. Scarpine,[7]
P. Schlabach,[7] E. E. Schmidt,[4] M. H. Schub,[12] R. Schwitters,[6] A. Scribano,[11]
S. Segler,[4] Y. Seiya,[16] M. Sekiguchi,[16] P. Sestini,[11] M. Shapiro,[6] M. Sheaff,[18]
M. Shochet,[3] J. Siegrist,[9] P. Sinervo,[10] J. Skarha,[18] K. Sliwa,[17] D. A. Smith,[11]
F. D. Snider,[3] R. St. Denis,[6] A. Stefanini,[11] R. L. Swartz, Jr.,[7] M. Takano,[16]
K. Takikawa,[16] S. Tarem,[2] D. Theriot,[4] M. Timko,[15] P. Tipton,[9] S. Tkaczyk,[4]
A. Tollestrup,[4] G. Tonelli,[11] J. Tonnison,[12] W. Trischuk,[6] Y. Tsay,[3]
F. Ukegawa,[16] D. Underwood,[1] R. Vidal,[4] R. G. Wagner,[1] R. L. Wagner,[4]

© 1989 American Institute of Physics

J. Walsh,[10] T. Watts,[14] R. Webb,[15] C. Wendt,[18] W. C. Wester. III,[9]
T. Westhusing,[11] S. White,[13] A. Wicklund,[1] H. H. Williams,[10] B. Winer,[9]
A. Yagil,[4] A. Yamashita,[16] K. Yasuoka,[16] G. P. Yeh,[4] J. Yoh,[4] M. Yokoyama,[16]
J. C. Yun,[4] F. Zetti[11]

[1] *Argonne National Laboratory, Argonne, Illinois 60439*
[2] *Brandeis University, Waltham, Massachusetts 02254*
[3] *University of Chicago, Chicago, Illinois 60637*
[4] *Fermi National Accelerator Laboratory, Batavia, Illinois 60510*
[5] *Laboratori Nazionali di Frascati, Istituto Nazionale di Fisica Nucleare, Frascati, Italy*
[6] *Harvard University, Cambridge, Massachusetts 02138*
[7] *University of Illinois, Urbana, Illinois 61801*
[8] *National Laboratory for High Energy Physics (KEK), Tsukuba-gun, Ibaraki-ken 305, Japan*
[9] *Lawrence Berkeley Laboratory, Berkeley, California 94720*
[10] *University of Pennsylvania, Philadelphia, Pennsylvania 19104*
[11] *Istituto Nazionale di Fisica Nucleare, University and Scuola Normale Superiore of Pisa, I-56100 Pisa, Italy*
[12] *Purdue University, West Lafayette, Indiana 47907*
[13] *Rockefeller University, New York, New York 10021*
[14] *Rutgers University, Piscataway, New Jersey 08854*
[15] *Texas A&M University, College Station, Texas 77843*
[16] *University of Tsukuba, Ibaraki 305, Japan*
[17] *Tufts University, Medford, Massachusetts 02155*
[18] *University of Wisconsin, Madison, Wisconsin 53706*
[19] *Visitor*

ABSTRACT

A search for new heavy quarks in proton-antiproton collisions at the Fermilab Tevatron Collider ($\sqrt{s} = 1.8$ TeV) is described. Primary emphasis is given to the search for final states in which the charged-current decays of a $t\bar{t}$ pair produce both an electron and a muon. The data taken during the first portion (2.0 pb^{-1}) of the 1988-1989 run is analysed. The top quark (as well as a fourth generation b' quark) is excluded at the 95% C.L. in the mass range 30 to 60 GeV/c^2. Other prospects for heavy quark physics at CDF are discussed.

INTRODUCTION

The strategies for a Top quark search at the Tevatron differ substantially from those at the $Sp\bar{p}S$ collider. This is because direct production of $t\bar{t}$ pairs dominates the production of t-quarks at $\sqrt{s} = 1.8$ TeV, whereas the principal production mechanism at $\sqrt{s} = 630$ GeV is the process $W \to t\bar{b}$. The $W \to t\bar{b}$ production mechanism has the advantage of a known cross section, but has very little sensitivity for Top quark mass $M_t > 60$ GeV/c^2[1]. Direct $t\bar{t}$ pair production at the Tevatron has the advantage of a much higher rate (especially at high values of M_t), and a variety of detectable final states.

The most copious decay channels open to the $t\bar{t}$ pair are hadronic. These produce a multi-jet final state which is difficult to observe in the presence of a QCD background which is orders of magnitude larger. An improvement in the signal-to-noise can be made by requiring that at least one of the t-quarks decay semileptonically ($t \to bl\nu_l$), which proceeds via either a real or virtual W according to whether M_t is above or below the W threshold. With the requirement of a single high P_T lepton and two or more detected jets, the dominant background is from W+jet(s) production and is still several times larger for $M_t > 50 GeV/c^2$. This background can be eliminated only by using kinematic cuts which require a fairly detailed knowledge of the response of the detector to hadronic jets. Despite this, the (single lepton + jets) signature remains attractive due to its relatively high branching ratio (15%).

A further enhancement of the signal-to-noise is possible by requiring the semileptonic decays of both the t and $\bar{t}$. This effectively eliminates the W + jet backgrounds but retains backgrounds from such sources as (Drell-Yan + jets) and (mismeasured Z)+jets. These backgrounds are elimated if one considers only events in which the leptons are of different families, e.g. muon-electron. The remaining sources of background are the semileptonic decays of lighter b-quarks and the decays $Z^0 \to \tau\tau \to \mu e$. These backgrounds are characterized by their soft lepton P_T distributions and their back-to-back (or collinear) event topology.

The basic $\mu - e$ Top signature is then an opposite-sign muon-electron pair, with a sufficiently high P_T(out of the "back-to-back" axis) to remove background from $b\bar{b}$ or τ pairs. In practice for our current level of statistics, simple cuts on the P_T of each lepton suffice to eliminate all background. Additional requirements such as acollinearity, missing transverse energy, lepton isolation, and jet activity remain powerful (unused) handles to separate backgrounds from any potential signal.

The $\mu - e$ signature has the benefit of being relatively free of theoretical uncertainties. Semileptonic decays of heavy quarks are directly calculable in the Standard Model. In particular, the P_T spectrum of the resulting leptons (on which the efficiency of this analysis chiefly depends) is well known[2]. Since one can eliminate backgrounds without making cuts on the jet properties of the event, no detailed understanding of the efficiencies for identifying low-E_T clusters, the hadronization of the B-jets in the simulation, etc. is required. The key points to be addressed for this analysis involve our knowlege of the efficiencies of the trigger and offline selection requirements for muon and electron candidates.

The branching ratio of $t\bar{t}$ into $\mu - e$ is 2.5% in the standard model. In actuality, the expected yield of μe pairs above a given P_T cut is increased by typically 20% due to "feed-down" from second generation decays involving τ leptons, lighter quarks, etc.

THE CDF DETECTOR

The CDF detector has been described in detail elsewhere [3]. Here we summarize the features relevant to this analysis. A Vertex Time Projection Chamber (VTPC) provides tracking information up to a radius of 28 cm from the interaction point for $|\eta| < 3.25$, where η is the pseudorapidity, $\eta = -\ln(\tan(\theta/2))$, with θ being the polar angle relative to the proton beam direction. At larger radius, an 84 layer drift chamber (CTC) measures charged particle momenta for $|\eta| < 1.5$, in a 1.4 Tesla magnetic field with a precision of $\Delta P_T / P_T^2 = 0.0011(\text{ GeV}/c)^{-1}$ [4]. Outside the tracking chambers, electromagnetic (EM) and hadronic calorimeters are arranged in a fine-grained, projective tower geometry covering most of the 4π solid angle. Strip chambers (wire chambers with cathode strips perpendicular to the wires) are located at a depth of six radiation lengths in the EM calorimeters in the region ($|\eta| < 1.1$), near the peak of an EM shower, allowing an accurate measurement of the particle position. In the central region ($|\eta| < 0.65$), drift chambers for muon detection are installed outside of the hadron calorimeter.

TRIGGERING

The CDF trigger is a multi-level system with an input rate of 50,000 inelastic $\bar{p}p$ collisions per second and an output rate of 2 Hz onto tape. It consists of:

- a "Level 0" trigger which requires a number of in-time hits in "beam-beam" trigger counters located near the beam pipe on both sides of the interaction region. These counters are sensitive to the beam jets produced in any hard-scattering process, and are essentially 100% efficient for events which deposit energy in the central calorimeter.

- a "Level 1" trigger which requires either: (1) any calorimeter element to be above an E_T threshold of typically 3 GeV/c^2, or (2) a prompt trigger signal from the muon chambers.

- a "Level 2" trigger consisting of 1) energy clustering logic which identifies localized depositions of electromagnetic and hadronic energy in the calorimeter, 2) a track processor which identifies high-momentum tracks in the central tracking chamber, and 3) a custom microcoded processor which correlates the calorimeter, tracking, and muon information to produce lists of leptons, jets, etc. and forms a trigger decision based on event classifications. There were more than 30 event classifications in the trigger mix by the end of the run.

- a "Level 3" trigger consisting of a processor farm, using a combination of offline code and special-purpose trigger code to sharpen trigger thresholds and reduce the volume of data written to tape.

Needless to say, there are a large number of things which can go wrong in such a system, and they have. A substantial fraction of the running time has been devoted to verification of the efficiency of various components of the trigger system. As an example of such a trigger efficiency check, we consider the track processor. The low threshold lepton triggers needed for Top searches at CDF rely on a track processor which identifies high P_T tracks in the Central Tracking Chamber before the next beam crossing. Without tracking information in the trigger, the thresholds for electrons (which are set by tape-writing speed and the bandwidth into the processor farm) would have to be raised from 12 GeV to 20 GeV. Similarly, without tracking information the muon thresholds would be raised from 9 GeV to infinity since even the highest muon threshold would require writing 300 Hz of events to tape. Therefore, verification of the track processor efficiency is crucial. One test sample used was a set of 816 $W \to e\nu$ events which were triggered by either the "Missing-Et" or "EM Cluster" triggers which do not rely on track processor information. The track processor efficiency measured from this sample was 99 ± 0.5%, which is consistent with estimates from dead wires and single-wire inefficiencies.

The results of this and many similar trigger efficiency tests give good confidence in our ability to selectively trigger on both muons and electrons.

Triggers for Muon-Electron Events

Each muon-electron event has four opportunities to trigger:
- As a single muon ($P_T(\mu) > 9$ GeV/c)
- As a single electron ($E_T(e) > 12$ GeV/c^2)
- As a low-threshold $\mu - e$ ($P_T(\mu) > 3$ GeV/c; $E_T(e) > 5$ GeV/c^2)
- Missing E_T (Missing $E_T > 20$ GeV/c^2)

The Missing E_T trigger is sensitive to both neutrinos and muons, and is efficient mainly for high M_t.

Trigger overlaps allow us to use $\mu - e$ candidate events themselves to study the threshold behavior of the inclusive electron and muon triggers. We find for this (and other) event samples that both the single muon and single electron triggers are fully efficient 3 GeV above their nominal thresholds. Thus the fiducial range chosen for the μe analysis (M_t limits) will be:

$E_T(e) > 15 GeV/c^2$, $P_T(\mu) > 12 GeV/c$.

In this range BOTH the electron and the muon should have satisfied their "single lepton" triggers so that trigger efficiency is doubly assured. The efficiency of the lower-threshold lepton triggers is still under study, so that any conclusions relating to low P_T μe events ($b\bar{b}$ production, etc.) are subject to this caveat.

MUON IDENTIFICATION

Offline muon identification begins by requiring a high-quality track in the Central Tracking Chamber, with an extrapolated impact parameter less than 0.5

cm in the transverse plane and 5 cm along the beam axis. These cuts discriminate against muons produced by decays-in-flight of π and K mesons.

Two classes of muons are identified offline according to whether they produced track segments in the Central Muon (CMU) Chambers. Both classes are important since the CMU chambers have only 70% coverage inside $|\eta| < 0.65$. The acceptance of the μe channel is nearly doubled by accepting muons inside $|\eta| < 1.2$ which miss the CMU Chambers but appear as high-momentum tracks pointing to a minimum-ionizing signal in the calorimeters. The thickness of the calorimeters is typically 6 λ_{ABS} in the region not covered by the CMU chambers. Events of this type will not trigger as muons but will be detected by the Electron or Missing Et trigger.

CLASS 1: (Central Muon Chamber Muons). These require: 1) a track with $P_T > 5$ GeV/c which extrapolates to within 10 cm of a track segment in the CMU chambers, 2) that the track slope observed in the CMU chambers matches that of the extrapolated Central track within 0.1 radian in the transverse plane, and 3) that the total calorimeter energy in a cone of R = $\sqrt{(\Delta\phi)^2 + (\Delta\eta)^2} < 0.13$ be less than 8 GeV.

CLASS 2: (Calorimeter Muons). These require: 1) a track with $P_T > 10$ GeV/c within $|\eta| < 1.2$, 2) that the total calorimeter energy in a cone of R < 0.4 be less than 13 GeV, and 3) that the extrapolated track does not point into "cracks" in the calorimeters. In CDF, these occur primarily at the boundaries of the calorimeter modules, every 15° in ϕ and at 90° in θ. The overall geometric acceptance for calorimeter muons is is 90% inside $\eta < 1.2$.

The non-geometric efficiency for muon identification is $95 \pm 3\%$. This has been verified using $Z^0 \rightarrow \mu^+\mu^-$ and cosmic rays, and is in agreement with test beam measurements.

ELECTRON IDENTIFICATION

The offline identification of an electron of transverse energy $E_T(e)$ requires the following:

(1) a calorimeter energy cluster with with transverse energy $\geq E_T(e)$,
(2) a ratio of hadronic energy to EM energy in the cluster less than 0.05,
(3) a ratio of the cluster energy to track momentum less than 1.5,
(4) a transverse shape of the energy cluster consistent with a single
    electron shower,
(5) a strip chamber energy profile in both ϕ (azimuth) and z (along the
    beam direction) views consistent with a single electron shower,
(6) a distance between the strip chamber shower position and the extrapolated
    track position of < 1.5 cm in ϕ and < 3.0 cm in the z direction, and
(7) fiducial cuts to avoid cracks between calorimeter modules.

Electrons from photon conversions and Dalitz decays of π^0's are rejected by requiring electron candidates to have a matching VTPC track, and rejecting candidates with a second nearby oppositely charged CTC track forming a low e^+e^- effective mass.

The electron fiducial volume covers $74 \pm 2\%$ of the solid angle for the region $|\eta| < 1.1$. Inside this fiducial region, the efficiency of the electron selection criteria is 0.77 ± 0.03, and is well modeled by the Monte Carlo. The efficiency was determined using both test beam data and a sample of $Z^0 \to e^+e^-$.

EVENT SIMULATION AND ACCEPTANCE

The ISAJET [5] Monte Carlo program is used to generate samples of $b\bar{b}$ and $t\bar{t}$ events with full detector simulation [6] for comparison with the data and determination of acceptance. The uncertainty in the determination of the expected rate was estimated by comparing the results of the ISAJET and PAPAGENO [7] Monte Carlo calculations, as well as other published calculations of the $t\bar{t}$ production cross sections[1]. The uncertainty varies from 30% at $M_t = 70$ GeV/c² to a factor of two at $M_t = 30$ GeV/c². This uncertainty is mainly due to reasonable variations of structure functions, scale, etc. in the $t\bar{t}$ production cross sections. The uncertainties due to the t quark and lepton P_T distributions, detector acceptance, and lepton detection efficiency are smaller. For the purpose of deriving a limit (below), we choose the ISAJET predictions, but divide the number of expected events by a factor of 1.9 to allow for theoretical uncertainties.

HEAVY QUARK MASS LIMITS

In order to separate possible $t - \bar{t}$ events from sources of low P_T background, we define a signal region (see fig. 1) to be:

$E_T(e) > 15$ GeV/c², $P_T(\mu) > 12$ GeV/c.

No events in this range are observed in the data. Table 1 indicates the number of events expected in this region for an integrated luminosity of 2.0 pb⁻¹, as a function of M_t. In this Table, the number of expected events has been divided by the factor of 1.9 mentioned above to allow for systematic and theoretical uncertainties. On the basis of this partial data set we are able to exclude values of M_t in the range 30 to 60 GeV/c². For updated limits based on the full CDF data sample from the 1988-89 run, as well as a more complete enumeration of systematic errors, the reader may consult [9] for an analysis of the $\mu - e$ channel and [8] for an analysis based on e+jets events.

APPLICABILITY OF THE $\mu - e$ LIMIT TO A b' QUARK

Our limit on the t quark mass is also applicable for a fourth generation charge $-\frac{1}{3}$ b' quark, providing that the b' is lighter than the top quark, that it decays predominantly via the charged current interaction into a virtual W and a light quark (u or c), and has a decay lifetime sufficiently short ($< 10^{-11}$ sec) that

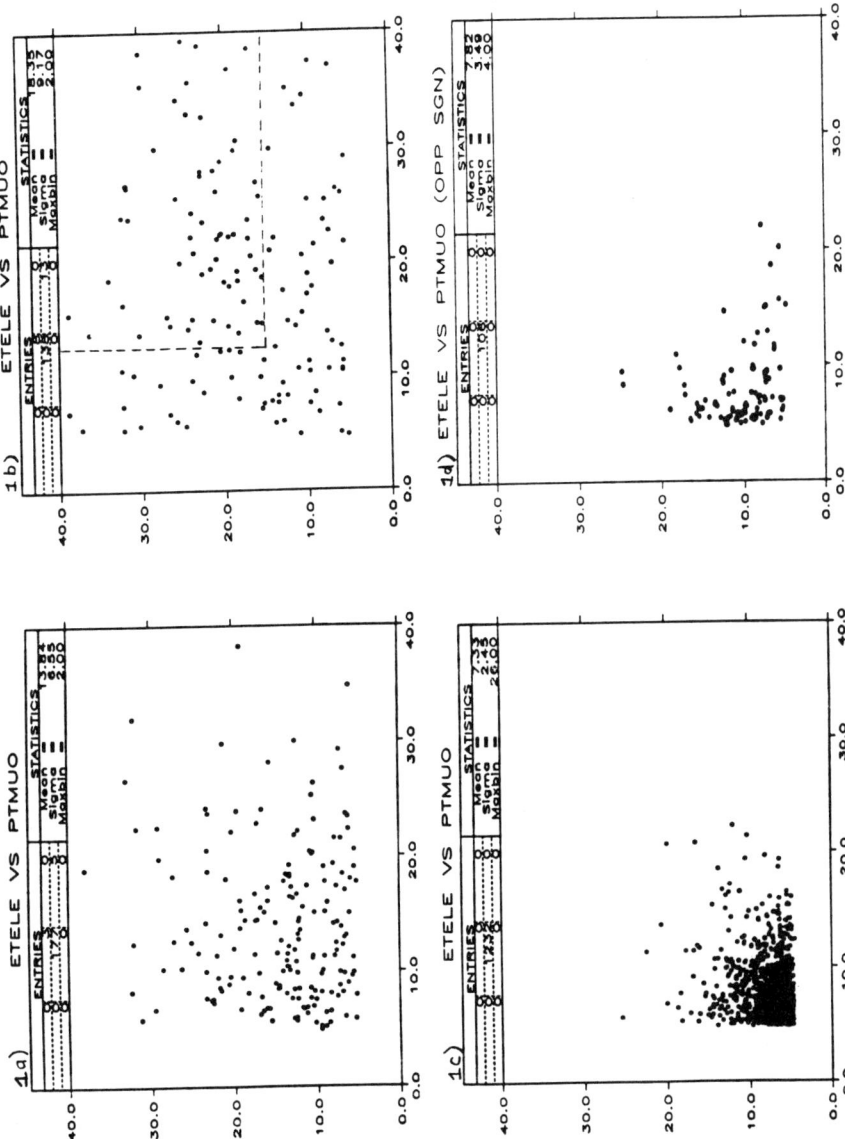

Figure 1 Scatter plots of $E_T(e)$ vs. $P_T(\mu)$ for the data and for simulated $t\bar{t}$ and $b\bar{b}$ events. 1a) 40 GeV simulated $t\bar{t}$, 9 pb^{-1}. 1b) 60 GeV simulated $t\bar{t}$, 60 pb^{-1}. 1c) Simulated $b\bar{b}$, 10 pb^{-1}. 1d) Data events from 2.0 pb^{-1}. Only events with leptons of opposite sign are included in this figure. Note that the t quark decays generate leptons with large transverse momentum, while the leptons from both b quark decay and the data are concentrated at much lower P_T. The dashed lines indicate the region used to obtain M_t limits.

Table 1: Expected Cross Section and Number of Events as a function of M_t

Top Mass (GeV)	30	40	60	70
$\sigma * B * \epsilon$ (pb)	18	7.8	2.5	1.5
N($\mu - e$) expected in 2.0 pb^{-1} after allowance for systematics	19	8.1	2.6	1.6

its decay products appear to come from the interaction vertex [10]. This last point is not necsssarily assured since the b' decay would be skipping one or two generations and the relevant mixing matrix element may be exceedingly small. With the assumptions stated above, b' quarks in the mass range 30 to 60 GeV/c^2 are also excluded at the 95% confidence level.

SO, WHAT ARE THESE $e - \mu$ EVENTS?

At this point we reiterate out caveat that the trigger and reconstruction efficiencies for low P_T leptons (below the M_t limit region) have not been fully studied. The tools to study these exist in our data (pure samples of low-P_T leptons from Υ, J/ψ, etc.) but the main emphasis to date has been on the high-P_T leptons for W and Z physics.

The process $Z^o \to \tau\bar\tau \to \mu e \nu \bar\nu \nu \bar\nu$ produces opposite sign μe pairs, back-to-back in ϕ (due to the small Q-value in τ decay), and a very "Clean" event (small total E_T in underlying event) as is typical of Z^0 production. The average P_T of the leptons is considerably lower than the $\sim$ 30 GeV/c transverse momenta from $Z \to e^+e^-$. We expect a cross section of 0.3 pb in the reduced momentum range $P_T(e)$, $P_T(\mu) > 10$ GeV/c. In 2.0 pb-1 of data we then expect 0.6 event. There is one such candidate in the data; however, the electron lies outside our fiducial η cuts.

Standard-model W^+W^- and WZ production could produce high-P_T μe events. We expect < 0.1 event from these sources in our data sample.

The main source of leptons in the range $5 < P_T < 15$ GeV/c at the Tevatron is the decay of $b\bar b$ pairs. The contribution from lighter quarks in this momentum range is negligible. In the case of $b\bar b \to \mu e + X$, we find good agreement between these data and the Monte Carlo predictions (c.f. fig. 1) for the shapes of distributions of several kinematic variables, including missing transverse energy, the azimuthal angular difference between the two leptons, and lepton isolation. In particular, the kinematical distributions show evidence for both the "Gluon Splitting" mechanism (which produces opposite sign μe pairs correlated in both η and ϕ) and "Direct $b\bar b$ Production" (which produces leptons back-to-back in ϕ but not strongly correlated in η).

A main focus of preliminary work on b physics at CDF is to estimate the purity of various leptonic event samples. A significant result in this regard is the evidence for D signal in the "single electron" trigger sample. Preliminary indications are that the normalization of this signal is consistent with this "single electron" data sample consisting of approximately 80% $b\bar{b}$ events. When this and other work is completed, it is expected that data from this run should allow a measurement of the b production cross-section at $\sqrt{s} = 1.8$ Tev. Other promising channels include single lepton, dilepton, J/Ψ and (Single Lepton + D meson).

A Silicon Vertex tracking device will be installed for the upcoming run in 1991. The device itself does not provide a trigger, but can be used to tag leptonic events which have secondary verticies (e.g. b and τ decays) using the online Level 3 "Processor Farm". We are encouraged by preliminary indications that purity of the lepton samples is such that we should be able to write essentially as many vertex-tagged $b\bar{b}$ events as we can tolerate onto tape.

CONCLUSIONS

CDF has been able to search for the t quark in both the (e + jets) and the muon-electron channel[8][9]. At the time of this conference, we have excluded the region $30 < M_T < 60$ GeV/c^2 using the first 2.0 pb^{-1} from a 4.6 pb^{-1} data sample. The outstanding performance of the Tevatron during the 1988-89 run (the delivered luminosity exceeded the design goal by a factor of two, and exceeded expectations by about a factor of 5) allows the $\mu - e$ channel to be fully competitive with the (e + jets) mode for Top searches, despite a six times smaller branching ratio. In future runs, the low background of the $\mu - e$ channel should allow each factor of two in integrated luminosity to buy about 15 GeV in M_t sensitivity. This situation is likely to persist until the standard-model background from $W^+W^- \to \mu\nu_\mu e\nu_e$ is reached. This occurs for $M_t \sim 160$ GeV and at integrated luminosities of ~ 100 Pb^{-1}.

The near term prospect for b physics at CDF seem good, with the data for this run being sufficient to measure production cross sections and identify a $b\bar{b}$ signal in several channels. The inclusion of the Silicon Vertex detector for next run should allow us to write a large number of vertex-tagged b events on tape.

References

[1] G. Altarelli, M. Diemoz, G. Martinelli and P. Nason, Nucl. Phys. B308, 724 (1988) and references therein. G. Costa *et al.*, Nucl. Phys. B297, 244 (1988).

[2] J. Rosner, Aspect of Top Quark Searches in TeV Hadron Collisions, EFI-89-02 IASSNS-HEP 89/02, Jan., 1989.

[3] F. Abe *et al.*, Nucl. Instrum. Methods A271, 387 (1988).

[4] F. Abe *et al.*, Phys. Rev. Lett. 63, 720 (1989).

[5] F. Paige and S.D. Protopopescu, BNL Report No. BNL 38034, 1986 (unpublished).

[6] J. Freeman, Proceedings of the Workshop on Detector Simulation for the SSC, edited by L. E. Price, Argonne National Laboratory (1987), p. 190.

[7] I. Hinchliffe, Private Communication.

[8] F. Abe *et al.*, Results of a Top Quark search in the electron + jets channel at $\sqrt{s} = 1.8$ TeV, to be published in PRL.

[9] F. Abe *et al.*, A Search for New Heavy Quarks in Electron-Muon Events at the Fermilab Tevatron Collider, to be published in PRL.

[10] E. Eichten and J. Rosner, Private Communication. V. Barger and R. J. N. Phillips, University of Wisconsin - Madison Preprint MAD/PH/508 (1989).

PREDICTIONS AND PROSPECTS FOR HQ AND CP VIOLATIONS

PREDICTIONS FOR CP VIOLATION*

Frederick J. Gilman
Stanford Linear Accelerator Center,
Stanford University, Stanford, CA 94309

ABSTRACT

Predictions for CP violation in the three-generation Standard Model are reviewed, especially as they pertain to the K and B meson systems.

INTRODUCTION

It is now 25 years since the initial discovery of CP violation and we are still faced with the question of its origin and its ultimate significance:
- Is it a curiosity? Could it be physics from a much higher mass scale, at which we are allowed only a peek—a tiny remnant of new physics beyond the Standard Model?

or

- Is it a cornerstone? Does it originate inside the Standard Model? Indeed, is it the signal that there are three or more generations, all quark masses unequal, and all weak mixing angles nonzero? Is it then the single statement summarizing all of this, and yielding a characteristic pattern of CP violation which is tied to quark flavor?

These are the basic questions which we seek to answer experimentally, and then to delineate the details of whatever is the mechanism of CP violation. To do so, we need to know how CP violation is manifested in the Standard Model.

CP VIOLATION IN
THE THREE-GENERATION STANDARD MODEL

The matrix[1] that describes the mixing of three generations of quarks has three real angles and one nontrivial phase. Any difference of rates between a given process and its CP-conjugate process (or of a CP-violating amplitude) always has the form:

$$\Gamma - \bar{\Gamma} \propto s_1^2\, s_2\, s_3\, c_1\, c_2\, c_3\, \sin\delta_{KM} = s_{12}\, s_{23}\, s_{13}\, c_{12}\, c_{23}\, c_{13}^2\, \sin\delta_{13} \quad , \quad (1)$$

where we express things first in the original parameterization of the quark mixing matrix[1] and then in the "preferred" parameterization adopted by the Particle Data Group,[2] using the shorthand that $s_i = \sin\theta_i$ and $c_i = \cos\theta_i$. Our present experimental knowledge assures us that the approximation of setting the cosines to unity, which we often adopt in the following, induces errors of at most a few percent. In that case, the combination of angle-dependent factors in Eq. (1), involving the invariant measure of CP violation,[3] becomes the

---

* Work supported by Department of Energy contract DE-AC03-76SF00515.

approximate combination,

$$s_1^2 \, s_2 \, s_3 \, \sin \delta_{KM} = s_{12} \, s_{23} \, s_{13} \, \sin \delta_{13} \quad , \tag{2}$$

which was recognized earlier as characteristic of CP-violating effects in the three-generation Standard Model.[4] Equation (1) shows us immediately that all three generations of quarks are necessary for CP violation; in particular, none of the angles can be zero, nor can any of the Kobayashi–Maskawa (K–M) matrix elements.

The K–M factors in Eq. (1) define the "price of CP violation" in the Standard Model. This "price" must be paid somewhere. It could be paid in a specific process by having many of these factors in both Γ and $\overline{\Gamma}$, corresponding to a very small branching ratio for that process; then when we form the asymmetry,

$$A_{CP\ violation} = \frac{\Gamma - \overline{\Gamma}}{\Gamma + \overline{\Gamma}} \quad , \tag{3}$$

the smallness of the denominator results in a large asymmetry. On the other hand, the price could be paid by having few of these factors in Γ and $\overline{\Gamma}$ separately (and hence in their sum), but only in their difference; the asymmetry is correspondingly small. There is, therefore, a very rough correspondence between rarer decays and bigger asymmetries. This rule of thumb is only that—it can be mitigated or exacerbated by other factors: hadronic matrix elements, dependence of one-loop amplitudes upon internal quark masses, and the possible presence of K–M factors in addition to those demanded by Eq. (1). A prime example of luck in this regard is provided by CP-violating effects which depend on $B - \overline{B}$ mixing, where the large top quark mass allows fairly big asymmetries between B and $\overline{B}$ decays to occur in modes which are themselves not suppressed in rate by K–M factors.

THE UNITARITY TRIANGLE

In principle, measurement of just the magnitudes of the K–M matrix elements could tell us about the phase, δ_{13}, as well as the "rotation angles" θ_{12}, θ_{23}, and θ_{13} in Eq. (1). This is most easily seen for the case at hand, where the "rotation angles" are small, by using the unitarity of the matrix as applied to the first and third columns to derive that (c_{ij} have been set to unity):

$$1 \cdot V_{ub}^* - s_{12} \cdot V_{cb}^* + V_{td} \cdot 1 \approx 0 \quad . \tag{4}$$

This equation is represented graphically in Fig. 1 in terms of a triangle in the complex plane, the lengths of whose sides are $|V_{ub}^*|$, $|s_{12} \cdot V_{cb}^*|$, and $|V_{td}|$, and the nontrivial phase in different parameterizations is the indicated interior or exterior angle. This triangle appears explicitly in Ref. 4, and has been commented on by many people,[5] but has been particularly emphasized by Bjorken.[6]

According to an ancient theorem, perfect measurements of the lengths of all three sides could determine a nontrivial triangle, thereby completely fixing the mixing matrix, including the phase. Alternately, a set of measurements of the lengths could show that the triangle can not exist, forcing us beyond three generations. As a special case, the triangle could collapse to a line, and we must go beyond the three-generation Standard Model for an explanation of CP violation. Unfortunately, given our present experimental knowledge and our limited theoretical ability to compute hadronic matrix elements, the three sides are not known with sufficient accuracy to discriminate between these situations, let alone determine the value of δ_{13}. For now, to get information on the phase we are forced to consider a CP-violating quantity and assume it can be understood within the three-generation Standard Model.

Note that twice the area of the triangle is:

$$s_1^2 s_2 s_3 \sin \delta_{KM} \approx s_{12} s_{23} s_{13} \sin \delta_{13} \ . \tag{5}$$

This is "the price of CP violation," and reaffirms that if the triangle degenerates to a line, then CP is conserved.

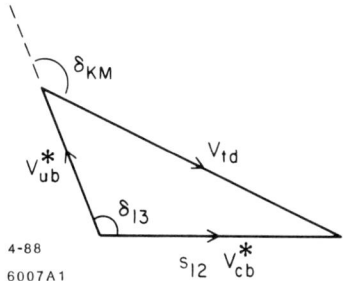

Fig. 1. Representation in the complex plane of the triangle formed by the K–M matrix elements V_{ub}^*, $s_{12} \cdot V_{cb}^*$, and V_{td}.

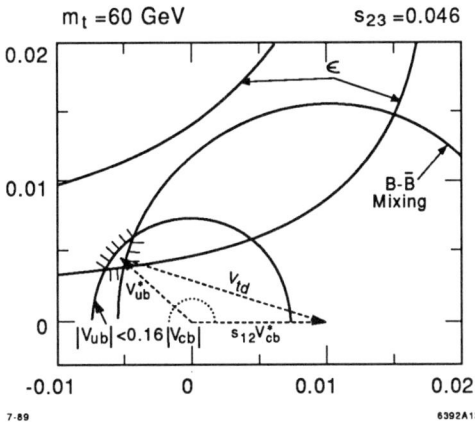

Fig. 2. Constraints on "unitarity triangle" for $m_t = 60$ GeV and $|V_{cb}| = 0.046$.

With this representation of the ill-determined parameters of the K–M matrix, it is possible to see more directly the interplay of various pieces of experimental information. In Figs. 2 to 6, we have placed[7] the side $s_{12} V_{cb}^*$ along the horizontal and taken $|V_{cb}|$ at its central value[2] of 0.046, so that one vertex is at the origin and a second vertex is very near the point $(0.010, 0)$. Constraints on the position of the third vertex follow from[8]

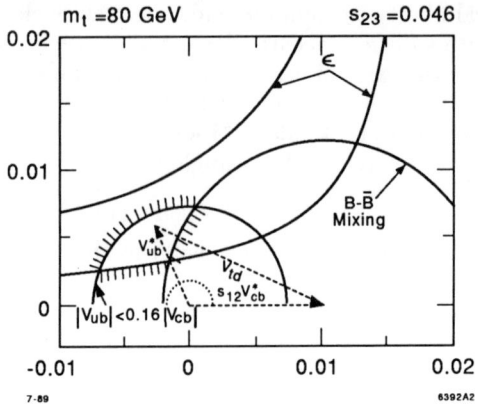

Fig. 3. Constraints on "unitarity triangle" for $m_t = 80$ GeV and $|V_{cb}| = 0.046$.

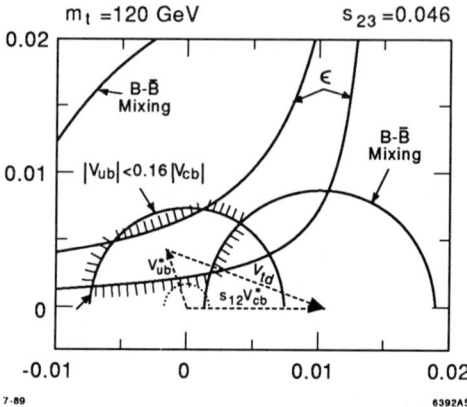

Fig. 4. Constraints on "unitarity triangle" for $m_t = 120$ GeV and $|V_{cb}| = 0.046$.

- $|V_{ub}|$: An upper limit on this quantity forces the third vertex to lie inside a circle about the origin. A lower limit, taken here to be $|V_{ub}| > 0.04|V_{cb}|$, is implied by data indicating $b \to u$ transitions presented to this conference.[9]
- $B - \bar{B}$ Mixing: The combination of the experimental value of $\Delta M/\Gamma$ and an upper and lower limit on the hadronic matrix element[8] forces the third vertex to lie outside and inside, respectively, circles drawn with the second vertex as an origin.
- ϵ: Imposing the constraint of obtaining the experimental value of $|\epsilon|$ along with upper and lower limits on the hadronic matrix element forces the third vertex to lie between hyperbolas.

Figure 2 shows the situation for $m_t = 60$ GeV, where the position of the third vertex is quite limited by the solid curves indicating the various constraints. The dotted circle represents the lower limit on $|V_{ub}|$ from the observation of $b \to u$ transitions. A sample unitarity triangle is indicated by the dashed lines.

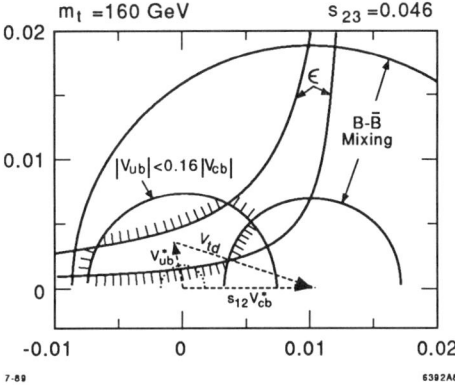

Fig. 5. Constraints on "unitarity triangle" for $m_t = 160$ GeV and $|V_{cb}| = 0.046$.

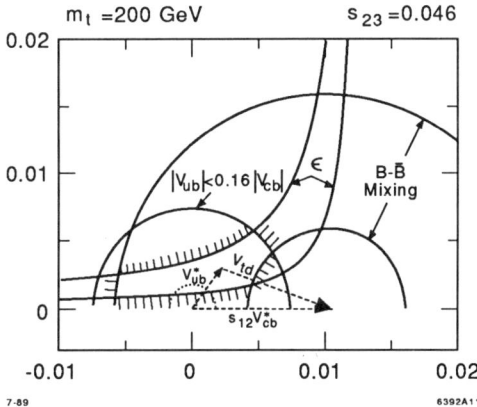

Fig. 6. Constraints on "unitarity triangle" for $m_t = 200$ GeV and $|V_{cb}| = 0.046$.

For still lower values of m_t, the inner limiting circle due to $B - \overline{B}$ mixing moves outward and eventually becomes incompatible with the other constraints; this is precisely how a lower limit of around 50 GeV for m_t came about after the observation of large $B_d - \overline{B}_d$ mixing.

As we move to a top quark mass of 80 GeV in Fig. 3, the region permitted for the third vertex opens up. Values of $m_t = 120, 160$, and 200 GeV in Figs. 4, 5, and 6, respectively, show a progressively longer and lower allowed region, as both the upper and lower limits from $B - \overline{B}$ mixing and from $|\epsilon|$ enter the picture. Note in addition that the base of the triangle, $s_{12}V_{cb}^*$, is itself only moderately well determined: Figs. 7 and 8 show what happens for $m_t = 200$ GeV when values of 0.036 and 0.056 are used for $|V_{cb}|$.

The new lower limit we are using for $|V_{ub}|$ plays little role, except for the heaviest top masses, once the "ϵ constraint" is imposed. Of course, the latter assumes that CP violation originates in the K–M matrix; it is very important

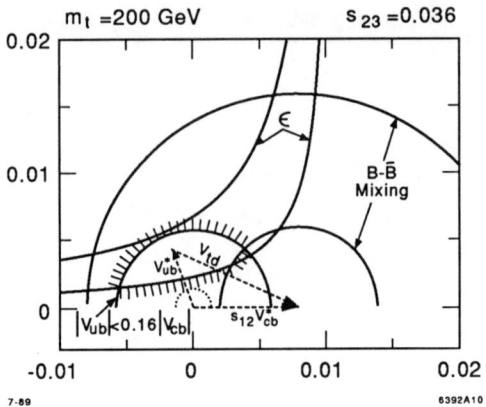

Fig. 7. Constraints on "unitarity triangle" for $m_t = 200$ GeV and $|V_{cb}| = 0.036$.

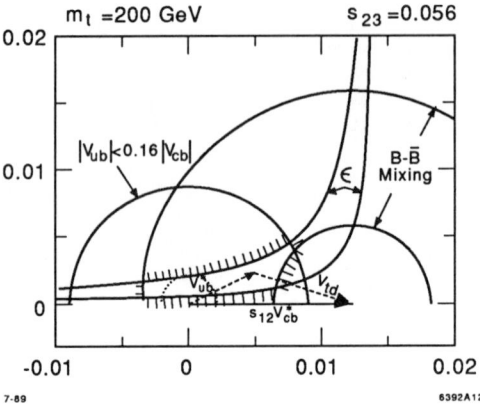

Fig. 8. Constraints on "unitarity triangle" for $m_t = 200$ GeV and $|V_{cb}| = 0.056$.

to ascertain without any such assumption that $|V_{ub}|$ is nonzero, and eventually, to pin down its value.

When viewed from the point of view of the "price of CP violation," i.e., twice the area of the unitarity triangle, it is the altitude times the base that matters. This quantity clearly has a large range, especially once we have allowed m_t to vary all the way up to 200 GeV. A ballpark figure for $s_1^2 s_2 s_3 s_\delta$ is several times 10^{-5}, which means that $s_2 s_3 s_\delta$ is of order 10^{-3}.

STATUS OF CP VIOLATION IN THE STANDARD MODEL

Given this "price of CP violation," we can "naturally" understand why

$$|\epsilon| \approx 2.28 \times 10^{-3} \quad . \tag{6}$$

is so small and CP seems to come so close to being a symmetry in K decays. When all the factors are put in, the size of $|\epsilon|$ is roughly governed by that of

$s_2 s_3 s_\delta$. This is "naturally" of the right size, in the technical sense that to have $s_2 s_3 s_\delta$ of order 10^{-3} does not require any angle to be fine-tuned to be either especially small or especially large.

This same factor of $s_2 s_3 s_\delta$ pervades all CP-violation observables in the K system, so it is then not so surprising that after 25 years the total evidence for CP violation in Nature consists of a nonzero value of ϵ, and one statistically significant measurement[10] of a nonzero value of the parameter $\epsilon'/\epsilon = 3.3 \pm 1.1 \times 10^{-3}$, representing CP violation in the $K \to \pi\pi$ decay amplitude itself. Experiments at Fermilab[11] and at CERN[10] are continuing with the aim of reducing the statistical and systematic errors. The value of ϵ' from Ref. 10 is consistent[12-14] with the three-generation Standard Model. Unfortunately, this is not a very strong statement. Other values of ϵ' would be consistent as well, because of our lack of knowledge both on the experimental and theoretical fronts:

- The hadronic matrix elements of the penguin operators, upon which the prediction of ϵ' depends, are fairly uncertain. Definitive results will presumably come from lattice QCD calculations, which still seem several years away.
- The predictions depend on the value of $s_2 s_3 s_\delta$, which in turn depends (aside from another hadronic matrix element) on m_t through imposing the constraint of obtaining the experimental value of ϵ. Very roughly, as m_t goes up, the range allowed for $s_2 s_3 s_\delta$ goes down, and so does the prediction for ϵ'.
- Also as m_t rises, the contributions from "Z penguin" and "W box" diagrams begin to be significant. For sufficiently large m_t, a recent calculation[15] contends that most of the usual (strong) penguin contribution to ϵ' can be cancelled in this way.

Experimental and theoretical progress over the next few years should clarify these points. But even if the situation becomes that the value of ϵ' is in significant accord with the three-generation Standard Model, this single number is unlikely to be regarded as conclusively establishing that the origin of CP violation lies in the K–M matrix. We would demand additional evidence: A <u>single</u> set of K–M angles (including the phase) must be able to fit several different processes which exhibit CP-violating effects, providing a redundant check on the theory.

There are two main avenues being pursued in order to get this additional evidence. One is to look for CP-violating effects in the B meson system. Here the CP-violating asymmetries potentially can be very large: of order 10^{-1} or more. The second way is to consider other K decays where CP-violating effects, although very small, may occur with a different weighting (from that in $K \to \pi\pi$) between effects originating in the mass matrix and in the decay amplitude. Possible K decays which come to mind include $K \to 3\pi$, $K \to \gamma\gamma$, and $K \to \pi\pi\gamma$,[16-18] and especially $K_L \to \pi^0 \ell^+ \ell^-$ and $K_L \to \pi^0 \nu \bar{\nu}$. We take up K decays in the next section, saving the B system for last.

CP VIOLATION IN RARE K DECAYS

The late 1960s and early 1970s marked a peak in experiments on K decays, sparked by the discovery of CP violation.[19] This effort tailed off as many important measurements were completed and new areas of physics opened up in the 1970s at electron-positron and hadron machines.

Then, in the late 1970s and early 1980s, both theoretical and experimental developments led to a "rebirth" of K physics. On the experimental side, great strides were made to create high flux beams, handle high data rates, incorporate "smart triggers," improve detectors (especially for photons), and be able to analyze enormous data samples. These matched, at least to some degree, the requirements in precision and rarity being demanded by the theory for incisive tests of the Standard Model. The last few years have seen the beginning of a parade of results which are the culmination of a decade of work in perfecting and performing the needed experiments. Much more is yet to come, and one can see the opportunity to make use of the beams and detectors which are already in existence, or are being developed, to attack the rare K decays which will give additional insight into CP violation.

On the theoretical side, the establishment of gauge theories for the strong and electroweak interactions provided a well-defined basis for calculations. The three-generation Standard Model could be used to make predictions of what, by definition, was inside, and, by its complement, outside the Standard Model. The question of "who ordered the muon" was generalized to "who ordered three generations with particular values of masses and mixing angles," and attention was directed at interactions which would connect quarks and leptons of different generations, producing flavor-changing neutral currents. It was realized that not only did the three-generation Standard Model provide an origin for CP violation in the nontrivial phase in the quark mixing matrix, but that CP violation should affect the K^0 decay amplitude as well as the $K^0 - \overline{K}^0$ mass matrix, resulting in values of ϵ'/ϵ in the 10^{-3} to 10^{-2} range.[20] There were also predictions for short-distance contributions to a number of other rare K decay amplitudes induced at one-loop, both CP-conserving and CP-violating.[21]

There has also been an associated experimental development which has important theoretical consequences: the rise of the top quark. Over the past decade, the "typical" or "best" value of the top quark mass used in theoretical papers has risen monotonically, somehow always remaining one step, or maybe one and a half steps, ahead of the experimental, then-current, lower bound. Values of 15, 25, 30, 45... GeV have been used in various papers (some of them mine), and subsequently fallen by the wayside as experiments have been able to search at higher and higher masses. The present lower limit is around 60 GeV, below which a top quark is said[22] to be "unlikely." It seems that limits even higher than this will be quoted at high confidence within a month or two, as the analysis of the present round of collider data (which is still being taken as I speak) is completed. An upper limit of around 200 GeV follows from analysis of neutral and charged current data and the measured W and Z masses (i.e., consistency of the ρ parameter with unity).[23] I suspect that we are headed for a lower limit in the neighborhood of M_W later this year.

The rise of the top quark mass has important consequences when we go to calculate one-loop contributions. For the penguin diagrams in Fig. 9 involving a top and charm quark and a virtual photon (the "electromagnetic penguin"); the conserved nature of the current demands that a factor of q^2, the square of the four-momentum carried by the virtual photon, be present in the numerator of the amplitude. This cancels the $1/q^2$ from the photon propagator; the leading term for small (compared to M_W^2) top mass in the coefficient of the

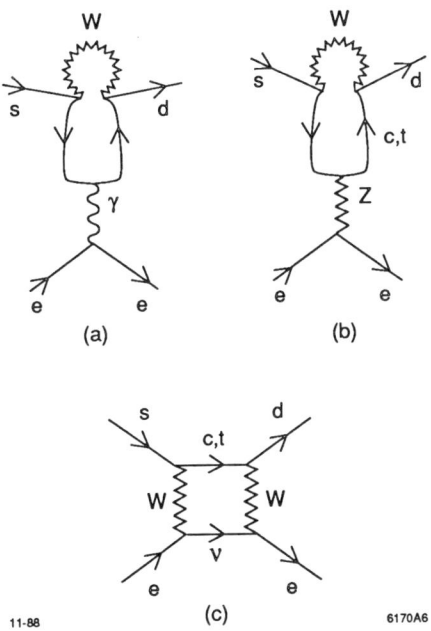

Fig. 9. One-loop diagrams giving short distance contributions to K decays, and in particular, to the process $K \to \pi \ell^+ \ell^-$: (a) the "electromagnetic penguin;" (b) the "Z penguin;" (c) the "W box."

appropriate operator behaves as $\ln(m_t^2/m_c^2)$. By contrast, the "Z penguin" or "W box" involve nonconserved currents: the factor q^2 in the numerator is replaced by the square of the quark mass in the loop and the propagator by $1/(q^2 + M_Z^2) \approx 1/M_Z^2$ or $1/M_W^2$. The corresponding coefficient behaves like $[(m_t^2/M_W^2)\ln(m_t^2/M_W^2) - (m_c^2/M_W^2)\ln(m_c^2/M_W^2)]$ when the top mass is small. In the days when $m_t^2 \ll M_W^2$, it was completely justified to throw away the Z penguin and W box contributions to such amplitudes in comparison to that of the electromagnetic penguin. Not so any more. The various graphs give comparable contributions, as we will see below in a specific example. Moreover, the contributions from the top quark become the dominant ones to various rare K decays when $m_t^2 \gg M_W^2$. In the three-generation Standard Model, as m_t rises farther and farther above M_W, more and more of one-loop K physics is top physics and we are in the interesting situation where those working at the highest-energy hadron colliders are pursuing another aspect of the same physics as those working on the rarest of K decays at low energies.

Let us illustrate a number of the above remarks by looking in more detail at one particular rare K decay in which it is possible to observe CP violation and which has emerged as the object of concentrated theoretical and experimental study.

$K_L \to \pi^0 e^+ e^-$

If we define K_1 and K_2 to be the even and odd CP eigenstates, respectively, of the neutral K system, then $K_L \to \pi^0 e^+ e^-$ has three contributions:

(1) Through a two-photon intermediate state:

$$K_2 \to \pi^0 \gamma\gamma \to \pi^0 e^+ e^- \ .$$

This is higher order in α, but is CP conserving. With two real photons, there are two possible Lorentz invariant amplitudes for $K_L \to \pi^0 \gamma\gamma$. One is the coefficient of $F^{(1)}_{\mu\nu} F^{(2)}_{\mu\nu}$, which corresponds to the two photons being in a state with total angular momentum zero. Consequently, it picks up a factor of m_e when contracted with the QCD amplitude for $\gamma\gamma \to e^+ e^-$, as the interactions are all chirality conserving, and its contribution to the $K_L \to \pi^0 e^+ e^-$ decay rate is totally negligible.[24] The other invariant amplitude is the coefficient of a tensor which contains two more powers of momentum and one might hope for its contribution to be suppressed by angular momentum barrier factors. In chiral perturbation theory, an order-of-magnitude estimate[25] for the resulting branching ratio of $K_2 \to \pi^0 e^+ e^-$ is 10^{-14}. However, a vector dominance, pole model predicts[26] a much bigger result: a branching ratio of order 10^{-11}, roughly at the level as that arising from the CP-violating amplitudes (see below). The experimental upper limit on the branching ratio for $K_L \to \pi^0 \gamma\gamma$ has very recently been considerably improved,[27] and now is only a few times larger than some of the predictions.[26,25] In the future, we might have not only a measurement of the branching ratio, but a Dalitz plot distribution which could help distinguish between models. The final answer for this contribution remains to be seen both theoretically and experimentally.

(2) Through the small (proportional to ϵ) part of the K_L, i.e., K_1, due to CP violation in the mass matrix:

$$K_L \approx K_2 + \epsilon K_1$$

$$K_1 \to \pi^0 \gamma_{virtual} \to \pi^0 e^+ e^- \ .$$

We call this "indirect" CP violation and may calculate its contribution to the decay rate once we know the width for the CP-conserving process $K_1 \to \pi^0 e^+ e^-$. Eventually, there will presumably be an experimental measurement of $\Gamma(K_S \to \pi^0 e^+ e^-)$, which will take all the present theoretical model dependence away. For now, equating this width to the measured one for $K^+ \to \pi^+ e^+ e^-$ gives the estimate:

$$B(K_L \to \pi^0 e^+ e^-)_{indirect} = 0.58 \times 10^{-11} \ . \tag{7}$$

(3) Through the large part of the K_L, i.e., K_2, due to CP violation in the decay amplitude:

$$K_2 \to \pi^0 \gamma_{virtual} \to \pi^0 e^+ e^- \ .$$

We call this "direct" CP violation, and the amplitude for it arises from the diagrams shown in Fig. 9. For values of $m_t \ll M_W$, it is the "electromagnetic penguin" that gives the dominant short-distance contribution to the amplitude, which is summarized in the Wilson coefficient, C_{7V}, of the appropriate operator,

$$Q_{7V} = \alpha \left(\bar{s}\gamma_\mu(1-\gamma_5)d\right)\left(\bar{e}\gamma^\mu e\right) .$$

Values of $m_t \sim M_W$ allow the "Z penguin" and "W box" contributions to become comparable to that of the "electromagnetic penguin," and bring in another operator,

$$Q_{7A} = \alpha \left(\bar{s}\gamma_\mu(1-\gamma_5)d\right)\left(\bar{e}\gamma^\mu\gamma_5 e\right) .$$

The QCD corrections are substantial for the "electromagnetic penguin" contribution and have been redone for the case[28,29] when $m_t \sim M_W$. In contrast, the top quark contributions from the "Z penguin" and "W box" live up at the weak scale and get only small QCD corrections. Still, the coefficient C_{7V} comes largely from the "electromagnetic penguin," even after its reduction from QCD corrections. On the other hand, the "electromagnetic penguin" cannot contribute to C_{7A}, and here it is the "Z penguin" which gives the dominant contribution. The overall decay rate due to the "direct" CP-violating amplitude can be obtained by relating the hadronic matrix elements of the operators Q_{7V} and Q_{7A} to that which occurs in K_{e3} decay. Then we find that

$$B(K_L \to \pi^0 e^+ e^-)_{direct} \approx 1 \times 10^{-5} (s_2 s_3 s_\delta)^2 [|\tilde{C}_{7V}|^2 + |\tilde{C}_{7A}|^2] . \quad (8)$$

The last factor, shown in Fig. 10, ranges[28] between about 0.1 and 1.0. As $s_2 s_3 s_\delta$ is typically of order 10^{-3}, the corresponding branching ratio induced by this amplitude alone for $K_L \to \pi^0 e^+ e^-$ is around 10^{-11}. Note that when $m_t \gtrsim 150$ GeV, the contribution from C_{7A} overtakes that from C_{7V}, and it is the "Z penguin" and "W box," coming from the top quark with small QCD corrections, which dominate the decay rate.

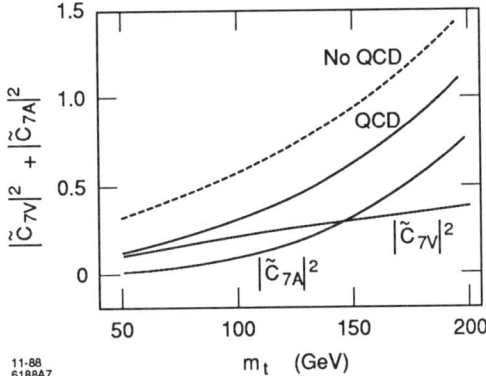

Fig. 10. The quantity $|\tilde{C}_{7V}|^2 + |\tilde{C}_{7A}|^2$, which enters the branching ratio for the CP-violating decay $K_L \to \pi^0 e^+ e^-$, as a function of m_t for $\Lambda_{QCD} = 150$ MeV, from Ref. 28.

Thus, it appears at this point that the three contributions from (1) CP-conserving, (2) "indirect" CP-violating, and (3) "direct" CP-violating amplitudes could all be comparable. The weighting of the different pieces in $K_L \to \pi^0 e^+ e^-$ is entirely different from that in $K \to \pi\pi$. The present experimental upper limit[30,31] is 4×10^{-8}, with prospects of getting to the Standard Model level of around 10^{-11} in the next several years. Hopefully, the CP-conserving and "indirect" CP-violating amplitudes will be pinned down much better by then, permitting an experimental measurement of this decay to be interpreted in terms of the magnitude of the "direct" CP-violating amplitude.

CP VIOLATION IN B DECAY

The possibilities for observation of CP violation in B decays are much richer than for the neutral K system. The situation is even reversed, in that for the B system the variety and size of CP-violating asymmetries in decay amplitudes far overshadows that in the mass matrix.[32]

- To start with the familiar, however, consider the phenomenon of CP violation in the mass matrix of the neutral B system.

Here, in analogy with the neutral K system, one defines a parameter ϵ_B. It is related to p and q, the coefficients of the B^0 and $\overline{B}^0$, respectively, in the combination which is a mass matrix eigenstate by

$$\frac{q}{p} = \frac{1-\epsilon_B}{1+\epsilon_B} \ . \tag{9}$$

The charge asymmetry in $B^0\overline{B}^0 \to \ell^\pm \ell^\pm + X$ is given by[33]

$$\frac{\sigma(B^0\overline{B}^0 \to \ell^+\ell^+ + X) - \sigma(B^0\overline{B}^0 \to \ell^-\ell^- + X)}{\sigma(B^0\overline{B}^0 \to \ell^+\ell^+ + X) + \sigma(B^0\overline{B}^0 \to \ell^-\ell^- + X)} = \frac{|p/q|^2 - |q/p|^2}{|p/q|^2 + |q/p|^2}$$
$$= \frac{\text{Im}(\Gamma_{12}/M_{12})}{1 + \frac{1}{4}|\Gamma_{12}/M_{12}|^2} \ , \tag{10}$$

where we define $<B^0|H|\overline{B}^0> = M_{12} - \frac{i}{2}\Gamma_{12}$. The quantity $|M_{12}|$ is measured in $B-\overline{B}$ mixing to be comparable in magnitude to the total width, while Γ_{12} gets contributions only from channels which are common to both B^0 and $\overline{B}^0$, i.e., K–M suppressed decay modes. This causes the charge asymmetry for dileptons most likely to be in the ballpark of a few times 10^{-3}, and at best 10^{-2}. For the foreseeable future, it is inaccessible experimentally.

- Now we turn to where the excitement is: CP violation in decay amplitudes.

In principle, this can occur whenever there is more than one path, with different K–M factors, to a common final state. For example, let us consider the all-time favorite and paradigm: decay of a neutral B to a CP eigenstate, f, such as ψK_s^0 or $D^+ D^-$. Since there is substantial $B^0 - \overline{B}^0$ mixing, one can consider two decay chains of an initial B^0 meson:

$$\begin{array}{c} B^0 \to B^0 \searrow \\ \qquad\qquad\qquad f \ . \\ B^0 \to \overline{B}^0 \nearrow \end{array}$$

The second path differs in its phase because of $B^0 \to \overline{B}^0$ mixing, and because the decay of a $\overline{B}$ involves the complex conjugate of the K–M factors involved in B decay. The strong interactions, being CP invariant, give the same phases for the two paths. The amplitudes for these decay chains can interfere and generate nonzero asymmetries between $\Gamma(B^0(t) \to f)$ and $\Gamma(\overline{B}^0(t) \to f)$. Specifically,

$$\Gamma(\overline{B}^0(t) \to f) \sim e^{-\Gamma t} \left(1 - \sin[\Delta m\, t] Im\left(\frac{p}{q}\rho\right)\right), \tag{11a}$$

and

$$\Gamma(B^0(t) \to f) \sim e^{-\Gamma t} \left(1 + \sin[\Delta m\, t] Im\left(\frac{p}{q}\rho\right)\right). \tag{11b}$$

Here, we have neglected any lifetime difference between the mass matrix eigenstates (thought to be very small), set $\Delta m \equiv m_1 - m_2$, the difference of the eigenstate masses, and $\rho \equiv A(B \to f)/A(\overline{B} \to f)$, the ratio of the amplitudes, and then used the fact that $|\rho| = 1$ when f is a CP eigenstate in writing Eqs. (11). From this, we can form the asymmetry:

$$A_{\text{CP Violation}} = \frac{\Gamma(B) - \Gamma(\overline{B})}{\Gamma(B) + \Gamma(\overline{B})} = \sin[\Delta m\, t] Im\left(\frac{p}{q}\rho\right). \tag{12}$$

Moreover, in the particular case of decay to a CP eigenstate with one combination of K–M factors contributing to the decay amplitude, the quantity

$$Im\left(\frac{p}{q}\rho\right) = Im\, e^{2i\Phi}$$

is given entirely by the K–M matrix and is independent of hadronic amplitudes, which cancel out in the ratio, ρ. Remarkably, the angles Φ turn out to be nothing but those of the unitarity triangle, as shown in Fig. 11, where the angles are labelled by examples of the neutral B decays to CP eigenstates whose asymmetries they govern.[34] Figure 12 shows the potential size of the time-dependent

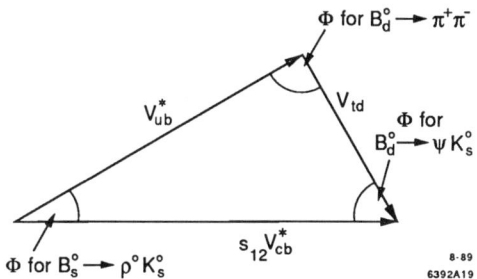

Fig. 11. The angles Φ, where $|\sin 2\Phi|$ is the magnitude of the asymmetry for $B_d \to \psi K_s$, $B_s \to \rho K_s$, and $B_d \to \pi^+\pi^-$, associated with the angles of the "unitarity triangle."

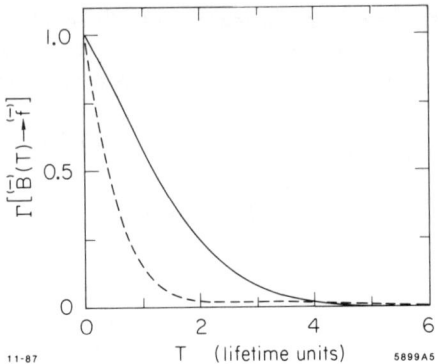

Fig. 12. The time dependence for the process $B_d \to \psi K_s^0$ (dashed curve) in comparison to that for $\overline{B}_d \to \psi K_s^0$ (solid curve) for $\Delta m/\Gamma = \pi/4$, a value consistent with that measured experimentally.

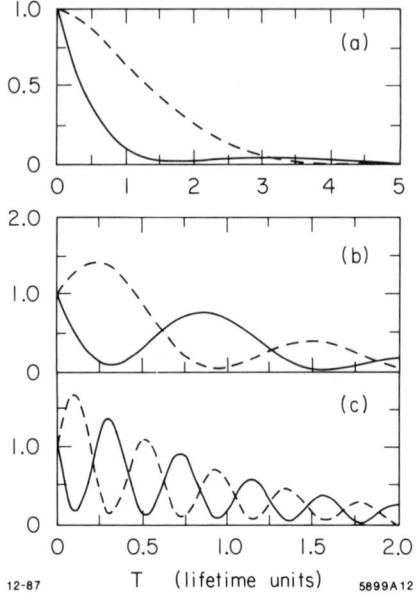

Fig. 13. The time dependence for the process $B_s \to \rho K_s^0$ (solid curve) in comparison to that for $\overline{B}_s \to \rho K_s^0$ (dashed curve) for values of (a) $\Delta m/\Gamma = 1$, (b) $\Delta m/\Gamma = 5$, and (c) $\Delta m/\Gamma = 15$, from Ref. 37.

differences[35] between B_d and $\overline{B}_d$ decaying[36] to the same (CP self-conjugate) final state, ψK_s^0. The likely situation for B_s mixing is shown[37] in Fig. 13(c). The oscillations are so rapid that even with a very favorable difference in the time dependence for an initial B_s <u>versus</u> an initial $\overline{B}_s$, the time-integrated asymmetry is quite small. Measurement of the time dependence becomes a necessity for CP-violation studies in this case.

We can also form asymmetries where the final state f is not a CP eigenstate. Examples are: $B_d \to D\pi$ compared to $\overline{B}_d \to \overline{D}\overline{\pi}$; $B_d \to \overline{D}\pi$ compared to $\overline{B}_d \to D\overline{\pi}$; or $B_s \to D_s^+ K^-$ compared to $\overline{B}_s \to \overline{D}_s^- K^+$. There is a decided disadvantage here in theoretical interpretation, in that the quantity $Im((p/q)\rho)$ is now dependent on hadron dynamics.

In all the above cases, to measure an asymmetry one must know if one starts with an initial B^0 or $\overline{B}^0$, i.e., one must "tag." This is one of the main difficulties experimentally, as the tagging efficiency is generally fairly low.[32]

A second path to the same final state could arise in several other ways besides through mixing. For example, one could have two cascade decays that end up with the same final state, such as:

$$B_u^- \to D^0 K^- \to K_s^0 \pi^0 K^-$$

and

$$B_u^- \to \overline{D}^0 K^- \to K_s^0 \pi^0 K^- \ .$$

Another possibility is to have spectator and annihilation graphs contribute to the same process.[38] Still another is to have spectator and "penguin" diagrams interfere.[39] These routes to obtaining a CP-violating asymmetry have the advantage that they do not require one to know whether one started with a B or $\overline{B}$, i.e., they do not require "tagging." These decay modes are in fact "self-tagging" in that the properties of the decay products (through their electric charges or flavors) themselves fix the nature of the parent B or $\overline{B}$. Their disadvantage, which is theoretical, is that they generally bring poorly known hadronic matrix elements into the interpretation of an asymmetry, and so the association with specific combinations of K–M angles is not clean.

CONCLUSION

In a sense, after 25 years we are still at the beginning of the study of CP violation; most CP-violating phenomena have yet to be explored, even those predicted by the Standard Model. The main thrusts in high energy physics are:

- K Decays: A strong effort is already underway at BNL, CERN, Fermilab, and KEK to pursue rare K decays. It includes measurements of ϵ'/ϵ and CP violating effects in $K_L \to \pi^0 e^+ e^-$ and other K decays. With a number of groups proposing to get to sensitivity levels corresponding to the Standard Model, we are almost guaranteed interesting results over the next few years.

- B Decays: We have seen that there are many manifestations of CP violation to look at in the B system. It appears that one needs of order 10^7 B's to begin to see the large asymmetries that are predicted by the Standard Model in some channels, but I would be the last to tell someone not to look for such effects if they had, say, 10^6 B's. Any nonzero asymmetry is important, and part of the signature of the Standard Model is the flavor dependence of the effects, with generally much larger CP-violating asymmetries characteristic of B's than of K's. We want to know if this pattern is correct. Ultimately, CP violation in the B system is the way to measure the K–M angles in a redundant way. However, unlike the situation in K decays, we do not have the likelihood of significant results in the next few years. The prospects are longer term, but it seems clear what we must do: Learn how to detect B's that are produced at hadron machines, and build electron-positron B factories.

REFERENCES

1. M. Kobayashi and T. Maskawa, Prog. Theor. Phys. **49**, 652 (1973).
2. Particle Data Group, Phys. Lett. **204B**, 1 (1988).
3. C. Jarlskog, Phys. Rev. Lett. **55**, 1839 (1985); Z. Phys. **29**, 491 (1985).
1. M. Kobayashi and T. Maskawa, Prog. Theor. Phys. **49**, 652 (1973).
2. Particle Data Group, Phys. Lett. **204B**, 1 (1988).
3. C. Jarlskog, Phys. Rev. Lett. **55**, 1839 (1985); Z. Phys. **29**, 491 (1985).
4. L.-L. Chau and W.-Y. Keung, Phys. Rev. Lett. **53**, 1802 (1984).
5. C. Jarlskog and R. Stora, Phys. Lett. **208B**, 268 (1988); J. L. Rosner, A. I. Sanda, and M. P. Schmidt, Proceedings of the Workshop on High Sensitivity Beauty Physics at Fermilab, Fermilab, November 11–14, 1987, edited by A. J. Slaughter, N. Lockyer, and M. Schmidt (Fermilab, Batavia, 1988), p. 165; C. Hamzaoui, J. L. Rosner and A. I. Sanda, ibid. p. 215.
6. J. D. Bjorken, lectures, private communication; Fermilab preprint (1988), unpublished.
7. C. O. Dib, I. Dunietz, and F. J. Gilman, to be published.
8. We use $|V_{ub}| < 0.16|V_{cb}|$, $B_B^{1/2} f_B = 150 \pm 50$ MeV, $m_c = 1.4$ GeV, and $1/3 < B_K < 1$ for the parameters that enter the various constraints in Figs. 2–8.
9. M. Procario, these proceedings, describes data from CLEO for the inclusive lepton spectrum indicating $b \to u$ transitions at the 2.2σ level of significance.
10. H. Burkhardt et al., Phys. Lett. **206B**, 169 (1988).
11. M. Woods et al., Phys. Rev. Lett. **60**, 1695 (1988); B. Winstein, invited talk at the Conference on CP Violation in Particle Physics and Astrophysics, Blois, France, May 22–26, 1989, reported a central value consistent with zero within the statistical error bars of $\pm 1.4 \times 10^{-3}$, based on 20% of the data from Fermilab experiment E731.
12. M. A. Shifman, Proceedings of the 1987 International Symposium on Lepton and Photon Interactions at High Energies, Hamburg, July 27–31, 1987, edited by W. Bartel and R. Ruckl (North Holland, Amsterdam, 1988), p. 289.
13. F. J. Gilman, International Symposium on the Production and Decay of Heavy Flavors, Stanford, September 1–5, 1987, edited by E. Bloom and A. Fridman (New York Academy of Sciences, New York, 1988), vol. 535, p. 211.
14. G. Altarelli and P. J. Franzini, CERN preprint CERN-TH-4914/87 (1987), unpublished.
15. J. M. Flynn and L. Randall, Phys. Lett. **224B**, 221 (1989).
16. L.-F. Li and L. Wolfenstein, Phys. Rev. **D21**, 178 (1980).
17. L.-L. Chau and H.-Y. Cheng, Phys. Rev. Lett. **54**, 1768 (1985); Phys. Lett. **195B**, 275 (1987); J. O. Eeg and I. Picek, ibid. **196B**, 391 (1987).
18. G. Ecker, A. Pich, and E. de Rafael, Nucl. Phys. **B303**, 665 (1988).
19. J. H. Christenson, J. W. Cronin, V. L. Fitch, and R. Turlay, Phys. Rev. Lett. **13**, 138 (1964).
20. F. J. Gilman and M. B. Wise, Phys. Lett. **83B**, 83 (1979); Phys. Rev. **D20**, 2392 (1979).

21. See, for example, the recent review of J. S. Hagelin and L. S. Littenberg, MIU–THP–89/039 (1989), to be published in Prog. Part. Nucl. Phys.
22. UA1, UA2, and CDF, any talk in early 1989.
23. U. Amaldi et al., Phys. Rev. **D36**, 1385 (1987); G. Costa et al., Nucl. Phys. **B297**, 244 (1988).
24. J. F. Donoghue, B. R. Holstein, and G. Valencia, Phys. Rev. **D35**, 2769 (1987).
25. G. Ecker, A. Pich, and E. de Rafael, Phys. Lett. **189B**, 363 (1987); Nucl. Phys. **B291** 691 (1987); and Ref. 18.
26. L. M. Sehgal, Phys. Rev. **D38**, 808 (1988); T. Morozumi and H. Iwasaki, KEK preprint KEK–TH–206 (1988), unpublished; J. Flynn and L. Randall, Phys. Lett. **216B**, 221 (1989).
27. V. Papadimitriou et al., Phys. Rev. Lett. **63**, 28 (1989).
28. C. Dib, I. Dunietz, and F. J. Gilman, Phys. Lett. **218B**, 487 (1989); Phys. Rev. **D39**, 2639 (1989).
29. Other recent work on the subject is found in J. Flynn and L. Randall, LBL preprint LBL–26310 (1988), unpublished.
30. L. K. Gibbons et al., Phys. Rev. Lett. **61**, 2661 (1988).
31. G. D. Barr et al., Phys. Lett. **B214**, 1303 (1988).
32. K. J. Foley et al., Proceedings of the Workshop on Experiments, Detectors, and Experimental Areas for the Supercollider, Berkeley, July 7–17, 1987, edited by R. Donaldson and M. G. D. Gilchriese (World Scientific, Singapore, 1988), p. 701, review CP violation in B decay and give references to previous work.
33. A. Pais and S. B. Treiman, Phys. Rev. **D12**, 2744 (1975); L. B. Okun et al., Nuovo Cim. Lett. **13**, 218 (1975).
34. This is correct if, as stated in the text, there is just one combination of K–M factors contributing. This appears to be an excellent approximation for $B_d \to \psi K_s$ and a good one for $B_s \to \rho K_s$, while there may be significant corrections for $B_d \to \pi^+\pi^-$. See M. Gronau, Max Planck Institute preprint MPI–PAE/PTh–27/89 (1989), unpublished, and B. Grinstein, Fermilab preprint FERMILAB–PUB–89/158–T (1989), unpublished.
35. The importance of the time dependence has been particularly emphasized by I. Dunietz and J. L. Rosner, Phys. Rev. **D34**, 1404 (1986).
36. This graph was constructed by R. Kauffman, in accord with the paper of Dunietz and Rosner, Ref. 35, but with somewhat different parameters: $s_1 = 0.22, s_2 = 0.09, s_3 = 0.05$ and $\delta_{KM} = 150^0$.
37. I. Dunietz, University of Chicago Ph.D. thesis (1987), unpublished.
38. This possibility has been particularly emphasized by L. L. Chau and H. Y. Cheng, Phys. Lett. **165B**, 429 (1985).
39. This has been first emphasized by M. Bander, D. Silverman, and A. Soni, Phys. Rev. Lett. **43**, 242 (1979).

EXPERIMENTAL PROSPECTS FOR OBSERVING CP VIOLATION IN B-MESON DECAYS
$- e^+e^-$ EXPERIMENTS $-$

T. Nakada
Paul Scherrer Institute(PSI)
CH-5232 Villigen-PSI, Switzerland

ABSTRACT

The prospects of observing CP violation in B-meson decay are considered for different e^+e^- experiments. For boosted $\Upsilon(4S)$, optimization for the amount of boost needed for the observation of CP violation in $B \to \Psi K_s$ is tried. Also discussed are experiments with B^*B and Z^o.

INTRODUCTION

The Cabibbo-Kobayashi-Maskawa [1] mass mixing matrix (CKM-matrix) for the three quark families can be parametrized by three rotation angles and one phase. In order to explain CP violation within the framework of the standard model, all four parameters must be neither 0 nor 180 degrees [2].

The CKM matrix elements can be directly obtained from the weak decay of various hadrons. Using the presently available experimental data, only two rotation angles can be extracted [3]. The third angle and the phase are so far indirectly determined from the magnitude of the CP violation in K^o-$\overline{K^o}$ oscillations and from the mass difference between the two neutral B-meson weak eigenstates, Δm.

Once all four parameters are obtained, a theoretical prediction can be made for the magnitude of CP violation in the K^o decay amplitude [4], $\mathrm{Re}(\epsilon')$, which is generated by the phase difference between the isospin 0 and isospin 2 decay amplitudes. The results published by the NA31 experiment [5] which indicate agreement with this theoretical prediction, must be confirmed by other experiments [6].

Although the magnitude of CP violation in K^o-$\overline{K^o}$ oscillations is well measured [7], the extraction of accurate values of the CKM matrix parameters from the data is not possible [8]. This is due to the difficulty in calculating hadronic matrix elements. A similar difficulty introduces a large uncertainty in the predicted value for $\mathrm{Re}(\epsilon')$.

© 1989 American Institute of Physics

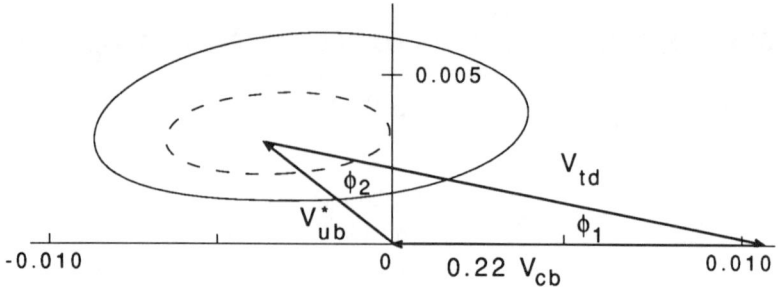

Figure 1: Unitarity Triangle. The curves correspond to one and two standard deviations

(A theoretical calculation for Δm of the neutral K-meson system suffers from even larger problems.)

In the neutral kaon system, the dominant contribution at the tree and the one loop level comes only from the quarks in the first two families. Therefore the relevant CKM matrix elements belong to a two by two sub-matrix, which can always be made real. This explains the observed smallness of CP violation in the kaon system. In the case of the neutral B-meson system, the dominant contribution in the box diagram comes from the top quark. The dominant decay of the b-quark is into the c-quark. The relevant CKM matrix elements are V_{tb}, V_{td} and V_{cb}, which cannot be reduced into a two by two sub-matrix. Therefore we expect that CP violation in the neutral B-meson system can be large.

Moreover, it was realized [9] that a particular type of CP violation which is generated by the interference between the decay and the oscillation amplitude is described only with the CKM matrix elements. Then, no uncertainty due to the hadronic matrix element obscures the predicted magnitude of CP violation. Examples are

$$\Gamma(B^\circ \to \Psi K_s) \neq \Gamma(\overline{B^\circ} \to \Psi K_s)$$
$$\Gamma(B^\circ \to \pi^+\pi^-) \neq \Gamma(\overline{B^\circ} \to \pi^+\pi^-).$$

A visual description of the magnitude of CP violation within the framework of the standard model is the so called unitarity-triangle [10] (fig.1). The magnitude of CP violation in the neutral B-meson decay is related to $-\sin(2\phi_1)$ for the ΨK_s final state and to $\sin(2\pi - 2\phi_2)$ for the $\pi^+\pi^-$ final state [11]. If we had precise measurements for $|V_{bu}|$ and $|V_{td}|$, the triangle would be completely determined.

Fig.1 also summarizes our present knowledge on the triangle [12] obtained including B°-$\overline{B^\circ}$ oscillation and CP violation in the neutral kaon system. Detailed discussion can be found elsewhere [13]. The curves correspond to the one and two standard deviation contours. They indicate that the magnitude of CP violation in the $B \to \Psi K_s$ decay can be estimated with a factor of six uncertainty within the framework of the standard model.

In order to improve this prediction, a direct measurement of V_{bu} is necessary. An indication of a non zero V_{bu} from the lepton spectrum of the semileptonic B decay is presented in this symposium [14]. However, the theoretical uncertainty of extracting a value for V_{bu} is as big as a factor of two [3]. For the reduction of the theoretical uncertainty, a detailed study of exclusive semileptonic decays such as $B \to \rho \ell \nu$ is necessary. With a sufficient number of reconstructed B mesons from the $\rho \ell \nu$ final state, say ~ 100 events, one can study the decay kinematics which allows to examine predictions from various theoretical models. An investigation [15] shows that one may need as much as $\sim 10^7 \cdot B^+ B^-$ pairs for such a study. This will be one of the important motivations for a B-meson factory.

A better determination of V_{td} requires not only more accurate measurements of Δm from B^o-$\overline{B^o}$ oscillation, but also the B-meson decay constant f_B [16]. Theoretical predictions for f_B vary by roughly a factor of two [3] and one wishes to measure it experimentally. This can be done by measuring the branching ratio for $B \to \tau \nu_\tau$, once we know the value of V_{bu}. The branching ratio can be written as

$$B(B \to \tau \nu_\tau) = 7.7 \cdot 10^{-5} \left(\frac{f_B}{120 \text{MeV}} \right)^2 \left(\frac{|V_{bu}|/|V_{cb}|}{0.16} \right)^2$$

which can be as small as 10^{-6}. The reconstruction of this decay is very difficult and an efficient tagging technique is required. Even with a B-meson factory, the study of $B \to \tau \nu$ will be difficult.

The real goal for the observation of CP violation in the B system is a consistency check of the standard model. A measurement of CP violation in $B \to \Psi K_s$ determines the angle ϕ_1 in fig.1. But, without an accurate knowledge on V_{bu} and V_{td}, no consistency check can be made. The measurement of these two elements is possible by studying the B-meson decays. Therefore, it is important to note that the aim of a B-meson factory is not just the discovery of CP violation.

In this article, we discuss the prospects of observing CP violation in B decays with e^+e^- experiments. A similar work was done at the Snowmass workshop [17] in summer 1988 and it serves as a master reference.

CP VIOLATION IN ΨK_s FINAL STATE

In the Snowmass workshop, the CP violation study was concentrated on the decay of $B \to \Psi K_s$. This can be justified by the following reasons:

- The theoretical prediction involves little complication due to the hadronic matrix elements. Hence the prediction is relatively reliable.

- The prediction for the CP asymmetry is expected to be large, $\sin(2\phi_1) = 0.1 \sim 0.6$ [18].

- The decay has been seen by ARGUS and CLEO [19] with a branching ratio of $\sim 5 \cdot 10^{-4}$.

- The final state is easy to reconstruct if it is restricted to $\Psi \to \ell^+\ell^-$ and $K_s \to \pi^+\pi^-$.

Disadvantages for this decay mode are:

- Since only $\sim 14\%$ of Ψ decays into the dilepton final state [7] and a realistic reconstruction of K_s can be done only from the $\pi^+\pi^-$ final state, the reconstruction efficiency of the ΨK_s final state is less than 9%.

- Since both B^o and $\overline{B^o}$ can decay into ΨK_s, flavour tagging is needed by the decay of the other B-meson which introduces a further loss in efficiency and some complication in the observation which is described later.

Keeping these difficulties in mind, let us discuss three different possibilities:

1. B^o-$\overline{B^o}$ pairs produced at the $\Upsilon(4S)$ resonance.

2. B^o-$\overline{B^o}$ pairs coming from B^*B pairs produced somewhat above the $\Upsilon(4S)$ energy.

3. Neutral B-mesons produced from Z^o's.

We do not discuss the possibility at e^+e^- center of mass energies around 14GeV.

EXPERIMENTS AT $\Upsilon(4S)$

There are several advantages for studying B-meson decays on the $\Upsilon(4S)$ resonance. Firstly, the B-meson cross section is large (~ 1.2nb) compared with higher center of mass energies. Since $\Upsilon(4S)$ decays exclusively into a B-$\overline{B}$ pair, by reconstructing one B the quantum number of the other B-meson becomes known (some complication appears for the neutral B-meson due to oscillation). The energy of the produced B-meson is also known, in particular with a storage ring having a small beam energy smearing. This can be used to reduce the combinatorial background in the reconstructed B-mesons.

For a symmetric e^+e^- collider running on $\Upsilon(4S)$, the B-meson is produced almost at rest ($p \approx 300$MeV/c) and the mean flight path, $\beta\gamma c\tau_B$, where τ_B is the B-meson lifetime (~ 1.13ps [12]), is only $\sim 20\mu$m. An even more serious problem is that only the sum of the two B-meson flight paths can be measured due to the absence of the primary vertex. As discussed below, this makes the observation of CP violation in the B-meson decay into the ΨK_s final state impossible for $\Upsilon(4S)$ produced at rest.

It was pointed out by P. Oddone [20] that moving $\Upsilon(4S)$ resonances produced by colliding e^+ and e^- with unequal energies stretch the flight paths of the B-mesons. Later, it was realized [21] that such moving $\Upsilon(4S)$ made the observation of CP violation possible. A study by R. Alexan et al. [22] triggered various activities [23] to investigate different possibilities of observing CP violation with moving $\Upsilon(4S)$.

When the $B^\circ \overline{B^\circ}$ pair is produced with odd relative angular momentum, the probability distribution for one of the neutral B-mesons to decay as a B° at $t = t_1$ and the other to decay into ΨK_s at $t = t_2$ is written as [24]

$$P[B^\circ(t_1), \Psi K_s(t_2)] = \Gamma^2 e^{-\Gamma(t_1+t_2)} \left[1 + \lambda \sin\{\Delta m \cdot (t_2 - t_1)\}\right] \quad (1)$$

where Γ is the B-meson total decay width, Δm is the mass difference between the two neutral B weak eigenstates and λ is the CP violation parameter [9]

$$\lambda = -\sin(2\phi_1).$$

The parameters t_1 and t_2 denote proper time. The CP conjugate of this function is

$$P[\overline{B^\circ}(t_1), \Psi K_s(t_2)] = \Gamma^2 e^{-\Gamma(t_1+t_2)} \left[1 - \lambda \sin\{\Delta m \cdot (t_2 - t_1)\}\right]. \quad (2)$$

It is easily seen that the time integrated rates

$$n = \frac{N}{2} \int_0^\infty dt_1 \int_0^\infty dt_2 P[B^\circ(t_1), \Psi K_s(t_2)] \quad (3)$$

and

$$\overline{n} = \frac{N}{2} \int_0^\infty dt_1 \int_0^\infty dt_2 P[\overline{B^\circ}(t_1), \Psi K_s(t_2)] \quad (4)$$

are the same, therefore this time integrated measurement does not yield any visible CP violation. Note N is the total number of events. If the sum of the decay times $T = t_1 + t_2$ is measured by reconstructing the B decay vertices, this does not lead to an observable CP asymmetry either. Due to the lack of a visible primary vertex and the beam spot sizes at the interaction point which are much larger than the B flight length, no individual decay time can be measured.

A visible CP asymmetry can be obtained if the decay time difference $\Delta t = t_2 - t_1$ is measured. The probability distribution of the time difference can be obtained from eq.1 as

$$P(B^\circ, \Psi K_s, \Delta t) = \frac{\Gamma}{2} e^{-|\Gamma \Delta t|} \{1 + \lambda \sin(\Delta m \cdot \Delta t)\}$$

and its CP conjugate from eq.2 as

$$P(\overline{B^\circ}, \Psi K_s, \Delta t) = \frac{\Gamma}{2} e^{-|\Gamma \Delta t|} \{1 - \lambda \sin(\Delta m \cdot \Delta t)\}.$$

We can combine these two probability functions by replacing Δt by $-\Delta t$ for the $\overline{B^\circ}$-ΨK_s final state events and obtain

$$f(\Delta \tau) = \frac{1}{2} e^{-|\Delta \tau|} \{1 + \lambda \sin(x \cdot \Delta \tau)\} \quad (5)$$

where the time is replaced by the normalized time, $\tau = t\Gamma = t/\tau_B$ and $x = \Delta m/\Gamma$.
A visible CP asymmetry can be now obtained by using time integrated rates

$$n_1 = N \int_0^\infty d\Delta\tau \, f(\Delta \tau)$$

and
$$n_2 = N \int_{-\infty}^{0} d\Delta\tau\, f(\Delta\tau)$$

as
$$AS = \frac{n_1 - n_2}{n_1 + n_2}$$
$$= \frac{x}{1+x^2}\lambda. \tag{6}$$

In a real experiment, the time difference can be measured only with a finite resolution. Let us assume that the resolution for the normalized time difference is described by a Gauss function with a width $\sigma_{\Delta\tau}$. Then the probability function eq.5 is replaced by

$$g(\Delta\tau) = \frac{1}{2\sqrt{2\pi}\sigma_{\Delta\tau}} \int_{-\infty}^{\infty} d\Delta\tau'\, e^{-\frac{(\Delta\tau-\Delta\tau')^2}{2\sigma_{\Delta\tau}^2}}\, e^{-|\Delta\tau'|}\{1 + \lambda\sin(x\cdot\Delta\tau')\}$$

and the CP asymmetry eq.6 becomes

$$AS = \alpha\lambda$$

where
$$\alpha = \frac{1}{2\sqrt{2\pi}\sigma_{\Delta\tau}} \int_{0}^{\infty} d\Delta\tau \int_{-\infty}^{\infty} d\Delta\tau'\, e^{-\frac{(\Delta\tau-\Delta\tau')^2}{2\sigma_{\Delta\tau}^2}}\, e^{-|\Delta\tau'|} \sin(x\cdot\Delta\tau'). \tag{7}$$

It can be shown [25] that the error for the CP violation parameter λ, σ_λ measured with this method for a given number of B°-$\overline{B^\circ}$ pairs, N is

$$\sigma_\lambda^2 = \frac{1}{N\cdot I_i} \tag{8}$$

where
$$I_i = \frac{\alpha^2}{1 - \lambda^2\alpha^2}. \tag{9}$$

This can be evaluated by integrating eq.7 numerically.

A better method is to fit eq.5 to the observed time difference distribution taking λ to be a parameter. For an asymptotic case, the error on λ can be expressed [25] as

$$\sigma_\lambda^2 = \frac{1}{N\cdot I_f} \tag{10}$$

where
$$I_f = \int_{-\infty}^{\infty} d\Delta\tau\, \frac{(\partial g(\Delta\tau)/\partial\lambda)^2}{g(\Delta\tau)} \tag{11}$$

which can be calculated numerically. Using the calculated value of $1/I_f$ eq. 11, eq.10 allows us to determine the necessary number of B-$\overline{B}$ pairs for the observation of CP violation with a given significance. Fig.2 shows the value of $1/I_f$ as a function of

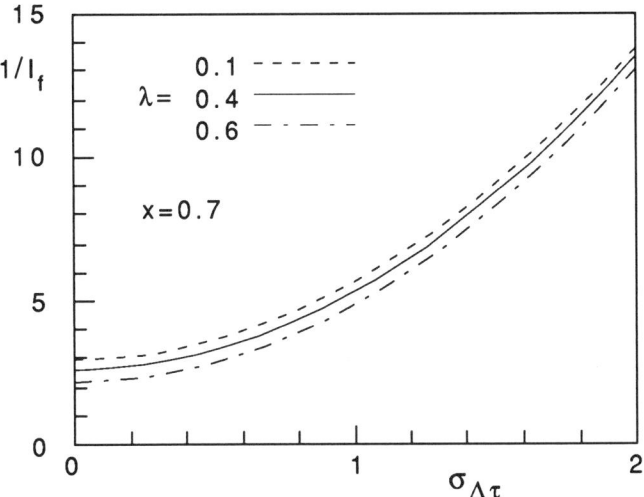

Figure 2: The parameter $1/I_f$ as a function of the normalized time difference resolution. The definition of I_f can be found in the text.

the normalized time difference resolution, $\sigma_{\Delta\tau}$ for three different values of the CP violation parameter λ. The mixing parameter x is chosen to be 0.7 [26].

Let us discuss how to measure the normalized time difference. The boost of the $\Upsilon(4S)$ can be obtained by colliding e^+ and e^- beams with unequal energies. The energies of the two beams, E_1 and E_2 must satisfy

$$E_1 \cdot E_2 = \frac{m^2_{\Upsilon(4S)}}{4} \approx 28 \; [\text{GeV}^2].$$

The decay time difference between the two B-mesons produced from the boosted $\Upsilon(4S)$ can be expressed as

$$\begin{aligned} \Delta t &= t_2 - t_1 \\ &= \frac{L_2}{\beta_2 \gamma_2 c} - \frac{L_1}{\beta_1 \gamma_1 c} \end{aligned}$$

where L is the length of the flight path. Due to the finite momenta of the B-mesons in the centre of mass system ($\sim 300 \text{MeV}/c$), B and $\overline{B}$ have different Lorentz factors β and γ. The first approximation we make is

$$\beta_1 \gamma_1 \approx \beta_2 \gamma_1 \approx \beta \gamma$$

where

$$\beta\gamma = \frac{E_1}{m_{\Upsilon(4S)}} - \frac{m_{\Upsilon(4S)}}{4E_1}$$

and E_1 is the energy of the more energetic beam.

The second approximation is to replace the difference between the two flight path lengths $L_2 - L_1$ by the difference of the z coordinates of the B decay vertices, $z_2 - z_1$ where the z axis is along the beam direction. Finally we obtain

$$\Delta t \approx \frac{z_2 - z_1}{\beta \gamma c}.$$

The validity of this approximation is examined with a simulation program using

$$m_{\Upsilon(4S)} = 10.58 \text{GeV}/c^2, \quad m_B = 5.281 \text{GeV}/c^2$$

and a beam energy resolution of $\sigma_E/E = 7 \cdot 10^{-4}$. No detector resolution is considered. It shows that the distribution of the difference between the real and approximated Δt has a full width half maximum (FWHM) of ~ 0.5ps for a beam of 6.00GeV colliding against a beam of 4.67GeV. With increasing asymmetry of the beam energies, the FWHM decreases rapidly. Although we take this effect into account in the analysis below, comparing with the vertex resolution of the detector considered, the error introduced by this approximation is negligible. Therefore, the normalized time difference resolution can be related to the z vertex distance resolution $\sigma_{\Delta z}$ as

$$\sigma_{\Delta \tau} \approx \frac{\sigma_{\Delta z}}{\beta \gamma \tau_B c}.$$

The vertex resolution depends on the spatial resolution of the vertex detector, on the distances between the vertex and the measured points of the tracks on the vertex detector, on the amount of the material through which the tracks must pass before reaching the last layer of the vertex detector and on the opening angles between the tracks. By increasing the boost of $\Upsilon(4S)$, both these distances and the thickness of the material increase. The opening angles between tracks become smaller.

Therefore, the B decay vertex resolution reconstructed from the ΨK_s final state, which is determined by two leptons from the Ψ decay, becomes worse for larger boost. Due to the clean signature of the final state, we do not expect any loss in the vertex reconstruction efficiency for the low boost case which is introduced by the difficulty in associating tracks to the right vertex.

For the tagging side, there are mostly two vertices from B and D decays. For those events where a clear separation between the two vertices is possible, one expects to have a better vertex resolution with the lower boost. However, with the low boost the average distance between the two vertices is small and a clear separation of the D and B vertices is not always possible. In such a case, one can reconstruct one common vertex which provides somewhat worse z vertex resolution [27] but minimizes the loss of the vertex reconstruction efficiency.

In summary, we assume that both the resolution of the z distance between the two B decay vertices and the vertex reconstruction efficiency stay constant for different beam configurations. It must be noted that a constant vertex resolution leads to an improvement of the normalized time difference resolution by increasing the boost.

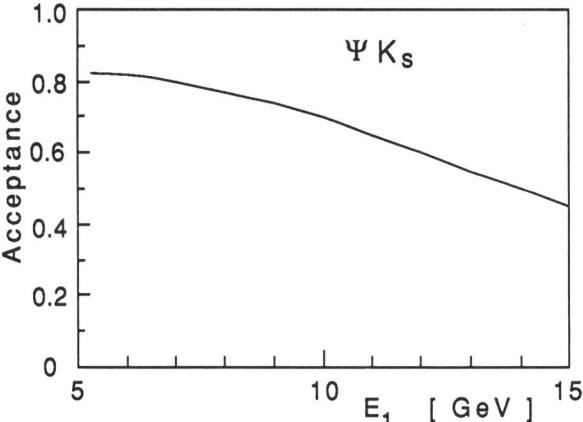

Figure 3: The acceptance for the ΨK_s final state as a function of the more energetic beam.

Various studies [28] show that a B decay vertex distance resolution of $\sigma \approx 40 \sim 60 \mu m$ can be achieved with a thin beam pipe with a small diameter of $\sim 1.5 cm$ and with a silicon vertex detector system. The vertex reconstruction efficiency can be as high as $\epsilon_{vertex} \approx 90\%$.

Loss of particles which go very close to the beam is inevitable. By increasing the boost, the loss due to the limited detector coverage in the forward region increases. An accurate momentum measurement is important for the reconstruction of $B \to \Psi K_s$ with small background, from e.g. $B \to \Psi K^*$, which has a higher branching ratio. In order to obtain the necessary momentum resolution, the acceptance of the detector cannot be extended beyond a minimum polar angle θ in the laboratory frame with present technology. We assume therefore, that the acceptance of the detector is limited to $\cos(\theta) \leq 0.95$ for all beam configurations.

Fig.3 shows the acceptance for the ΨK_s final state as a function of the boost. For the tagging B side, we require at least two particles to be detected. The acceptance for this is estimated by assuming them to be kinematically uncorrelated.

The determination of the flavour for the tagging B can be done by the semileptonic decay of the B-meson using the charge of a lepton with a momentum $\geq 1.3 GeV/c$. A lepton with lower momentum can be used if on requires a kaon with the same charge. For the hadronic B decays, the charge of the kaon from the sequential D decay can be used. However, events with additional kaons have to be rejected in order to suppress wrong tags from Cabibbo suppressed decays and $B \to D_{(s)}\overline{D}$. We assume that the flavour tagging efficiency is unchanged for different boost. Simulation studies [29] show that the flavour tagging efficiency, without taking the acceptance into account is $\epsilon_{flavour} \sim 50\%$ with $w \approx 10\%$ for the wrong flavour tagging.

The necessary number of $\Upsilon(4S)$ resonances produced for the observation of CP

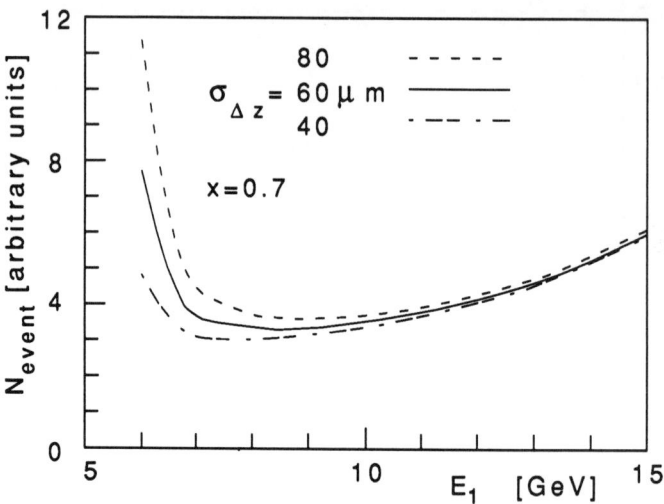

Figure 4: Number of necessary events in order to see CP violation with a given significance as a function of the higher beam energy.

violation with a significance s can now be calculated as

$$N_{\Upsilon(4S)} = \frac{s^2}{\lambda^2 \cdot I_f \cdot \eta_{tag} \cdot \eta_{\Psi K_s} \cdot B(B \to \Psi K_s) \cdot (1-2w)^2}$$

where

$$\eta_{tag} = \epsilon_{vertex} \cdot \epsilon_{flavour} \cdot \epsilon_{acceptance\ tag}$$
$$\eta_{\Psi K_s} = \epsilon_{vertex} \cdot B(\Psi \to \ell^+\ell^-) \cdot B(K_s \to \pi^+\pi^-) \cdot \epsilon_{acceptance\ \Psi K_s}$$

and a cross section ratio of

$$\frac{\sigma_{B^0\bar{B}^0}}{\sigma_{\Upsilon(4S)}} = \frac{1}{2}$$

is assumed.

Fig.4 shows the necessary number of B-$\bar{B}$ pairs produced in order to observe CP violation with equal significance, as a function of beam energy asymmetry. Three values for the resolution of the z decay vertex difference $\sigma_{\Delta z}$, 40μm, 60μm and 80μm are considered. In all cases, a modest asymmetry of the beam energies give the best result. Below 7GeV against 4GeV, the necessary number of events increases sharply due to the deterioration of the decay time resolution and above 10GeV against 2.8GeV, the loss in the acceptance becomes significant.

AT B^*B THRESHOLD

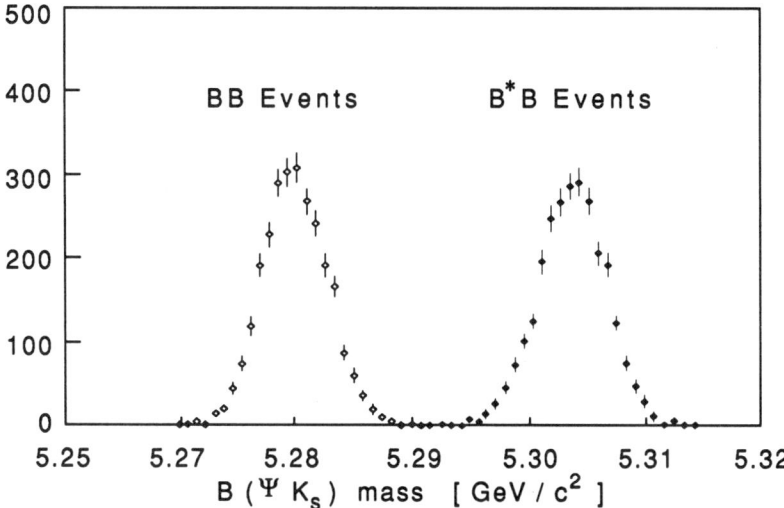

Figure 5: Reconstructed B-meson mass from the ΨK_s final state using the beam energy constraint for $B\overline{B}$ events and for B^*B events.

A potential model [30] predicts the mass of the spin one B-meson, B^* to be $\sim$ 5.338GeV/c^2. B^*-mesons decay electromagnetically into $B + \gamma$. We expect the production of B^*B pairs somewhat above the $\Upsilon(4S)$ energy. Note that the wave function of B^*B state is a superposition of $B^*\overline{B}$ and $B\overline{B^*}$ states. A coupled channel analysis [31] predicts that at $\sqrt{s} \approx 10.64 GeV$ the production of B^*B dominates with a cross section of 0.3nb and almost no $B\overline{B}$ production takes place.

The production of the B^*B pair can be detected [32] by fully reconstructing the mass of the B-mesons from their hadronic decay modes. The B-mesons from the reaction $e^+e^- \to B^*B$ have smaller energy than the B mesons produced from $e^+e^- \to B\overline{B}$. If the mass of the B-meson is determined from its hadronic final state assuming the energy of the B-meson to be equal to the beam energy, the determined B mass becomes smaller than the nominal B mass for the B-mesons coming from the former process. Fig.5 shows the reconstructed B mass from the ΨK_s final state using the beam energy constraint. It shows a clear separation between the B-mesons from B^*B and from $B\overline{B}$. The resolution of the invariant mass is determined by the beam energy spread, $\sigma_E/E = 7 \cdot 10^{-4}$ in this case, rather than by the detector resolution.

The B°-$\overline{B^\circ}$ pair from B^*B is in an even relative angular momentum state. The time evolution of a B°-$\overline{B^\circ}$ pair, eq.1 and eq.2 is modified [9] as

$$P^*[B^\circ(t_1), \Psi K_s(t_2)] = \Gamma^2 e^{-\Gamma(t_1+t_2)}[1 + \lambda \sin\{\Delta m \cdot (t_2 + t_1)\}]$$
$$P^*[\overline{B^\circ}(t_1), \Psi K_s(t_2)] = \Gamma^2 e^{-\Gamma(t_1+t_2)}[1 - \lambda \sin\{\Delta m \cdot (t_2 + t_1)\}].$$

In this case, we can obtain a visible integrated CP asymmetry from the time integrated rates eq.3 and eq.4

$$AS^* = \frac{n^* - \overline{n^*}}{n^* + \overline{n^*}}$$
$$= \frac{2x}{(1+x^2)^2}\lambda$$

where

$$n^* = \frac{N}{2}\int_0^\infty dt_1 \int_0^\infty dt_2 P^*[B^\circ(t_1), \Psi K_s(t_2)]$$
$$\overline{n^*} = \frac{N}{2}\int_0^\infty dt_1 \int_0^\infty dt_2 P^*[\overline{B^\circ}(t_1), \Psi K_s(t_2)].$$

This method does not require any vertex reconstruction, although it may help to reduce the background. The measurement can be done with a storage ring of symmetric beam energy configuration [33]. It should be noted that the measured value of $x = 0.7$ gives the coefficient $2x/(1+x^2)^2 \approx 0.63$ which is closed to its maximum value ~ 0.65 for $x \sim 0.58$.

The statistical error on the CP parameter λ is similar to eq.8 and eq.9 and given by

$$\sigma_\lambda^2 = \frac{1}{NI_i^*}$$

where

$$I_i^* = 0.424$$

for $x = 0.7$ and $\lambda = 0.4$.

Although the reconstruction of the vertex is not required, some losses are inevitable for the reconstruction of $B \to \Psi K_s$ due to the cut in the invariant mass. We assume that the total reconstruction efficiency for the ΨK_s can be expressed as

$$\eta_{\Psi K_s}^* = \epsilon_{\text{mass cut}} \cdot B(\Psi \to \ell^+\ell^-) \cdot B(K_s \to \pi^+\pi^-) \cdot \epsilon_{\text{acceptance}}^* \,\Psi K_s$$
$$\sim 6.6 \cdot 10^{-2}$$

where

$$\epsilon_{\text{mass cut}} = \epsilon_{\text{vertex}}.$$

For the flavour tagging, no invariant mass cut is required. Therefore, we assume

$$\eta_{\text{tag}}^* = \epsilon_{\text{flavour}} \cdot \epsilon_{\text{acceptance tagg}}^* \sim 45\%.$$

Assuming that half of B^*B are neutral B pairs, the necessary number of B^*B for a 3σ significance is, for $\lambda = 0.4$,

$$N_{B\cdot B} = \frac{9}{\lambda^2 \cdot I_i^* \cdot \eta_{\text{tag}}^* \cdot \eta_{\Psi K_s}^* \cdot B(B \to \Psi K_s) \cdot (1-2w)^2}$$
$$\approx 1 \cdot 10^7$$

where w is the wrong tag of 10%. This corresponds to about four years of data taking with a storage ring of a luminosity $1 \cdot 10^{33} \text{cm}^{-2}\text{s}^{-1}$.

COMPARISON BETWEEN BOOSTED $\Upsilon(4S)$ AND B^*B

The ratio of the necessary integrated luminosity for observing CP violation with the same significance can be expressed as

$$\frac{\mathcal{L}_{\text{boosted } \Upsilon(4S)}}{\mathcal{L}_{B^*B}} = \frac{I_i^* \cdot \sigma(B^*B) \cdot \eta_{\Psi K_s}^* \cdot \eta_{\text{tag}}^*}{I_f \cdot \sigma(\Upsilon(4S)) \cdot \eta_{\Psi K_s} \cdot \eta_{\text{tag}}}. \quad (12)$$

If we choose 40μm for the vertex z distance resolution, the optimal beam energy configuration is roughly 8GeV against 3.5GeV and we obtain

$$\frac{\mathcal{L}_{8+3.5\text{boosted } \Upsilon(4S)}}{\mathcal{L}_{B^*B}} \approx \frac{1}{3}. \quad (13)$$

Therefore, the luminosity of asymmetric storage rings working at $\Upsilon(4S)$ must be larger than one third of the luminosity of symmetric storage rings operating in the B^*B region.

AT Z^o

The biggest advantage of studying B-meson decay at the Z^o resonance is the large $b\bar{b}$ cross section of 6.3nb. B-mesons produced from the b-quarks are expected to carry a large fraction of the b-quark energy. A jet simulation study [34] suggests that in average as much as $\sim 90\%$ of the b-quark energy is taken by the B-meson. The multiplicity of particles in the same jet but not from the B-meson decay can be estimated using the measured charge multiplicity of the e^+e^- interactions at $\sqrt{s} = 10$GeV [35]. This gives about 5 charged tracks and 5 γ's which is equivalent to the average multiplicity from one B-meson decay. Therefore, the background situation within the same jet is very similar to that at the $\Upsilon(4S)$ resonance.

Another advantage is the large flight path length of the B-meson which is almost 3mm. Since the primary vertex can be reconstructed, one should be able to measure the decay time for $B \to \Psi K_s$ with an accuracy of $0.1 \sim 0.2\text{ps}$[34] or better.

The flavour tagging is considerably more difficult. In particular, a b jet can produce a charged or neutral B-meson or Λ_b. Therefore, one has to know the fraction of B^o in order to take the effect of oscillation into account properly.

The flavour tagging is largely improved if the electron beam can be longitudinally polarized [36]. This allows to tag the b flavour by just separating the $B \to \Psi K_s$ final states produced in the backward and forward hemispheres. However, a very good polarization is required for an effective tagging. More details can be found elsewhere [17].

CP VIOLATION OTHER THAN ΨK_s

Although controversial theoretically [37], there could be relatively large CP violation of up to 10% in the decay amplitude itself [38]. A typical example is

$$\Gamma(B^° \to K^+\pi^-) \neq \Gamma(\overline{B^°} \to K^-\pi^+). \tag{14}$$

Since the final state itself tags the flavour, no additional flavour tagging is required. Once the kaon is correctly identified, the reconstruction is very easy.

It is also important to look for CP violation at places where the standard model prediction is very small, e.g. B_s-$\overline{B_s}$ oscillation. Unlike the $B^°$ system, the B_s system can have some difference in the lifetimes between the two weak eigenstates, $\Delta\Gamma/\Gamma \approx 10\%$ [39]. This allows new physics to generate a sizable effect in CP violation in the oscillation.

Consequently, we should keep our eyes open to various possibilities and not concentrate too much on one channel. The goal of B physics after all is a precision consistency test of the standard model, which requires more than just a three standard deviation effect in CP violation in the ΨK_s final state.

ACKNOWLEDGEMENT

R. Eichler, H.-J. Gerber, R. Horisberger, L. Rivkin and K. R. Schubert are acknowledged for various useful discussions. The author thanks K. Gabathuler for his numerous comments and suggestions on this manuscript.

References

[1] N. Cabibbo, Phys. Rev. Lett. 10 (1963) 531
    M. Kobayashi and K. Maskawa, Prog. Theor. Phys. 49 (1972) 282

[2] Various useful articles on CP violation can be found in "CP Violation", ed. C. Jarlskog, World Scientific, Singapore

[3] See for example
    G. Altarelli, Contribution to this symposium

[4] See for example
    A. Buras, Proceedings for PSI Spring School on Heavy Flavour Physics, Zuoz, April 1988

[5] H. Burkhardt et al., Phys. Lett. B 206 (1988) 169

[6] Experiment E731 at FNAL is expected to present results soon. Another ongoing experiment is PS195 at CERN.

[7] Particle Data Group, Phys. Lett. B 204 (1988) 1

[8] See for example Ref. 3 and 4

[9] I. I. Bigi et al. in Ref. 2 for further references.

[10] See ref. 2.

[11] See for example
A. I. Sanda, Proceedings of the XXIV International Conference on High Energy Physics, München, August, 1988, p. 517

[12] K. R. Schubert, Invited talk at the Conference on Phenomenology on High Energy Physics, Trieste, July, 1988, IEKP-KA/88-4 (Uni. Karlsruhe)

[13] See for example
Ref. 3, 12 and K. R. Schubert, Contribution to this symposium

[14] M. Procario (CLEO), Contribution to this symposium

[15] T. Ruf, IEKP-KA/89-2 (Uni. Karlsruhe)

[16] See for example
A. Buras, W. Slominski and H. Steger, Nucl. Phys. B238 (1984) 529 and B245 (1984) 369

[17] High Energy Physics In The 1990's, Snowmass, June-July, 1988

[18] See for example
F. J. Gilman, Contribution to this symposium

[19] H. Albrecht at al. (ARGUS), Phys Lett. B 199 (1987) 451
I. Kim (CLEO), Talk given at the XXIV International Conference on High Energy Physics, München, August, 1988

[20] P. Oddone, Proceedings of the UCLA Workshop on Linear Collider B-Factory Conceptual Design, Los Angels, January, 1987

[21] I. Dunietz and T. Nakada, Z. Phys. C 36 (1987) 503

[22] R. Alexan et al., Phys. Rev. D39 (1989) 1283

[23] H. Nesemann, W. Schmit-Parzefall and H. Willeke, Contribution to the European Accelerator Conference, Rome, June 1988
R. Eichler et al., PSI-PR-88-22, 1988 (PSI)
D. Hitlin, T. Nakada and A. Sanda, Proceedings for Ref. 17
H. Nesemann et al., DESY 89-080, 1989 (DESY)
see also an earlier work by
I. Peruzzi, Proceedings for Workshop on Heavy-Quark Factory and Nuclear-Physics Facility with Superconducting Linacs, Courmayeur, December 1987

[24] See for example Ref. 21

[25] T. Nakada, PSI/TN-2/89 and PSI/TN-3/89, 1989 (PSI)

[26] H. Albrecht et al. (ARGUS), Phys. Lett. B 192 (1987) 245
   M. Artuso et al. (CLEO), Phys. Rev. Lett. 62 (1989) 2233

[27] T. Nakada, PSI/TN-3/88, 1988 (PSI)

[28] See Ref. 22 and Ref. 27

[29] See Ref. 22, Ref. 27 and H. Nesemann et al. in Ref. 23

[30] E. Eichten, Phys. Rev. D 22 (1980) 1819

[31] S. Ono, A. I. Sanda and N. A. Törnquvist, Phys. Rev. D 34 (1986) 186

[32] T. Nakada, PSI/TN-1/88, 1988 (PSI)
   D. G. Cassel et al., Proceedings for Ref. 17

[33] See Ref. 9
   S. Stone, Mod. Phys. Lett. A 3 (1988) 541
   R. Eichler et al. in Ref. 23

[34] See for example
   P. Roudeau, Talk given at XXIV the Recontres de Moriond, Les Arcs, March, 1989

[35] See for example
   W. Hofmann, "Jet of Hadrons", Springer, New York-Berlin-Heidelberg, 1981

[36] W. B. Atwood, I. Dunietz and P. Grosse-Wiesmann, Phys Lett. B 216 (1989) 227

[37] J.-M. Gerard and W.-S. Hou, MPI-PAE/PTh-26/88, 1988 (MPI-München)

[38] See for example Ref. 9

[39] A. Datta, E. A. Paschos and U. Turke, Phys. Lett. B 196 (1987) 382

PROSPECTS FOR CP VIOLATION EXPERIMENTS IN HADRON BEAMS

M.S. Witherell
Univ. of California, Santa Barbara, CA 93106

ABSTRACT

Some of the most important questions in particle physics can be addressed by very sensitive studies of B meson decay. With the next generation of hadron colliders, experiments with up to 10^{12} produced B mesons will be possible. This paper addresses the question of whether it will be possible to observe CP violation at the expected level with such experiments.

INTRODUCTION

One of the most important problems in elementary particle physics is the origin of CP violation. There is a natural explanation for CP violation in the Standard Model with three generations of quarks, and the unique opportunity to test this explanation is in decays of B mesons. I will not try to reproduce the excellent summary of the situation given by Fred Gilman,[1] but will quote the main result. Although the expected CP violation asymmetries are large (10%), they occur in decay modes with branching fractions of less than 10^{-4}. The required sensitivity is orders of magnitude beyond that achieved by existing experiments.

The large production of B mesons in hadron collisions raises the prospect that the necessary sensitivity can be reached at hadron accelerators. Certainly the experience with charm physics is encouraging, as we have heard at this conference. The advent of silicon vertex detectors and powerful computing engines has made it possible to obtain large, clean charm signals in fixed target experiments. As of 1990, groups will be collecting data samples with 30,000 events in the best modes, such as $D^+ \to K^-\pi^+\pi^+$.

The contrasting situation in bottom physics is striking. Production of B mesons has been observed and measured, as detailed in the review at this conference.[2] On the other hand, there has been no B decay physics done at hadron machines, for fairly straightforward reasons. At fixed-target energies, the charm cross section represents about 10^{-3} of the total cross section, but the fraction of events with bottom production is about 3×10^{-7}. At very high energy colliders, the bottom cross section is large, but all existing experiments have been designed to study high-p_T physics.

In this paper I will review the prospects for making dramatic progress in B decay physics at hadron machines. I will start with a brief review of the status of fixed target experiments in the near future. I will then describe proposals for collider experiments designed to study B decays. Finally I will attempt to assess the long-term prospects for CP violation experiments using B decays. In particular, I will try to detail what are the major experimental problems that must be overcome to reach the desired sensitivity.

FIXED TARGET EXPERIMENTS—1990

As mentioned above, the difficulty in pursuing B physics in the fixed target environment comes from the fact that the bottom cross section is about

3×10^{-7} of the total cross section. This improves for high-A nuclei, but the high-p_T background also increases. There are three problems: (1) How can one build a sufficiently selective trigger? (2) How can one suppress the backgrounds enough to see clean signals? and (3) What interaction rate can the experiment stand?

There are proposed solutions to the first two problems. One is to use specific decay modes, such as $B^0 \to \pi^+\pi^-$ or ψK_S, which have distinctive signatures. There is also the possibility of using photoproduction, in which bottom production is a relatively high fraction of the cross section, about 10^{-5}. I will give examples of these approaches from experiments running at Fermilab in 1990, but I will make no attempt to give an exhaustive list.

Fermilab experiment E771 is designed to study the process $B \to \psi + X$, $\psi \to \mu^+\mu^-$. At $\sqrt{s} = 40$ GeV, the bottom cross section is expected to be about 10μb. The $\psi \to \mu^+\mu^-$ decay gives a signature which is both clean and easy to trigger on. About 10^{-3} of the ψ's come from B decays, and silicon vertex detectors will be used to separate the ones that do. At least initially in 1990 the experiment will run with an interaction rate of 2×10^6 Hz, corresponding to 5×10^{12} interactions total. This corresponds to 2.5 million $b\bar{b}$ events, of which 3500 include the process $B \to \psi X$, $\psi \to \mu^+\mu^-$. With acceptance and efficiency of about 10%, this would lead to 350 observed B events, which would be enough to measure the cross section and lifetimes. Of this inclusive sample, about 10 should be identified as $B \to \psi K_S$, $K_S \to \pi^+\pi^-$, to name one example. This experiment will give very valuable experience in operating a silicon vertex detector with MHz interaction rate. The collaboration will explore how to operate a single muon trigger, so that other decay modes can be observed in the future.

The Fermilab experiment E789 is designed to study the two-body decays $B^0 \to \pi^+\pi^-$, $K^+\pi^-$, etc. This experiment uses the two stage spectrometer built for E605 to study high mass pairs, such as $\Upsilon \to \mu^+\mu^-$. The first stage has a 15 meter long magnet which selects high-mass pairs; the second stage has particle identification, etc. The active elements are shielded from the full intensity of produced particles, except for the silicon vertex detector (SVD). Even the silicon will not cover the region near the beam direction. The desired interaction rate is 5×10^8 Hz, for a total of 6×10^8 $b\bar{b}$ events in one run. This would give 10^4 $B^0 \to \pi^+\pi^-$ events if the branching ratio is 2×10^{-5}, of which 60 would be observed.

The first question for this experiment is whether the SVD can operate successfully at a 5×10^8 Hz interaction rate. The second is whether the vertex cuts can reduce the dihadron background by a factor of 10^4–10^5. An additional physics benefit from this experiment should be good limits on the decay modes $B^0 \to ee$, $e\mu$, and $\mu\mu$, where the backgrounds are very low.

The Fermilab experiment E687 will be operating with a high intensity photon beam in the 1990 run. Its major goal will be to collect 10^5 reconstructed charm events, or 10 times the sample collected by E691. Two search strategies for B mesons come to mind. Of the 10^5 charm events, about 100 will be due to B decay, of which about 20 are semileptonic decays. This should give a clear signal using good lepton identification and vertex cuts. In addition, there will be about 250 events in which the two B mesons decay to $e^-\mu^+$ or $e^+\mu^-$. It might be possible to identify 20 or so events of this type.

Although photoproduction does provide a less dilute source of bottom quarks than does hadroproduction, it may not be possible to increase these rates substantially. The flux and energy of the photon beam are the limiting factors, since the production of the beam is a four-step process. Direct photon beams are possible, but they have the problem of large backgrounds from neutral hadrons.

In summary, there will be a large number of heavy quark experiments running in a high-rate fixed target environment soon. Experiments such as E687 and E791 will observe <u>charm</u> samples 10 times as large as E691, and there are plans for another order of magnitude in subsequent experiments. There will also be first results from the first fixed-target experiments designed to study B-decay physics at high rates. In two years or so we will know more about the ultimate limits of such experiments.

HADRON COLLIDERS

An important fact from strong interactions is that production of B mesons is large at hadron colliders. The $b\bar{b}$ cross sections have been calculated to order α_s^3 and they are large: $500\mu b$ at the SSC and $50\mu b$ at the Tevatron[3] The differential cross sections are well known, too. The rapidity distribution is flat to about a rapidity of 5 at the SSC and 3 at the Tevatron. The average p_T is about equal to the b quark mass, 5 GeV. The average rapidity gap between b and $\bar{b}$ is 1-2 units. Figure 1 shows the p_T distribution, taken from a paper by Ed Berger,[4] and the rapidity dependence. The important lesson is that to attain good acceptance, one must accept tracks down to $p_T \simeq 1$ GeV/c, over a very wide range in rapidity.

The Bottom Collider Detector (BCD), which has been proposed to Fermilab, is shown in Fig. 2. It covers the central region, $|\eta| < 3$, where η is the pseudorapidity, $\eta = -\ln(\tan\theta/2)$. The spectrometer is designed to detect low multiplicity decays into all-charged modes, such as $\pi^+\pi^-$, $K^+\pi^-$, and ψK_s. The plan is to use a fast trigger based on an electron and/or tracks with $p_T \gtrsim 2$ GeV/c, followed by a second level trigger which uses massive processor farms. The collaboration submitted a letter of intent, along with a proposal for research and development. Besides successful results from the R&D efforts, the experiment depends on the main injector project at Fermilab and the construction of a third low-β interaction region.

There is also a proposal for the CERN SPS collider to build a forward B spectrometer. The experiment covers angles up to 600 mrad using 2 large magnets, a quadrupole at 3 meters from the interaction point, and a dipole at 9 meters. This proposal is being considered by the advisory committee at CERN.

What are the expected rates for BCD at the Tevatron? A favored decay process is $B \to \psi K_s$, with $\bar{B} \to D(D^*)e^-\bar{\nu}$. If we assume $\mathcal{L} = 2 \times 10^{31}$ cm^{-2}s^{-1}, the number of $b\bar{b}$ events produced in a Snowmass year of 10^7 live seconds is 10^{10}, of which 2×10^4 decay into these modes. In addition one needs to take into account acceptance (tracks must have $p_T > 1$ GeV/c), trigger efficiency, vertex efficiency, reconstruction efficiency, efficiency for identifying the electron, and acceptance of the K_s decay. An order-of-magnitude estimate of the product of these factors is 10^{-3}, which corresponds to 20 events observed, although the error is an order of magnitude or more.

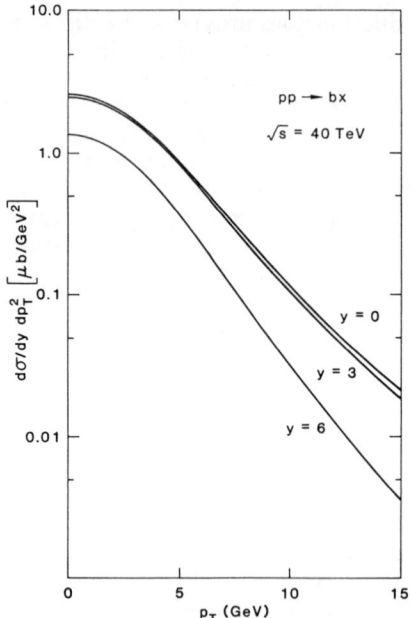

Fig. 1. The expected p_T distribution for production of B's at the SSC. The only major differences at lower energy machines is the rapidity range covered.

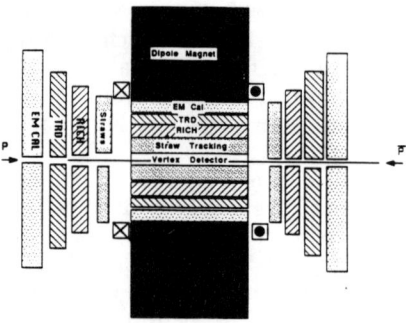

Fig. 2. The proposed Bottom Collider Detector at Fermilab.

PROSPECTS FOR CP VIOLATION

Finally I come to the main topic, the prospects for observing CP violation in B mesons produced using hadron collisions. Table 1 lists the relevant rates for producing B's in fixed target experiments, and colliders from the SPS to the SSC. The range of effective luminosities for the fixed target experiments is $10^{32} - 2 \times 10^{34}$ cm^{-1}s^{-1}, corresponding to interaction rates from 2×10^6 to 4×10^8 Hz.

Table I Rates for Bottom Production

Accelerator	$\sqrt{s}$ (TeV)	$\sigma(\mu b)$	L (cm^{-2}s^{-1})	$b\bar{b}$/year	$\sigma_{b\bar{b}}/\sigma_{tot}$
FNAL Fixed Target	0.04	0.01	10^{32}–2×10^{34}	3×10^6–10^9	10^{-6}
SPS Collider	0.6	10	10^{30}	10^8	2×10^{-4}
FNAL Collider	1.8	50	10^{31}–5×10^{31}	10^{10}	10^{-3}
SSC	40	500	10^{32}	5×10^{11}	5×10^{-3}

There is a preliminary caution that should be mentioned, which is relevant to all methods of studying B decays. The fact that CP violation is "large" in the B system does not mean that it is easy to measure. For comparison, if there were CP violation in the D^0 system at the level expected for B decays, we would still be at least three orders of magnitude away from being able to observe it. (We now observe D decay modes at a rate of 5×10^4 times the branching ratio.)

There are many advantages to the SSC for a B-decay experiment, relative to the Tevatron. The cross section is a factor of 10 larger, and the luminosity is a factor of 2-10 more, so there is a much greater production rate and a richer sample. The interaction region at the SSC has an r.m.s. deviation along the beam direction of 7 cm, which is a factor of 5 smaller than that of the Tevatron. This makes the vertex detector easier to build, and the vertexing more efficient. Finally the larger value of maximum rapidity means that a larger fraction of the cross section is in the forward region, $|y| > 1$, where muons are possible to detect. This is important because muons provide one of the most straightforward trigger signatures. The one possible drawback is that at SSC energies, there may be more jets of energy 5 GeV or so, which would complicate the fast triggers.

There is by now a long history to the idea of a B spectrometer for the SSC.[5-7] There are a few items of acquired wisdom from these various workshops. 1) The operating luminosity should be about 10^{32} cm^{-2}s^{-1}. At this rate the problems of triggering and radiation damage are already difficult. Also, a long clear space is needed along the beam for the detector to handle the rate in the forward direction, and theat requires lower luminosity than in the low-β regions. 2) The rates depend only on R_T, the transverse distance to the beam. Typical guesses at the minimum distance for good operation are 40 cm for calorimetry and 15 cm for straw tubes. 3) One needs the longest continuous acceptance in rapidity possible. The acceptance for observing both parts of the $B\bar{B}$ pair is roughly proportional to $(\Delta\eta - 1.5)$, so adding to the rapidity range gains a great deal.

The forward spectrometer roughly designed at the Berkeley workshop in 1987[6] consisted of a forward, two-stage spectrometer covering angles $\theta < 20°$. The forward spectrometer has a large dipole magnet, drift chambers, two ring-imaging Cherenkov counters, a transition radiation detector (TRD), electromagnetic calorimeter, and a muon filter. It covers from 6 to 57 mrad, and extends 75m from the interaction region. By the end of the Berkeley workshop, two major shortcomings were identified. The coverage extending to $\eta \simeq 6$ causes the transverse size to be ±4m, which makes the spectrometer too big and too costly. The second is that the magnet restricts θ_{max} to be 20°, and this causes too large a loss in acceptance.

The solutions to these problems were worked out at Snowmass in 1988.[7] The forward angle was increased to 14 mrad, corresponding to $\eta_{max} = 5$, which reduces the length and transverse size of the forward arm by a factor of two. The solution to the large angle acceptance was a monster cyclotron magnet, adapted by the BCD group for the Tevatron experiment from the Fermilab Workshop of 1987. The acceptance is naturally divided into three regions, a barrel region from 30°–150°, the intermediate region from 100mr to 30°, and the forward region from 14mr to 100mr. The spectrometer has good coverage from $\eta = +5$ to $\eta = -3$.

Figure 3 shows the sketch of this arrangement. The intermediate and forward regions are similar to the Berkeley version, and the barrel region is very similar to the BCD letter of intent. The major difference is in the vertex detector, which consists of barrel silicon detectors for angles greater than 45°, and disk detectors for angles less than 45°. Because the interaction region is short, the region of interlaced barrel and disk detectors only extends 14 cm from the central point, while the disk detectors extend to about 3m. In the Tevatron experiment, the overlapping region covers most of the acceptance of the vertex detectors.

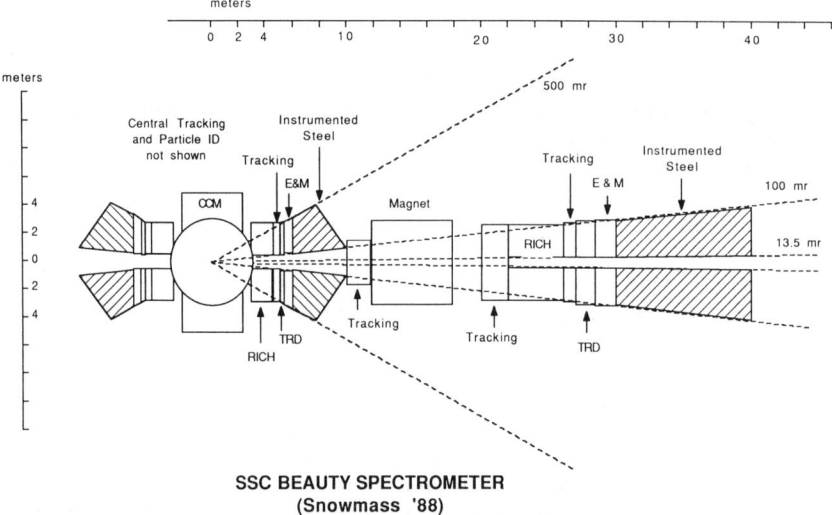

Fig. 3. The SSC Beauty Spectrometer, as seen by the group at Snowmass 1988 (reference 7).

The benchmark process for measuring CP violation is the one discussed earlier for the Tevatron, $B \to \psi K_s$ or $\pi^+\pi^-$, and $B^- \to D(D^*)e^-\nu$. The increased luminosity of 10^{32} cm^{-2}s^{-1} and the increased cross section of 500μb correspond to a production rate of 5×10^{11} bb events per Snowmass year. In addition, both muons and electrons are observable in the somewhat forward region where most of the acceptance lies. This experiment would see about 4000 events, in these modes, with opposite side tagged, even with an acceptance times efficiency of about 10^{-3}.

It will be possible to see clean signals, even in these rare B decay modes. The long B lifetime and the relatively rich $b\bar{b}$ production, 1/2% of the total cross section, are more favorable than charm in fixed target experiments. In addition, these are doubly tagged events, which further suppresses background. The question is, what is the cost in efficiency of obtaining the clean signal?

There are many difficult experimental problems to solve for this experiment. Some of the most important are the fast trigger; the data acquisition system, including the processor farms; the vertex detector; and the high rate capability required of all detectors. The general scheme for the trigger is to reduce the 10^7 Hz interaction rate to 10^5 Hz with a fast trigger, which is then fed into the numeric processors for a reduction to the tape-writing rate of 10^3 Hz. All of those numbers are merely estimates, however. The signature for the fast trigger might be high p_T electrons, or muons, or pions for the $B^0 \to \pi^+\pi^-$ decay. Perhaps one could use a dilepton trigger for ψK_s and a single lepton plus one high-p_T pions for the $\pi^+\pi^-$ mode. The processing capacity required for the second level trigger is about 10^3 that of the present farms and in CDF or D^0.

In summary, the performance required for the B spectrometer at the SSC is beyond that of any operating experiment. There needs to be a large R&D effort on about eight different systems to be able to show the viability of the experiment. No experiment of this type has operated at a high energy hadron collider, at any luminosity. The experience needed to design this experiment will come from R&D for the BCD and SPS experiments, from high rate fixed target experiments, and from dedicated R&D projects.

CONCLUSIONS

The B production rate at the SSC is about 5×10^{11} $b\bar{b}$ events per year, which represents 1/2% of the cross section. It is quite plausible that one can collect enough tagged B decays into rare CP eigenstates to measure CP violation, throughout the expected range. There are a number of problems to be solved, and they will require a large R&D effort. On the other hand, the physics of CP violation in the B system is so important that all reasonable paths should be explored; it is worth the effort.

REFERENCES

1. F.J. Gilman, in this volume.

2. R.J. Morrison, in this volume.

3. P. Nason, S. Dawson, and R.K. Ellis, *Particle Phys.* **303**, 607 (1988).

4. E.L. Berger, "Heavy Flavor Production," Argonne report ANL–HEP–CP–88–26.

5. J.W. Cronin, and M.S. Witherell, two chapters in "Proceedings of the 1984 Summer Study on the Design and Utilization of the Superconducting Super Collider, Snowmass—1984," Ed. by R. Donaldson and J.G. Morfin; Initial SSC Experimental Equipment, SSS–SR–1023, June 1986; and B. Cox, F.J. Gilman, and T.D. Gottschalk in "Physics of the Superconducting Supercollider, Snowmass—1986," Ed. by R. Donaldson and J. Marx.

6. K.J. Foley, et al., in "Experiments, Detectors, and Experimental Areas for the Supercollider," Ed. by R. Donaldson and M.G.D. Gilchriese (World Scientific, Singapore, 1988).

7. M.P. Schmidt, in the proceedings of Snowmass, 1988 (to be published).

B FACTORY PLANS AT CORNELL

Karl Berkelman
Cornell University, Ithaca, NY 14853

CESR HISTORY, STATUS, AND NEAR-TERM UPGRADE

Since the Cornell Electron Storage Ring (CESR) began operating in 1979 for e^+e^- experiments in two interaction regions occupied by the CLEO and CUSB detectors, the performance of the storage ring and the detectors have been frequently upgraded. The most ambitious improvements have come through a capital upgrade project started in 1985 and just now being completed.

CESR was modified in two ways that will become part of any e^+e^- B factory design.

(a) Microbeta. Permanent magnet quadrupoles were installed 66 cm from the interaction point to make $\beta_y^* = 1.5$ cm. This required that the RF accelerating voltage be increased to shorten the bunches to a similar value, $\sigma_z = 1.7$ cm.

(b) Multibunch. Instead of single bunches of e^+ and e^-, CESR now circulates 7 bunches in each beam. The beam orbits are separated by electric fields except in the CLEO and CUSB interaction regions.

As a result of these and other changes in the CESR design, the peak luminosity was raised a factor of five to a record value for e^+e^- machines: $L_{pk} = 1 \times 10^{32}$ cm^{-2}s^{-1}.

The CLEO detector has also been rebuilt with a new 1.5 T superconducting magnet and a high resolution electromagnetic calorimeter made of cesium iodide scintillators. CLEO-II will be the first large-solid-angle detector with energy resolution better than 2% for both charged particles and photons over most of the particle momentum range.

Even at the improved performance levels the potential for important discoveries in b-quark physics will be limited by the rate at which B mesons can be produced. We have therefore worked out a plan, called CESR-Plus, to increase further the CESR luminosity in the next two years. When the CUSB experiment is completed in 1990, CESR will be converted to a single-I.R. machine and the number of circulating bunches will be increased to 14 per beam. The increased damping per interaction and the increased number of bunches will enable us to store higher beam currents and achieve higher luminosity. The goal is $L_{pk} = 5 \times 10^{32}$ cm^{-2}s^{-1}.

In the past decade a number of important physics discoveries have been made by the CLEO and CUSB collaborations at CESR. Upsilon spectroscopy was enriched by the discovery of the $\Upsilon(4S)$, $\Upsilon(5S)$, and $\Upsilon(6S)$ $^3S_1$ resonances in the e^+e^- cross section and by the discovery of the intermediate ^3P states, $\chi_b(1P)$ and $\chi_b(2P)$, in photon transitions. The discovery of the pseudoscalar B (and vector B*) mesons above the flavor threshold led to the discovery that the b quark decays dominantly to the c quark; many modes were first observed at CESR.

PHYSICS GOALS

Since the b quark is the lighter member of the heaviest doublet of quarks, its decays involve all the lighter quarks, but only through off-diagonal couplings in the Cabibbo-Kobayashi-Maskawa matrix. The resulting decay rate suppression allows rare processes to compete and makes the study of b quark decays especially sensitive to new physics beyond the Standard Model. These rare processes often involve branching ratios beyond the sensitivity of existing e^+e^- machines. The following is a partial list of the phenomena one would like to be able to study if higher luminosities were available.

(a) Rare B-decay branching ratios and limits:
   charmless ($b \to u$),
   flavor-changing neutral currents ($b \to s\gamma$, $b \to sg$),
   nonspectator,
   Standard-Model-forbidden decays.
(b) Charge-tagged branching ratios:
   separate semileptonic branching ratios for B^+ and B^0,
   $B^- \to \tau \nu$.
(c) Branching ratios for B_s, B_c, Λ_b, ... decays.
(d) $B_s^0 - \bar{B}_s^0$ mixing.
(e) CP violation in B (and B_s) decays:
   mass matrix--
   ex.- $B\bar{B} \to \ell^+\ell^+$ vs $\ell^-\ell^-$
   flavor-specific decay asymmetry--
   ex.- $B^+ \to K^+\rho^0$ vs $B^- \to K^-\rho^0$
   neutral decay asymmetry through interference with mixing--
   ex.- $B^0\bar{B}^0 \to \psi K_s + \ell^+ X$ vs $\psi K_s + \ell^- X$
   CP-violating double-eigenstate decay--
   ex.- $e^+e^- \to B^0\bar{B}^0 \to \psi K_s + \psi K_s$

The total number of produced B mesons required to study these phenomena can be estimated from branching ratios and asymmetries predicted by the Standard Model with the known or guessed values of the quark masses and CKM mixing angles. For example, if $|V_{ub}/V_{cb}| = 0.1$, then the inclusive $b \to u$ branching ratio is about 1%, and the branching ratio for B to decay to any readily detectable exclusive charmless mode is at most 10^{-5}. Another example is the leptonic decay $B \to \tau\nu$; the branching ratio should be about 3×10^{-4}, but the tagging efficiency will cost another factor of 10^{-2} to 10^{-3}. CP violation can be observed through any of the four techniques listed, although the second (flavor-specific decay asymmetry) and third (neutral decay asymmetry through interference with mixing) are probably possible with fewer produced B's. If the violation comes from the phase parameter in the CKM matrix with three quark families, then the asymmetries can be larger than in K^0 decays, but the relevant branching ratios will be 10^{-4} or lower.

LUMINOSITY REQUIREMENT

The prospect of a measurement of CP violation in B decay has generated worldwide interest in building an e^+e^- collider in the CESR/DORIS energy range, but with higher luminosity than is now available. Although one would not build such a machine to study only CP violation, it is instructive to see what the two most promising CP violation scenarios require.

The measurement of the CP violation asymmetry is straightforward in the case where the decay mode is flavor-specific. For example, starting with $\Upsilon(4S) \to B^+B^-$, one simply compares the total numbers of $B^+ \to K^+\rho^0$ and $B^- \to K^-\rho^0$ observed. The flavor (B^+ or B^-) is obvious from the detected final state; no tagging is required and decay length measurements are not needed. The CP violation comes from a small interference between two diagrams in the decay amplitude, say spectator and annihilation or penguin. The theoretical prediction for the asymmetry in any such mode therefore involves not only the CKM parameters but also strong interaction dynamics. In our present state of ignorance this would make the interpretation of a measured asymmetry somewhat uncertain.

In the other scenario the CP violation comes through the interference between the decay of a neutral B directly to some final state (say ψK_s, or $\pi^+\pi^-$) and the decay through an intermediate stage where the B has oscillated to its antiparticle. If the final state happens to be a CP eigenstate, the asymmetry is likely to have little or no dependence on strong interaction dynamics. The expected asymmetry tends to be larger, although this advantage is at least partially offset by the necessity of tagging the initial flavor (B^0 or $\bar{B}^0$) by observing the decay of the other B produced in coincidence. A more serious difficulty is the fact that if the original $B^0\bar{B}^0$ pair is produced directly as $e^+e^- \to B^0\bar{B}^0$, say at the $\Upsilon(4S)$ resonance, then the asymmetry is always zero if one integrates over all decay times. In order to make an observation of a nonzero CP asymmetry one has to either (a) use $B^0\bar{B}^0$ pairs produced through $e^+e^- \to B\vec{B}^* \to B\bar{B}\gamma$, or (b) for each tag pair determine which B decays first. Option (a) requires that one run immediately above the BB^* threshold, thus losing about a factor of 4 in production rate relative to running at the $\Upsilon(4S)$ resonance. One can retain the advantage of running at the $\Upsilon(4S)$ in option (b), but only by producing the $B\bar{B}$ in motion with unequal e^+ and e^- energies, and measuring the decay lengths.

Even if we knew how to calculate the relevant branching ratios and asymmetries, the prediction of the number of produced $B\bar{B}$ pairs needed to make a significant measurement of the CP asymmetry will obviously depend on which scenario, decay mode, and energy ($\Upsilon(4S)$ or above) are picked, and how good the detector efficiency and resolution are. For favorable modes and detector performance assumptions one estimates that about 10^8 $B\bar{B}$ pairs will be needed, whichever scenario and energy are chosen. Since CESR is now capable of producing about 10^6 $B\bar{B}$ pairs per year at the $\Upsilon(4S)$, we require about two orders of magnitude more luminosity, that is $L_{pk} = 10^{34}$ cm^{-2}s^{-1}.

THE IDEAL CESR B FACTORY

In addition to luminosity there are several other requirements that we would like to impose on a B factory to be built at Cornell. Ideally we would like to ask for each of the following.

1. $L_{pk} = 10^{34}$ cm^{-2}s^{-1} (at $E_{cm} = 10.6$ GeV), in order to study CP violation and the other listed physics goals.

2. E_{cm-max} = 15 GeV, above the threshold for B_c, at least as an upgrade option with some sacrifice in luminosity.

3. Should be able to run symmetrically or (at the $\Upsilon(4S)$) asymmetrically, e.g., 5.3 + 5.3, 3.8 + 7.5, or 7.5 + 7.5 (GeV).

4. Should make use of the present CESR tunnel, injector, and detector, appropriately upgraded.

Since the luminosity we are asking for is two orders of magnitude beyond the present state of the art, and no one has real experience with asymmetric colliders, it is reasonable to ask whether such a machine is possible.

CESR-B STUDY

At Cornell we have begun an R&D effort to try to answer by early 1990 the question of whether it is possible to build a B-factory with the specifications given above. If we are not discouraged by what we learn, we intend to make a conceptual design and preliminary cost estimate.

For the case of symmetric collisions of flat beams one can write the luminosity (units of 10^{34} cm^{-2}s^{-1}) in terms of the total current per beam I (Amps), the vertical tune shift ξ, and the vertical β^* function at the I.R. (cm):

$$L = 11.5\ I\xi/\beta^*.$$

If one makes reasonable assumptions for tune shift and β^*, a luminosity of 10^{34} cm^{-2}s^{-1} requires currents of the order of 3 Amperes. For comparison, CESR now collides 70 mA per beam; several hundred mA have been achieved under special conditions. This prodigious beam current requirement leads to a number of problems which demand a serious research effort before a credible B-factory design is possible. Single-beam stability considerations demand that the number of bunches be as high as possible.

1. <u>RF system</u>. It must couple megawatts of power into the beams to make up for synchrotron radiation losses. The accelerating volatage must be much higher than normal in order to keep the bunch length at 1 cm or less. The cavities must have a minimum contribution to the ring impedance, which can limit the stable beam current. We are working on a solution involving single-cell superconducting RF cavities.

2. <u>Separation</u>. The two beams must be kept separate except at the experimental I.R., and the longitudinal spacing between bunches must be minimized without sacrificing focusing power at the I.R. Of the various schemes (electrostatic, delay line, RF, magnetic, and crossing angle), all have serious problems.

3. <u>Beam profile</u>. Should the beams be round at the I.R. instead of flat, as in existing machines? Simulations indicate a luminosity advantage for round beams. We will be making a test soon in CESR.

4. <u>Vacuum system and beam pipe</u>. Poor vacuum will hurt beam lifetime, exacerbate ion trapping, and make intolerable experimental backgrounds. It will be a challenge to make a quality vacuum in the face of megawatts of synchrotron radiation power and heating due to beam-induced currents, while keeping the ring impedance low.

5. **Asymmetric energies.** In addition to the above problems that must be faced for any high luminosity e^+e^- collider, an asymmetric collider has its own special problems. A way must be found to optimize the parameters for each beam for maximum current, while optimizing the beam-beam interaction for high luminosity. The increased length of RF cavities in the high energy ring should not force too high a ring impedence and limit the maximum stable current. The high and low energy I.R. focusing elements must be kept separate without compromising the achievement of low β^* or making intolerable backgrounds for the experiment. The backgrounds at the experimental I.R. must be low enough to enable <30μm tracking accuracy at about 1 cm from the beam line.

There are a number of choices to be made which are strongly conditioned by the physics goals and by detector considerations. We hope to get community input on these at a workshop to be held at Syracuse University in September, 1989.

CONCLUSIONS

1. Continued experimental progress in B physics will be important to our understanding of the Standard Model and may give clues to the more general theory beyond it.

2. An e^+e^- collider operating at or near the $\Upsilon(4S)$ resonance with a luminosity of the order of 10^{34} $cm^{-2}s^{-1}$ is likely to be the most useful facility for the study of B physics, and for CP violation in particular.

3. Our best bet for a successful e^+e^- B factory is to build on the experience at Cornell in making high luminosity.

4. The CESR operations group is pursuing an R&D program to resolve the acclerator physics questions that have to be answered before making a credible design. Assuming that the present R&D program is encouraging, a conceptual design and a proposal will follow.

DETECTOR FOR THE B FACTORY AT PSI

J.Chauveau

L.P.N.H.E. Universites Paris VI et VII. FRANCE

In the proposal[1] for the B factory at PSI, now known as the BETA project, a study of the accompanying universal detector was included[2]. We summarize here its main points. We start [§1] by describing the constraints for such a detector. We mention without any details the machine design and outline the physics goals which we, as of now, think will be of relevance in the years 1995 and beyond. Then we list the various detector elements which constitute the design and explain the choices that have been made [§2]. We go on by describing the subdetectors which are the most characteristic of the apparatus, namely the silicon vertex detector[§3], and the ring imaging Cherenkov detector (RICH) [§4]. We close [§5] by giving some examples of the detector performance.

1. THE CONSTRAINTS FROM THE PHYSICS AND THE MACHINE

B factories are proposed to try to discover CP violation in the decay of the b quark. This is predicted in the standard model[3] to give rise to rare and subtle effects (mainly asymmetries of various kinds). The paucity of events explains the need to construct high luminosity colliders (factories), the experimental difficulties require a sophisticated universal detector to be installed at such an accelerator.

The BETA factory, planned at PSI, has been described elsewhere[1,4]. Here we just recall it is a two ring $e^+ e^-$ collider which has been optimized to reach a maximum peak luminosity of 10^{33} cm^{-2}s^{-1} when producing the $\Upsilon(4S)$ resonance at rest. We hope this figure can be improved by a factor 5 after a few years experience. The maximum beam energy is 7 GeV and this enables an asymmetric mode of operation leading to a maximum boost of the $\Upsilon(4S)$: $\beta\gamma=0.28$. In this configuration which has been advocated to study specific CP violating decay modes of the B_d meson[5], the maximum peak luminosity is reduced, perhaps by a factor 3. The means by which the collider obtains the high luminosity affect the detector geometry and operation. The mini-β quadrupoles (see Fig.1) will be as close as 70 cm from the interaction point. The bunch frequency (10 MHz) is a challenge for the trigger and data acquisition. The high beam currents will create unprecedented levels of background for an $e^+ e^-$ machine. Note however the advantage of the two-ring concept in which the interaction point can be offset from the plane of the hard bending magnets and shielded from the fiercest synchrotron background.

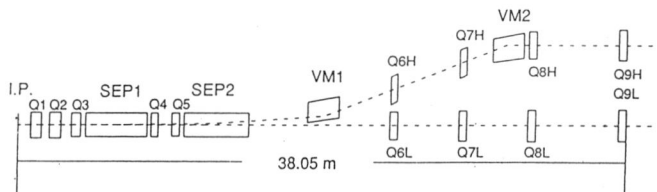

Figure 1 : The design of the intersection region. Two horizontal separator, two vertical bending magnet and quadrpole magnets are shown

Apart from CP violation, the main physics topic for BETA will presumably be B_s mixing. Less spectacular but mandatory to narrow down the standard model quantitative predictions for CP violation and mixing, one will need precision measurements of experimental quantities like the CKM matrix element V_{ub}, or the B mesons decay constants. At the $\Upsilon(4S)$ the events are exclusively $B\overline{B}$ pairs [1] and the main problem one faces is the combinatorial background since the the decay products of the B and the $\overline{B}$ are mixed in the detector. The known features of B decays[6] (charged and neutral multiplicities, predominance of charm in the final state with the corresponding strangeness enhancement) have been used to design a better detector than the present experiments (in particular CLEO II which is just turning on) in secondary vertex detection, charged particle identification and π^0 reconstruction. We will explain how this is achieved in the next section.

Note that, in such an experiment, a lot of other physics topics will be addressed like the rare decays[1] of the B mesons and the τ lepton. In fact, BETA will be a more than decent τ-charm factory.

2. DETECTOR OVERVIEW

The constraints outlined in the previous section have been combined to set forth the following goals for the detector design: π/K separation up to 3 GeV/c; secondary (charm) vertex reconstruction capability; state of the art electromagnetic calorimetry; hermeticity and 4π coverage. The result of the study (Fig.2) is a general purpose detector which, on top of the usual tracking and calorimetric devices, incorporates a silicon vertex detector and a RICH. The latter two subdetectors are quite new in a medium energy e^+e^- experiment and therefore will be described in separate sections. In this one, we give a brief account of the other more conventional detector elements. The cost of the apparatus is estimated to be 50 millions Swiss francs.

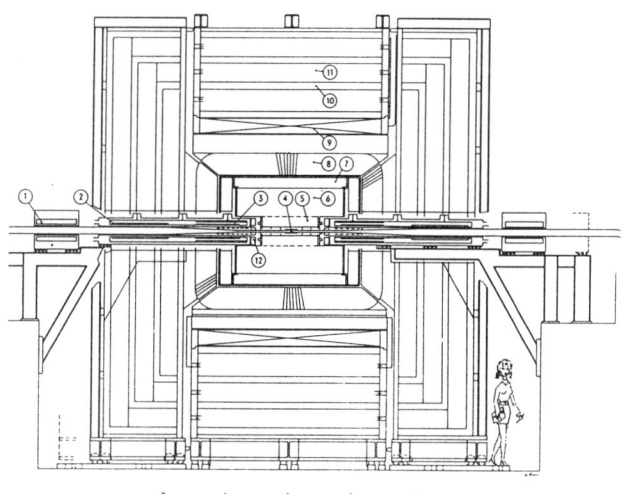

Fig.2 Detector side view showing the SVD(4), the PTC(5), the MTC(6), the RICH(7), the shower counter(8), the coil(9), and the muon filter(10,11).

---

[1]this is not exactly true for center of mass energies above the resonance but this extra complexity can be safely neglected

Tracking. All subdetectors except the muon chambers are immersed in a 1.5 T solenoidal magnetic field. Three devices contribute to the measurements of the momentum and angles of charged tracks. The momentum resolution for a 90^0 track is:

$$(\sigma_p/p)^2 = (0.0019p)^2 + (0.0048)^2.$$

The silicon vertex detector (SVD) [§3] provides r-ϕ and z measurements at 2 radii: 1.7 and 5 cm.

The precision tracking chamber (PTC) occupies the space between 6 and 20 cm from the beam line. The technology of this gaseous chamber has yet to be decided upon. It will provide accurate momentum and angular measurements ($\sigma_{r\phi}$, $\sigma_z \sim 100\ \mu m$ on a handful of layers), the capability to match tracks with the SVD points, a fast first level trigger mainly based on the r-z view, high efficiency tracking for low p_t tracks ($\epsilon \geq 90\%$ for p_t=50 MeV/c), and efficient secondary vertex reconstruction for K_s and Λ particles.

The main tracking chamber (MTC) takes over for radii between 20 and 65 cm. Its total length is 170 cm which means that the solid angle for tracks seing the whole depth of this device is 79% of 4π. It is a jet chamber split in two arrays of 30 and 60 tilted sectors respectively. 11460 (2430 sense and 9030 field) wires provide 52 layers with $\sigma_{r\phi} \sim 80\ \mu m$ and amplitude measurements for dE/dx (see below). Coarse z measurements will be obtained by charge division. However an array of cathode strips in the outer skin of the chamber will provide a precise enough space determination for matching to the RICH. With the slow gas (80/20 CO_2/isobutane) used and the chosen cell geometry the maximum drift time is 4 μs. The MTC takes part in the second level of the trigger (r-ϕ).

Particle Identification. Charged particle identification is obtained by means of up to 52 dE/dx measurements in the MTC, the RICH [§4], the electromagnetic calorimeter and the muon filter[1]. The precision on the specific ionization expected from the MTC is 7% rms. The sensitivity of this device is pictured on Fig.3a where the usual drop of resolving power for momenta close to 1 GeV/c is obvious. The increased leverage brought by the RICH is shown on Fig.3b where it can be seen that pions and kaons can be distinguished with a significance better than 3 σ over most of the relevant momentum range for a run on the $\Upsilon(4S)$ at rest. Our Monte Carlo study of π/K separation in $B\overline{B}$ events shows that 95% of the kaons are correctly identified and that the sample is 99% pure.

Calorimetry. To try to approach for π^0's the momentum precision achieved for charged particles, a shower counter of the best kind has to be built. We have chosen to surround the RICH with a total absorption calorimeter consisting of 12160, 18 radiation length (34 cm) CsI(Tl) crystals. This technique is already employed in several experiments (CLEO II and the Crystal Barrel at LEAR) and will not be described here. The barrel starts at a radius of 86 cm. With the endcaps the solid angle coverage is 96% of 4π. This is increased to 98% with a coarse precision forward calorimeter[1] and even more if the luminosity monitor can be used for this purpose. The expected performance can be summarized by :

- the energy resolution

$$(\sigma/E)^2 = (1\%/\sqrt{E} + 1\%)^2 + (\sigma_{noise}/E)^2 + \sigma_{rl}^2 + \sigma_{nc}^2; E\ in\ GeV$$

where σ_{rl} comes from the fluctuations of the shower leakage at the back of the crystals and σ_{nc} from the particles which traverse the photodiodes. The first two figures could be improved by a factor two if the presently underway R&D is totally successful.

- the angular resolution $\sigma_\theta(E) = (3mrad/\sqrt{E} + 1mrad)\sin\theta$ (the $\sin\theta$ factor is absent for the endcaps i.e. $\theta \leq 20^0$)

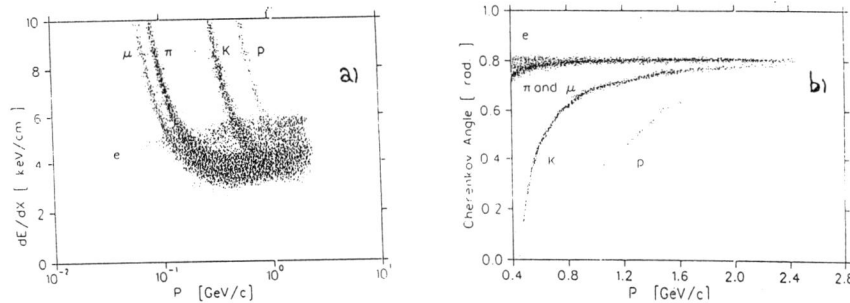

Fig.3 Particle identification: a) dE/dx vs p and b) θ_{ch} vs p.

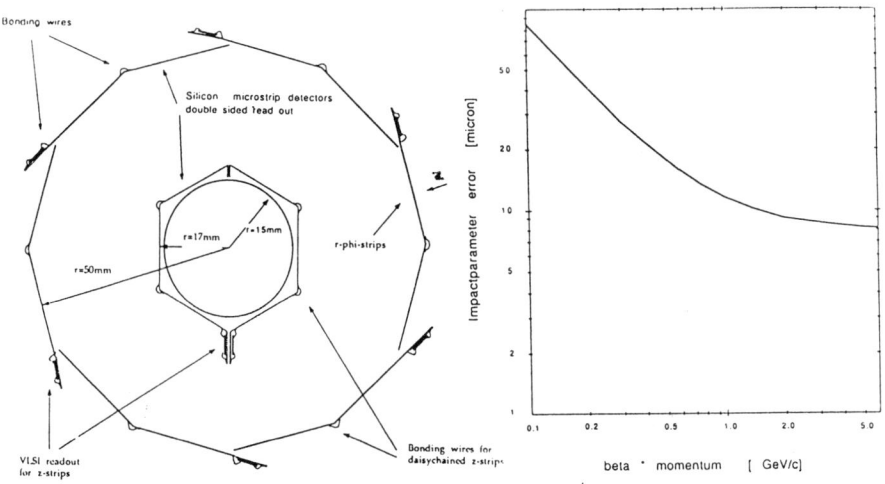

Figure 4a : r-φ view of Silicon Vertex Detector

Figure 4c   Impact parameter resolution for the Silicon Vertex Detector(SVD)

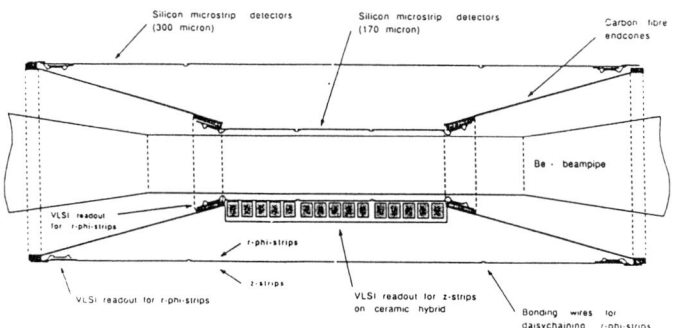

Figure 4b : r-z view of Silicon Vertex Detector

- A noise level of 100-150 keV per slab. This is mainly due to a recent breakthrough in the light collection part of the readout process[7].

<u>Trigger and Data Acquisition.</u> The physics rate at the $\Upsilon(4S)$ is expected to be 7 Hz, not including the 2γ processes (15 Hz). The average event buffer length is estimated to be 15 kbytes. We plan to use triggering methods pioneered by the HERA collaborations to keep the background trigger rate no higher than ~ 10 times the physics rate. This implies a multilevel trigger scheme including programmable processors. For data acquisition VLSI circuits on the very detector will be the standard.

3. THE SILICON VERTEX DETECTOR

In the symmetric mode of operation of the collider at the $\Upsilon(4S)$, a charmed particle coming fom B decay flies 110 μm on the average and a B meson flies 25 μm. In the asymmetric mode, the latter will have a decay length of the order of 60 μm. Thus an accuracy of ~ 20 μm on impact parameters is called for.

The proposed layout of the main part (SVD) is schematized on Figs. 4a and b. The machine allows a very thin (190 μm of Be) and small radius (1.5 cm) beam pipe. The SVD consists of two polygonal assemblies of Si detectors. The inner layer is a hexagon made of $170\mu m$ wafers at a minimum radius of 1.7 cm. Its total length is 11.6 cm. Because the distance from the interaction point is short and the cumulated thickness of the beam pipe and silicon small, the effect of multiple scattering is minimized. The outer layer is a 12 face polygon made of $300\mu m$ wafers at a minimum radius of 5 cm and its length is 31 cm. On each layer there are strips for $r - \phi$ and z measurements to an accuracy of ~ 5 μm. The detector is complemented in the forward directions with vertical counters, the silicon forward detector (SFD), which we do not describe. The overall solid angle coverage is 99.5 % of 4π.

The large number of channels (56064 for the SVD, and 32768 for the SFD) require the use of VLSI electronics for the readout. A circuit based of CMOS technology and incorporating, for 128 channels, low noise preamplification, sample and hold, and, for the first time, analog multievent storage, is under developpement.

From Monte Carlo, the achievable impact parameter resolution is:

$$\sigma_\delta^2 = (8.3\mu m/\beta p_t)^2 + (8\mu m)^2$$

(Fig.4c) which corresponds to standard deviation errors of 16 and 26 μm respectively for the $J/\psi K_s$ and $l^\pm X$ vertices for the golden events[5] looked for in the asymmetric mode.

Evidently the above design is not final. We already know that the inner layer is not obvious to build and some R&D incorporating bipolar transistor structures within the detector silicium is currently investigated[8].

4. THE RICH

π/K separation helps reduce the combinatorial background in B reconstruction since the CKM favored decay chain is b $\to$ c $\to$ s. It is espicially crucial for distinguishing two-body rare decays like $B_d \to \pi\pi$ or $K\pi$ which have high momentum prongs. It is mandatory to study the CP violating asymmetry in the rates $B_d \to \pi^+ K^-$ and $B_d \to \pi^- K^+$. It provides the means to study rare τ decays involving kaons.

The proposed RICH (Fig.5) enables a 3σ separation up to ~ 2.5 GeV/c. It is a proximity focused imaging counter which occupies the space between the MTC (R=65 cm) and the shower counter (R=85 cm). The Cerenkov light is produced by fast charged particles in a 1 cm thick sodium fluoride radiator (the transparent solid with the lowest refraction index in the vacuum ultraviolet region) and detected in a multiwire proportional chamber (MWPC) with pad readout which is made photosensitive by the inclusion of TEA vapor in its gas mixture. The geometry is such that the image reconstructed from the pad pattern are narrow enough to provide the required particle identification. The barrel consists of 24 sectors like the one pictured on Fig.5. With the endcap the solid angle covered by the RICH is 95 % of 4π.

Compared to the LEP RICH or the SLC CRID, this counter is fast. This is desirable at a high luminosity machine, makes the mechanical construction much simpler and lighter but has the drawback of requiring the readout of 380k channels. For this purpose a VLSI circuit is presently under study at Rutherford Lab. Another advantage of this design is the improved stability of the photon detector because TEA gives much less photoelectron feedback than TMAE. The MWPC enables also a fast charged multiplicity trigger to be devised from its wires. The RICH thickness amounts to 20% of a radiation length in front of the electromagnetic calorimeter and will somewhat degrade the π^0 reconstruction efficiency. This is currently being studied in Monte Carlo. A prototype for the barrel RICH is under construction and should be in operation in the summer of 1990.

Recent measurements on the TEA quantum efficiency[9] and the chromatic variation of the NaF index of refraction[10] have shown that the expected performances described in ref.[1](see Fig.3b) were a little optimistic. We estimate now that the 3 σ separation limit between pions and kaons will be at 2.5 GeV/c instead of 2.8. An exhaustive study is in progress. This however should not affect drastically the results obtained in ref.[1] for heavy flavor reconstruction which showed, for instance a reduction of a factor 3 in the combinatorial background under the D^0 mass peak(Fig.6) in representative samples from $B\overline{B}$ events.

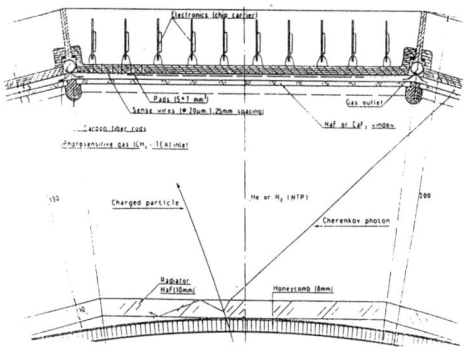

Figure 5. Cross section of the fast RICH counter (r vs φ view)

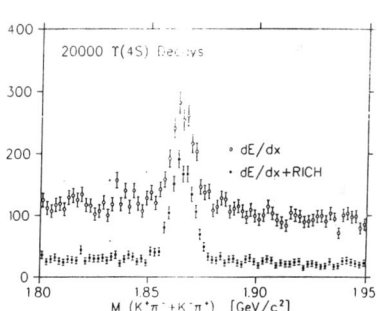

Fig.6 Inclusive D^0 mass for the $K\pi$ channel for $B\overline{B}$ with (•) and without (o) using the RICH.

5. A FLAVOR OF THE DETECTOR PERFORMANCE

We have simulated the detector performance on B meson reconstruction and some specific reactions of interest like $B_d \to J/\psi K_s$ and exclusive semileptonic decays. The efficiencies to reconstruct exclusive final states with ≤ 1 π^0 have been found above 40% with small background. The overall reconstruction efficiency is not yet known since a lot of decay chains with small branching ratios have to be added. For the chain: $\overline{B_d} \to \pi^-(D^{*+} \to \pi^+(D^0 \to$

$K^-\pi^+$)) we find 1 standard deviation resolutions of ~ 5.5 and 0.3 MeV/c² for the D^0 mass and the D^*-D mass difference respectively. The relative merits of an asymmetric run at the $\Upsilon(4S)$ and a symmetric operation above the threshold for $B\overline{B}^*$ production have been compared for the study of the $(J/\psi K_s)(l^\pm X)$ events. Partial reconstruction methods for b → u semileptonic exclusive decays with efficiencies above 10% and background levels smaller than 10^{-4} have been developped.

6. CONCLUSIONS

A brief account of the initial studies of the universal detector to be installed at the BETA facility at PSI has been given. Simulation and detector developpement are currently being actively pursued and should materialize next as a detector proposal shortly after the accelerator is approved.

REFERENCES

[1] Proposal for an Electron Positron Collider for Heavy Flavour Particle Physics and Synchrotron Radiation, PSI PR-88-09.
[2] Physicists from Aachen, CERN, Dortmund, Erlangen, Geneve, Heidelberg, Karlsruhe, Krakow, Munich, Neuchatel, Orsay, Paris, Siegen, Strasbourg, Villigen (PSI), and Zurich were involved in the detector study.
[3] see e.g. I.Bigi, in CP Violation C.Jarlskog ed. (World Sc. 1988).
[4] T.Nakada, PSI PR-89-07, to be published in the Proc. XXIVth Renc. Moriond.
[5] I.Dunietz and T.Nakada, Z. Phys C36, 503 (1987).
 R. Aleksan et al., Phys. Rev. D39, 1283 (1989).
[6] K.R.Schubert, these Proceedings.
[7] E.Lorenz, Max Planck Report MPI-PAE Exp. E1, 147 (1984).
[8] R.Horisberger, The Bipolar Silicon Microstrip Detector, to be published in the Proc. of a Conf. held in Munich (1989).
[9] R.Arnold et al., private communication.
[10] M.Hempstead et al., in preparation.
[11] T.Nakada, these Proceedings.

The Status of the TPC/2γ Detector at PEP

Helmut Marsiske
Stanford Linear Accelerator Center
Representing the TPC/2γ Collaboration

Abstract

The TPC/2γ detector and the PEP e^+e^- storage ring have been upgraded for high-luminosity running. We describe the performance of the new configuration during a test run in fall 1988. Prospects for progress in τ physics, based on a large data sample to be recorded with the TPC/2γ detector, are reviewed.

Storage Ring and Detector

Since 1986 the TPC/2γ (see Fig. 1) has been the only detector at the PEP storage ring [1]. The central Time Projection Chamber (TPC) is used to track charged particles over 87% of 4π sr, and to identify them by the simultaneous measurement of ionization energy loss (dE/dx) and momentum in a 1.325 T magnetic field. The dE/dx resolution for a track in a hadronic event is 3.4% and the momentum resolution was $(\sigma_p/p)^2 = (0.015)^2 + (0.007\,p)^2$, with p in GeV. The central detector has recently been upgraded by inserting a 14 layer straw vertex chamber (VC) with a beryllium beam pipe, starting at 4 cm radius from the beam [2]. The VC has an impact parameter resolution of 90 μm at 1 GeV and 40 μm at 10 GeV, and we expect it to improve the momentum resolution to $(\sigma_p/p)^2 = (0.011)^2 + (0.003\,p)^2$. Photons are detected in a 10.4 radiation length (r.l.) thick electromagnetic calorimeter which surrounds the 0.87 r.l. superconducting coil. The calorimeter is highly segmented (8 mrad projective strips) and achieves an energy resolution of 17%/$\sqrt[4]{E/\text{GeV}}$ below 5 GeV and about 14% for Bhabha events. The detector is surrounded by muon chambers which identify muons above 1-2 GeV. To adopt the new mini-β scheme of the PEP machine (see below) the forward spectrometers were removed, retaining however the NaI arrays for tagged two-photon physics. During the test run in fall 1988 all major detector components performed well. (For further details see the following talk by Gerry Lynch.)

Also the reconfigured PEP machine performed well despite its 2.5 year shutdown. The instantaneous luminosity of PEP has been increased by a factor of about 2.5 by moving the face of the first quadrupole magnets to 3.5 m from the interaction point, thereby reducing β_y^* from 12 cm to 5.5 cm. During the run in fall 1988 we achieved typical peak luminosities of $\mathcal{L} = 6 \times 10^{31}$ cm^{-2}s^{-1} with peak total currents of $I = 42$ mA. (See Fig. 2 for a plot of $\mathcal{L}$ versus I for a typical fill.) The total current was limited by higher order mode losses which caused unacceptable heating of the VC. A further reduction of β_y^* was prevented

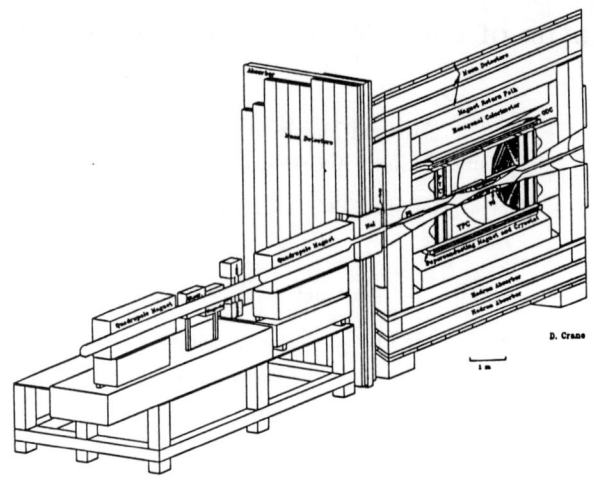

Figure 1: The TPC/2γ detector.

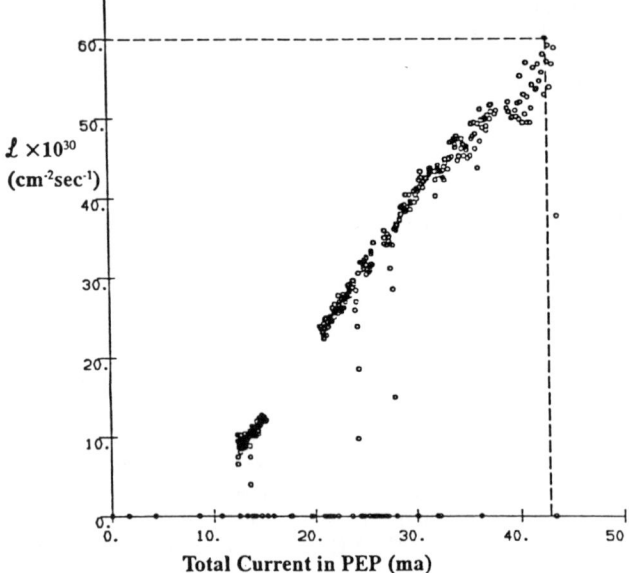

Figure 2: Instantaneous luminosity versus total current in PEP.

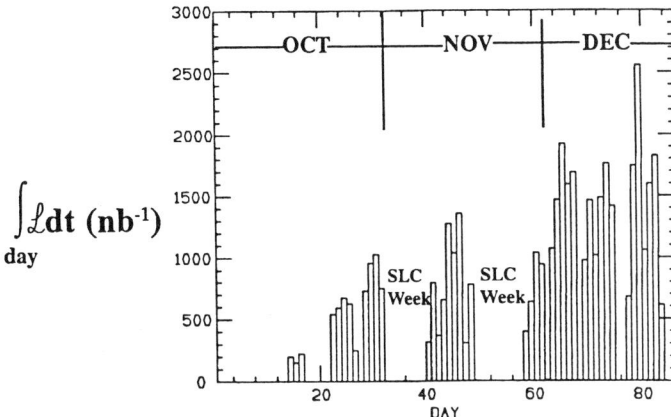

Figure 3: Daily integrated luminosity delivered by PEP.

by high radiation background originating at a mask 37 m from the interaction point that interfered with the 'beam stay clear'. After removing the mask and watercooling the VC we expect to reduce β_y^* to 4.0 cm and to increase the total current to 52 mA, which will result in an instantaneous luminosity of $10^{32}\,\mathrm{cm}^{-2}\mathrm{s}^{-1}$.

In order to improve the integrated luminosity, the injection into PEP has been upgraded by using the low emittance beams from the SLC damping rings and by adding an energy feedback system. Fig. 3 shows the daily integrated luminosity delivered by PEP during the fall run. It increased considerably as we gained experience in operating the machine. We reached an average of $\geq 1\,\mathrm{pb}^{-1}$ per day for the last three weeks of the run, the maximum being 2.5 pb^{-1} per day. The total integrated luminosity delivered was 42 pb^{-1}.

For future running PEP has been promised 20% of the Linac time to fill. Based on our expectations for switching between SLC operation and storage ring filling, we project $\geq 250\,\mathrm{pb}^{-1}$ of data per six month of running. This number will improve as we gain operational experience and as the SLC stability improves, and we hope to collect a total of 1000 pb^{-1} over the next three years.

The TPC/2γ Physics Program

Given a large data sample the TPC/2γ collaboration has a rich physics program in various fields:

- two-photon physics
    - quark/gluon structure of $C = +1$ states
    - QCD tests in exclusive and inclusive measurements
    - missing mass search for new states
- hadronization
    - flavor dependent cross sections and correlations
    - gluon fragmentation
- B physics
    - mixing for B_d and B_s
    - charge and flavor tagged lifetimes
- τ physics
    - τ decay branching ratios
    - τ lifetime
    - ν_τ mass
    - structure of the τ-W-ν_τ vertex

We will focus here on the prospects in τ physics. (For recent reviews of the field see Ref. [3].) At 27 GeV center-of-mass energy the cross section for $e^+e^- \to \tau^+\tau^-(\gamma)$ is 156 pb, resulting in a sample of 156k $\tau^+\tau^-(\gamma)$ events in the anticipated 1000 pb^{-1} of data.

τ Decay Branching Ratios

Measuring topological and exclusive branching ratios will be one of the major points in the TPC/2γ τ physics program. This is motivated by our failure to solve the problem of missing 1-charged particle decay modes ('1-prong problem'). If only experimental results are used (see Tab.1 in Ref. [4]) there appears to be no problem in accounting for every decay, mainly because only relatively poor upper limits exist for 1-prong decays with multiple π°'s or η's. The 1-prong problem shows up however, when other experimental data (*e.g.* $e^+e^- \to hadrons$ in the 1-2 GeV region) and conventional theoretical concepts (conserved vector current and isospin invariance) are used to tighten these upper limits. Then $(6.1 \pm 1.2)\%$ of the 1-prong decays are left unaccounted for. We do not know if this is caused by experimental errors, by misuse of the theory, or by unknown phenomena in τ decay.

Recently, a set of measurements of most τ decay channels performed by the CELLO collaboration has been published [5] that points to a conventional solution of the problem. They measure $B_1 = (84.9 \pm 0.5)\%$, which is nearly three standard deviations (s.d.) lower than the previous world average of $(86.6 \pm 0.3)\%$. They also measure B_e, $B_{\pi 2\pi^\circ}$, $B_{\pi 3\pi^\circ}$, B_3

and $B_{3\pi}$ somewhat higher than the world averages, thus being able to account for all 1-prong decays [1].

Existing TPC/2γ measurements [6], [7] agree very well with the CELLO values for B_1, B_e, $(B_{\pi 2\pi^\circ} + B_{\pi 3\pi^\circ})$ and B_3, however, they have larger statistical errors. ($B_{3\pi}$ has not yet been measured by TPC/2γ.) Given 1000 pb^{-1} of data and the capabilities of the TPC/2γ detector we will also be able to measure most of the τ decay channels (due to the TPC's superior particle identification we will also be able to measure channels with kaons), and thus do a conclusive test of CELLO's proposed solution of the 1-prong problem.

One of the key channels in this context is [2] $\tau^- \to \nu_\tau \pi^- \pi^\circ \pi^\circ$, which unfortunately is very difficult to measure because of the high photon multiplicity. This is the only substantial 1-prong decay channel for which no firm theoretical prediction exists. However, it is bound by isospin invariance to the corresponding all-charged channel [8]: $B_{\pi 2\pi^\circ} \leq B_{3\pi}$. Calculating a formal average for $B_{\pi 2\pi^\circ}$ yields $B_{\pi 2\pi^\circ} = (8\text{-}9 \pm 0.8)\%$, depending on which of the statistically correlated Mark II measurements is used (see Tab. 2 in Ref. [9]). The formal average for $B_{3\pi}$ is $B_{3\pi} = (6.8 \pm 0.4)\%$, even somewhat *lower* than $B_{\pi 2\pi^\circ}$:

$$(8-9 \pm 0.8)\% = B_{\pi 2\pi^\circ} \stackrel{isospin}{\leq} B_{3\pi} = (6.8 \pm 0.4)\%.$$

Note however, that there is considerable disagreement among the $B_{3\pi}$ measurements, which seem to cluster into two groups. TPC/2γ, which has already proven its ability to measure the difficult channel $\tau^- \to \nu_\tau \pi^- \pi^\circ \pi^\circ$ and which can easily handle the all-charged channel, will redo all these measurements to clarify the subject [3].

τ Lifetime

The τ lifetime has been determined by various experiments using a vertex chamber to measure the τ impact parameter or the τ decay length. The world average is [4]:

$$\tau = (3.03 \pm 0.09)\, 10^{-13}\, \text{s}.$$

The two most precise measurements to date, both using a decay length method, come from HRS [10] at PEP ($\tau_\tau = (2.99 \pm 0.15 \pm 0.10)\, 10^{-13}$ s) and from ARGUS [11] at DORIS ($\tau_\tau = (2.95 \pm 0.14 \pm 0.11)\, 10^{-13}$ s). Although the mean τ decay length at DORIS is almost a factor of three smaller than at PEP (see Tab. 1), ARGUS reaches the same precision as HRS due to a much larger data sample and a superior VC. Note that both measurement errors are still dominated by the statistical error.

Given the large expected 1-3 event sample and the excellent properties of the TPC/2γ VC, and scaling from the HRS and ARGUS results, we expect to measure τ_τ with better precision than today's world average.

---

[1] Note that B_1 and B_3 are anti-correlated since they are constrained to add up to $1 - B_5 = 99.85\%$.

[2] Throughout the text τ^- is used for both charge states of the τ.

[3] Note that due to various detector upgrades the amount of material in front of the electromagnetic calorimeter has been cut in half since its previous $B_{\pi 2\pi^\circ}$ measurement.

Table 1: Comparison of quantities relevant for a lifetime measurement. E_b denotes the beam energy, $<l>$ the mean decay length, δ the impact parameter resolution and $(\Delta\tau_\tau)_{stat}$ the statistical error on the τ lifetime.

Experiment	E_b (GeV)	$<l>$ (μm)	δ (μm)	1-3 events	$(\Delta\tau_\tau)_{stat}/\tau_\tau$ (%)
HRS	14.5	725	140	1311	5
ARGUS	5	250	95	5696	5
TPC/2γ	13.5	675	40	9000	1

τ Neutrino Mass

At present, the τ neutrino mass is limited to < 35 MeV at 95% C.L. [12]. This limit has been obtained by the ARGUS collaboration using the invariant mass distribution of 11 events from the decay channel $\tau^-\to\nu_\tau\pi^-\pi^+\pi^-\pi^+\pi^-$, for which they measure a branching ratio [12] of $B_{5\pi} = (0.064 \pm 0.025)\%$. Another hadronic final state, even better suited for such a study because of the high rest masses of the particles involved is $KK\pi$. Its branching ratio has been measured [13] to be: $B(\tau^-\to\nu_\tau K^+K^-\pi^-) = (0.22^{+0.17}_{-0.11})\%$, in good agreement with the theoretical expectation of 0.24%. Due to the unique particle identification of the TPC this channel can be selected (in the 1-3 topology) with an efficiency of about 8% and negligible background. From 1000 pb^{-1} of data we thus expect about 50 1-3 events with $K^+K^-\pi^-$ which will enable us to derive a ν_τ mass limit of

$$m_{\nu_\tau} < 15 \text{ MeV} \quad \text{at 95\% C.L.}$$

Structure of the τ-W-ν_τ Vertex

The exact form of the τ coupling to the charged current is far from being sufficiently constrained by existing data, mainly because of lack of large data samples. Sofar, only the decay parameter ρ has been determined using the lepton energy spectrum in $\tau\to\nu l\nu$. Its value, $\rho_\tau = 0.70 \pm 0.05$ [14], is in agreement with the $V-A$ expectation of 0.75. The lepton energy spectrum can also be used to determine the low energy shape parameter η.

Whereas ρ and η are best measured near $\tau^+\tau^-$ threshold (due to the smaller Lorentz boost distortion of the energy spectrum), the decay asymmetry parameters ξ and δ will most likely be determined at higher energies. ξ and δ describe the decay asymmetry with respect to the spin of the mother lepton and their determination thus requires knowledge of the τ polarization. This is possible (without polarized beams) because the spins of two τ's produced in e^+e^- annihilation are strongly correlated if their β is close to one. (At high energies the τ helicities prefer to be opposite to each other [15].) opposite to each other [15].) Using a whole set of measurements (see Ref. [16]) yields ξ_e^2, ξ_μ^2 and $h_{\nu_\tau}^2$, the ν_τ helicity, and their relative signs. However, with the anticipated data sample it will not be possible to determine the sign of h_{ν_τ}. Without longitudinally polarized beams this measurement needs $\geq 10^7$ produced τ pairs and thus might be performed at a future B factory.

Conclusions

The PEP storage ring and the TPC/2γ detector have been successfully upgraded for high-luminosity running.

During a test run in fall 1988, PEP reached instantaneous luminosities of $\mathcal{L} = 6 \times 10^{-31}$ cm^{-2}s^{-1} and average integrated luminosities in excess of 1 pb^{-1} per day. With minor modifications we expect to reach $\mathcal{L} = 10^{-32}$ cm^{-2}s^{-1} in the near future.

The TPC/2γ detector has been greatly improved by insertion of a high precision straw vertex chamber. The chamber performed well during the fall 1988 run.

Given a large data sample, projected for the next three years, TPC/2γ has an interesting and competitive physics program in the fields of two-photon physics, hadronization, B physics and τ physics. Particularly in τ physics, we want to measure τ decay branching ratios and the τ lifetime to a higher precision, improve the limit on m_{ν_τ} and further explore the structure of the τ-W-ν_τ vertex.

Acknowledgements

The author thanks the members of the TPC/2γ τ working group for useful discussion. This work was supported by the Humboldt Foundation.

References

[1] *Proceedings of the Workshop on e^+e^- Physics at High Luminosities*, SLAC-283 (1985).

[2] M.T. Ronan et al., in Proc. IEEE Nuclear Science Symposium, San Francisco, USA (1987).

[3] B.C. Barish and R. Stroynowski, Phys. Rep. 157(1988)1;
C. Kiesling, in High Energy Electron Positron Physics, eds. A. Ali and P. Söding, World Scientific, Singapore, 1988, p. 716.

[4] K.G. Hayes and M.L. Perl, Phys. Rev. D38(1988)3351.

[5] C. Kiesling, in Proc. XXIV Rencontre de Moriond, Les Arcs, France (1989).

[6] H. Aihara et al., Phys. Rev. D35(1987)1553.

[7] H. Aihara et al., Phys. Rev. Lett. 57(1986)1836.

[8] F.J. Gilman and S.H. Rhie, Phys. Rev. D31(1985)1066.

[9] H. Marsiske, in Proc. Tau-Charm Factory Workshop, SLAC-343 (1989).

[10] S. Abachi et al., Phys. Rev. Lett. 59(1987)2519.

[11] H. Albrecht et al., Phys. Lett. B199(1987)580.

[12] H. Albrecht et al., Phys. Lett. B202(1988)149.

[13] G.B. Mills et al., Phys. Rev. Lett. 54(1985)624.

[14] H. Janssen et al., DESY 89-054 (1989) and SLAC-PUB-4958 (1989), submitted to Phys. Lett. B.

[15] Y.S. Tsai, Phys. Rev. D4(1971)2821.

[16] *Proposal for an Electron Positron Collider for Heavy Flavour Particle Physics and Synchrotron Radiation*, PSI PR-88-09 (1988).

Tau Charm Factory Physics*

Walter H. Toki

Stanford Linear Accelerator Center
Stanford University, Stanford, California 94309

Abstract

Physics from a Tau Charm Factory is presented.

INTRODUCTION

Tau Charm Factories proposed for future machines will provide powerful and unique facilities to study a variety of physics topics: the tau lepton, charm mesons, charmonium and the J/ψ decays. These topics cover the physics of the members of the first and second quark doublets and the third lepton doublet. The number of events produced in a running year of 5,000 hours is shown in table 1. This represents a factor 100-1000 increase over previous data samples. A workshop held at Stanford Linear Accelerator Center[1] reviewed the physics,[2] the machine[3] and the detector[4] for such a facility. In this paper,

Table. 1 Tau Charm Factory particle yields at 10^{33}

e+e−	Particle	Produced events
J/ψ	J/ψ	10^9/month
3.680	ψ'	5×10^8/month
2m(τ)+2 MeV	$\tau^+\tau^-$	4×10^6 pairs/year
3.67 GeV	$\tau^+\tau^-$	4×10^7 pairs/year
ψ'' (3.77)	$D^0\bar{D}^0$	4×10^7 pairs/year
ψ'' (3.77)	D^+D^-	5×10^7 pairs/year
4.03 GeV	Ds^+Ds^-	10^7 pairs/year
4.14 GeV	$Ds\,Ds^*$	2×10^7 pairs/year

Invited talk presented at the meeting of the 1989 International Symposium on Heavy Quark Physics at Cornell University, Ithaca, N.Y., June 13-17, 1989

*Work supported in part by the National Science Foundation and by the Department of Energy contracts DE-AC03-76SF00515, DE-AC02-76ER01195, DE-AC03-81ER40050, DE-AC02-87ER40318 and DE-AM03-76SF000324

© 1989 American Institute of Physics

highlights of this meeting will be reviewed. We will begin with a short sketch of the machine issues and then briefly describe topics in tau, charm and charmonium-J/ψ physics.

STORAGE RING

The machine for the Tau Charm Factory was studied extensively by accelerator physicists.[3] The main conclusion was; the machine is difficult to build but well within present day technologies. The schematic layout is shown in Fig. 1. The general layout plan is to have

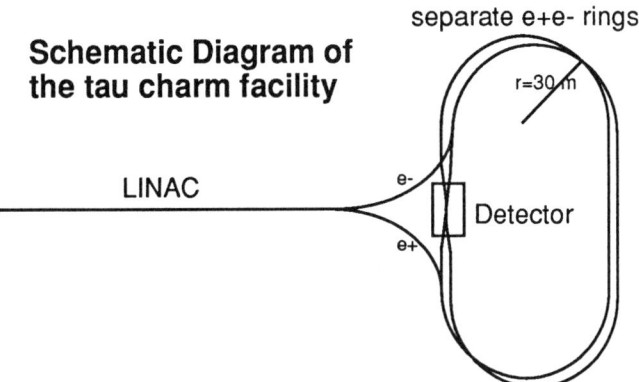

Fig. 1 Layout of a Tau Charm Factory

separate rings, a dedicated e^+ and e^- injector and probably a single interaction region. The operating range was chosen in the energy range 3-4.2 GeV. The peak luminosity to be obtained is $L=10^{33}$ cm^{-2}s^{-1} at E=4.0 GeV and at lower energies the luminosity should scale quadratically downward. It is expected to achieve high integrated luminosity with frequent filling of the rings, "topping off", which would occur every 30-60 minutes. The start up luminosity is expected to a few times 10^{33} cm^{-2}s^{-1} and in a few years achieve routine 10^{33} running.

The design parameters considered were 24 bunches per beam, a current of 0.5 ampere and beta values of β_x^*=20 cm and β_y^*=1 cm to achieve a luminosity of 10^{33} cm^{-2}s^{-1}. The main technical challenges to handle ampere sized currents include the vacuum system design and beam loading. The only technical hurdles for the detector are the requirements of close in mini-beta quad magnets, a 50 ns beam crossing and up to a 1 Khz signal rate for the J/ψ decays.

TAU PHYSICS

Tau physics in a Tau Charm Factory has several important advantages:
- Well defined low energy kinematics for tagging
- Low momentum tracks for high resolution mass measurements
- Small backgrounds (no charm nor beauty decays)
- Possibility to run below $\tau^+\tau^-$ threshold to study backgrounds

In the following physics examples,[5] the data sample is based on two years of running at the design luminosity. The branching ratio measurements and Michel parameter measurements were studied very near threshold to take advantage of tau's being almost at rest. The tau neutrino measurement was done at a center mass energy of 4.2 GeV.

Tau neutrino mass

The tau neutrino mass may be measured from the decay $\tau \to 5\pi\nu$. The basic technique is to measure the end point mass spectrum of the five pions. The key

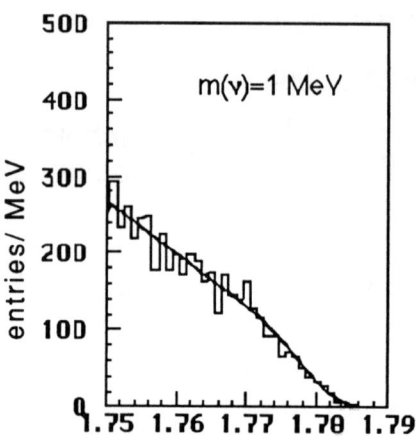

Fig. 2 Five pion endpoint mass

benefits are the very high statistics and the excellent mass resolution attainable due to the low momentum (~300 MeV/c) of the pions from a tau which is nearly at rest in the lab. The end point spectrum for a one MeV tau neutrino mass is shown in Fig. 2. For a two year run at a center mass energy of 4.2 GeV, there will be ~1000 events near the end point, $m(5\pi) > 1750$ MeV. This limit will approach 3 MeV which will improve the current limit of 35 MeV by a factor of ten.

Tau absolute branching ratios

The precise determination of absolute tau branching ratios are important to study the one prong problem.[6-8] A unique advantage of the measurement of the tau branching ratios near threshold is the ability to separate by kinematics the one prong tau decays of $\tau \to \pi\nu$ and $K\nu$ from the $\tau \to e\nu\nu$ and $\mu\nu\nu$. Near threshold, the momentum spectrum of the pion and the kaon in the one prong tau decay, appears as a narrow peak near the kinematic limit. This enables

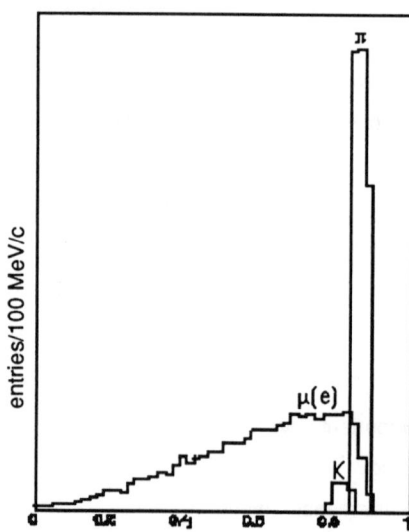

Fig. 3 One prong momentum spectrum

a useful separation from the leptonic decays where the electron and the muon momentum spectrum will be spread over the allowable momentum range. This is shown in Fig. 3. By measuring the number of events where both tau's decay into the $\pi\nu$ modes ("double tag") and the number of events where one of the two tau's decays into the $\pi\nu$ mode ("single tag"), we can obtain the total number of produced tau pairs in the sample and the branching ratio $B(\tau \to \pi\nu)$. Using a similar technique for the electron, muon and kaon modes we can achieve a fractional error of 0.4-0.5% on the leptonic modes and 1% on the kaon mode. This result is a factor ten better than the current fractional error on the pion modes which is 10%.

Michel Parameters

The Michel parameters[9], ρ and η, are the constants that determine the shape of the energy distribution of the electron or muon in the leptonic decay of the tau. These parameters are determined by fitting the energy spectrum of the electron or the muon. The measurement of these parameters from tau's produced near threshold is advantageous due to the tag selection of the pion mode. Selecting the e-π and μ-π decay modes will provide a clean sample. In particular, the low energy lepton energy spectrum will not be contaminated by the $\pi-\pi$ mode which is the case when this is studied using tau's produced at higher center of mass energies. The resulting study indicates the fractional error on the ρ parameter will be 0.4% which is a factor ten improvement over current values and comparable to the measurement for the muon.

Other measurements not discussed here but included in the workshop are: study of the multiple photon decays,[10] effects of charged Higgs in tau decays,[11] anomalous magnetic moments,[12] search for 2nd class currents,[13] electric dipole form factors,[14] supersymmetry affects in tau decays,[15] precision measurement of the tau mass,[16] QCD tests in tau decays[17,18] and the search for rare tau decays[19].

CHARM PHYSICS

Charm physics in a Tau Charm Factory has several important advantages:
- Large well established $D°$, D^+ and Ds^+ cross sections
- Exclusive, associated production of charm meson pairs to provide uniquely defined kinematics and low background contamination

- Well understood light quark continuum backgrounds and the ability to run below $c\bar{c}$ threshold
- Hadronic calorimetry can be used to veto background events that feed into the event selections with a missing neutrino

The main limitations in further progress in understanding charm meson decays is limited statistics.[20] The Mark III obtained a combined tag sample of charged and neutral D mesons of less than 10^4. The Tau Charm Factory in two years of running will amass nearly 10^7 tags. This very high statistics plus the detector improvements in neutral calorimetry resolution for the η and π° modes and hadronic calorimetry to reject K_L backgrounds should provide a major improvement in charm meson physics. The following examples are based on two years of running on the ψ''.

Leptonic Decays

Measurement of the leptonic decays of the Ds and the D^+ will provide a measurement of the pseudoscalar decay constant which is related to the branching ratio in the following formula;

$$B(D^+ \to \mu^+ \nu) = \frac{G_F^2}{8\pi} f_D^2 \tau_D M_D m_\mu^2 |V_{cd}|^2 \left(1 - \frac{m_\mu^2}{M_D^2}\right)^2$$

The pseudoscalar decay constant has only been measured for the pion and the kaon. It is basically a measurement of the wavefunction overlap or the annihilation of the quark-antiquark pair. Its importance lies not only in testing various models[21] but also in the model extrapolation for f_B which is a parameter in the box diagrams from $B\bar{B}$ mixing. If the values of the heavy quark decay constants are found to be anomalously too large or small, our understanding of the top quark mass and $B\bar{B}$ mixing may change.

The measurement[22] of the D^+, $Ds^+ \to \mu^+ \nu$ requires the use of fully reconstructed events where the exclusive pair production is achieved and the only missing track is the neutrino. This allows the use of the constraint of the missing mass of the neutrino. In addition, hadronic calorimetry will be important to reject backgrounds with K_L decays that could feed down into this channel. Using the expected tag sample in a Tau Charm Factory, the measurement of f_D and f_{Ds} should be achievable with a few percent error. The $\tau\nu$ mode was also studied. Depending on the backgrounds, this mode could also provide a large rate and could provide another independent measurement. Thus far there is only an upper limit for f_D from the Mark III group.

Exclusive Semileptonic Decays

Measurement[23] of the semileptonic decays of D→Kev and D→πev in all possible modes will provide important measurements of the KM matrix elements.[24] The relation between the KM matrix element and the partial semileptonic rate is;

$$\Gamma(D^0 \to \pi^- e^+ \nu) = \frac{G_F^2 M_D^5}{192\pi^3} |V_{cd}|^2 \int |f_+^\pi(t)|^2 p_\pi^3 dt$$

The measurement will provide a measurement of the KM parameters, V_{cd} and V_{cs} and the form factor for the hadronic transition;

$$f_+^{K,\pi}(t) = f_+^{K,\pi}(0) \times \left[\frac{M_{D^*,F^*}^2}{M_{D^*,F^*}^2 - t} \right]$$

The predicted number of detected events are at a few times 10^5 level for the Cabbibo allowed D decays and a few times 10^4 level for the Cabbibo suppressed modes. The measurement is performed by detecting a D decaying into a Kev or πev recoiling against a D tag. An important technique in this analysis is the use of the exclusive nature of the measurement where all the tracks are measured except the missing neutrino. The missing mass of the neutrino is a powerful constraint that other non-exclusive measurements will not have. These measurements will provide a 1% determination of the KM parameters and a precise measurement of the form factors. In a Tau Charm Factory the projected number of πev (Kev) events is 4×10^4 (3×10^5) whereas the current existing modest sample is 7 (55) from the Mark III.

$D^0 \bar{D}^0$ Mixing

Mixing in the charm sector is predicted from the Standard Model to be;

$$2r_D = \left(\frac{\Delta m}{\Gamma}\right)^2 + \left(\frac{\Delta \Gamma}{2\Gamma}\right)^2$$

where r_D is the ratio of mixed to unmixed events and $\Delta\Gamma$ and Δm are the mass mixing parameters. Expected rates from the Standard Model are $r_D \sim 10^{-4}$-10^{-5} and any unexpected larger rates could be the case for new physics.[25] The mixing can be measured by comparing the charge of the leptons in events where $D°\overline{D}°$ is produced and both of the $D°$'s have semileptonic decays. The second but more problematical technique is to use the case where both of the D's have the non-leptonic hadronic decays and compare the strangeness by the sign of the charged kaon. This last case may be plagued by double Cabbibo suppressed (DCSD) decays[26] where one D has a Cabbibo suppression at each W connection with a quark line. Although this rate is small, $\sim \tan^4 \theta_C$, these events cannot be separated by time evolution measurements in a Tau Charm Factory and they may mask the real mixing rate. It has been pointed out[27] that Bose statistics forbids the decay of both of the D's into identical hadronic states that are in a p-wave. Hence the decay $\psi' \to D°\overline{D}° \to K^+\pi^-K^+\pi^-$ is only possible from mixing and not from DCSD, whereas the decays $e^+e^- \to \gamma D°\overline{D}° \to \gamma K^+\pi^-K^+\pi^-$ or $\psi' \to D°\overline{D}° \to K^+\pi^-\pi°K^+\pi^-$ may be attributed only to DCSD or in combination with mixing. By systematically measuring decays that are s-wave (DD*) and p-wave (ψ') and several hadronic modes that decay into identical pairs the question of mixing and DCSD can be disentangled and it is possible to measure $\Delta m/\Gamma$ and $\Delta\Gamma/2\Gamma$ separately. The studies[28] indicate that mixing may be detected to a level of $r_D \approx 10^{-4}$ and by measuring several modes in different techniques the level of $r_D \approx 10^{-5}$ may be obtained.

Other charm physics topics covered in the workshop included a measurement of the absolute branching ratios of the D^+, $D°$ and Ds mesons by use of the double tag method, measurement of inclusive semileptonic branching ratios,[29] search for CP violation by measuring differences in the $D/\overline{D}$ rates[30,31] and the search for rare D decays[32] such as ee, $\mu\mu$, μe,[33] penguin decay modes[34] and CP violating decays.[31]

J/ψ AND CHARMONIUM PHYSICS

J/ψ and charmonium physics in a Tau Charm Factory will benefit enormously from the ultra statistics and improvements from better detector capabilities which include;
- High resolution electromagnetic calorimetry
- Uniform charged and neutral track acceptance
- Very large solid angle and very forward angle acceptance

The very high statistics enables precise measurements of branching ratios of the J/ψ by use of the decay mode $\psi' \to \pi^+\pi^- J/\psi$ where the J/ψ events can be tagged by detecting missing mass recoil of the $\pi^+\pi^-$ system. One third of the ψ' decays into this mode. Another major improvement will be the inclusive photon capability that a high resolution crystal calorimeter will allow. This resolution will enable "Crystal Ball" like inclusive photon resolution. This permits precise measurements of absolute branching ratios of radiative decays such as the η_c, iota, and theta as the luminosity errors will cancel out when the exclusive and inclusive

rates are measured in the same experiment.

Charmonium decays

Among the charmonium decays, several measurements[35,36] can really put the model to a more confining test. These include the two photon decay of the η_c and the three photon decay of the J/ψ. They are simply the two(three) photon diagram divided by the two(three) gluon diagram of the η_c(J/ψ).

$$\frac{\Gamma(J/\psi \to \gamma\gamma\gamma)}{\Gamma(J/\psi \to e^+e^-)} = \frac{4\alpha e_q^4 (\pi^2 - 9)}{3\pi} \qquad \frac{\Gamma(\eta_c \to \gamma\gamma)}{\Gamma(\eta_c \to gg)} = \frac{8}{9}\left(\frac{\alpha}{\alpha_s}\right)^2 \left(\frac{e_q}{2/3}\right)^4$$

The three photon decay of the J/ψ can be measured via the transition $\psi' \to \pi^+\pi^- J/\psi$, J/$\psi \to \gamma\gamma\gamma$. This channel is not affected by the $e^+e^- \to \gamma\gamma\gamma$ background. This decay channel can be easily normalized as the total J/ψ events can be counted from studying the missing mass from the $\pi^+\pi^-$ tags. The η_c decays can be measured in J/$\psi \to \gamma\eta_c$, $\eta_c \to \gamma\gamma$. The normalization can be obtained by use of the inclusive photon measurement of the radiative photon from the η_c. Because the inclusive measurement is performed in the same experiment the luminosity errors will cancel out. This of course will enable precise measurements of the branching ratios B($\eta_c \to K\bar{K}\pi$) and B($\eta_c \to p\bar{p}$) which are needed to normalize two photon and antiproton gas jet production of the η_c and the χ states. These relations predict a rate of 10^{-5} for the three photon decay of the J/ψ and 10^{-3} for the two photon decay of the η_c. In a Tau Charm Factory there will be 2,550 detected three photon decays of the J/ψ and 4,500 detected two photon decays of the η_c from a radiative decay of the J/ψ. From the existing world data there is a limit for the J/ψ decays and for the η_c there are all combined 150 events so far detected and a wide range of branching ratio values from the two photon experiments.

Search for Glueballs and Hybrids

The central prediction of the QCD lattice gauge theories is the existence of the lowest lying scalar glueball. The radiative decays of the J/ψ are the seminal hunting ground for these particles. Thus far the search in J/$\psi \rightarrow \gamma\pi\pi$, $\gamma\eta\eta$ and γKK has been negative. The possible explanations include a small branching ratio and/or the signal is hidden by overlapping $q\bar{q}$ resonances. Recent theory[37] has suggested that the scalar mass may be near 1.5 GeV instead of of 1 GeV and that the tensor mass is a factor 1.5 higher. If this is correct the glueball may be hidden among the backgrounds of $\rho\pi$ and f(1270) in the $\pi\pi$ mode and the f(1525) in the $\eta\eta$ and KK modes. To disentangle[38] [39] the signal from other resonances these channels requires a partial wave analysis requiring high statistics and a uniform and very forward acceptance. The current samples from the Mark III and DM2 are based on 5-10 million events and the forward acceptances of these detector were limited to |cosθ|<0.8. The proposed Tau Charm Factory will produce 10^9 J/ψ events in a month of running and the acceptance is designed to achieve cosθ measurements up to 0.95. These measurements will allow a careful search for underlying resonances but also a complete measurement of many decay modes to verify the flavor decay of the candidates which are important tests of the glueball predictions.

Rare decays

The ultra high number of J/ψ events allows the possibility of observing weak decays. We can estimate the rate from the ratio of the lifetime of the D mesons and the width of the J/ψ;

$$\frac{2\hbar/\tau(D)}{\Gamma(J/\psi)} = 5 \times 10^{-7}$$

This rate is could easily be enhanced by non-spectator effects such as exchange diagrams. These decays[40] could appear either as a spectacular signature in the direct decay into a charm meson, J/$\psi \rightarrow$Ds+X, or into C or CP violating decays such as J/$\psi \rightarrow$KsKs or $\phi\phi$. These rare decays follow similar physics proposed for very high statistics machines such a super LEAR or SATURNE[41] which are proposed to study rare decays of light quark mesons which are expected at the level of 10^{-13} because of their wide widths. The OZI suppressed nature of the J/ψ causes a relatively narrow width and this provides an advantage in terms of the larger branching ratios of these rare decays. What is interesting in the Tau Charm case is the possibility of observing these decays is within reach and the decay modes are striking and unmistakable.

Other studies not discussed here include the study of light quark spectroscopy,[42] search for the η_c' and the $^1p_1$,[35] study of the iota[43,44], E and theta[45] mesons, search for hybrids and 4-quark states[37,46], study of the $\rho\pi$ puzzle in ψ' decays[47] and study of η_c hadronic decays[48].

SUMMARY

The main conclusions of the Tau Charm studies from the workshop were;
- Precision tau and charm measurements testing the Standard Model are possible with a very high luminosity, $\mathcal{L}=10^{33}$ cm^{-2}s^{-1} machine
- Machine designs understudy, capable of this luminosity, are challenging but well within present technologies
- Detector designs[49] for Tau Charm physics are easily within present daytechnologies and represent substantial improvements in physics capabilities over previous detectors.
- The scope and breath of the physics is very broad, a Tau Charm Factory represents a major facility studying a wide range of particles: tau leptons, charm mesons, J/ψ and ψ' decays and their secondaries.

REFERENCES

[1] Tau Charm Factory Workshop, May 23-27, 1989, Stanford Linear Accelerator Center, Stanford University, Stanford, CA., organized by K. Brown, D. Coward, J. Kirkby, M. Perl, A. Seiden, R. Schindler and W. Toki.
[2] A. Seiden, *Proceeding of the Tau Charm Workshop*, SLAC-REPORT-343, Vol. 1 and 2.
[3] J. Jowett, *Proceeding of the Tau Charm Workshop*, SLAC-REPORT-343, Vol. 1 and 2.
[4] J. Kirkby, *Proceeding of the Tau Charm Workshop*, SLAC-REPORT-343, Vol. 1 and 2.
[5] J.J. Gomez-Cadenas and A. Seiden, *Proceeding of the Tau Charm Workshop*, SLAC-REPORT-343, Vol. 1 and 2.

[6] F. Gilman and S. Rie, Phys. Rev **D31**, 1066 (1985).
[7] K. Hayes, *Proceeding of the Tau Charm Workshop*, SLAC-REPORT-343, Vol. 1 and 2.
[8] J. Repond, *Proceeding of the Tau Charm Workshop*, SLAC-REPORT-343, Vol. 1 and 2.
[9] W. Fetscher, *Proceeding of the Tau Charm Workshop*, SLAC-REPORT-343, Vol. 1 and 2.
[10] T. Skwarnicki, *Proceeding of the Tau Charm Workshop*, SLAC-REPORT-343, Vol. 1 and 2.
[11] Y. Tsai, *Proceeding of the Tau Charm Workshop*, SLAC-REPORT-343, Vol. 1 and 2.
[12] D. Silverman, *Proceeding of the Tau Charm Workshop*, SLAC-REPORT-343, Vol. 1 and 2.
[13] K. K. Gan, *Proceeding of the Tau Charm Workshop*, SLAC-REPORT-343, Vol. 1 and 2.
[14] W. Bernreuther, O. Nachtmann, *Proceeding of the Tau Charm Workshop*, SLAC-REPORT-343, Vol. 1 and 2.
[15] R. Stroynowski, *Proceeding of the Tau Charm Workshop*, SLAC-REPORT-343, Vol. 1 and 2.
[16] J.J. Gomez-Cadenas, *Proceeding of the Tau Charm Workshop*, SLAC-REPORT-343, Vol. 1 and 2.
[17] E. Braaten, *Proceeding of the Tau Charm Workshop*, SLAC-REPORT-343, Vol. 1 and 2.
[18] A. Pich , *Proceeding of the Tau Charm Workshop*, SLAC-REPORT-343, Vol. 1 and 2.
[19] B. Barish, *Proceeding of the Tau Charm Workshop*, SLAC-REPORT-343, Vol. 1 and 2.
[20] R. Schindler, *Proceeding of the Tau Charm Workshop*, SLAC-REPORT-343, Vol. 1 and 2.
[21] N. Paver, *Proceeding of the Tau Charm Workshop*, SLAC-REPORT-343, Vol. 1 and 2.
[22] P. Kim, *Proceeding of the Tau Charm Workshop*, SLAC-REPORT-343, Vol. 1 and 2.
[23] J. Izen, *Proceeding of the Tau Charm Workshop*, SLAC-REPORT-343, Vol. 1 and 2.
[24] B. Ward, *Proceeding of the Tau Charm Workshop*, SLAC-REPORT-343, Vol. 1 and 2.
[25] M. Shin, M. Bander and D. Silverman, *Proceeding of the Tau Charm Workshop*, SLAC-REPORT-343, Vol. 1 and 2.
[26] L.L. Chau, *Proceeding of the Tau Charm Workshop*, SLAC-REPORT-343, Vol. 1 and 2.
[27] I. Bigi, *Proceeding of the Tau Charm Workshop*, SLAC-REPORT-343, Vol. 1 and 2.
[28] G. Gladding, *Proceeding of the Tau Charm Workshop*, SLAC-REPORT-343, Vol. 1 and 2.
[29] D. Pitman, *Proceeding of the Tau Charm Workshop*, SLAC-REPORT-343, Vol. 1 and 2.
[30] L. L. Chau, *Proceeding of the Tau Charm Workshop*, SLAC-REPORT-343, Vol. 1 and 2.
[31] U. Karshon, *Proceeding of the Tau Charm Workshop*, SLAC-REPORT-343, Vol. 1 and 2.
[32] R. S. Willey, *Proceeding of the Tau Charm Workshop*, SLAC-REPORT-343, Vol. 1 and 2.
[33] I. Stockdale, *Proceeding of the Tau Charm Workshop*, SLAC-REPORT-343, Vol. 1 and 2.
[34] T. Browder, *Proceeding of the Tau Charm Workshop*, SLAC-REPORT-343, Vol. 1 and 2.

[35] R. Mir and T. Burnett, *Proceeding of the Tau Charm Workshop*, SLAC-REPORT-343, Vol. 1 and 2.
[36] N. Isgur, *Proceeding of the Tau Charm Workshop*, SLAC-REPORT-343, Vol. 1 and 2.
[37] F. Close, *Proceeding of the Tau Charm Workshop*, SLAC-REPORT-343, Vol. 1 and 2.
[38] T. Bolton, *Proceeding of the Tau Charm Workshop*, SLAC-REPORT-343, Vol. 1 and 2.
[39] W. Dunwoodie, *Proceeding of the Tau Charm Workshop*, SLAC-REPORT-343, Vol. 1 and 2.
[40] W. Toki, *Proceeding of the Tau Charm Workshop*, SLAC-REPORT-343, Vol. 1 and 2.
[41] B. Nefkens, preprint SACLAY-DPHN-2562, June 1989
[42] B. Ratcliff, *Proceeding of the Tau Charm Workshop*, SLAC-REPORT-343, Vol. 1 and 2.
[43] M. Burchell, *Proceeding of the Tau Charm Workshop*, SLAC-REPORT-343, Vol. 1 and 2.
[44] J. Weinstein, *Proceeding of the Tau Charm Workshop*, SLAC-REPORT-343, Vol. 1 and 2.
[45] F. Liu, *Proceeding of the Tau Charm Workshop*, SLAC-REPORT-343, Vol. 1 and 2.
[46] B. Li, *Proceeding of the Tau Charm Workshop*, SLAC-REPORT-343, Vol. 1 and 2.
[47] S. Brodsky, *Proceeding of the Tau Charm Workshop*, SLAC-REPORT-343, Vol. 1 and 2.
[48] C. Heusch, *Proceeding of the Tau Charm Workshop*, SLAC-REPORT-343, Vol. 1 and 2.
[49] J. Kirkby, *Proceeding of the Tau Charm Workshop*, SLAC-REPORT-343, Vol. 1 and 2.

Multidimensional analysis : b-tagging at LEP

Ch. de la Vaissière and S. Palma-Lopes

(L.P.N.H.E. – Universités Paris VI-VII –)

At the Z^0, the cross-section for $e^+e^- \to b\bar{b}$ is large (6.5 nb), as is the fraction of hadronic events leading to $b\bar{b}$ (22%). A jet topology allows to distinguish naturally the products of the b and $\bar{b}$ fragmentation and decays. The Z^0 looks therefore an attractive place to pursue B physics. Techniques previously used at PEP and PETRA to tag the b-flavour, have provided reasonable b-purities, at the cost of poor efficiencies. A first technique originally proposed to measure the b-lifetime was to use leptonic decays, but the corresponding branching ratios are at the 10 % level. At Z^0 energies, P. Roudeau [1] shows that a 91 % purity and 6 % efficiency can be obtained. The TASSO collaboration [2] was the first to use a vertex detector for b-enrichment. They achieved a b-purity of about 68 %, with a 16 %-efficiency. The best way to increase these low yields is to improve the resolution of vertex detectors on impact parameters. DELPHI [3] will be equipped with a silicon microstrip vertex detector [4] which will provide an asymptotic accuracy of 20 μm on impact parameters in the plane transverse to the beam, to be compared with the 150 μm quoted by TASSO. However this 20 μm, combined with limited coverage, can not disentangle the multiple decays occuring in a $b\bar{b}$ event. In this intermediate situation multidimensional analysis may provide tagging of $b\bar{b}$ events with high purity and good efficiency.

1. Multidimensional analysis and class likelihoods

The principle of multivariate analysis is to maximize the use of the information and analyse events in the space of their characteristics (observables). In this space, which could have many dimensions, different elementary processes will populate hopefully clusters with little overlap. Criteria of class membership are needed to classify events. Several analysis techniques in high-dimensioned spaces are available [6], [7],[8]. Classification algorithms require the knowledge of the clusters, which have therefore to be modelled. A second function of the Monte-Carlo generation is, before the LEP start-up, to simulate raw data where every event has a known origin. A correct strategy should be a *double model* mode where a first M.C. defines a metric in the space of observables. Then a second M.C. simulates real data, in which the origin of events are ignored. The tagging process uses the characteristics of the clusters defined from the "model" sample. The two Monte-Carlo can be chosen slightly different to emulate nature.

In this analysis, the tagging was based on class likelihoods (other criteria are discussed in [9]). Class likelihoods are defined as follows. At the level of a single variable V_l, the probability distribution for the class c, $p(V_l : c)$, is approximated by the histogram of this variable for a sample of events belonging to that class in the "model sample". The $p(V_l : c)$ are poorly estimated in the tails of the histograms. The effect of statistical fluctuations is reduced by fitting the histograms by suitable functions. Neglecting correlations, the joint probability to observe the vector $\vec{V} = (V_1, ..., V_l, ... V_L)$ is the product of individual $p(V_l : c)$. This probability should be multiplied by the weight of the class $W_c = N_c/N_{tot}$ where N_c is the population of the class and N_{tot} the number of events in the data. N_c/N_{tot} can be set to a theoretical value in the case of $Z^0 \to q\bar{q}$. Finally, the class likelihood is :

$$L_c(\vec{V}) = Ln W_c + \sum_{l=1}^{L} Ln(p(V_l : c)) \quad (1)$$

The advantage of the "class likelihood" is that it accepts all types of distributions. The choice of the best class likelihood leads to the attribution of the event to the corresponding class. The quality of the tagging can be improved by reducing ambiguities between two classes. This is done by requesting that

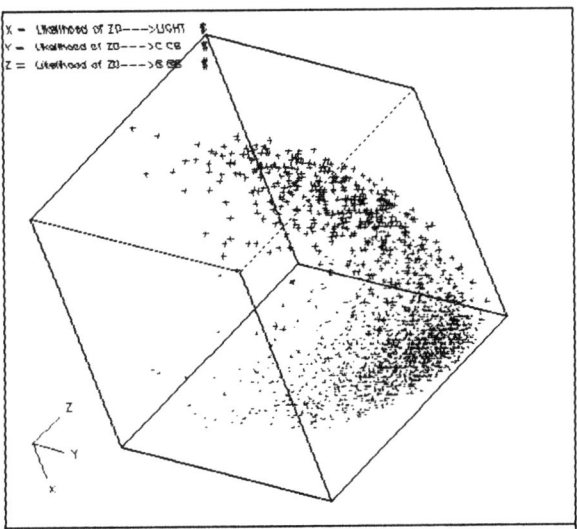

Figure 1: : *Display of the cube representing the "likelihoods" of the 3 classes in this analysis. The cluster of b-events symbolized by crosses can be seen clearly away from the 2 other classes represented by dots.*

the difference between the best likelihood and the second best is large. Next, the event can be classified as "untagged", if the best value lies below a quality threshold. An unexpectedly large amount of untagged events in the data means either that clusters have long tails or a clue for new physics.

2. Data simulation

Simulation of the data sample

The data sample is made of 2000 events simulated at the Z^0 peak. Among these, 1208 belong to the $u\bar{u}$, $d\bar{d}$ and $s\bar{s}$ flavours, 319 to the $c\bar{c}$ flavour, 412 to the $b\bar{b}$ flavour. The sample contains in addition 61 $t\bar{t}$ events. An option allows to switch off these events. The mass of the top quark has been set to 40 GeV. A model sample made of 500 events/class was created without the top class and with similar generator and acceptances. Details of the simulation can be found in ref [9]. We have used the release 6.3 of the Lund model [10] based on the standard model at Z^0 with the production of extra quarks and gluons up to second order in α_s. Except for the mass of the t quark and the choice of the second order α_s, all parameters are set to default values. The LUND package models also the decays of c and b hadrons. The modelling is tuned to reproduce essentially the multiplicities and semi-leptonic ratios. The major source of systematic uncertainty is the knowledge of the B-lifetimes, since half of the variables rely on impact parameters. The lifetimes may vary between various b hadrons, as they do for charm. This dependence is watered down by the b hadrons cascade into charmed hadrons.

The simulation of the DELPHI detector has been simplified. Emphasis has been put on the reconstruction of the trajectories of charged particles seen by the vertex detector. The extrapolation length in the beam tube from the first layer of the detector is 9 cm. No emphasis has been put on particles identities.

Description of variables

This analysis is based on 15 variables. The 8 first variables, calculated from all detected particles, are: The

	A - Generated class			B - Generated class			
TAG	purity	L	C	B	purity	L	C	B
L	89.7 ± 1.	913.	93.	12.	87.4 ± 1.0	1051.	133.	18.
C	34.2 ± 2.1	288.	179.	57.	40.9 ± 2.7	149.	138.	50.
B	86.4 ± 1.7	7.	47.	343.	86.0 ± 1.7	8.	48.	344.
Loss in %		34.4 ± 1.2	43.9 ± 2.8	16.7 ± 1.8		13.0 ± 1.0	56.7 ± 2.8	16.5 ± 1.8

Table 1: *Classification of 1939 Z^0 Monte-Carlo events between 3 classes : $L(q = u,d,s)$, C ($q = c$), $B(q = b)$ based on likelihoods: a) when class populations are ignored ($W_c = 1$); b) when class populations are taken into account.*

average $|cos\theta|$, the logarithms of sphericity and aplanarity, the number of reconstructed jets, the hadronic visible energy, the missing momentum in the plane perpendicular to the beam, the total charged (visible) multiplicity and a Fox-Wolfram moment.

The 7 next variables are linked to the micro-vertex detector. We assume that the interaction point has been found and fitted. The first two variables represent, respectively, the contributions of the χ^2 of the fitted vertex in the plane perpendicular to the beam and along the beam. The third variable represents the sum (over all charged particles) of the distances of closest approach to the vertex (normalized to the error) projected on the reconstructed jet axis associated with each particle. This quantity takes positive values for decays occurring "downstream" of a jet. Furthermore, the micro-vertex detector allows to select and count "secondary" tracks that "miss" the interaction vertex. Four more variables are calculated from these secondary tracks: a secondary multiplicity, an average P_t of secondaries (w.r.t jet axis), the sum of the momenta of the "secondaries", and the number of secondary kaons and leptons.

This set of 15 variables carries information about the topology of the event and exploits the micro-vertex possibilities, but is not optimal. Correlations exist that can be difficult to avoid. For example, when most particles travel along the beam pipe and are undetected, all micro-vertex variables become useless.

3. Experimental results

Classification and b-tagging

The table 1-a shows the classification obtained from likelihoods with class populations ignored ($W_c = 1$). The overall efficiency is 74.0 ± 1.0 %. Among the 26 % mistagged events, 15 % are u,d,s events tagged charm. The purity of the b-sample is 86.4 ± 1.7 %, with a loss of 16.7 ± 1.8 %. The display of the likelihood cube shows a horseshoe structure, with a $b\bar{b}$ cloud distinct from the uds and $c\bar{c}$ clouds which begin to separate (fig.1). Classifications are improved if class populations are accounted for, as can be seen from the table 1-b. The 288 u+d+s events unduly tagged charm shrink to 149 events, while well tagged charm events diminish only from 179 to 138. Weighting according to class population has little influence on b-tagging. This suggests that the b-cluster is separated from the others. The purity of the b-sample can be improved by removing evemts found ambiguous between two classes. With a criteria more elaborate than likelihoods, it has been found that a purity of 98 % can be reached at the cost of a 30 % loss [9].

Losses and contaminations in b-tagging

The main contamination is due to charm. An event-by-event scan indicates a first group of events, where one or both charmed hadrons have a long range (around 1 cm and even more) combined to a large charged decay multiplicity (3/4 prongs). One observes a close competition between the charm and beauty tags. A second group contains poorly detected events.

Important for the physics analysis are losses. Monte-Carlo allows to trace back to the generated b-decays and to plot the decay-length and $\cos\theta$ of the parents B hadrons (fig.2). It appears that b-events mistaken as u, d or s flavour have both B's characterized by very short projected decay paths. The direction of these B's is clearly aligned with the z-axis. The main loss, due to a charm tag, has less clear features. The event scan suggests two main causes of misclassification: 1) decays poorly detected because of acceptance problems; 2) one B hadron with a very short decay range associated to a second B hadron with no charged

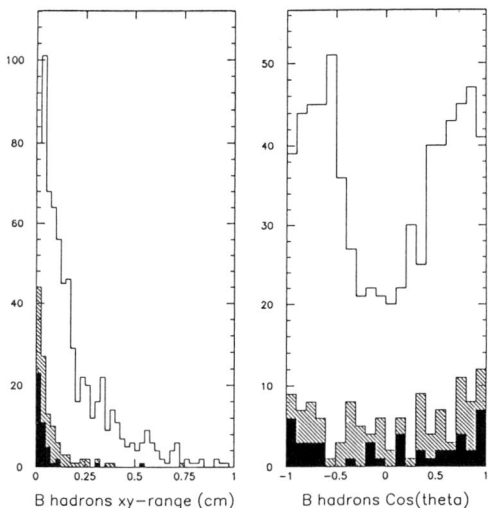

Figure 2: : *Decay range and production angle of B hadrons in events well classified as $b\bar{b}$ and events erroneously attributed to other flavours (hatched area). Losses affect short decays, mainly for b-events classified as uds (black area)*

visible secondaries (either genuine or due to reconstruction uneffeciencies).

A better accuracy and coverage of the micro-vertex detector, together with a smaller beam-pipe radius is obviously the best way to improve charm and beauty separation. But with the present detector, the amount of losses is already close to an irreductible core of low acceptance events.

Sensitivity to variables and influence of micro-vertex

To measure the influence of the microvertex variables, we have compared efficiencies and purities in classifications involving the 8 topological variables only, the 7 microvertex variables alone, and all the variables (Table 2). The classification based on topological variables is already at the level of 47 % in the case of b-tagging, with a loss of 38 %. The microvertex variables perform, as expected, an already excellent b-separation of its own.

Criteria	Variable set		
	General	microvertex	all variables
Overall eff.	63.4 ±1.	79.0 ±0.9	79.1 ±0.8
b-purity	47.0 ±2.1	83.5 ±1.8	86.0 ±1.7
b-efficiency	61.7 ±2.4	82.5 ±1.9	83.5 ±1.8

Table 2: *Comparison of efficiencies and purities in classifications based on general variables only, microvertex variables only and all variables.*

Combining all variables improves the efficiencies and purities, but the improvement for b-tagging appears small. As mentioned in the previous section, we are approaching the irreductible core of low acceptance events. However, the fact that the tagging has already begun with the general variables allows to tag outside the coverage of the microvertex detector. On the other hand, if the weight of the microvertex is dominant in the separation, b-tagging would be more sensitive to uncertainties in the lifetimes of b hadrons or knowledge of the vertex detector than to the Monte-Carlo generator and fragmentation scheme.

Sensitivity to the model

Results were found to be stable with respect to acceptance or reconstruction efficiencies. Both of these have been taken more severe than in our first published analysis [11], without significant change on the results. The same was observed for the accuracies of the impact parameters and the angular coverage of the microvertex detector. This robustness has is origin in the summation made on many particles for the chosen variables.

The *double model* mode allows to investigate the effect of discrepancies between the model and data files. An example is the impact of unexpected physics. Data at the Z^0 may contain unexpected events. Such a situation, where a class of events is not modelled but is present in the data, can be emulated, by switching on the 61 top events in the data file. Among these 61 events, 7 were found classified uds, 8 $c\bar{c}$, and 48 $b\bar{b}$. These events are expected to have low uds, c and b likelihoods. A "no tag" condition is introcuced by requesting that the best likelihood L_{max} is below a threshold. 72 events (40 top and 32 non-top) pass the condition $L_{max} < -47.5$, which corresponds to a 7 σ excess. These low likelihoods events can be furthermore asked to have at least 9 detected charged particles detected i.e to fulfill a quality condition. The remaining sample contains 39 top events among 54. To summarize, with a statistics limited to 2000 events only, 3 % of unexpected events could easily be enriched to 70 %. The implicit condition is that the tails are well understood.

Conclusion

With a micro-vertex detector and multidimensional analysis, one can expect to achieve 85 % purity with 10-20 % losses in b-tagging. Quasi pure b samples can be achieved at the cost of 30 % losses. Therefore, one may expect to analyse, at LEP energies, samples of 10^5 to 10^6 b-events with high purity. The solidity of the classification suggests that the b-cluster has little overlap with the others in the space of variables. B-tagging requires a good understanding of the micro-vertex detector but is not really model dependent. In order to go beyond the stage of tagging, it appears important to increase the angular coverage of the micro-vertex detector, to get accuracy in the z coordinate along the beam, and to reduce the beam pipe radius.

References

[1] P. Roudeau - Rapport LAL 86/22

[2] TASSO collaboration, Report DESY 88-112 (August 1988) and J.M.Pawlak, TASSO collaboration, Report DESY F1-88-01 (September 1988).

[3] DELPHI Technical proposal DELPHI 83-66/1.
DELPHI Progress report DELPHI 84-60/GEN -11.

[4] DELPHI microvertex Detector, Addendum to the technical proposal, DELPHI 86-86 GEN-2 (Oct 1986)

[5] I.Pless, J.E.Brau et al. 1971. Phys.Rev.Lett. 27:1481

[6] J.Friedman, SLAC-PUB-3841 (STAN-LCS 18) Dec. 85

[7] J.M.Romeder, 1973. Méthodes et programmes d'analyse discriminante. Dunod Paris; and bibliography therein

[8] P.Lutz. How to recognize the hard elementary process in hadronic events at LEP energies? Rapport College de France LPC 84-26 and D.Delkaris, P.Lutz - Fisher Discriminant Analysis for Heavy Quark Production - (Contribution to the "New Quarks and Leptons" LEP 200 Working Group - 6th January, 1987)

[9] Ch. de la Vaissière and S. Palma-Lopes DELPHI Progress report DELPHI 89-32/PHYS-38.

[10] The Lund Monte-Carlo program for jets fragmentation and e^+e^- physics, LU TP 86-22 (Release JETSET 6.3)

[11] Ch. de la Vaissière and S. Palma-Lopes In proceedings of Int. Symposium on production and decay of heavy flavours, stanford 1987, p.50-59.

AUTHOR INDEX

A

Alam, M. S., 226
Anjos, J. C. et al., 172

B

Barberis, D., 285
Berkelman, K., 409
Besson, D., 311
Bigi, I. I., 18
Botner, O. et al., 268

C

Chauveau, J., 414

D

DeJongh, F., 197
de la Vaissiere, Ch. et al., 440
Ducati, M. B. Gay, 274

F

Foster, G. W., 354

G

Gilman, F., 367
Gladding, G., 157
Gläser, R., 116
Golutvin, A., 111

I

Isgur, N., 3

J

Jensen, T., 325

K

Katayama, N., 103
Kim, P. C., 203
Kubota, Y., 142
Kutschke, R., 191
Kwan, S., 262

L

Lipkin, H. J., 72
Lipton, R., 255
Lynch, G., 130

M

Marsiske, H., 421
Moneti, G., 179
Morrison, R., 239

N

Nakada, T., 385
Neubert, M., 52
Nir, Y., 35

O

O'Donnell, P., 46

P

Paul, S., 280
Pimiä, M., 339
Procario, M., 122
Punkar, G., 210

S

Sakamoto, H., 345
Schäfer, M., 97
Schubert, K., 79
Sharpe, S., 59
Spaan, B., 136

T

Toki, W., 317, 428
Tuts, M., 299

W

Waldi, R., 331
Witherell, M. S., 401
Wormser, G., 219

AIP Conference Proceedings

		L.C. Number	ISBN
No. 140	Boron-Rich Solids (Albuquerque, NM, 1985)	86-70246	0-88318-339-0
No. 141	Gamma-Ray Bursts (Stanford, CA, 1984)	86-70761	0-88318-340-4
No. 142	Nuclear Structure at High Spin, Excitation, and Momentum Transfer (Indiana University, 1985)	86-70837	0-88318-341-2
No. 143	Mexican School of Particles and Fields (Oaxtepec, México, 1984)	86-81187	0-88318-342-0
No. 144	Magnetospheric Phenomena in Astrophysics (Los Alamos, 1984)	86-71149	0-88318-343-9
No. 145	Polarized Beams at SSC & Polarized Antiprotons (Ann Arbor, MI & Bodega Bay, CA, 1985)	86-71343	0-88318-344-7
No. 146	Advances in Laser Science–I (Dallas, TX, 1985)	86-71536	0-88318-345-5
No. 147	Short Wavelength Coherent Radiation: Generation and Applications (Monterey, CA, 1986)	86-71674	0-88318-346-3
No. 148	Space Colonization: Technology and The Liberal Arts (Geneva, NY, 1985)	86-71675	0-88318-347-1
No. 149	Physics and Chemistry of Protective Coatings (Universal City, CA, 1985)	86-72019	0-88318-348-X
No. 150	Intersections Between Particle and Nuclear Physics (Lake Louise, Canada, 1986)	86-72018	0-88318-349-8
No. 151	Neural Networks for Computing (Snowbird, UT, 1986)	86-72481	0-88318-351-X
No. 152	Heavy Ion Inertial Fusion (Washington, DC, 1986)	86-73185	0-88318-352-8
No. 153	Physics of Particle Accelerators (SLAC Summer School, 1985) (Fermilab Summer School, 1984)	87-70103	0-88318-353-6
No. 154	Physics and Chemistry of Porous Media—II (Ridge Field, CT, 1986)	83-73640	0-88318-354-4
No. 155	The Galactic Center: Proceedings of the Symposium Honoring C. H. Townes (Berkeley, CA, 1986)	86-73186	0-88318-355-2
No. 156	Advanced Accelerator Concepts (Madison, WI, 1986)	87-70635	0-88318-358-0

No. 157	Stability of Amorphous Silicon Alloy Materials and Devices (Palo Alto, CA, 1987)	87-70990	0-88318-359-9
No. 158	Production and Neutralization of Negative Ions and Beams (Brookhaven, NY, 1986)	87-71695	0-88318-358-7
No. 159	Applications of Radio-Frequency Power to Plasma: Seventh Topical Conference (Kissimmee, FL, 1987)	87-71812	0-88318-359-5
No. 160	Advances in Laser Science–II (Seattle, WA, 1986)	87-71962	0-88318-360-9
No. 161	Electron Scattering in Nuclear and Particle Science: In Commemoration of the 35th Anniversary of the Lyman-Hanson-Scott Experiment (Urbana, IL, 1986)	87-72403	0-88318-361-7
No. 162	Few-Body Systems and Multiparticle Dynamics (Crystal City, VA, 1987)	87-72594	0-88318-362-5
No. 163	Pion–Nucleus Physics: Future Directions and New Facilities at LAMPF (Los Alamos, NM, 1987)	87-72961	0-88318-363-3
No. 164	Nuclei Far from Stability: Fifth International Conference (Rosseau Lake, ON, 1987)	87-73214	0-88318-364-1
No. 165	Thin Film Processing and Characterization of High-Temperature Superconductors	87-73420	0-88318-365-X
No. 166	Photovoltaic Safety (Denver, CO, 1988)	88-42854	0-88318-366-8
No. 167	Deposition and Growth: Limits for Microelectronics (Anaheim, CA, 1987)	88-71432	0-88318-367-6
No. 168	Atomic Processes in Plasmas (Santa Fe, NM, 1987)	88-71273	0-88318-368-4
No. 169	Modern Physics in America: A Michelson-Morley Centennial Symposium (Cleveland, OH, 1987)	88-71348	0-88318-369-2
No. 170	Nuclear Spectroscopy of Astrophysical Sources (Washington, D.C., 1987)	88-71625	0-88318-370-6
No. 171	Vacuum Design of Advanced and Compact Synchrotron Light Sources (Upton, NY, 1988)	88-71824	0-88318-371-4
No. 172	Advances in Laser Science–III: Proceedings of the International Laser Science Conference (Atlantic City, NJ, 1987)	88-71879	0-88318-372-2

No. 173	Cooperative Networks in Physics Education (Oaxtepec, Mexico 1987)	88-72091	0-88318-373-0
No. 174	Radio Wave Scattering in the Interstellar Medium (San Diego, CA 1988)	88-72092	0-88318-374-9
No. 175	Non-neutral Plasma Physics (Washington, DC 1988)	88-72275	0-88318-375-7
No. 176	Intersections Between Particle and Nuclear Physics (Third International Conference) (Rockport, ME 1988)	88-62535	0-88318-376-5
No. 177	Linear Accelerator and Beam Optics Codes (La Jolla, CA 1988)	88-46074	0-88318-377-3
No. 178	Nuclear Arms Technologies in the 1990s (Washington, DC 1988)	88-83262	0-88318-378-1
No. 179	The Michelson Era in American Science: 1870–1930 (Cleveland, OH 1987)	88-83369	0-88318-379-X
No. 180	Frontiers in Science: International Symposium (Urbana, IL 1987)	88-83526	0-88318-380-3
No. 181	Muon-Catalyzed Fusion (Sanibel Island, FL 1988)	88-83636	0-88318-381-1
No. 182	High T_c Superconducting Thin Films, Devices, and Application (Atlanta, GA 1988)	88-03947	0-88318-382-X
No. 183	Cosmic Abundances of Matter (Minneapolis, MN 1988)	89-80147	0-88318-383-8
No. 184	Physics of Particle Accelerators (Ithaca, NY 1988)	87-07208	0-88318-384-6
No. 185	Glueballs, Hybrids, and Exotic Hadrons (Upton, NY 1988)	89-83513	0-88318-385-4
No. 186	High-Energy Radiation Background in Space (Sanibel Island, FL 1987)	89-083833	0-88318-386-2
No. 187	High-Energy Spin Physics (Minneapolis, MN 1988)	89-083948	0-88318-387-0
No. 188	International Symposium on Electron Beam Ion Sources and their Applications (Upton, NY 1988)	89-084343	0-88318-388-9
No. 189	Relativistic, Quantum Electrodynamic, and Weak Interaction Effects in Atoms (Santa Barbara, CA 1988)	89-084431	0-88318-389-7
No. 190	Radio-frequency Power in Plasmas (Irvine, CA 1989)	89-045805	0-88318-397-8
No. 191	Advances in Laser Science–IV (Atlanta, GA 1988)	89-085595	0-88318-391-9

No. 192	Vacuum Mechatronics (First International Workshop) (Santa Barbara, CA 1989)	89-045905	0-88318-394-3
No. 193	Advanced Accelerator Concepts (Lake Arrowhead, CA 1989)	89-045914	0-88318-393-5
No. 194	Quantum Fluids and Solids—1989 (Gainesville, FL 1989)	89-081079	0-88318-395-1
No. 195	Dense Z-Pinches (Laguna Beach, CA 1989)	89-046212	0-88318-396-X

MAY 1 5 1990